Sandia National Laboratories
The Postwar Decade

THE WHITE HOUSE
WASHINGTON

May 13, 1949

Dear Mr. Wilson:

I am informed that the Atomic Energy Commission intends to ask that the Bell Telephone Laboratories accept under contract the direction of the Sandia Laboratory at Albuquerque, New Mexico.

This operation, which is a vital segment of the atomic weapons program, is of extreme importance and urgency in the national defense, and should have the best possible technical direction.

I hope that after you have heard more in detail from the Atomic Energy Commission, your organization will find it possible to undertake this task. In my opinion you have here an opportunity to render an exceptional service in the national interest.

I am writing a similar note direct to Dr. O. E. Buckley.

Very sincerely yours,

Harry Truman

Mr. Leroy A. Wilson,
President,
American Telephone and Telegraph Company,
195 Broadway,
New York 7, N. Y.

Sandia National Laboratories

The Postwar Decade

Necah Stewart Furman

University of New Mexico Press
Albuquerque

DISCLAIMER

AT&T welcomes the publication of this history. It recounts the experiences, opinions, and first-hand observations of many of the individuals who directly participated in the matters and events discussed. As such, all statements, opinions, and observations expressed in this history, unless otherwise noted, are solely those of the author. AT&T did not participate in the writing of this history and has exercised no right of approval over the style, form, or content of this book.

Published 1990 by the University of New Mexico Press. All rights reserved. First edition.

Design by Tonimarie Stronach.

Library of Congress Cataloging-in-Publication Data

Furman, Necah Stewart.
  Sandia National Laboratories: the postwar decade / by Necah Stewart Furman.
     p. cm
     Includes bibliographical references.
     Includes index.
     1. Sandia Laboratories—History. 2. Manhattan Project (U.S.)—History. I. Title.
QC792.8U6S264   1989
623.4′5119′072073—dc20   89-16726

CONTENTS

PREFACE xi

AUTHOR'S NOTE xv

ACKNOWLEDGMENTS xix

Prologue
BATTLE OF THE LABORATORIES: The Secret War 3

Part I.
HERITAGE 25

    Chapter I.
    **TRINITY: The Field Test That Revolutionized
    the World** 27

        Future Sandians View Trinity; Countdown Zero: July 16, 1945

    Chapter II.
    GENESIS: The Laboratory on the Hill 45

        Los Alamos; Thin Man and Fat Man; A Shift in Emphasis: Implosion; Reorganization; Jumbo; Selection of a Test Site; Pioneering Field Testing Techniques

    Chapter III.
    DESTINATION: Hiroshima and Nagasaki 89

        Inyokern; Wendover; Sandy Beach; Training the 509th; Project Silverplate; The Glory Boys; Operation Centerboard

    Chapter IV.
    THE POSTWAR YEARS: Z Division 119

        Oxnard Field; Formation of Z Division; Sandia Base, Albuquerque

v

CONTENTS

Chapter V.
ORIGINS OF AN ETHOS: Field Testing 147

    Los Lunas Test Range; Selection of Salton Sea Site; Testing at Salton Sea; Improvement of Facilities; GAFCO

Chapter VI.
CROSSROADS: Impact and Legacy 181

    Z–Division Personnel to Overseas Duty; Staging the Crossroads Colossus; Tests Able and Baker; Fallout

Chapter VII.
MANAGING THE NUCLEAR ARSENAL: The Atomic Energy Act 207

    The Quest for Control; The Great Debate; The Political Machinery; Revitalization

Chapter VIII.
PRODUCT OF POSTWAR READINESS 231

    The AEC Visits Sandia Base; The Manley Evaluation; Interaction With the Military; The Custody Issue; The Readiness Problem Defined

Part II.
NUCLEAR ORDNANCE ENGINEER FOR THE NATION 249

Chapter IX.
MEETING THE CHALLENGE: Sandia Laboratory— An Independent Branch 251

    Paul Larsen Appointed Director; Advances on the Administrative Front; "A Rube Goldberg Affair"; Sandia Laboratory Adds an Applied Physics Department; Organizations for Interaction; Weapons Development, A Project Group Orientation

Chapter X.
SANDSTONE: A New Realm in Weapons Development 283

    Enewetak Selected as Test Site; Sandia Branch Personnel Arrive at Sandstone; A Dry Run; Zero Hour; Impact on the Sandia Branch Laboratory

CONTENTS

Chapter XI.
"Let's Get the Show on the Road" 307

    The "Road" Department; Origins of the Integrated Contractor Complex; "Road" Becomes Production Engineering; Stockpile Surveillance and Training Liaison; Quality Assurance; Alert Status, Site "Easy"; Project "Water Supply"; Sandians on Site

Chapter XII.
SANDIA CORPORATION: From University to Industrial Management 329

    "This Whole Sandia Matter . . . Seems to Have Gotten Out of Hand"; Mervin Kelly Recommends Changeover to Industrial Management; The AEC Selects AT&T; President Truman Intercedes

Chapter XIII.
ON THE BASIS OF "GOOD INDUSTRIAL PRACTICE" 349

    Larsen Responds to the Kelly Report; Bridging the Gap; Prime Contract Relations; Contractual Modifications

Chapter XIV.
MEETING THE CHALLENGE: The Production Era, 1950–1952 365

    George A. Landry: First President; Expansion of Production Facilities; Engineering for Manufacture and Production; TWA Engineering; Materials and Standards

Chapter XV.
A SHIFT IN EMPHASIS: Toward Research and Development 395

    Formation of the Research Directorate; Origins of the Systems Analysis Function; Aerodynamics, The Mk 4 Case Controversy, and Timer Problems; 4N Program; Other Marks Meet Emergency Capability Requirements; Project Greenfruit

CONTENTS

Chapter XVI.
UNIONIZATION: Aftermath of Incorporation 427

Formation of Metal Trades Council and Sandia Employees' Association; Federal Mediators and the Davis Panel Intercede; Impact of Unionization on Nonunion Employees; Sandia Security Guards Unionize

Chapter XVII.
MEETING THE ADMINISTRATIVE CHALLENGE: Expansion of Services 453

Financial, Personnel, and Business Operations; Wage and Salary Administration; Legal, Medical, and Safety; Purchasing "Keeps the Wheels of Progress Turning"; Public Relations

Chapter XVIII.
A MISSILE REVOLUTION IN THE MAKING 485

From German Rockets to Redstone Arsenal; Sandia Establishes Directorate for Warhead Development; Warheads for Guided Missiles; Environmental Test Facilities Expanded to Meet Missile Needs; Reorganization

Chapter XIX.
PARAMETERS FOR INTERACTION: DOD-AEC-Sandia 527

Donald A. Quarles, Second President, Sandia Corporation; The Development and Production Controversy; The Fuzing Controversy; The DOD Assumes Responsibility for Fuzing; Implementation of the 1953 Agreements; Sandia Strives to Improve Organizational Effectiveness

Chapter XX.
ERA OF THE SUPERBOMB: Politics to Prooftesting 559

Steppingstones to the Super; Politics of the Thermonuclear Program; Ranger; The Greenhouse Operation; The Second Lab Controversy

CONTENTS

Chapter XXI.
WEAPONS TESTING ON THE CONTINENT 595

Buster–Jangle; Tumbler–Snapper; Upshot–Knothole; Teapot

Chapter XXII.
ADVANCING TECHNOLOGICAL FRONTIERS 629

McRae Becomes Sandia's Third President; The McRae Committee; Emergency Capability Programs; The Laydown Bomb; The Quest for the Wooden Bomb; Toward Fundamental Research; The Westward Migration; In Retrospect

NOTES 685

BIBLIOGRAPHY 779

PHOTO CREDITS 833

INDEX 837

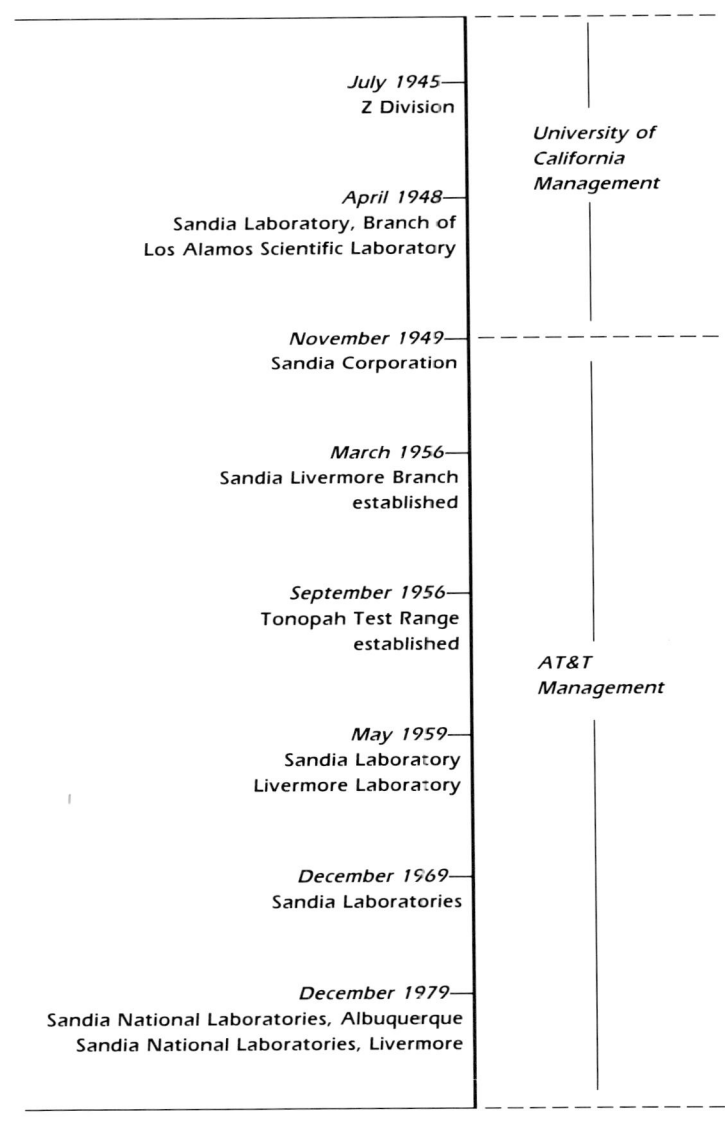

PREFACE

On May 13, 1949, President Harry Truman wrote to the President of AT&T, announcing that the Atomic Energy Commission planned to ask the company to accept direction of the Sandia Laboratory in Albuquerque, New Mexico. At stake was not only the successful operation of a small, nuclear engineering and assembly facility located in the Southwestern outback, but also the growth and reliability of the nation's nuclear stockpile. In Truman's opinion, this was an opportunity to render "an exceptional service in the national interest."

AT&T accepted the task and the opportunity, beginning the Laboratory's forty-year association November 1, 1949. Throughout its history, Sandia National Laboratories would respond to technological challenges in the defense of the nation. Recognized by the technical community for its long-standing responsibility for weaponization of nuclear explosives, Sandia has pioneered techniques through research, development, design, and testing to assure reliable, safe, and effective nuclear weapons. In so doing, it has become a center of excellence in many areas: arming, fuzing, and firing; radiation effects, simulation, and hardening against countermeasures; devices to assure nuclear safety and controlled use of nuclear weapons; and a host of other weapons-related technologies. Sandia is also known for research and development in treaty verification technologies, nonnuclear weapons, and energy. Yet, because of the sensitive nature of its mission, this Laboratory has been an enigma to the general public. While many books have been written on its progenitor, Los Alamos Scientific Laboratory, this is the first public history of the Laboratory referred to as "the nuclear ordnance engineer for the nation."

Although Sandia has retained its primary identity as an engineering research and development laboratory, it is also recognized for its strong basis in fundamental research. The story of this evolution—how and why it occurred, major technological developments, and the people who made them possible—is the focus of this first volume of the history of Sandia.

Sandia National Laboratories: The Postwar Decade is organized into two major sections: Part I: "Heritage" and Part II: "Nuclear Ordnance Engineer for the Nation." Within this general structure, the narrative connects a series of subjects, beginning with a Prologue that surveys the national and international events leading to development of the atomic bomb and establishment of the national laboratory complex under the Manhattan Engineer District. Chapters 1–8 trace the genesis of Sandia Laboratory from its engineering–ordnance origins in 1945 as the Z Division of Los Alamos, through its reorganization in 1948 as Sandia Laboratory, Branch of Los Alamos Scientific Laboratory. The story, told from the perspective of those individuals who would become "Sandians," starts with Trinity and proceeds to detonation of the atomic bombs over Hiroshima and Nagasaki, full-scale test operations in the Pacific, and interactions with the military at the Army Air Field on the desert outskirts of Albuquerque, New Mexico. Throughout this formative period, the trend is toward growing independence as an organizational entity, while the underlying ethos of the Lab is that of a "can–do" response to the nuclear ordnance needs of the nation.

In Part II, Chapters 9–15 describe how the Laboratory, through strong leadership, an improved organizational structure, and the 1949 changeover to industrial management under AT&T, contributed to the growth of the nation's stockpile. These chapters address weapons development activities, establishment of the integrated contractor complex, and the gradual shift in emphasis from the production orientation of the Laboratory toward research and development.

Chapters 16–20 analyze the immediate impact of incorporation and the attempts of new management to meet administrative and technical challenges of the time. These chapters recount Sandia's participation in the guided missile revolution, establishment of the parameters for interaction among the DOD–AEC and

PREFACE

Sandia, response to the emergency capability programs of the thermonuclear era, and development of advanced weapon concepts and basic research activities. The underlying theme of Chapters 20-22 is expansion and maturation—both technologically and organizationally. While Los Alamos Scientific Laboratory was being recognized as "a bastion of big Science," Sandia Laboratory within the span of a decade had earned its place as "the nuclear ordnance engineer for the nation."

Overall, the history of Sandia National Laboratories during the postwar decade is the story of an amazing group of mission-oriented people, who, in the Cold War temper of the times, considered their work to be essential to the national defense.

AUTHOR'S NOTE

In his history of AT&T, *Telephone, The First Hundred Years*, John Brooks makes a valid point: "Because books about corporate affairs are . . . subsidized," he says, "the author of a corporate history owes it to both his craft and his readers to set forth plainly at the outset the essential terms and conditions under which he has done his work." Brooks explains that this approach prevents the reader from being deceived and provides future historians the information necessary to evaluate properly the book as source material. Accordingly, I would like to comment about the writing of this history.

Sandia National Laboratories, a long-time proponent of the low profile, has never before allowed production of an official history for external publication. For this reason, among others, this book has been a trailblazer of sorts, and the path to publication has led through uncharted ground.

Major turning points or pendulum swings in Sandia's history have occurred almost every ten years; therefore, it was decided that the general timeframe for this first volume should be the postwar decade. This approach permits certain threads in the organization's history to be carried to a logical culmination or transition point, rather than being artificially constrained by specific years, such as 1945–1955 exclusively.

At the inception of the project, the decision was made to direct the history of the Laboratories to a general audience as well as to Sandia personnel. Suggestions were also made regarding methodology. Although the case study or topical essay format would have been easier, a narrative, chronological style seemed better suited to the overall goal of explaining the genesis and growth of the Laboratory as a product of the defense posture of the nation and political decision-making. When possible, the

AUTHOR'S NOTE

simple statement of events has been expanded to deal with the hows and whys of technological innovations and their resulting impact.

Another challenge established early on was the decision to include human interest anecdotes and the stories of individuals responsible for the growth and development of the Laboratory and its technologies. The pitfalls to such an approach are obvious, but so are the rewards. In an organization numbering over 8,000 in population, it is impossible always to give credit where credit is due. The available documentation, or the lack thereof, simple judgment calls, or the inability to weave in a personality at a certain point in the story are all factors that determined who was included and who was not. I apologize in advance to those who were slighted.

The fact that this book involves the history of a national laboratory dealing in the highly secret business of the weaponization of nuclear explosives has influenced the manuscript's content. The constraints of Classification, the policy of "need-to-know," instruction to "neither confirm nor deny" certain weapons development data, or even the location of storage sites, as well as legal considerations, limited coverage of technological accomplishments and some Laboratory activities. While this may be frustrating from the scholarly perspective, from the standpoint of national security, it is essential.

Similar limitations also affected treatment of full-scale test activities at the Pacific Proving Grounds and the Nevada Test Site.* Such parameters are a fact of life for some corporate historians, particularly in a government setting. On the other hand, the writing of the Sandia history was enriched by my having access to many private collections, correspondence, and the confidences of personnel that would have been unavailable to the outside scholar. Within these bounds, I have attempted to be as candid as possible in presenting the growing pains of the organization, the trauma associated with management

---

*For a more in-depth presentation of the consequences of nuclear testing—a theme not included in this book—one will want to read the forthcoming third volume of the AEC histories, *Atoms in War and Peace*, by Richard Hewlett and Jack Holl.

AUTHOR'S NOTE

change, and certain aspects of the military–civilian controversy. Free access to numerous classified and unclassified sources from government and corporate archives has permitted a solid basis for interpretation, although technological detail has been restricted.

This then is the Sandia story. In the telling, it is hoped that this first volume will add a new chapter to the historical annals of the atomic era.

Necah Stewart Furman
May 1989

SANDIA NATIONAL LABORATORIES

HISTORY REVIEW BOARD
May 1987–November 1988

Chairman
Robert L. Peurifoy, Jr.
Vice-President, Technical Support

Glenn A. Fowler
Vice-President, Systems
(retired)

Robert W. Henderson
Executive Vice-President,
Weapons Development
(retired)

Richard W. Lynch
Director, Nuclear Waste
Management and
Transportation

Leon D. Smith
Director, Monitoring Systems
(retired)

* * * * * * * * * *

HISTORY REVIEW BOARD
December 1988–Present

Chairman
Richard W. Lynch
Director, Nuclear Waste
Management and
Transportation

Glenn A. Fowler
Vice-President, Systems
(retired)

Robert W. Henderson
Executive Vice-President,
Weapons Development
(retired)

William C. Myre
Director, Monitoring Systems

L. Herbert Pitts
Director, Information and
Communications Services

Leon D. Smith
Director, Monitoring Systems
(retired)

ACKNOWLEDGMENTS

The assignment to write the history of Sandia National Laboratories was a challenging one, but the assistance of many people made the task much easier. Scholarly groundwork laid by the following authors proved to be of immense value: Roger Anders, *Forging the Atomic Shield*; John Brooks, *Telephone*; Paul Johnson, *Modern Times*; Gregg Herken, *The Winning Weapon*; Richard G. Hewlett and Oscar Anderson, Jr., *The New World*; Richard G. Hewlett and Francis Duncan, *Atomic Shield*; Richard G. Hewlett and Jack Holl, *Atoms for Peace and War*; Daniel Kevles, *The Physicists*; William Manchester, *The Glory and the Dream*; George T. Mazuzan and J. Samuel Walker, *Controlling the Atom*; Peter Pringle and James Spigelman, *The Nuclear Barons*; Richard Rhodes, *The Making of the Atomic Bomb*; and Ferenc Szasz, *The Day the Sun Rose Twice*, among many others.

As this list indicates, numerous secondary sources existed to provide a foundation for the Sandia history; however, internal documentation proved to be another matter. When the Sandia History Project began five years ago, no real archives existed, and the majority of the Laboratories' documents resided in a mountainside depository in uncataloged form. Fortunately, Frederick C. Alexander, Jr., Sandia's first historian and Supervisor of Special Projects (1960–1969), had retained a safe of well-organized classified documents that he had used as resource materials for the writing of a monograph on the history of Sandia. The brevity of Alexander's end product, *History of Sandia Corporation through Fiscal Year 1963*, published in that same year, scarcely indicates the extensive amount of research that went into the work, nor its value to this historian as a concise summary and point of embarkation. Equally as valuable were the special projects—a number of weapons histories—produced by Alexander and staff members Phillip Owens and Daniel

ACKNOWLEDGMENTS

Freshman. Beryl Hefley later added several oral history interviews to the *Alexander Collection*, as it is now designated. To all these individuals, I owe a debt of gratitude.

After Alexander's retirement, Sandia *Lab News* Editor and Acting Historian John Shunny, with the assistance of *Lab News* personnel, published an excellent pictorial history called *Sandia, Looking Back*. This brochure and Alexander's history were both internal publications directed largely toward Sandia employees. The current volume, therefore, represents a distinct departure in Sandia's traditional low-profile policy.

Others deserving of credit for laying the groundwork for this first externally published history of the Labs are: Larry Lopez, who assisted in the collection and organization of Z-Division materials, and William A. "Bill" Stevens, who served as the Project's Technical Advisor until his retirement in 1985. Stevens' extensive collection of draft papers and notes, ranging from Research to Quality Assurance to Components and the Characteristics of Early Nuclear Weapons, has served as invaluable guideposts into Sandia's technological past. A portion of Stevens' Sandia-published reports, "On the Division of Responsibilities Between the AEC and DOD for Fuzing of Nuclear Warheads Used on Guided Missiles and Rockets 1950–1953" and "On the Evolution of the AEC/DOD Agreement on Development, Production, and Standardization of Atomic Weapons of March 1953," have been incorporated liberally into the writing of Chapter 19 of the current volume of the Sandia history. Perhaps more important, it was Bill Stevens who patiently guided this historian through the initial learning and organizational stages; who generously shared his valuable insights into personalities as well as programs; and who even after retirement provided an excellent critique of many chapters.

For fifteen years, from 1969 to 1984, no formal history project existed at the Laboratories. Then certain events occurred to generate a renewed interest. In 1982, during the upper management retreat held at the scenic Inn of the Mountain Gods in Ruidoso, New Mexico, Vice-President Glenn A. Fowler spoke of the factors that had influenced the culture of the Laboratory and

ACKNOWLEDGMENTS

stressed the value of capturing the corporate memory while it was still possible. He noted especially how industrial, as opposed to academic management of the Laboratories, had shaped the character of the Labs and recommended that serious consideration be given to the writing of the Laboratories' history.

External support for the idea came the following year when The American Institute of Physics' Center for the History of Physics and the Department of Energy Headquarters' History Office, as a result of a joint study, strongly encouraged the Laboratories to establish archival programs and write their histories. Subsequently, in April 1984, President George C. Dacey formally established the Sandia History Project and appointed a Corporate Historian. The Project, positioned organizationally in the Technical Information Services Directorate, received the strong support of Director Henry M. "Hank" Willis, who held a true appreciation for the value of corporate history and the multiple uses for historical research and analysis. Without the foresight of these two people, this history would not have been written.

Established according to Department of Energy guidelines, the Sandia History Project soon grew to include an extensive oral history collection, the beginnings of a corporate archives, and the capability for response to special projects. A unique aspect of the Project is a network of representatives from all of Sandia's thirty-four directorates, who assist in the history-gathering and reporting process. This approach, initiated by President Dacey, resulted in an enthusiastic team effort and increased historical awareness throughout the Laboratory.

Because of the large number of people involved and changes in organizational appointments, I regret that it is not possible to acknowledge by name all those who have served as history representatives over the five-year period from 1984 to the present. However, it has been a pleasure coordinating with you; your contributions of reports, photographs, and resource materials have greatly simplified a complex task. This institutional approach to the writing of history has also had its fringe benefits, namely, directorate collections that can be transformed into

ACKNOWLEDGMENTS

individual histories of the Laboratories' major organizational threads. Internal publication of a history of the Sandia Livermore Weapons Development directorate has already established the precedent.

With George Dacey's retirement, President Irwin Welber continued support of the Sandia History Project by encouraging the directorate representatives to provide annual reports for the growing history archives and by establishing a Review Board to provide technical review of the manuscript. The original Board, chaired by Robert L. Peurifoy, Jr., and including members Glenn A. Fowler, Robert W. Henderson, Richard W. Lynch, and Leon Smith, provided the historian with excellent technical and editorial assistance and more than a modicum of moral support. In 1989, two new members were added to the Board, William C. Myre and L. Herbert Pitts, under whose directorate the History Project resides. At this time Dick Lynch was named the new Chairman of the Board. To Lynch has fallen the task of ushering the manuscript through the extensive corporate and governmental review and approval processes. Review Board members have donated many long hours without compensation to the review and editing of this manuscript. For your expertise, forebearance, and friendship, I am most appreciative.

I would also like to express my appreciation to Albert Narath, who succeeded Irwin Welber as President of Sandia National Laboratories in April 1989, for his personal review of the manuscript.

For serving as an exofficio Advisory Board and for providing expert and valuable professional reviews of the manuscript, I would like to acknowledge the following members: Ferenc Szasz, Professor of History at the University of New Mexico and author of the prize-winning book, *The Day the Sun Rose Twice*; Dana Asbury, Editor, University of New Mexico Press; Marcy Goldstein, Archivist-Historian and Department Manager, AT&T Archives; Jack Holl, former Chief Historian DOE, currently Assistant Dean, Kansas State University; J. Samuel Walker, Chief Historian for the U.S. Nuclear Regulatory Commission; and Roger Meade, Archivist-Historian, Los Alamos National Laboratory.

ACKNOWLEDGMENTS

The following institutions and individuals have also assisted in making this volume possible: Jack Holl as Chief Historian for the Department of Energy History Office provided guidance and support for the Project from its inception; Roger Anders, Alice Buck, Sheila Convis, and Skip Gosling, also of the DOE History Office, who expertly assisted with the location of important documents, information, and photographs; Gil Ortiz and Roger Meade (co-author of *Critical Assembly*) of the Los Alamos Archives and Records Center, and his predecessor Allison Kerr, who with their gracious staffs assisted in locating hard-to-find documents, often on short notice; Joan Warnow and Spencer Weart of the American Institute of Physics' Center for the History of Physics, for providing access to their valuable oral history collection; and the Bancroft Library, also for providing interviews of the Manhattan Project pioneers; Frank Shelton of Kaman Sciences Corporation, who shared his manuscript "Reflections of a Weaponeer"; Robert Seidel, author of *A History of the Lawrence Berkeley Laboratory, Volume I: Lawrence and His Laboratory*, and Director of the Bradbury Science Museum, who extended the opportunity to introduce the history of Sandia to the International Congress of the History of Science; Lillian Hoddeson, co-author of the Los Alamos history *Critical Assembly*, for her leads to various resources; Norma McCormick, AT&T Archives Assistant, for contributing resource material and photographs; and Beth Hadas, the enlightened and talented Director of the University of New Mexico Press.

I am especially grateful to Manhattan Project pioneers Norris E. Bradbury, John Manley, Carson Mark, and the numerous individuals who took the time to be interviewed or to read a portion of the manuscript. Others to whom I am indebted include: Bill Jack Rogers of Los Alamos, for providing many of the excellent photographs included in the manuscript; Richard Ray of the National Atomic Museum on Kirtland Air Force Base, for loaning his excellent collection of news articles from the fifties; James Carothers of Lawrence Livermore National Laboratory and staff, for introducing me to his computerized archival system and for providing photographs and information; Edward

ACKNOWLEDGMENTS

Reese of the Modern Military History Branch of the National Archives, for his expert research assistance.

Two Sandians in particular deserve special recognition for their long-standing and valuable contributions to the production of this volume: W. Lee Garner, under whose division the History Project resided for the longest period of time, and Tonimarie Stronach, Technical Publications Coordinator and Research Assistant. Garner, talented writer and artist, donated personal time to edit the entire manuscript. Tonimarie Stronach, who joined the History Project in December 1984, has been responsible for compositing and formatting the entire text and bibliography. Her diligent efforts in locating documents and photographs from various depositories around the country have added immensely to the interest and value of the book. Equally important, her skillful coordination of production details and assistance in the acquisition of resources, combined with a dedication to "getting the job done" within rigid time constraints have made her a valuable and loyal contributor to the publication of this volume.

Part-time staff member Arthur Wright, who worked with the History Project during two summers, contributed by providing research assistance and a draft paper on components, and compiling a preliminary bibliography. Alice Matlock, the Project's Youth Opportunity Trainee, has assisted in the operation of the History Office and Archives.

Other Sandians deserving of special recognition include Richard Craner and the Classification staff, who carefully perused each chapter as it was produced and also reviewed the manuscript in its entirety; Jan Willis and ATEX personnel, for patient support and assistance in seeing us through the various production crunches; E. Roberta "Bobbi" Voelker, for providing a final edit of the volume; Shawkeet Hindi and his helpful staff, for printing and reproducing numerous photographs, for providing photolithographic support, and for printing related historical materials for the Project; and Jim Hayes, Bob Colgan, and the talented Tech Art staff.

ACKNOWLEDGMENTS

It is a pleasure to express my appreciation to Barry Schrader and Cindy English of the Sandia, Livermore, Public Relations office, for help in setting up numerous interviews and in gathering data on the early history of the Livermore Laboratory; Gersedon "Gerse" Martinez of the Sandia *Lab News* and photographer Bill Laskar, now retired, for their photographic expertise; former *Lab News* editor Bruce Hawkinson for sharing personnel exit interviews; Carmen de Souza and her cheerful assistants, for helping with production of review drafts of the manuscript; the supportive staff of the Sandia Technical Library, especially Edna Baca, Central Technical File Librarian, for her persistence in tracking down hard-to-find documents; Connie Souza, Books Librarian, who assisted in the acquisition of numerous secondary sources; and Information Specialists Walter Roose and Marge Meyer. Last, but not least, I would like to thank the many other generous Sandians who have volunteered their time, talents, and documents to this Project. The production of this first volume of the history of Sandia has indeed been a cooperative venture.

NSF

Sandia National Laboratories

The Postwar Decade

Jubilant employees on Wall Street celebrate end of war in Europe.

Prologue

BATTLE OF THE LABORATORIES:
The Secret War

On August 6, 1945, President Harry Truman announced the dawn of a new era—the Atomic Age. "Sixteen hours ago," he said, "an American airplane dropped one bomb on Hiroshima. . . . It is a harnessing of the basic power of the universe. The force from which the sun draws its power has been loosed against those who brought war to the Far East." Continuing, Truman declared the development of the bomb to be "the greatest achievement of organized science in history."

Fearing that the Germans were working feverishly to develop the bomb, the United States and Great Britain by 1942 had pooled their scientific knowledge and entered the race. With the blinding flash over Hiroshima August 6, 1945, Truman asserted: "We have won the Battle of the Laboratories as we have won other battles."[1]

The victory of the Battle of the Laboratories involved individuals and institutions ranging from Europe to an isolated hideaway in the high desert of the southwestern United States. It is, in effect, the story of a war within a war, a secret war, or S-1, as it was officially known to a select few.

In 1939 the world stood on the threshold of an Elizabethan Age in science. At this time, the race to harness the atom began in earnest. As history has proven repeatedly, and for some reason not totally comprehensible, syntheses of creative thought and intelligence seem to burst forth from different individuals and from different parts of the world almost simultaneously. This was the situation in 1939 when international advances in science and technology approached a breakthrough.

Three decades after Albert Einstein had introduced his theory of relativity, it appeared conceivable that theory might become reality. Significant discoveries relevant to harnessing the atom were being made in a half-dozen countries. Pioneering in these investigations were German physical chemists Otto Hahn and Fritz Strassmann, Danish Nobel Laureate Niels Bohr, discoverer of the neutron James Chadwick at Cambridge, the Frederick Joliot-Curies in Paris, and in Italy, Enrico Fermi, who was to receive the Nobel Prize for his work with neutrons.

Until 1939, scientists and physicists had enjoyed an openness and sharing that allowed them to advance on the ladders of learning and expertise provided by others—a situation that would soon cease. The interplay and exchange among them, in effect, had made each person stronger and wiser within the group than individually. Hahn's report on atom splitting, for example, published openly in *Die Naturwissenschaften*, created not only an intellectual furor of speculation, but also a fever of humanitarian anxiety among scientists who understood its ramifications. If multiple neutrons could be liberated in the disintegration of uranium, a chain reaction might be possible. As Hungarian-born physicist Leo Szilard wrote to Joliot-Curie February 2, 1939: "In certain circumstances this might then lead to the construction of bombs, which would be extremely dangerous in general and particularly in the hands of certain governments."[2]

There was no doubt as to the identity of one "certain government." For some time scientists had been fearful that Hitler would attain the means to realize his goal of world domination. Already some scientists had been forced to flee. Among them was Lise Meitner, an Austrian Jew and head of the physics department at the Kaiser Wilhelm Institute in Berlin, where she worked with Otto Hahn and Fritz Strassman. Together they had been successfully bombarding uranium with neutrons and recording exciting results when she found herself a prime candidate for a Nazi concentration camp. Despite intercession by her respected colleagues, the Fuhrer issued orders for her arrest. Making a hasty escape disguised as a tourist, she fled to Stockholm, Sweden, where she stayed at a boardinghouse near Göteborg. At Christmastime, she was joined by her nephew Otto Frisch, who became fascinated with the new calculations that the small, energetic woman had been working on with Hahn.

From her experiments, Meitner theorized that the radioactive isotope, resulting from the bombardment of uranium with neutrons, signified nuclear fission. Simply stated, as the nucleus split into two lighter elements, energy was released and neutrons as well, making a chain reaction feasible. After sharing her ideas with Niels Bohr, founder of the institute for Theoretical Physics, Meitner and Frisch agreed to develop an experiment to verify the hypothesis.

Fearful of the specter of the swastika, other scientists soon joined Meitner in a general European exodus. Edward Teller went to George Washington University, and Victor F. Weisskopf became a member of the University of Rochester faculty. Enrico Fermi, instead of returning to Fascist Italy from the Nobel Prize ceremonies in Sweden, traveled to New York and Columbia University.

Meanwhile, Bohr left Copenhagen for the Institute for Advanced Study in Princeton. When he reached New York January 16, 1939, he found a cable from Meitner and Frisch informing him that the experiment had been successful—the process of atom splitting had freed 200 million electron volts of electricity. On January 25, 1939, scientists at Columbia University, with Fermi as adviser, again confirmed the experiment with the same positive results. With a simple push of the button, the oscilloscope registered "precisely 200 million volts."

Still unencumbered by the shroud of secrecy, Niels Bohr in January discussed the new developments at the Fifth George Washington Conference on Theoretical Physics. At the spring meeting of the American Physical Society, also in Washington, Bohr gave a report in which he stated in very plain terms that "a projectile armed with a tiny fragment of U-235 under bombardment from slow neutrons could blow up most of the District of

Columbia."[3] This graphic pronouncement caused a stir among scientists in the audience—including a young American named J. Robert Oppenheimer.

Surprisingly, however, the scientific breakthrough of nuclear fission, although widely reported in the press, initially made little national impact. Several reasons can be identified. First, uranium was a rare commodity, and the majority of it was U-238, which was too stable for fission. Second, no bomb had been designed, although the German refugee, Klaus Fuchs, would soon be involved in the task. Third, the creation of an actual device would be an expensive undertaking. Only one man in the United States had such an amount of money under his control—President Franklin D. Roosevelt.

Despite these handicaps, several events occurred to spur on the active interest of scientists in the realization of an atomic device. Dr. S. Flugge, an anti-Nazi physicist, learned that the Germans were constructing a "uranium machine." Believing that the scientific community should be aware of these developments, he published a report on uranium chain reaction in *Die Naturwissenschaften* and also in a conservative German newspaper. American scientists, fearful that Flugge was revealing only a small part of what he knew, became even more alarmed when the Germans placed a ban on the exportation of uranium ore from Czechoslovakia and a blackout on information pertaining to uranium experiments. These factors, combined with knowledge of Operation U (as the German uranium control system was labelled) and its positioning under the Army Weapons Department in Berlin, motivated scientists in the United States to action.

After futile attempts by Szilard, Teller, and Fermi to warn Washington of the consequences of national lethargy, they decided to enlist the aid of the best known of them all—Albert Einstein. As William Manchester points out in his colorful style: "Albert Einstein was so famous that when he decided to wear his hair long, he added a phrase ['long hairs'] to the American idiom."[4] The scientists reasoned accurately that if any name could catch the attention of the President, it would be Einstein's.

Szilard and Eugene Wigner contacted the scientist at his summer home at Peconic, Long Island. Einstein quickly recognized the significance of a chain reaction in uranium, and, at the suggestion of Szilard and Wigner, agreed to write a letter that would be hand-carried to Franklin D. Roosevelt. After some decision-making regarding approach, Szilard returned to Long Island August 2, with Edward Teller, who translated Einstein's letter from German to English. Alexander Sachs, financier and economic adviser to the President, was to play courier.

On October 11, Sachs met with the President and, to make sure that it was not filed away and forgotten, read the Einstein letter aloud. The letter informed Roosevelt that the work of Joliot-Curie in France and Fermi and Szilard in America made possible the generation of a nuclear chain reaction that could result in "vast amounts of power." The new phenomenon, Einstein warned, would lead to the construction of "extremely powerful bombs." In view of this probability, he urged the Administration to take action "to secure a supply of uranium ore for the United States" and "to speed up experimental work" by obtaining the cooperation of industrial laboratories. Roosevelt, however, became bored and indicated that he felt "government intervention might be premature." Quickly assessing the President's reaction, Sachs wisely requested that the meeting be continued the next morning at breakfast.

That night Sachs tried to come up with a new approach—a way to catch and hold Roosevelt's interest. The following morning he successfully dramatized the issue by reminding Roosevelt that Robert Fulton initially offered his steamship design to Napoleon Bonaparte. Napoleon, not realizing the importance of the invention, rejected it as impractical—in effect, discarding the means with which to invade England and attain victory.

The dramatization worked. Suitably, the President poured two glasses of Napoleon brandy and lifted his in a toast to Sachs:

"Alex," he said, "what you are after is to see that the Nazis don't blow us up."

"Precisely," Sachs replied.

Albert Einstein
Old Grove Rd.
Nassau Point
Peconic, Long Island

August 2nd, 1939

F.D. Roosevelt,
President of the United States,
White House
Washington, D.C.

Sir:

    Some recent work by E.Fermi and L. Szilard, which has been communicated to me in manuscript, leads me to expect that the element uranium may be turned into a new and important source of energy in the immediate future. Certain aspects of the situation which has arisen seem to call for watchfulness and, if necessary, quick action on the part of the Administration. I believe therefore that it is my duty to bring to your attention the following facts and recommendations:

    In the course of the last four months it has been made probable - through the work of Joliot in France as well as Fermi and Szilard in America - that it may become possible to set up a nuclear chain reaction in a large mass of uranium,by which vast amounts of power and large quantities of new radium-like elements would be generated. Now it appears almost certain that this could be achieved in the immediate future.

    This new phenomenon would also lead to the construction of bombs, and it is conceivable - though much less certain - that extremely powerful bombs of a new type may thus be constructed. A single bomb of this type, carried by boat and exploded in a port, might very well destroy the whole port together with some of the surrounding territory. However, such bombs might very well prove to be too heavy for transportation by air.

The United States has only very poor ores of uranium in moderate quantities. There is some good ore in Canada and the former Czechoslovakia, while the most important source of uranium is Belgian Congo.

In view of this situation you may think it desirable to have some permanent contact maintained between the Administration and the group of physicists working on chain reactions in America. One possible way of achieving this might be for you to entrust with this task a person who has your confidence and who could perhaps serve in an inofficial capacity. His task might comprise the following:

a) to approach Government Departments, keep them informed of the further development, and put forward recommendations for Government action, giving particular attention to the problem of securing a supply of uranium ore for the United States;

b) to speed up the experimental work, which is at present being carried on within the limits of the budgets of University laboratories, by providing funds, if such funds be required, through his contacts with private persons who are willing to make contributions for this cause, and perhaps also by obtaining the co-operation of industrial laboratories which have the necessary equipment.

I understand that Germany has actually stopped the sale of uranium from the Czechoslovakian mines which she has taken over. That she should have taken such early action might perhaps be understood on the ground that the son of the German Under-Secretary of State, von Weizsäcker, is attached to the Kaiser-Wilhelm-Institut in Berlin where some of the American work on uranium is now being repeated.

Yours very truly,
A. Einstein
(Albert Einstein)

HERITAGE

Albert Einstein and Leo Szilard reenact the writing of the 1939 letter to President Roosevelt. The letter led to development of the first atomic bomb and establishment of the United States' atomic weapons program.

Summoning his military aide, Roosevelt handed General Edwin ("Pa") Watson the Einstein letter and the accompanying documents.

"Pa," he said, "this requires action!" With this statement, Roosevelt declared the Secret War to build the bomb.[5]

By 1940, America was on the brink of *open* war as well. As early as 1937, there had been ominous signs in Europe. In that year, James Byrnes, Roosevelt's delegate to the Inter-Parliamentary Union meeting in Paris (and later to be Director of Economic Stabilization), was in Nuremberg, Germany, ostensibly to evaluate Germany's program for economic recovery. What he saw, however, was "a program not for economic recovery but for armed aggression." Byrnes recalled the scene:

> Twelve thousand troops marched past our stand, accompanied by the rumble of scores of tanks and motorized weapons, while 450 planes swept across the skies above the stadium. As the demonstration reached its climax, Hitler rode across the field in an open automobile standing erect and with his right arm uplifted in the Nazi salute. An exultant cry arose from the multitude. But I was frightened.[6]

Ironically, on August 28, 1939, Hitler concluded a *non*aggression pact with the Soviet Union. Two days later, Nazi armies invaded Poland. Denmark and Norway were the next to fall as German Panzer divisions methodically plowed through lines of the Allies, and Stuka dive bombers slaughtered soldiers and civilians alike. Holland surrendered in May, followed by Belgium and France. Churchill's magnificent oratory strengthened British morale and generated Allied support from across the Atlantic, while the Luftwaffe bombed London and battled the Royal Air Force. In the United States, the President mobilized American industry, listened to the pleas of interventionists and isolationists, and initiated Lend-Lease. On September 4, 1941, after a German submarine fired the first shot at the *Greer*, an American destroyer, Roosevelt changed his naval orders from "search and patrol" to "search and destroy." In October, a German U-boat torpedoed another destroyer, the *Reuben James*, creating a sensation in the American press. However, it was the infamous

Japanese attack on Pearl Harbor December 7, 1941, that provided the final impetus to unite the nation.

By 1942, with the United States at war with both Japan and Germany, the race to build the bomb intensified. Reports of consignments of uranium and heavy water being shipped to Germany's atomic physicists, in addition to the arrest of two German agents outside what would soon be known as Oak Ridge, Tennessee, further alarmed Allied scientists.

After receipt of the Einstein letter in 1939, Roosevelt had appointed the Advisory Committee on Uranium. In addition to committee members, scientists Leo Szilard, Eugene Wigner, and Edward O. Teller were included as well as Fred L. Mohler, Alexander Sachs, and Richard B. Roberts. Lyman Briggs headed the Committee as chairman; Colonel Keith F. Adamson represented the Army Ordnance Department; and Commander Gilbert C. Hoover, the Navy Bureau of Ordnance. Their recommendations mentioned both atomic power and an atomic bomb.

In June 1940, the National Defense Research Committee (NDRC) was established with Vannevar Bush as chairman, and the Uranium Committee reconstituted as a subcommittee of that group. After the summer of 1941, the Committee became known as the Uranium Section or S-1 of the NDRC. In October of that year, Roosevelt initiated a comprehensive research project relating to the development of atomic energy as a military weapon and appointed as director Dr. Vannevar Bush of the Office of Scientific Research and Development (of which the NDRC was a part) to conduct the study. James Conant then moved to the chairmanship of the NDRC. These appointments signified White House sanction for a full-scale effort directed toward the development of an atomic weapon. Secretary of War Henry L. Stimson served as the President's senior adviser on the military employment of atomic energy.

Two factors encouraged Bush and Conant to feel that such a development was a distinct possibility: First, at the University of California in Berkeley, Glenn T. Seaborg and Emilio Segrè discovered plutonium. Secondly, British scientists sent a report indicating their belief that uranium had a military significance

Members of the NDRC included (from left to right) E. O. Lawrence, A. H. Compton, Chairman Vannevar Bush, and James B. Conant.

"worth a major effort" and outlined a concrete program geared toward that objective. To implement such an effort, Bush set up a Planning Board consisting of a young chemical engineer, Eger V. Murphree, as chairman, and Ernest O. Lawrence, Harold C. Urey, and Arthur H. Compton as program chiefs. The Planning Board would be responsible for engineering, planning studies, and the supervision of a pilot plant and laboratory investigations. In essence, they were to ensure that the necessary plans were available when the production phase was reached.[7]

Thus, by September 1942, the Secret War had gained considerable momentum. The project, known as the Manhattan Engineer District, was transferred to the War Department under the dynamic direction of Brigadier General Leslie Groves in that same month, and the ability to let contracts to different laboratories was initiated. The appointment of J. Robert Oppenheimer as Director of the Los Alamos effort during the summer of 1943 provided additional leadership.[8]

Other laboratories established to carry out certain tasks necessary to the success of the project included the Clinton Engineer Works, which concerned itself largely with separation of U-235; Hanford Engineering Works in Richland, Washington, established for plutonium production; and the Los Alamos facility in New Mexico, where the bomb itself was designed and built. Contracts were also granted to Columbia University, Princeton University, Standard Oil Development Company, Cornell University, Carnegie Institute, the University of California, Rice, Purdue, and the Massachusetts Institute of Technology. Early in 1943, the Argonne Lab was constructed outside Chicago, and a pilot plant for production of plutonium was built at the Clinton (Oak Ridge) site in Tennessee. This network of laboratories, organized under the auspices of the United States, represented a revolutionary relationship between science and government established in the interests of national defense.[9]

A stupendous undertaking, the Manhattan Engineer District Project, as Anthony Cave-Brown has so aptly phrased it, was "a gamble on the theoretical calculations of a group of scientists and economists."[10] Organized and carried out in great secrecy—

from the public, cabinet members, and administrative personnel—the project had to be concealed from enemies and allies alike. In line with this policy, Roosevelt and Churchill attempted to exclude the Soviets throughout the duration of the war. Subsequent events, however, proved exclusion to be an exercise in futility.

As in the United States, Germany, and Great Britain, Soviet interest in the development of nuclear fission became public knowledge in 1939 when A. I. Brodsky published an article dealing with the separation of uranium isotopes. Two of Brodsky's colleagues, in fact, had conducted fission experiments in the shaft of a Moscow subway, but the project was terminated by Stalin "on the grounds that it had no practical value." Furthermore, Soviet physicist Igor Kurchatov and associates, spurred on by the announcement of the German success in splitting the atom, made investigations leading to the discovery of spontaneous fission. Kurchatov then wired a telegram reporting his findings to the American journal *Physical Review*. When publication of the lengthy telegram in July 1940 roused little reaction among American scientists, the Russians inferred that the United States was already involved in a secret project to build the bomb. This conclusion prompted the Soviets to implement an espionage campaign that would keep them well abreast of developments in the American and British atomic program.

Among the espionage agents operating in the United States was a Philadelphia chemist of Russian extraction, Harry Gold, and a German scientist, Klaus Fuchs, who worked first for the British atomic program and later for the United States as a trusted member of the Manhattan Project. Working together, Fuchs and Gold, known by the Russians as "the candy man," would manage to pass numerous atomic secrets to the U.S.S.R. by filtering the information out of the country through the New York spy ring directed by Soviet Vice-Consul Anatoli Yakovlev. Donald MacLean, a British "Yakovlev" and a member of the Combined Policy Committee, was also privy to Anglo-American atomic policy and dutifully kept the Russians informed about the Manhattan Project.

By June 1941, Soviet studies on uranium had to be curtailed because of the German invasion, but investigations were resumed as soon as possible. In the interim, Soviet agents kept Stalin abreast of Anglo-American efforts in the atomic field. The resulting information leaks would have increasing significance as the end of the war approached.[11]

Meanwhile, scientific advances in the United States continued. At the University of California, Ernest O. Lawrence initiated conversion of the cyclotron into a giant mass spectograph; and on December 2, 1942, Enrico Fermi and assistants, working in an abandoned squash court within the ivy-covered walls of the University of Chicago, achieved the first controlled nuclear chain reaction. There, beneath the west stands of Stagg Field, Fermi and his staff assembled "a [U-235] pile of unprecedented size." To reduce the risk of losing control, they inserted seven strips of cadmium and three rods of boron steel through the pile to consume neutrons. By sliding them in and out, they hoped to have control over the incredible force they were creating. For added safety, two young physicists volunteered to act as a "suicide squad." Standing on a scaffold that extended over the pile, they held buckets of cadmium salt solution to hurl as a final fail-safe mechanism. On December 1-2, 1942, the intense clicking of the counters signalled that a self-perpetuating chain reaction was a reality; while at Berkeley, J. Robert Oppenheimer's research team, conducting an investigation of an atomic bomb, began to realize that their theoretical studies were fast approaching the experimental stage.[12]

By 1943, it was apparent that a revolutionary new "industry" was in the making—one that would ultimately determine the victor in the battle of the laboratories. Suitably, perhaps, for such a prime example of American pioneering spirit, ingenuity, and inventiveness, two wilderness sites were selected—one in the red soil of eastern Tennessee, the other in mountainous terrain near Santa Fe, New Mexico.

The Tennessee group, or the Clinton Engineer Works as the site was formally known, became the headquarters for the Manhattan Engineer District (MED) Project and General Leslie

Groves' "home away from home." The mushrooming community that developed to service the production sites being established on more isolated parts of the reservation became known as Oak Ridge.

To turn the fissionable material being produced at Oak Ridge into a weapon required a special laboratory for that purpose. Until 1942, the S-1 project had been compartmentalized into units to prevent the enterprise from being leaked to the enemy. In the interest of more rapid progress, however, this approach began to change toward a coordinated effort. As the newly appointed director of Project Y, J. Robert Oppenheimer spoke out strongly in favor of integration. He personally found the "disjointed character" of the project detrimental. Oppenheimer realized, too, that ordnance experiments required an isolated proving ground.[13]

The result was the selection, after some debate, of another wilderness site that could be tightly secured by the Army. Toward the end of October 1942, Major John H. Dudley of the MED began his search, eventually narrowing his quest to northern New Mexico. With the assistance of the United States Engineering Office in Albuquerque, the choice was further reduced to two locations north of the city—Jemez Springs (fifty miles away) and the Los Alamos Ranch School, a private academy for boys near Otowi Crossing (ninety miles away).

Oppenheimer, accompanied by Major Dudley and Edwin M. McMillan, set out on horseback to investigate the Jemez Springs site. Located in a canyon, limited in space, and vulnerable to flooding, the area proved unsuitable. General Groves, who arrived later in the day, had misgivings about dispossessing the owners of the many small farms in the region and agreed with the conclusion.

The selection team, proceeding by automobile to the Los Alamos Ranch School, took a picturesque route along the rim of a gigantic extinct volcano that formed a huge caldera covered with lush, green grass grazed by cattle. Popularly known as El Valle Grande, after the best known of the valleys, the terrain is impressive in its beauty. The public road, winding along mountainsides marked by canyons carved by centuries of rain and snow, led to the area's second most prominent geological feature, the Pajarito Plateau. Lying at an altitude of between 6,300 and 7,300 feet above sea level, the Plateau is a volcanic bench, approximately two and one-half to three miles wide. On this plateau, forested with piñons, juniper, ponderosa pine, fir, spruce, aspen, and oak, an easterner named Ashley Pond in 1917 established the Los Alamos Ranch School for boys.

Oppenheimer, who owned a summer home across the valley near the headwaters of the Pecos River east of Santa Fe, had discovered the area some years earlier while on a pack trip. The setting, he felt, would meet the criteria for a secret laboratory site. Other members agreed, and on December 7, 1942, the first anniversary of Pearl Harbor, Secretary of War Henry L. Stimson sent a brief communiqué informing officials of the Los Alamos School that the government would be expropriating the property for a special project. A contract for constructing laboratory buildings and temporary living quarters was let in December 1942, and the school was closed in mid-February.

A letter of intent from the Office of Scientific Research and Development enabled the University of California and Oppenheimer to begin recruitment of personnel and the acquisition of equipment for Los Alamos. The first formal contract between the University of California and the Manhattan Engineer District of the War Department, made retroactive to January 1, 1943, was not signed until April 20, 1943.

The isolation of the area not only made construction difficult, but also created problems in the recruitment of technical personnel and administrative staff. Oppenheimer was able to bring a few individuals from the University of California; and James Conant, as the new chairman of the National Defense Research Committee, recruited a few others. By April 1943, there were about a hundred scientists working on Project Y in addition to supporting personnel. The laboratory complex to build the bomb was well under way. Only twenty-eight months after the first scientific contingent arrived at Los Alamos, the world's first manmade atomic explosion took place near Alamogordo, New Mexico.[14]

The Los Alamos Ranch School for Boys, founded by easterner Ashley Pond in 1917, was expropriated by the government in December 1942 for use in Project Y. Shortly after the last class graduated and the school closed its doors in February 1943, nearly 100 scientists relocated to the area to take part in the secret project. Pond's daughter, Margaret Hallet Pond, had three sons—Ted, Hugh, and Allen Church. Ted would join the Electronic Engineering Group of Z Division in 1947. Younger brothers Hugh and Allen would become employees of Sandia Laboratory.

Ironically, before the first successful test of an atomic bomb at Trinity Site on July 16, 1945, the war against Germany had been won by conventional means. Victory in Europe, however, did not mean victory in Japan. Japanese kamikazes continued to guarantee their immortality by diving into U.S. ships, while entrenched ground troops stained the Pacific islands with the blood of American soldiers (the eight square miles of volcanic ash called Iwo Jima alone cost the lives of 5,000 Marines).

In the United States, Americans mourned the sudden death of President Roosevelt at Warm Springs, Georgia, April 12, 1945. His successor, the tough, feisty man from Missouri, Harry S. Truman, carried on. With the promise of Russian support, the Allies planned a November 1 invasion of the Japanese mainland—an operation expected to be costly in terms of American lives lost. Secretary of War Henry L. Stimson—the man who, with the concurrence of Acting Secretary of State James C. Grew and Secretary of the Navy James Forrestal, recommended that the bomb be used—explained the rationale for his decision in 1947: "I was informed that such operations might be expected to cost over a million casualties to American forces alone." General Douglas MacArthur thought the casualties would be higher, the "biggest bloodletting in history," in fact. After all, the Japanese army was digging in; thousands of tons of ammunition were being stowed in caves; and 5,350 aircraft awaited kamikaze action from underground hangars. Philosophically, for the Japanese the word *surrender* had no meaning. "As we understood it in July [1945]," Stimson recalled, "there was a very strong possibility that the Japanese government might determine upon resistance to the end."[15]

With the distinct possibility of having to face a lengthy guerrilla engagement, President Truman sailed for Potsdam. Truman, however, had one ace in the hole—preparation for testing of the atomic device had reached the final stages. On July 16, two coded messages were sent to Potsdam. The first, wired to the anxious President from Washington, read: "Operated on this morning. Diagnosis not yet complete but results seem satisfactory and already exceed expectations." The second message,

BATTLE OF THE LABORATORIES: THE SECRET WAR

President Franklin D. Roosevelt sanctioned the Manhattan Project after being convinced of the need to develop the atomic bomb.

Harry S. Truman, who became President of the United States following the unexpected death of Franklin D. Roosevelt, first learned of the bomb's existence when briefed by Secretary of War Stimson on April 25, 1945.

addressed to Churchill from Stimson, informed the prime minister: "Babies satisfactorily born," to which he responded: "This is the second coming, in wrath." A more detailed message arrived on July 18 indicating that the Trinity device was much more powerful than anticipated.[16]

After some discussion, Churchill and Truman decided that Stalin must be informed, but with little ado. When Truman casually told Stalin that the United States had developed "a weapon of unusual destructive force," the Russian leader showed little overt reaction, merely indicating that he hoped the Americans would make "good use of it against the Japanese." Stalin's private reaction, however, was quite different. To Marshall Zhukov and Foreign Minister Molotov, he showed some concern: "We've got to work on Kurchatov and hurry things up," he ordered. According to the analysis of Anthony Cave-Brown, this event marked the genesis of atomic diplomacy and the Russian decision to begin an atomic program in earnest.[17]

For Truman, the news of Trinity represented a twofold victory. On the one hand, the possession of an atomic bomb obviated the need for Soviet entrance into the war against Japan. From another standpoint, and of the most immediate importance, Trinity allowed the President to take a firm policy stand against Japan, including the demand for unconditional surrender and, ultimately, a swift end to hostilities.

For some time, American policy toward atomic energy had been directly related to Soviet Russia. In his book on the subject, Secretary of War Stimson disclosed that the relationship had been a critical issue since he first briefed Truman on the bomb, April 25, 1945, following Roosevelt's death. On June 6, the Interim Committee decided that information relating to S–1 should be kept secret from Russia "or anyone else" until it was used against Japan. By July, however, Stimson maintained that the United States "no longer cared about or presumably even wanted Soviet entry into the Pacific War." Churchill agreed: news of Trinity meant "the end of the war in one or two violent shocks." "Moreover," he said, "we should not need the Russians." Stimson explained the political impact of Trinity succinctly:

BATTLE OF THE LABORATORIES: THE SECRET WAR

> The news from Alamogordo . . . made it clear to the Americans that further diplomatic efforts to bring the Russians into the Pacific war were largely pointless. The bomb as a merely probable weapon had seemed a weak reed on which to rely, but the bomb as a colossal reality was very different.[18]

The decision to use the bomb against Japan, however, was not made until a "last chance" warning had been issued. In what was subsequently referred to as the Potsdam Declaration, Truman, Churchill, and Chiang Kai-shek broadcast to the Japanese people assurances of humane treatment, no recriminations, the retention of basic freedoms, industry, world trade, limited temporary occupation by Allied forces, and a new order based on peace, security, and justice. The capstone, of course, was the demand for unconditional surrender; the alternative, "prompt and utter destruction."

In Tokyo, there were a few who understood that Truman was offering an opportunity for the Japanese to determine their own form of government. However, the uncompromising attitude was epitomized by Admiral Baron K. Suzuki, who rejected the Potsdam Declaration as "unworthy of public notice, a rehash of old proposals, and beneath Japanese contempt." Despite Suzuki's response—and hoping for a last-minute change in position—Truman waited until August 2 to issue the final order that would introduce the world to the atomic bomb and a new dimension in military and political power.[19] With the bombing of Hiroshima on August 6, 1945, and Nagasaki three days later, the new era was born and the battle of the laboratories won.

The remarkable story of how the United States achieved this victory and continued to maintain its military supremacy involves not only the contributions of the better known laboratories within the complex, but also the contributions of personnel who would form the nexus of a small branch of Los Alamos, referred to as Z Division. In Z Division—the forerunner of Sandia National Laboratories—an ethos would be born that continues to this day in the tradition of rendering "exceptional service in the national interest."[20]

WAR IN PACIFIC OVER!

ALBUQUERQUE JOURNAL
NEW MEXICO'S LEADING NEWSPAPER
65th Year | Volume 265 Number 46 | Entered as second class matter Albuquerque, N. M., postoffice under act of Congress 1879 | Wednesday Morning, August 15, 1945 | Published Every Morning | * * | Price Five Ce

TODAY'S SMILE
ATLANTA, Aug. 14 (AP)—Sound of the times (from one filling station attendant to another):
"Hey, Eddie, wipe that windshield and check those tires—the war's over."

Good Morning
"Good Morning" Is Right. It' Best Morning the World Has Since 1939—War Is Ended, W All Hope, Forever.

Japs Accept Terms

War Manpower Controls Revoked As Fighting Ends

WMC Announces Seven-Point Program To Aid Reconversion

WASHINGTON, Aug. 14 (AP) — The government has revoked all war-time manpower controls, effective immediately, and set forth a plan aimed at speedy re-employment of veterans and released war workers.

In an action timed to coincide with Japan's surrender, the War Manpower Commission announced a seven-point program which it said would stimulate "reconversion activities and the speedy re-employment of displaced workers, at the same time restoring a free labor market."

Among the controls lifted are those providing for hiring through the United States Employment Service, employment ceilings to channel workers to essential industries, and the requirement for certificates of availability in changing jobs.

[illegible] At Once

Acting WMC Chairman Frank L. McNamee said regional directors have been instructed to put the new program into effect at once in the 1500 local USES offices throughout the country. Specifically, it provides that:

1. All manpower controls are to be lifted immediately. In their place voluntary community action to speed reconversion will be substituted.

2. The number of displaced workers and returning veterans in each community will be determined in co-operation with local management-labor groups. Action will be taken by the WMC and local USES offices in co-operation with the communities to speed reconversion and re-employment.

Eyes on Bottlenecks

3. Labor will be channeled by voluntary methods into civilian industries "especially into industries which may become reconversion bottlenecks and thus delay mass re-employment throughout the country as a whole."

4. Full facilities of the USES again will be made available to all employers, including those for whom services were restricted because of war requirements.

5. Extended services will be rendered to veterans in their re-adjustment to civilian employment.

6. Increased emphasis will be given to job counseling and other personalized services to assist job seekers to adapt their wartime experience to peacetime job opportunities.

7. Displaced war workers, many of whom have migrated during the war, will be assisted in finding employment in other communities where civilian production has expanded.

U. S. E. S. offices will continue to give preferential treatment to all reconversion activities, McNamee said.

Russia and Japan Wage Shortest War

By the Associated Press

Russia's war against Japan was the shortest ever waged between major powers. It lasted just 6 days, from midnight Aug. 9 when Russia's declaration of war took effect. Italy's "stab-in-the-back" war against France lasted 14 days, from the declaration effective June 11, 1940, until the armistice.

The Weather

ALBUQUERQUE AND VICINITY: Partly cloudy Wednesday and Thursday with scattered afternoon and evening thunder showers over mountains. Slightly lower afternoon temperature. High Wednesday 85.

NEW MEXICO: Partly cloudy with scattered thundershowers Wednesday. Cooler southeast and central portions. High 80 northeast, 85 southwest.

Yes! Sir! Gen. MacArthur

Hirohito and horse

Text of Truman Statement on Jap Surrender

WASHINGTON, Aug. 14 (AP)—Following is the text of President Truman's statement on the Japanese surrender:

"I have received this afternoon a message from the Japanese government in reply to the message forwarded to that government by the Secretary of State on Aug. 11.

"I deem this reply a full acceptance of the Potsdam declaration which specifies the unconditional surrender of Japan. In this reply there is no qualification.

"Arrangements are now being made for the formal signing of surrender terms at the earliest possible moment.

"General Douglas MacArthur has been appointed the supreme Allied commander to receive the Japanese surrender.

"Great Britain, Russia and China will be represented by high ranking officers.

"Meantime, the Allied armed forces have been ordered to suspend offensive action.

"The proclamation of V-J Day must wait upon the formal signing of the surrender terms by Japan."

Small Boy Keynotes Victory Celebration

All by himself, a little red-headed boy "key-noted" the spirit of the holiday by blowing loudly on his trumpet in front of a downtown shoe store.

At another business establishment, a restaurant, the owner came outside with a big butcher knife, dug the mud out of the flag holder in the sidewalk, inserted his flag. Then he looked all around him with a broad smile.

Petain Convicted Of Treason, Given Death Sentence

Court Recommends Clemency for Marshal, Head of Vichy Regime

PARIS, Wednesday, Aug. 15 (AP) — Marshal Henri Philippe Petain was convicted and sentenced to death early Wednesday by three judges and a 24-man jury who deliberated almost seven hours.

The high court of justice added it "hoped the sentence would not be executed."

(This recommendation for clemency presumably will be considered by Gen. DeGaulle, president of the French provisional government.)

Besides condemning the 80-year-old former chief of the Vichy state to death for "plotting against the internal safety of France," the court also sentenced him to national indignity and ordered confiscation of all his property.

The lengthy judgment, read by Judge Mongibeaux, president of the court, went over the acts of collaboration of the Vichy government with Germany point by point and laid their responsibility at Petain's feet. Mongibeaux said the Marshal instituted "a veritable regime of terror" in France.

Truman Calls For Two-Day Holiday

WASHINGTON, Aug. 14 (AP)—Wednesday and Thursday are days off for government workers, and holidays for pay purposes for workers in general.

And V-J Day, when it comes, will be a premium pay day too.

President Truman announced both rulings Wednesday night.

It was presumed, but not specifically stated, that government workers generally will be off on V-J day, too.

The White House said the next two days are to be regarded as legal holidays.

SECOND HOLIDAY A SURPRISE HERE

The President's proclamation of both Wednesday and Thursday as legal holidays came as a surprise to Albuquerque's leaders. They had planned on only Wednesday being a holiday. There was no decision Tuesday night as to whether Thursday would be a holiday here as well.

Among places announcing Wednesday closings are city, county and federal offices, university offices and classes, retail stores, and liquor stores.

HIROHITO'S NEW BOSS—General Douglas MacArthur has been named to accept the formal surrender of Japan. As supreme Allied Commander, he will, under the terms of the surrender, dictate the policies of the Japanese nation by using Emperor Hirohito as his "mouthpiece."

MacArthur Is Appointed To Direct Occupation of Jap Homeland

WASHINGTON, Aug. 14 (AP)—Now Emperor Hirohito—whom the Japanese believe descended from the sun—becomes a mouthpiece for the Allies.

Gen. Douglas MacArthur, appointed Supreme Allied Commander to receive the Japanese surrender, will tell Hirohito what to do. The Japanese understood this when they accepted the surrender terms. Nothing like this—taking orders from a white man or any foreigner—has ever before happened to a Japanese emperor.

Gen. MacArthur, who told the Filipinos "I will return" and did it, was assigned to govern the enemy he whipped on the road back.

President Truman's announcement specifically mentioned MacArthur only for the assignment to accept formal surrender, but it said he would be Supreme Allied Commander. The last Allied note to Tokyo said the supreme commander would give the emperor his orders.

He had been commanding general of American Army forces in the Pacific since April 5, 1945. MacArthur moved up from the southwest Pacific command which he had held since shortly after the war began.

Over six feet tall, wide shouldered, always spectacularly dressed, generally stern of face, with broad forehead, prominent nose and rather thin lips, MacArthur at 65 is still the complete soldier; he never leaves the role of commander and leader. His officers are fiercely loyal.

He knows the Japanese and by all accounts they respect and fear him. The general started learning about the Japanese shortly after his graduation from West Point, when he served a brief tour in that country.

Congress Ordered To Re-Convene Sept. 5

WASHINGTON, Aug. 14 (AP)—Congress, under the urgency of transforming the nation from war to peace, was called Tuesday to reconvene Sept. 5.

Senate Democratic Leader Barkley (D., Ky.) voiced the hope the legislative body would work with the same harmony in "the momentous transformation" that marked "the greatest victory ever won in a war for freedom."

Victory, Rain, Hurrah!

CARLSBAD, Aug. 14 (AP)—The V-J announcement and a heavy rain arrived simultaneously at Carlsbad Tuesday. Drouth-bitten citizens welcomed both. Fire and railroad whistles proclaimed the announcement and impromptu parades formed. Rain and din continued into the night.

GALLUP STORES CLOSE

GALLUP, Aug. 14 (AP)—Practically every store in Gallup closed Wednesday afternoon as fire sirens spread the V-J news. A general holiday will be observed Wednesday.

Albuquerque Celebrates Noisily; Stores, Public Offices Close Today

With the official announcement of Japan's surrender, Albuquerque went into its victory celebration late Tuesday afternoon.

The news, which came at 5 o'clock, was greeted with silence by persons in cafes and other downtown establishments with radios. For about five minutes it was as if the announcement had concerned a baseball score or some other everyday occurrence.

Auto Horn Sets It Off

Then a single auto horn sounded on Central Avenue, swelling quickly into a roar from hundreds of cars as pent up emotions, whetted by the many days of suspense while awaiting Japan's official capitulation, were loosened in a noisy demonstration that lasted far into the night.

Confetti began drifting into the streets, tossed out from upper story office windows but appearing to come out of thin air.

Excited persons smiled at one another, stranger to stranger. Sometimes there was a brief comment: "Wonderful, isn't it?" or "I'm still numb."

Whistles, Bells, Join Din

The crowds of homeward bound office workers slowed their usual hurried pace and groups of friends or strangers stood on the corners and even in the street. Slow moving automobiles wended their way along the streets, every car with its horn blasting full. Santa Fe Railway Shop whistles, locomotive whistles, sirens and church bells added to the din.

A shower which fell in the downtown area at the height of the early celebration failed to dampen it in the slightest. A clap of thunder added to the noises.

In residential districts, housewives opened their doors to call to passersby with joyful remarks.

Those with loved ones in Pacific service who hadn't received letters were calculating how soon it would be before they knew "for sure" that these were now safe, too, from battle.

Bars Shut 'Til Thursday

Within 30 minutes after the surrender announcement, liquor stores and bars were closing. Gus Bruskas, president of the Bernalillo County Liquor Dealers Assn. said they will remain closed until 7 a. m. Thursday.

Restaurants should remain open, Bruskas said, since their

Continued on Page Two

MacArthur Appointed Supreme Commander Of Occupation Army

Truman Announces Unqualified Acceptance of Potsdam Declaration; Arrangements Being Made for Formal Signing of Surrender; V-J Day Proclamation to Await Signatures on Official Document

WASHINGTON, Aug. 14 (AP)—The second World War, history's greatest fl of death and destruction, ended Tuesday night with Japan's unconditional surr der. Gen. Douglas MacArthur was designated Supreme Allied Commander to rece the surrender.

Formalities still remained—the official signing of surrender terms and a pro mation of V-J day.

But from the moment President Truman announced at 5 p. m. (Albuquer Time) that the enemy of the Pacific had accepted surrender, the world put as for a time woeful thoughts of the cost in dead and dollars and celebrated in w frenzy. Formalities meant nothing to people freed at last of war.

To reporters crammed into his office, showing now-useless war maps agains marble mantle, the President disclosed that:

Japan, without ever being invaded, had accepted completely and without res vation an Allied declaration of Potsdam dictating unconditional surrender.

There is to be no power for the Japanese empero although Allies will let him remain their tool. No lon will the war-lords reign, through him. Hirohito—or a successor—will take orders from MacArthur.

Allied forces were ordered to "suspend offensive tion" everywhere.

From now on, only men under 26 will be draft Army draft calls will be cut from 80,000 a month 50,000. Mr. Truman forecast that five to five and a h million soldiers may be released within 12 to 18 month

The surrender announcement set in motion a whole chain of events. Among them:

To a Japanese government which once had boasted it would dictate peace terms in the White House, Mr. Truman dispatched orders to "direct prompt cessation of hostilities," tell MacArthur of the effective date and hour, and send emissaries to the general to arrange formal surrender.

The War Manpower Commission terminated all manpower controls.

The Navy piled a $6,000,000,000 cancellation of contracts on top of a previous $1,200,000,000 cut in its shipbuilding program.

Congress was summoned back to work on Sept. 5, more than a month ahead of schedule.

History-Making News

Those developments and surrender capped a week packed with some of history's most stunning news:

The first atomic bomb on Hiroshima, Russia's declaration of war, another atomic bomb on Nagasaki, Japan's offer to surrender if she could have her emperor and his sovereign prerogatives, an Allied declaration he would become merely their instrument.

Surrender followed—at an instant when carrier planes of the mighty Pacific fleet were a few seconds from the targets in the Tokyo area. Pilots eager for a last lick at a weakening foe were reported to have gotten this word from Adm. William F. Halsey:

"It looks like the war is over. Cease firing, but if you see any enemy planes in the air shoot them down in friendly fashion."

Call For President

So there was reason for rejoicing. A war-wracked world made the most of it. At times President Truman had to come out on the White House porch to greet tremendous crowds—75,000 people by official estimate—who jammed the streets and parks around the executive mansion.

The Chief Executive spent half an hour dining with his staff. For him there was no personal celebrating, even with close friends.

But across the Potomac in the Pentagon building, nerve center of the Army's winning war, there wasn't any jubilation. There no one left except a couple bored public relations offi answering phones.

Sacrifices Rewarded

For millions of Americans, hundreds of millions of Al people, his surrender announ ment signified victory, peace the eventual return of loved o from war. To millions who sl beneath stark white crosses meant their sacrifices had been in vain.

For Japan, as for already v quished Germany, it meant end of savage conquest and gression, dismemberment of empire won by blood-spill disarmament and occupation.

Those were the terms of Allied declaration of Potsd decreeing unconditional surre er, to which Japan acceded.

Once the Japanese sphere stretched from Attu to Ti and Java and India. Once Ja kept a billion people ensla under iron rule, and threate to enfold another half billion.

Terrific Cost

Now she is defeated—with invasion—but at a terrific co

For the United States, the p of victory in Word War II more than a million casualt an expenditure of some $3 000,000,000.

The world counted its killed wounded at perhaps 23,000, plus millions more slaughtered air raids or dead of starvation

And even while Japan's s render note was on the way Washington, the crash of bo and clash of arms sounded on Pacific fronts, signalling de almost at the moment of peace enemies and Allies alike.

But for the moment the wo forgot the horrors of a conf which had its seeds in Japa grab of Manchuria in 1931.

In Washington and the rest the nation, screaming, whoop throngs let off steam in joy celebration.

The President made the histo announcement to a huge cro of reporters who had been vir ally living in the White Ho for days in anticipation of such a development.

Smiling and surrounded by staff, the President told the pr

Continued on Page Two

Loss of Cruiser, 883 Men, Occurs Near War's End

Indianapolis Is Sunk July 30 by Jap Sub In Philippines Sea

PELELIU, Palau Islands, Aug. 5 (Delayed) (AP)—The 10,000-ton cruiser Indianapolis was sunk in less than 15 minutes, presumably by a Japanese submarine, 12 minutes past midnight July 30—and 883 crew members lost their lives in one of the Navy's worst disasters.

She went down in the Philippines Sea, within 450 miles of Leyte while on an unescorted high speed run from San Francisco.

A total of 883 crew members lost their lives.

Survivors believe two underwater torpedoes smashed into the starboard side near the bow of the 14-year-old cruiser, setting off one of the 8-inch gun magazines.

Announcing this Tuesday the Navy said the famous vessel was lost shortly after completion of her last mission, sailing from San Francisco July 16 on a high-speed run to Guam to deliver essential atomic bomb material. She was lost after safely delivering her cargo.

Santa Fe Greets News With Jubilation, Prayers

SANTA FE, Aug. 14 (AP)—Sirens and church bells quickly joined by a chorus from scores of automobile horns Tuesday signaled to this ancient capital the news of Japan's surrender. Soon a crowd of several thousand persons was moving deliriously around the downtown section centering on the plaza.

Men and women soon had filled the historic cathedral of St. Francis to near capacity, streaming there to offer prayers of thanksgiving.

Drive Carefully

SANTA FE, Aug. 14 (AP)—Gov. Dempsey Tuesday night asked motorists to use extraordinary caution and to drive with reduced speeds during the period of celebration following Japan's surrender.

Part I.

HERITAGE

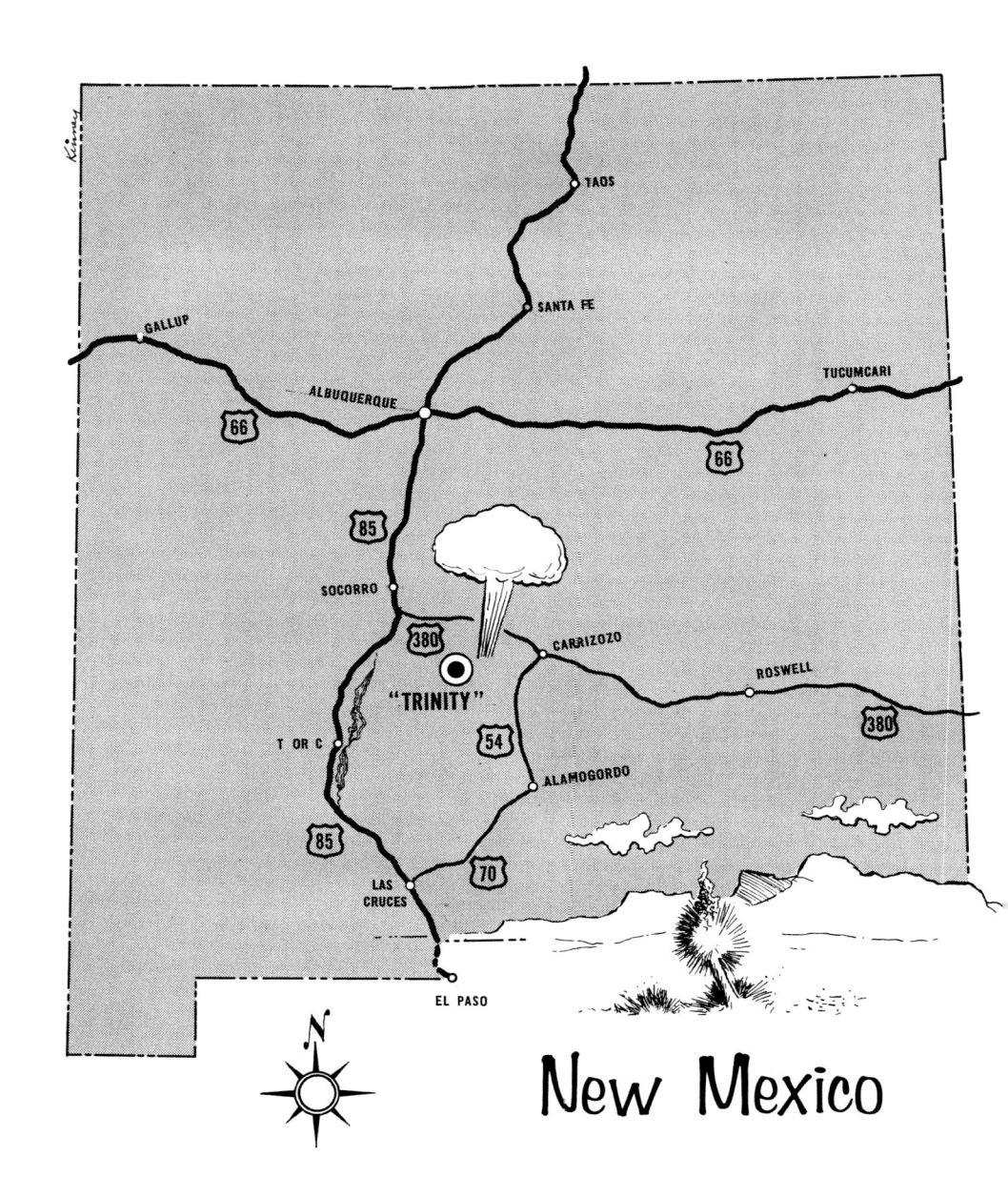

New Mexico, circa 1945

Chapter I.

TRINITY:

The Field Test That Revolutionized The World

And then without a sound, the sun was shining; or so it looked.

Otto Frisch

On July 16, 1945, when Americans sat down with their morning paper, they read that Harry Truman, Winston Churchill, and Joseph Stalin were in Berlin for the Potsdam Conference. Over the weekend, Truman planned a thirty-five-mile drive through some of the battered cities of Belgium. Elsewhere on the international scene, London "lit up" for the first time since 1939, and Britishers celebrated the end of blackouts as war emphasis shifted to the Pacific. In the United States, people watched the Braves topple the Cardinals and cheered at the track as "Pot Luck" won the Arlington Classic. In Albuquerque, New Mexico, readers noted that "Mrs. Ernie Pyle was presented a Jo Davidson bust of her famous war correspondent husband at ceremonies in the Pyle home, 700 South Girard." At the movies, the Kimo Theatre featured "G. I. Joe."

Despite the local-interest stories, war news of action on the various fronts occupied the majority of space. Indications were that "an early end to hostilities was, that week, unforeseen."[1] Not too far from Albuquerque, however, a momentous event was taking place that day that would hasten the end of the war.

HERITAGE

Future Sandians View Trinity

At an isolated site between Alamogordo and Socorro in south-central New Mexico, a flurry of secret activity involving scientists, engineers, and technicians gradually slowed as observers waited for the weather to clear. There, in an area known as the Jornada del Muerto (Dead Man's Route), thunder and lightning warned of a gathering storm and threatened delay of the most significant weapons test of the era.

For those most intimately associated with Project "Trinity," the tension was increasing. Timing was extremely important— and not only to those at the site. Across the ocean at No. 2 Kaiserstrasse in Potsdam, President Truman awaited news of the test while playing a game of international diplomacy with high stakes. Truman, who had just concluded a morning conference with Winston Churchill, Secretary of State James Byrnes, and British Foreign Minister Anthony Eden, expected to hear momentarily from Henry L. Stimson. *If* the test was on schedule and *if* the bomb was successful, Truman planned to issue a strong ultimatum to the Japanese demanding their unconditional surrender. On the other hand, if the test had to be postponed, it would mean a delay in Japanese reaction and, therefore, a delay in a decisive end to the war. Furthermore, from the technical viewpoint, postponement meant that exposure of electrical connections to excessive moisture increased chances of short circuits and a misfire.

Meanwhile, at the Trinity Site, twenty miles away from Ground Zero on a volcanic outcropping called Compañia Hill, a member of the Special Engineering Detachment (SED), Army Sergeant Leo M. "Jerry" Jercinovic, huddled in his poncho and waited for the rain to clear.[2]

Just three days earlier on Thursday, July 12, at the remote Los Alamos Laboratory thirty-five miles northwest of Santa Fe, he and other members of the High Explosives group, including Ira "Tiny" Hamilton and Charles R. Barncord proceeded with the assembly of the Trinity charge. By 3:00 P.M., the assembly was complete; by 4:00 P.M. all the holes in the case had been sealed

TRINITY: THE FIELD TEST THAT REVOLUTIONIZED THE WORLD

British Prime Minister Winston Churchill, U.S. President Harry Truman, and Soviet Marshall Joseph Stalin engage in a three-way handshake at the Potsdam Conference in Berlin in July 1945.

with tape. After loading the charge onto the truck, the crew drew straws to see who would spend the evening with the device before departure for Trinity. Jercinovic "won." At one minute past midnight, the truck, with its valuable cargo escorted by Army convoy, started on the 212-mile drive to Trinity. In defiance of superstition, George Kistiakowsky had insisted upon leaving Friday, July 13.[3]

By noon the convoy rumbled into the test area. With the truck in position at the base of the tower and a small canvas tent erected for protection from the environment, Arthur B. Machen, Lieutenant W. F. Shaffer, and Tech Sergeants Jercinovic and Alvin Van Vessem began to partially disassemble the charges to allow the nuclear core to be inserted. Under the careful scrutiny of Roger S. Warner, George Kistiakowsky, Norris Bradbury, and Harry Linschitz, the final assembly began at 1:00 P.M. that same day.[4]

For the assembly work of the plutonium core and initiator, the team appropriated a four-room house belonging to a ranch family named McDonald. In preparation the men cleaned the interior of the house, sealed and insulated the windows with plastic sheets, and applied masking tape to prevent dirt from contaminating the components. Robert Bacher and Marshall Holloway's nuclear assembly team made final checks on the plutonium core.

At 3:18 P.M., the call came from the High Explosives group at the tower. They were ready. Inside the ranch house Louis Slotin deftly positioned the plutonium hemispheres and inserted the initiator between them.[5] This action joined the hemispheres in a single sphere referred to as the "core," which was then placed in the spherical cavity. Together the initiator and plutonium made up the "plug." The assembled plug was then taken from the ranch to the tower. Once there, the scenario, according to Machen—who, as a member of Warner's assembly team, would perform similar duties at the Pacific island of Tinian—went like this:

> The polar cap was removed and replaced by a cover with a pentagonal hole (this was to protect the exposed H.E. blocks

and keep them from shifting). The charges were then removed through the hole and a brass funnel placed in the cavity to protect the sides of the adjacent charges. Now, with access to the cap in the pusher of the pit, a clamping spanner wrench was introduced into two holes in the cap, the cap unscrewed and removed. With the removal of the cap, the cylindrical hole in the tamper was now clear for the insertion of the plug. After insertion of the plug, the aluminum cap was screwed in, the funnel removed, and the two H.E. charges replaced.

After the charges were seated, the cover plate with its pentagonal hole was removed and the regular polar cap reinstalled. Tongs were then keyholed onto the trunions, the unit lifted and rotated to lug-up position and set on the steel stand. The stand was wired onto clip washers and the whole thing was ready for hoisting.[6]

There had been a tense moment during installation when the plug, which had been transported inside a case, momentarily became stuck because of expansion from the natural heat of the plutonium. After Bacher identified the expansion problem, the crew endured a few agonizing minutes while they waited to see if cooling would contract it adequately. The simple idea worked; the plug slipped smoothly into place; and the explosives team replaced the charges.

At 8:00 A.M. on July 14, the operation to lift the device to the tower top began. "One of my duties," Machen recalled, "along with Henry Linschitz was to string the detonator cables on the sphere and stake on the detonators as inspired by [Vincent] Caleca. This job put me up the ladder, and since I was going to be up there anyway, I had the honor of helping wrestle the trapdoor planks."[7]

Waiting for the men at the top was Robert W. Henderson, the tower designer and engineer in charge of tower operations during the test. With the device safely inside the shed, the men replaced the oak planks across the hole in the cab floor, completing the platform.

To test the adequacy of the tower and hoisting mechanism, the crew had first lifted a concrete block weighing fifty percent more than the bomb. As an additional insurance against some malfunction, a large pile of GI mattresses was brought in and stacked six feet high on the ground directly under the bomb during the lift.

In the shed the last of the instrumentation that had to be close to the surface of the device was installed and hooked up. The control panel from which the jumper cables went to the bomb firing set was checked but not connected to the gadget. The bomb assembly crew then installed the chemical explosive detonators and completed all of the wiring necessary except the final cable to the control panel, which would be installed at the last minute. This completed the test setup, but concern still existed over the inclement weather.[8]

With time to spare, Henderson took off for Mockingbird Gap to investigate a few old mines. Jercinovic and crew retired to Compañia Hill, where they waited in the cool morning air and speculated on the outcome. Would the gadget go off at all? Would the bomb incinerate the atmosphere or tilt the earth off its axis? The brilliant theoretician Hans Bethe had investigated the possibility and discounted it. Nevertheless, Nobel Laureate Enrico Fermi and some of his colleagues passed the time by wagering how quickly the earth would be incinerated in the event the bomb triggered a chain reaction in the atmosphere. Young engineer William Caldes listened intently. "To hear theoretical physicists talking like that," he said, "was a little hairy."[9]

On the Hill, the young SEDs rubbed shoulders with some of the biggest names in science: Sir James Chadwick, discoverer of the neutron; Edward Teller, who would come to be known as the father of an even mightier weapon; and Ernest O. Lawrence, inventor of the cyclotron. Just the day before, Major General Leslie R. Groves, the man in charge of the laboratory complex known as the Manhattan Engineer District, had arrived at the site to view the all-important demonstration in person.

Groves found the base camp, situated ten miles from Ground Zero, to be a center of activity—men coming in to eat, sleep, deposit supplies or pick them up, and hustle out again; but attention centered on the 100-foot tower, standing tall and stark on the desert landscape. Radiating from the tower base were three blacktop roads. Ten thousand yards out toward the north, south, and west stood a wood and concrete shelter covered with earth. Robert Wilson was the scientist in charge of North 10,000; John Manley had West 10,000; and Frank Oppenheimer, brother to Robert, was responsible for South 10,000. At the North station Walt Treibel operated the timing instrumentation he had helped to design and develop.[10]

One-half mile from Ground Zero sat Jumbo, the massive 214-ton steel containment vessel. Because of concern that the test explosion might turn out to be a nuclear dud, engineers Bob Henderson and Roy W. Carlson had designed the pressure vessel to withstand the detonation of the chemical high explosive [H.E.] charge that was to initiate the nuclear reaction. The small horde of invaluable plutonium could then be recovered later by acid washing the interior of the intact vessel. The decision not to use Jumbo was made only at the last minute.[11]

Countdown Zero: July 16, 1945

Shortly after his arrival, Groves met with J. Robert Oppenheimer, director of the Los Alamos Laboratory, to review the situation. It quickly became apparent to Groves that there was potential for trouble. The clouds rolling in from the Oscura Mountains to the east brought a blustery drizzle and flashes of lightning on the southern horizon. In addition to the unfavorable weather, Groves was alarmed by what he termed "a dangerous air of excitement" at the camp. Feeling that this was a time for calm deliberation, and noting that Oppenheimer was getting advice from all sides, the general suggested that he and Oppenheimer leave the base camp for the control bunker at South 10,000. There he knew that everyone would be too busy with last-minute cross checks to offer unnecessary advice. At the bunker Groves and Oppenheimer discussed the weather. Periodically they went outside to dodge the mud puddles and pace restlessly in the rain.

HERITAGE

Frequent dust storms were a problem at the McDonald ranch house (above). To prevent dirt from contaminating the units during assembly, the men cleaned the interior of the house, sealed and insulated the windows with plastic sheets, and applied masking tape. Below: Around 6 P.M. on July 12, T/3 Herbert Lehr carried initiators for the gadget into the assembly room at the McDonald ranch house. The note written on the door to the left emphasized the need for cleanliness: "Please use other door—Keep this room clean." The plutonium is loaded into the back seat of a government vehicle for transport to the base of the tower at the Trinity Site.

TRINITY: THE FIELD TEST THAT REVOLUTIONIZED THE WORLD

Sergeant Ben Benjamin and T/5 George Econnomu prepare instrumentation calibration charges on the hood of a military jeep for use during the 100-ton H.E. shot at Trinity.

HERITAGE

The base camp at Compañia Hill (above) looked deceptively calm until the day of the Trinity test. The camp then became a flurry of activity as scientists, engineers, and technicians made ready for detonation of the first atomic device. The 100-foot tower (left) stood tall and stark on the desert landscape of New Mexico.

Station South 10,000 was the main control point for the Trinity test and the bunker from which Groves and Oppenheimer witnessed the explosion of the first atomic bomb. The bunkers shown below protected an instrumentation station (on the right) and its power supply (left) for the Trinity shot.

HERITAGE

The first atomic bomb was detonated at the Trinity Site on July 16, 1945, at 5:30 A.M.

TRINITY: THE FIELD TEST THAT REVOLUTIONIZED THE WORLD

Among the scientists measuring radioactivity in the seared sand particles at Trinity two months after the test were (above, left to right) Kenneth T. Bainbridge, Joseph G. Hoffman, J. Robert Oppenheimer, Louis H. Hempelman, Victor Weiskopf, Robert F. Bacher, and Richard W. Dodson.

Oppenheimer and Groves (right) investigate the remains of the tower at Ground Zero. The charred remnants of one post are all that were left of the 100-foot tower.

Together, the two men created an odd pair. Groves, almost corpulent in his well-tailored Army uniform, provided an interesting contrast to the ascetic-looking Oppenheimer, who appeared emaciated in expensive but casual clothes and porkpie hat. During these last hours before countdown, Groves devoted himself to shielding the slight, intellectual physics professor from the excitement enveloping them. Groves realized that Oppenheimer needed to be able to consider the situation calmly since he had to make decisions based on his appraisal of the technical factors involved.[12]

Meteorologist Jack Hubbard was also feeling the pressure. He had spent several months securing wind and surface observations to determine the most favorable test time. "Then someone in Washington, D. C., selected July 16," he said, a date that was unfavorable "due to a deep tropical air mass, wrong wind directions, and thunderstorms." Furthermore, Groves had not minced words: "Your forecast had better be right or I'll hang you," he told Hubbard.[13]

At 3:00 A.M. Bob Henderson, accompanied by the man responsible for closing the safety switches, drove to the tower where the device reposed in eerie solitude. Henderson stood watching at the base of the tower in the drizzling rain while his companion climbed the tower, made the final electrical connections and determined that all was as it should be, then drove back to their designated observation points twenty miles away.

By 4:00 A.M. the rain stopped. Kenneth Bainbridge, test director for the project, anxiously awaited the weather report that would allow the test to proceed. Described by Groves as "quiet and competent," a man who "had the liking and respect" of his men, Bainbridge was a Harvard physics professor and member of the Steering Committee of the Radiation Laboratory, Massachusetts Institute of Technology.[14]

At 4:45 A.M. he learned that the wind conditions were stable and began the arming ritual. At 5:10 A.M., accompanied by the stirring strains of the National Anthem, Samuel Allison intoned the countdown. The weather caused the Trinity radio frequency to cross wavelengths with Station KCBA in Delano, California, and its Voice of America broadcast to Latin America.[15]

Earlier, at Kirtland Field in Albuquerque, two B-29 observation planes equipped with radar had readied for takeoff. With instructions from Oppenheimer to steer a course at least fifteen miles west of the detonation point, the pilots headed down the runway. Thunderstorm activity soon forced the planes to drop from 23,000 to 18,000 feet before circling Trinity Site. In the navigator's seat just behind the pilot sat Glenn Fowler, a young radar and ordnance expert, who was responsible for positioning the two aircraft using IFF ("identification of friend or foe") equipment. Fowler would soon have a grandstand seat on history.[16]

On the desert below as the countdown neared zero, observers were instructed to lie face down on the ground with feet toward the blast and eyes covered. At 5:30 A.M., July 16, 1945, the first manmade atomic blast lit the sky like a thousand suns. Through welder's goggles provided to protect their eyes, those present saw a sight they would never forget—a crimson fireball that boiled skyward, gradually changing into a gray, mushroom-shaped cloud surrounded by a strikingly luminous halo. In the intensity of the light, the desert growth glowed white in the flash. Shock waves churned up dust on the valley floor. At Ground Zero thousands of emerald beads suddenly glistened garishly, transforming the white sands of the desert into a strange sea of green glass soon to be known as "trinitite." For those witnessing the violence of the atom, emotions ranged from awe and fear to joy as the Trinity detonation indicated the successful culmination of twenty-eight months of intense and cooperative effort.[17]

Staff Sergeant Louis F. Jacot, who had been assigned to an evacuation outfit (in case of radioactive fallout), recalled seeing Nobel Prize winner Enrico Fermi coolly conduct his now famous experiment in which he assessed the yield of the blast by dropping a fistful of paper fragments and measuring the distance of their dispersal caused by the shock wave.[18]

Oppenheimer, his thin face tense and worried, held on to a post to steady himself. When the burst of light and the roar of the explosion indicated the success of the test, observers noted that "his face relaxed into an expression of tremendous relief."[19] George B. Kistiakowsky, who had bet Oppenheimer that the bomb would work, enthusiastically slapped him on the back. "Oppie, you owe me ten bucks!" he crowed.[20] Dr. James Conant of the Nuclear Defense Research Committee shook hands with Vannevar Bush, chairman of the Office of Scientific Research and Development. Both congratulated Groves and Oppenheimer.

Glenn Fowler (left), a young radar and ordnance expert, was aboard one of the B-29 observation planes that circled near detonation point during the Trinity test. The caption on the photograph reads: "Men From Mars." Bernard Waldman stands on the right.

Bob Henderson, who was to observe many other "shots," including Crossroads, Sandstone, and Greenhouse, recalled that Trinity made the most indelible impression. "I'll never forget my first atomic detonation," he said. "We were experiencing something that man had never seen before." For Glenn A. Fowler, there was the immediate realization that Trinity was "the start of a new era."[21] Trinity also represented a new era for that special branch of ordnance engineering called "Field Testing."

New Mexico State Highway 4 winding up "the Hill" toward Los Alamos.

Chapter II.

GENESIS:

The Laboratory on the Hill

> In every investigation, in every extension of knowledge, we're involved in action. And in every action we're involved in a kind of loss, the loss of what we didn't do. We find this in the simplest situations. . . . Meaning is always obtained at the cost of leaving things out. . . . In practical terms this means, of course, that our knowledge is always finite and never all encompassing. . . . This makes of ours an open world, a world without end.
>
> J. Robert Oppenheimer

To some, the label *Trinity* seemed almost a sacrilege. To the more analytical, however, the code name for the atomic bomb project was especially apropos, incorporating the joint efforts of a triad of pure science, applied science, and engineering.

Some claim that Trinity was an allusion to the rather frightening fact that only three such bombs existed. If the Japanese could have deciphered the code, the American position would have been weakened—with only two blasts left to end the war quickly.[1]

Robert Henderson explains that "Lex" Stevens named Trinity when he and Stevens surveyed ways to transfer the mammoth containment vessel called Jumbo from the rail siding to Ground Zero. Colonel Stevens, a devout Catholic, noting that the railroad stop was called Pope's Siding, remarked that it was indeed fortunate that the Pope had special intercession with the Trinity because the scientists would need all the help they could get to move Jumbo to its proper spot.[2]

A more popular explanation can be traced to a conversation between Kenneth Bainbridge and Oppenheimer. After General Groves approved Bainbridge's selection of the test site to meet the criteria of flatness, isolation, and proximity to Los Alamos, Bainbridge telephoned Oppenheimer. Reading a book of sonnets by John Donne when he received the call, Oppenheimer turned to the following verse:

> "Batter my heart, three-person'd God; for, you
> As yet but knock, breathe, shine, and seek to mend. . . ."

"Trinity," Oppenheimer felt, would be most appropriate. Nuel Pharr Davis, in his biography of Oppenheimer, agrees with this derivation of the name and quotes Oppenheimer as saying: "'This code name didn't mean anything. . . . It was just something suggested to me by John Donne's sonnets, which I happened to be reading at the time.'"[3]

More analytical, perhaps, is the explanation provided in a dissertation written by Marjorie Bell Chambers of Los Alamos. Genesis of the term *Trinity*, she maintains, is found in the Hindu, rather than Christian connotations of the word. Whereas the Christian Trinity is composed of the Father, the Son, and the Holy Ghost, the Hindu concept views the Trinity as Brahma, the Creator; Vishnu, the Preserver; and Shiva, the Destroyer. Self-taught in Sanskrit and well-versed in Hindu culture, Oppenheimer was aware of the Hindu belief in reincarnation, which posits that "whatever exists in the Universe is never destroyed; it is simply transformed."[4] Carried a bit further, then, the most likely analogy may be—atoms transformed into energy.[5]

Los Alamos

Regardless of the source of the name, credit for the success of the project must go to the talented cadre of physicists, scientists, and engineers who first gathered at the mountaintop laboratory of Los Alamos early in 1943. Arthur H. Compton, Nobel Prize winner and head of the Chicago Metallurgical

Laboratory, in his book *Atomic Quest* provides a genteel assessment: "Never . . ." he said, "has there been gathered together in one place . . . so large a group of competent men of science."[6] Referred to by Groves in more earthy terms as "the finest collection of crackpots the world has ever seen," the group began to gather the equipment necessary to support the brainpower. From Harvard, they borrowed a cyclotron; from the University of Illinois, they acquired a Cockroft–Walton accelerator; and from the University of Wisconsin came two electrostatic accelerators.[7]

Despite security questions relating to Oppenheimer's Communist friends, Groves personally selected him to direct the project. As Groves explained, "I felt that his potential value outweighed any security risk, and to remove the matter from further discussion, I personally wrote and signed the following instructions to the District Engineer on July 20, 1943":

> In accordance with my verbal directions of July 15, it is desired that clearance be issued for the employment of Julius Robert Oppenheimer without delay, irrespective of the information which you have concerning Mr. Oppenheimer. He is absolutely essential to the project.[8]

Oppenheimer soon showed surprising administrative ability in forging together his brilliant and diverse group. After learning that the Laboratory would be concerned "with the development and final manufacture of an instrument of war," he began to organize accordingly.[9] A small component of the huge complex of institutions devoted to the development of the atomic bomb, "Project Y," as Los Alamos was known, was itself to divide into even smaller units—Sandia, Wendover, and Tinian—to realize its ultimate mission.

In operation by April 1943, the Los Alamos Laboratory, under contract with the University of California, began to take shape.[10] Roads were suddenly carved out of mountainsides; and Pacific hutments, mobile trailers, and green laboratory buildings crowded the mesa in haphazard array. At this isolated site in the picturesque Jemez Mountains northwest of Santa Fe, some of the greatest minds of the day gathered to solve the mysteries of the atomic bomb, including bomb design, preparation of core materials, assembly, and delivery. Ray Powell, an engineer who became head of the Administrative Division of the Chemistry and Metallurgical Group, remembered Los Alamos as "the most supercharged intellectual atmosphere that had ever existed."[11]

Theoretically, the scientists envisioned two methods of assembly: the gun method, which was well understood by ordnance experts of the day, and the possible, but purely theoretical, implosion method. By far the simpler concept, the gun method involved the firing of a subcritical mass of fissionable material as a projectile at another subcritical mass of fissionable material. Upon contact, the two would form a supercritical mass. The implosion method, on the other hand, was based on the entirely new concept of having a slightly subcritical mass of fissionable material surrounded by high explosives that, upon detonation, would compress the fissionable material and render it supercritical. Initially, it was hoped that the gun method could be used for both the uranium and plutonium bombs.[12]

Before establishment of the Los Alamos Laboratory, the work conducted on the fast fission process had been theoretical, led by Oppenheimer and his staff of consulting physicists. "Nothing had been done," Groves recalled, "on such down-to-earth problems as how to detonate the bomb."[13] Even after the Laboratory began operation in April 1943, the first few months of endeavor were devoted to research.

To make the transition from theory and scientific investigation to the construction of a weapon for wartime use, Groves initiated an eclectic brand of engineering that would combine the theories of scientists with the expertise of engineers and explosives and ordnance personnel. Finding a competent ordnance expert to head the program was not an easy task. The individual had to be respected by ordnance people and scientists alike. Furthermore, it was desirable, from Groves' point of view, that the person be a regular Army officer. Unable to find such an individual within Army ranks, Groves had the problem solved for him when Vannevar Bush recommended Navy Commander William S. Parsons.[14]

Parsons, an Annapolis graduate with considerable ordnance and gunnery experience, had also spent several years working on the development and testing of the proximity fuze. In the early thirties, Groves had had dealings with Parsons while he was involved in the development of radar for the Navy. Parsons, promoted to captain, was assigned to the Manhattan Engineer District, and in May made a preliminary visit to Los Alamos. In June 1943, "Deak" [also referred to as "Deke"] Parsons, as he was called, returned as head of the Ordnance Division, which took up residence in three small rooms located in Building U behind Oppenheimer's office.[15]

Ordnance Engineering, classified as "E Division," would become the matrix from which Z Division, forerunner of Sandia Corporation, would evolve. Under Parsons' direction, E Division would also serve as the organizational vehicle to implement the field testing program leading to development of the first atomic bomb and to maintenance of national security through a strong nuclear arsenal.

In its original conformation, the new division had among its group leaders Edwin F. McMillan, who assumed responsibility for the Explosives Proving Ground (E-1), and Kenneth T. Bainbridge, in charge of Instrumentation (E-2), the man who would ultimately be assigned technical accountability for Trinity. One of the first tasks assigned to McMillan's group was the selection of an isolated area to serve as a proving ground for the gun project. For this purpose, the Pajarito Plateau, sloping eastward from the Jemez Mountains toward the Rio Grande, seemed especially suitable because of numerous canyons and gullies that served as natural protective barriers. The Anchor Ranch Proving Ground soon included sand butts, gun emplacements, a control room, bombproof magazines, and portable gun shelters that could be rolled away when firing took place.

Included among the cadre of group leaders was the head of E-3, Robert B. Brode, who would be instrumental in establishment of the Electronic Engineering Group of Z Division. Brode, assisted by E. B. Doll and S. J. Ratner, transformed the bomb into a sensitive, self-sufficient machine designed to detonate at the correct moment. Alan Ayers, who would eventually head his own division at Sandia, provided able assistance. Doll noted in the Daily Log that D. S. Dreesen, later a member of Sandia's Test Operations Division, also arrived all "dressed in a pretty new khaki suit," to join the ranks of the Scientific Engineering Detachment.[16]

After Parsons' initial visit to the Laboratory, he had added George Chadwick, long-time head engineer of the Navy Bureau of Ordnance, to lead the Engineering Group (E-6). Chadwick worked on the design and fabrication of the first experimental guns and helped establish the layout for the Anchor Ranch Proving Ground. Chadwick, however, served in a consultant capacity for only four months. After September 1943, he returned to Detroit to assist in the procurement of machinists and draftsmen. Subsequently, the Engineering Group was headed first by J. L. Hittell, then P. Esterline, and eventually by L. D. Bonbrake. Destined to grow rapidly, the Ordnance Division had expanded to include eleven groups by the fall of 1943.

Thus, during the first year of its existence, Los Alamos was characterized by the conversion of a primitive hilltop to a functional laboratory, organizational realignment, and a devotion to research. Design work moved confidently ahead, especially on the gun-type weapon, although work on implosion would prove to be slow and frustrating. During this early period, as Norman Ramsey, head of Delivery Group, reported: "Two external [bomb] shapes and weights were selected. . . . For security reasons Air Force representatives called these the Thin Man and the Fat Man, respectively."[17]

Thin Man and Fat Man

The major focus that first year was on the Thin Man, a gun-type device, seventeen feet long, with a maximum diameter of about two feet. The Thin Man was so named to imply a reference to the United States president, Franklin D. Roosevelt;

E DIVISION
(Ordnance)

Captain William S. Parsons, Leader

E-1, Proving Ground
  E. M. McMillan
  A. F. Birch

E-2, Instrumentation
  K. T. Bainbridge
  L. G. Parrott

E-3, Fuze Development
  R. B. Brode

E-4, Projectile, Target and Source
  C. L. Critchfield

E-5, Implosion Experimentation
  S. H. Neddermeyer

E-6, Engineering
  L. D. Bonbrake

E-7, Delivery
  N. F. Ramsey

E-8, Interior Ballistics
  J. O. Hirschfelder

E-9, High-Explosive Development
  K. T. Bainbridge

E-10, S-Site Maintenance, Construction, for Implosion and Plant Operation
  Major W. A. Stevens

E-11, RaLa and Electric Detonator
  L. W. Alvarez

HERITAGE

The technical area of the Los Alamos Scientific Laboratory during the 1950s: Note the enclosed walkways (center of photograph) connecting buildings on either side of the main street to protect workers going from one building to another during severe weather. Ashley Pond is visible on the left.

Wives also contributed to the Los Alamos workforce. Here, Marie Smith, wife of Leon Smith, works on the cyclotron.

GENESIS: THE LABORATORY ON THE HILL

Quonset huts and wooden military hutments were among housing available to residents working on the project.

whereas the Fat Man, an implosion-type assembly, nine feet long and five feet in diameter, was selected as an obvious reference to British Prime Minister Winston Churchill. With this linguistic subterfuge, telephone conversations concerning bombers to transport these weapons could be made to sound as if a plane was being modified to carry Churchill or Roosevelt.[18]

The aerodynamic qualities of scale models of the Thin Man, which was first tested at the Dahlgren Naval Proving Grounds in Virginia, proved discouraging. Because of the materials of its construction, the bomb fell in a flat spin when first tested. To add stability, the fin area was eventually increased and the center of gravity shifted forward. The Fat Man had its problems as well. As Bob Henderson recalled: "The Mk 3 (Fat Man) ballistics were horrible. It could fly backward as easily as forward."[19]

Manufacturers in Detroit were provided full-scale dimensions of the aerodynamic shapes, and the University of Michigan, which performed fuze work, collaborated with Los Alamos in solving the problem of adapting an Air Corps plane to drop the payload. By November 1943, under the priority code name "Silverplate," modifications of the B-29s selected to carry the bombs had started, and by February 28, 1944, these aircraft were being used in bombing tests at the Air Corps' Muroc field in California.[20]

Although torrential rains on the Mojave Desert combined with mechanical problems resulted in a four-week delay, the test series was completed, with mixed results. The Thin Man, for example, had a frustrating tendency to jam in the bomb bay at release but showed a stable flight pattern. On the other hand, the Fat Man wobbled like a poorly thrown football during the drop but released efficiently. On the last test, the Thin Man model released prematurely at 22,000 feet, seriously damaging the bomb bay doors of the aircraft.* This accident resulted in cancellation of the tests for three weeks while the bomb release mechanism was reworked.[21]

---

*A formerly secret teletype from Parsons to Groves dated 20 March states that, according to Norman Ramsey, the early release took place at 26,000 feet.

Testing also proved the proximity fuzes to be unreliable. Robert Brode's staff, in charge of arming and fuzing, had considered adapting currently available barometric switches or radio proximity fuzes, tail warning devices, radio altimeters, and even clocks as fuzing mechanisms. The series of mockup bombs dropped at Muroc revealed that the barometric fuze *might* be used as a feasible alternative to a more reliable method; however, news that the bomb could be detonated at altitudes as high as 30,000 feet precluded the use of the proximity fuze. Brode and company went back to the drawing boards. Fortuitously, the Radio Corporation of America had perfected a tail warning radar device (nicknamed Archie by the Los Alamos staff). In actual air drops, the Archies showed real possibilities as being the answer to the fuzing problems. Field testing resumed in June and continued throughout July of 1944.[22]

In midsummer a crisis developed when Emilio Segrè's tests on plutonium 239 from the Clinton pile revealed the presence of the isotope Pu 240, known to be a strong spontaneous fissioner. This discovery was to have serious ramifications for the gun program. As Norman Ramsey recalled: "It became apparent that Pu 239 could not be used in a gun [assembly] due to neutrons of Pu 240 [spontaneously emitted] almost certainly causing a predetonation."[23] If Pu 240 were to be used as the nuclear material for the gun–type weapon, discouragingly high–assembly velocities of the subcritical pieces would be required, pushing the state of the art of gun design.

These combined factors demanded the use of prohibitively heavy gun barrels; therefore, the decision was made to use uranium 235 in the gun–type weapon. The use of this element meant that assembly velocity could be significantly reduced and the weapon shortened to fit easily into the B–29 bomb bay. In this manner, the Thin Man was reduced in length and transformed into the "Little Boy."[24]

Problems, however, still existed with the Fat Man. Modification of the tail assembly from the original circular shroud to a square box tail fifty–eight inches on a side did not solve the instability of the bomb. Finally, David Semple, a bombardier working with the group at Muroc in California, and project members Norman Ramsey and Sheldon Dike, provided a solution by welding forty-five-degree baffle plates inside the shroud to create a parachute effect and increase the drag. "To everyone's surprise," Ramsey reported, "this modification was successful and resulted in an improved ballistic coefficient."[25]

The first tests of these models with a combat unit began at Wendover Air Corps Base (code name "Kingman" or "W-47") in Utah. Developments of this kind reflected the transition taking place at the Los Alamos facility. Weapon prototypes, and not mere laboratory devices, were now being created and tested.[26]

A Shift in Emphasis: Implosion

The decision to change the emphasis of the bomb program from the gun-type weapon to implosion was made at conferences held in Chicago, July 17, 1944. Subsequently, in August, General Groves presented a new weapons timetable to Chief of Staff George Marshall indicating that several implosion bombs of the Mk 3 variety would, he hoped, be ready by June 1945, and that a gun-assembly bomb would be ready by August 1, 1945.[27]

With the discovery that the plutonium gun was not feasible, the implosion program became all-important. Credited in numerous secondary sources as the unsung hero of implosion, Seth H. Neddermeyer, although he did not have the ability to carry it through administratively, did foster credence for the theory through his dogged support and experimentation.[28] According to David Hawkins, official chronicler of Project Y: "Neddermeyer had developed an elementary theory of high-explosive assembly."[29]

A quiet, almost inarticulate physicist from the National Bureau of Standards and former student of Oppenheimer's, Neddermeyer had attended the series of lectures presented at Los Alamos by the professor's close colleague, Robert Serber. The lectures, which began on the day the Laboratory formally opened, April 15, 1943, stemmed from a conference held earlier in Berkeley at which the state of the art in uranium and plutonium research had been discussed.[30]

At the Berkeley conference, held March 11–17, 1943, Oppenheimer, Serber, and Richard Tolman of the National Research Defense Committee had developed a report for General Groves and James Conant. It was this same secret report that formed the basis for the new staff lectures at Los Alamos. During the lectures, Serber introduced three different theories of bomb assembly—the gun method, the autocatalytic or self-assembly process, and implosion.[31]

Several significant theoretical considerations were presented regarding a detonating device. For example: "To obtain a suitable device for detonating a bomb when the desired supercritical configuration is reached, it is thought that capsules containing polonium and beryllium could be developed which would shatter at the right times and lead to neutron production through mixing of the two materials." And, concerning initiation of the reaction: "It might be possible to bring this about by explosive charges which would blow fragments of the shell into the interior. This latter possibility appears an interesting one to consider."[32]

The possibility did indeed appear interesting, particularly to Neddermeyer, whose imagination was captured by the implosion idea. Although he had difficulty expressing himself, he spoke out strongly in favor of the implosion configuration of the bomb, which he believed to be a viable alternative. Undaunted by the initial negative response, Neddermeyer convinced Oppenheimer that the idea was worth investigating. Subsequently, Neddermeyer went to the Navy's Explosives Research Laboratory at Bruceton, Pennsylvania, to study the techniques of high explosives. In the summer of 1943 he began experiments to prove the implosion theory.

By the Fourth of July, Los Alamos residents heard explosions on the mesa that sounded like fireworks. The sounds reverberated off the walls of the canyon near Anchor Ranch where Neddermeyer and a few others were playing a noisy and dangerous game with TNT. As leader of the implosion group, he was staging a demonstration for his skeptical boss, "Deak"

Parsons. Neddermeyer, attempting to achieve uniform compression, carefully placed boxes of tamped TNT around large, hollow steel cylinders (actually stove pipes), then detonated them. The result was an implosive force that transformed the pipes into solid blocks of metal. Applying the concept to bomb assembly, Neddermeyer believed that pieces of plutonium could be fuzed into a small ball by the exertion of symmetrical spherical pressure from the outside. According to this theory, a sphere of highly compressed plutonium could be created by implosion in millionths of a second.[33]

Parsons, who favored the gun method, felt that there were too many problems involved in refining implosion—that uniformity, for example, could not be achieved. Nevertheless, as early as September 1943, according to minutes of the Governing Board, Neddermeyer's method, along with the gun method and the idea proposed by John von Neumann and James Tuck for the use of shaped charges, was being seriously considered. In fact, it was duly noted that "encouragement should be given to the H.E. methods, particularly Neddermeyer's, which was felt to be short staffed." Oppenheimer wanted Neddermeyer to increase his efforts.[34]

The two individuals largely responsible for this shift in attitude were John von Neumann, renowned for his mathematical theory of games, and Robert F. Christy of the Theoretical Division. As Project Y historian David Hawkins indicated: "The decisive change in the picture of the implosion occurred with the visit of J. von Neumann in the fall of 1943."[35]

Recognized as an "extremely warm, friendly person" and a true genius, von Neumann had experience in the use of shaped charges for armor penetration. At Los Alamos he saw the possibility of using a shaped charge assembly for implosion and provided theoretical calculations that lent weight to the implosion concept. Furthermore, Christy's division in the summer of 1943 had discovered that the kinetic energy from implosion would create compression resulting in a situation where almost incompressible matter could, in fact, be compressed. The "Christy gadget," as the discovery became known, achieved criticality by extra compression rather than extra size.[36]

GENESIS: THE LABORATORY ON THE HILL

We, The Undersigned, in grateful appreciation of our joint association in the Los Alamos, New Mexico Atomic Bomb Project, hereby list below a lasting record of friendship spent in working on the World's Greatest Secret,

The Atomic Bomb

HERITAGE

Brigadier General Leslie Groves was named to direct the project known as the Manhattan Engineer District when the project was transferred to the War Department.

Captain William S. Parsons directed Los Alamos ordnance activities.

Director of the Los Alamos–based Manhattan Project, J. Robert Oppenheimer assumed administrative control in 1943.

Norris E. Bradbury became Director of Los Alamos Scientific Laboratory after Robert Oppenheimer resigned his position to return to academic life.

HERITAGE

Scientists at Los Alamos took part in weekly colloquia to exchange ideas. Among those present for a nuclear physics conference were (from left to right, front row) Norris Bradbury, John Manley, Enrico Fermi, and J. M. B. Kellogg. Robert Oppenheimer and Richard Feynman are in the back row.

At 8:00 A.M. on July 14, the operation to lift the device to the top of the tower began. In preparation for assembly, Art Machen, Bill Stewart, and Herbert Lehr (left to right) unload the shell of the device at the test site. Onlookers from the ground include Phil Dailey and Captain Wilbur Shaffer (at far left) and Norris Bradbury (fourth from left).

Norris Bradbury stands next to the fully assembled device atop the 100-foot tower. The device had to be hoisted through a trap door in the oak platform.

The gadget, adapted by February 1944, helped convince Oppenheimer of the feasibility of implosion. Moreover, several months earlier, Groves and Conant had given their support. In November, Oppenheimer had reported: "Both Groves and Conant seemed very much in favor of pushing the implosion method."[37] Groves had gone so far as to establish a six-month time scale for the project. The Governing Board, however, felt that this was "a little optimistic."[38]

Among others contributing important advances on the road to Trinity was James Tuck, a young British scientist, who supported the idea that uniformity of the explosive force or shock waves could be achieved by using lens charges comprising fast and slow detonating explosive components shaped to generate the desired detonation wave form. There was also Charles L. Critchfield. At the firing site, Critchfield, a member of G Division, tested a device about the size of a nut. Officially known as an "initiator" and dubbed the "urchin" by the scientists, it was composed of polonium and beryllium, two elements that produced neutrons when placed in contact with each other. The device, perfected in final form by Dr. Ruby Scherr, would trigger the neutrons when the gadget compressed.[39]

Meanwhile, Neddermeyer was unclear on complex problems associated with spherically converging shockwaves. The realization of a practical weapon "seemed as far off as ever." Despite his attempts to make implosion work, the group was "stagnating" under his leadership. A lack of manpower, poor management style, and personality clashes with Parsons and with Oppenheimer appeared to be the source of the problem. Oppenheimer was becoming increasingly concerned about Neddermeyer's "lone-wolf" attitude and, in particular, his lack of progress. Consequently, in June 1944, Oppenheimer asked NDRC explosives expert George Kistiakowsky to take charge of the implosion effort as Deputy Division Leader.[40]

Reorganization

Kistiakowsky, a physical chemist by training and Russian by birth, had the combustible personality of some of the explosives he worked with. Experimentation during the earlier years of the war had proven to his satisfaction that the process of detonation could be "completely described by rigorous laws of thermodynamics and hydrodynamics." This meant that Kistiakowsky believed explosives could be made into precision instruments that could be controlled by man.[41] His temerity, tempered by concern for the men who worked for him, led to advances that might not otherwise have taken place.

As the other Deputy Division leader, Oppenheimer assigned Edwin M. McMillan to head the gun program, while he retained Neddermeyer as a technical adviser to the Ordnance Division and as a member of the H.E. Group. Since Kistiakowsky was still reporting to Parsons, the crisis had yet to be solved.[42]

By July, a combination of the disillusioning news about Pu 240 and the decision to abandon the plutonium gun, capped by pressure from Groves and Conant, caused those in charge to realize that "implosion was the only hope, and from the current evidence not a very good one."[43] These factors encouraged Oppenheimer in August of 1944 to effect a sweeping reorganization. In continued efforts to make implosion work, Oppenheimer took the program out of Ordnance and divided responsibilities between two new divisions.[44]

One of these, the Gadget or Weapons Physics Division under Robert Bacher, was to be concerned primarily with investigation of implosion hydrodynamics. The other, the Explosives Division, was placed under Kistiakowsky and charged with the development of adequate H.E. components. Group X-2 of Kistiakowsky's division was to be headed by Kenneth T. Bainbridge, whose duties were to "make preparations for a field test." Robert Wilson headed the Experimental Physics or Research Division and was placed in charge of the North 10,000 shelter. John

Williams as Deputy Test Director became involved in Trinity Site preparations early on and served as right-hand man for Bainbridge. Ordnance remained under Parsons.[45]

By March 5, 1944, Kistiakowsky had developed a provisional work schedule for the implosion project. His final entry for November and December 1944, was humorously pessimistic: "The test of the gadget failed. Project staff resumes frantic work. Kisty goes nuts and is locked up!"[46] Admittedly, Kistiakowsky had numerous problems to face. G Division, for example, began to place incredible demands on Sawmill site (S-site) that interfered with production of castings for its own experimental use. The problem was compounded by the scarcity of molds in which to make the castings, despite the fact that Bob Henderson and Earl Long of X Division, according to Kistiakowsky, were "of tremendous help" in providing these precision fabrication tools, some of which weighed up to 100 pounds or so.[47]

When it became apparent that the Anchor Ranch Proving Ground could not accommodate the expansion of the implosion program, Kistiakowsky proposed the construction of another plant in Pajarito Canyon, but Parsons resisted and built an expansion of the S-site. Apparently more concerned about the sensitivity of the H.E. than he professed, Kistiakowsky felt that S-site was a particularly bad choice because the raw explosives had to be trucked across the mesa through the center of the Los Alamos project, where hundreds of "wild WACs and GIs" careened around in trucks and jeeps.[48] In operation by August of 1944, the S-site included a large casting plant and was staffed mostly by young soldiers with scientific training. Known as SED, the Special Engineering Detachment would play a large part in the project to build the bomb. Among this group were Charles R. Barncord and Ray Brin, who would later work on improved versions of the bomb as members of Sandia Corporation. At Los Alamos, Barncord and Brin helped install the H.E. casting kettles at S-site. Later, Barncord and another SED surveyed and laid out an area that would become part of S-site, where simulated airplane loadings and environmental tests were conducted.

Ira "Tiny" Hamilton (at right) poses with two of his buddies in the "assembly room" at T–D (Trap–Door) site with an early prototype of the housing for the Fat Man bomb. This particular unit, shown here suspended from a C-2 wrecker by an aircraft hook, used bathtub fittings and did not work very well. T–D site was out in the open, near what was to become S-site, and was frequently muddy; hence, the men who worked there wore galoshes.

In addition to the Anchor Ranch Proving Ground and S-site, some of the most dangerous work took place at a remote area called Omega site. There, Enrico Fermi supervised construction of the atomic reactor known as the "Water Boiler." There, also, Otto Frisch's Critical Assembly team conducted criticality tests called "tickling the dragon's tail." Assembly experts such as Louis Slotin and his assistant Harry Daghlian, both to become victims of nuclear accidents, performed experiments involving the plutonium cores.

In these bizarre experiments, Slotin would lower the top half of the Pu-239 sphere toward the bottom half to a point near criticality. On the day of his accident, the screwdriver he was using to separate the two halves somehow slipped, allowing the assembly to become supercritical. Throwing himself forward to protect others in the room, Slotin absorbed the piercing purple flash of radiation and died nine days later: May 30, 1946. Daghlian's accident occurred a year earlier when he somehow dropped a heavy "watercress" brick atop the tungsten carbide reflector cabin that shielded the plutonium sphere. The ionization resulted in Daghlian's death twenty-one days later.[49]

Jumbo

For some time, Los Alamos personnel heatedly debated whether or not the plutonium device should be field tested. General Groves, who had a healthy respect for Congressional investigating committees, was opposed to the idea. Since plutonium production had been very slow, there was a good chance that a field test of the implosion gadget would wipe out Groves' chance of meeting government demands for a combat-ready weapon before the end of the war.

Oppenheimer and Kistiakowsky objected. In their opinion, "there had to be a test because the whole scheme was so

HERITAGE

Jumbo, the massive 214-ton steel containment vessel, was designed to enclose the bomb to save the limited supply of nuclear material in the event the test failed. Transporting Jumbo thirty miles over the desert to Ground Zero at Trinity Site was no small task. Los Alamos engineers and members of the Eichleay Corporation moved the vessel from the 200-ton flat car on which it arrived at Pope's Siding south of Socorro and placed it atop a huge pneumatic-tired trailer with a 64-wheel base. Six caterpillar D-8 tractors, three pushing and three pulling, supplied the power to move the heavy cargo.

Although tremendous engineering and manufacturing resources were used to construct the vessel, Jumbo was never used.

uncertain."[50] If the bomb failed over enemy territory, for example, the enemy would have control of an atomic weapon—*gratis*—not to mention the embarrassment involved. In either case, Groves would probably land in front of a Congressional investigating committee. Furthermore, in 1944, plutonium production still dribbled forth, and prospects of obtaining a larger supply were slim. The scientists agreed that the active material could not be wasted in an unsuccessful test.[51]

To make sure that this did not happen, a water recovery method for salvaging the active material and a method of trying to salvage the material in sand were both investigated.[52] The ultimate compromise, however, came in the shape of a 214-ton cylinder-shaped tank called Jumbo. The idea was to construct a steel containment vessel to enclose the bomb so that the nuclear material would be saved in the event the test was a nuclear dud. This proposal seemed to reassure Groves. In February 1944, Oppenheimer announced officially that the implosion gadget would be field tested, explaining that "the test is required because of the incompleteness of our knowledge" and because the "pressures under which the gadget will operate are . . . unobtainable in the laboratory." In a letter to Groves, Oppenheimer promised that they would "attempt to have a container fabricated and completely assembled by September so that it . . . [could] play as useful a part as possible in the later stages of the implosion program." Oppenheimer also announced that Jumbo was to be classified "secret" and "should be treated so in conversations and written matter."[53]

The engineering, design, and procurement of such a vessel would lay additional stresses on the overburdened Ordnance Engineering division—a group that had already had its share of difficulties. The various reorganizations and attempts to make the group function more effectively had not resolved the situation. In a lengthy memo to Parsons, dated July 18, 1944, Kistiakowsky presented his perception of the problem: "In my opinion," he said,

> the difficulties have not been satisfactorily solved, and the fact that they exist, regardless of who is the Group Leader,

> suggests very strongly that the weakness lies not in the personality of the Group Leader but in the basic principle of the present organization. These difficulties in the organization are that the group is charged with providing design and engineering on such a great variety of unrelated items, ranging all the way from refined optical cameras to Jumbos, that no intimate understanding of the problems involved can be hoped for from the limited staff at the Project. It is not surprising, therefore, that mistakes are repeatedly made . . . .[54]

Difficulties such as these lend themselves to speculation about interactions between engineers and scientists and their ultimate impact. However, Robert W. Henderson, the engineer in charge of design at Los Alamos, discounts this as a significant factor, although he admits that there were the inevitable personality differences. Henderson essentially agrees with the analysis of Project Y historian David Hawkins, by identifying as the crux of the problem the fact that most of the engineering staff was recruited by Parsons from the Detroit "job shops"—facilities that specialized in tool and die design for the automobile industry; i.e., production engineers. The Los Alamos job, on the other hand, was original design work, totally unrelated to production tooling.[55] Nevertheless, and fortunately for the outcome of the project, certain individuals were exceptions to the general rule.

A memo from Kistiakowsky to Parsons, for example, commended the efforts of Bob Henderson and Roy W. Carlson. "The part of engineering which, to my knowledge, is functioning relatively very well," he said,

> is that part in which Henderson and Carlson have been intimately connected. As you remember, these latter men are in a somewhat anomalous position because they were brought in not to be our engineers but to act as liaison between our research and the engineering group. The not altogether satisfactory operation of the engineering group has led, however, to a situation where Henderson and Carlson take very active, in fact a leading, part in the actual engineering of the components related not only to Jumbo but also to the H.E. gadget as a whole.[56]

Robert W. Henderson, who would become the first Executive Vice-President of weapons for Sandia National Laboratories, had been recruited ("shanghaied," he terms it) for the Los Alamos project while working for Nobel Laureate Ernest O. Lawrence at the Radiation Laboratory in Berkeley. Lawrence had recruited Henderson from Paramount Studios, where he was assistant to the Chief Engineer and the recipient of an Academy Award in 1942 for his work in color process photography and special effects. At Berkeley, Henderson became acquainted with the weapons business and developed the highest personal regard for Lawrence.

Henderson's participation in the nation's first active full-scale field test, and ultimately in Sandia National Laboratories, can be traced to an urgent visit to Berkeley by two men referred to by E. O. Lawrence as "a couple of high-pressure artists." As it turned out, these visitors were none other than J. Robert Oppenheimer and George B. Kistiakowsky, who persuaded Bob that he was needed for a very important project in the mountains of New Mexico. Henderson indicated that he could get away in "perhaps a week." "No," Oppenheimer said, "how about this afternoon?" With that he pulled a plane ticket in Henderson's name from his pocket. Henderson called his wife and asked her to pack a bag for what he thought would be a four-day stay. He returned four months later to move his family.[57]

At Los Alamos, Henderson would not only serve as liaison between research and engineering, but would also head the engineering group assigned to locate and develop a site, facilities, and techniques for the field test of the first atomic bomb. One of the most unusual "facilities" was, of course, the containment vessel known as Jumbo.

The design and engineering problems posed by Jumbo proved much greater than expected. How could a containment vessel be constructed that would not rupture under the detonation shock of 5,300 pounds of high explosives? In the event of a nuclear reaction dud, how could the active material be recovered for reprocessing and reused in another experiment? And what about transportation?

To help with the project, Henderson, who was named Supervisor of Design in March 1944, asked Oppenheimer if he could try to persuade Roy W. Carlson to come to Los Alamos. Carlson, who had been a Civil Engineering professor of Henderson's at Berkeley, was a world authority on the design of mass concrete structures and had developed the unique strain/stress meters buried in all large concrete dams since Boulder. Used in conjunction with temperature gauges, these instruments could monitor the internal stress conditions of the concrete. Thus, when temperatures rise, cold water would be pumped through the pipes to equalize thermal stress and to prevent thermal expansion and the possibility of cracks in the dam. Carlson's instruments, which were installed in the Boulder Dam in the United States and the Itaipu Dam between Brazil and Paraguay— among others the world over—had earned him an international reputation. Carlson's ability to measure stresses, combined with Henderson's experience with Standard Oil from 1937 to 1939 on the design of high-pressure/high-temperature vessels used in oil refining, held promise for meeting the challenge of designing a pressure vessel capable of containing the high explosive of the atomic device.[58]

Carlson arrived at Los Alamos in 1944 and started work on the design of the vessel walls, while Henderson developed the closure system. As Henderson frankly admitted, the closure system was based upon "a German patent I had lifted from I. G. Farben Industry that was uniquely suited to the application we had in mind."[59]

Carlson, on the other hand, devoted attention to a design that would support the extremely high shock forces and sustained pressure. He recognized that the shape of the vessel should be changed from a sphere to an elongated shape "comprising a heavy, cylindrical central section with lighter, hemispherical ends, all of ductile steel and not cast."[60] Carlson also suggested that the cylinder be banded with concrete to provide the mass needed to absorb the momentum generated by the shock of the blast. According to theory, the container's steel walls would expand to contain the charge, then come to rest before rupture.[61]

Early Jumbino experiments: Bob Henderson and Roy Carlson first envisioned a spherical vessel (top, left), but soon realized that the triaxial stress pattern inherent in a sphere would not allow the vessel wall expansion required to absorb the shock. This idea was soon abandoned in favor of a cylinder with hemispherical heads. The top right photo shows a Jumbino chucked in a lathe to clean up the welds and provide a true surface over which three belly bands could be shrunk (below). Not shown is the concrete band that was poured around the cylindrical section to add mass to help in absorbing the initial shock.

Essentially, the design principle represented more of a problem in dynamics than in strength of material. As Henderson explained it:

> We would allow the walls of the vessel to be accelerated outward upon the outset of the explosive detonation at a speed slow enough so that the walls could be brought to rest by the tensile strength of the steel before the walls ruptured. ... In fact, the entire vessel was designed on the premise that the walls were completely fluid having only mass but no strength during the acceleration phase resulting from the shock wave.[62]

In other words, inertia was the phenomenon that would retard the outward motion of the vessel walls. Since inertia was the basis for the design, a cast-in-place concrete band was placed on the outside of the pressure vessel to give added mass when needed. The band would pulverize and fly off as sand under the impact of the shock wave, at which time the strength of the steel walls would come into play. The whole design was, therefore, based on dynamics.

Joseph Hirschfelder, known as a keen-minded man who wore a raccoon coat to the test site in midsummer, would later conduct studies on the spalling and breakup of Jumbo and the ballistics of fragments for internal explosions in the range of 50 to 500 tons. This information was needed for the design of shelters for personnel and the establishment of cleared areas, although Bainbridge estimated that "the absolute worst that could happen at the observation posts at a distance of 10,000 yards is for a six-inch cube fragment to descend almost vertically at 1,000 feet per second."[63]

To test the design principles on a small scale, approximately twenty models, twelve inches in diameter, were built. Feasibility experiments with the models proved Carlson's design concept and the ability to contain the chemical explosives. According to Carlson, Groves' response to this success was characteristic. "I thought that elephant was dead," he grumbled. And from that incident Jumbo supposedly received its name.[64]

After various steel companies refused to tackle the job of manufacturing Jumbo, the challenging task was offered to

Babcock and Wilcox. From his previous experience with Standard Oil Company, Bob Henderson knew of the firm's reputation and their unique capabilities in the manufacture of large high-pressure vessels of this type. Early in August 1944, the order for the design was placed officially.

When Vannevar Bush phoned to explain the importance of the containment vessel, Babcock and Wilcox officials responded enthusiastically, despite the stress that the order would place on the company's large metal-working facilities. As Henderson recalled, "The uniqueness of the design intrigued them, and they were able to add some good ideas."[65] The result of their efforts was the world's largest pressure vessel—twenty-five feet long, twelve feet in diameter, and 214 tons in weight. Constructed of banded steel walls fifteen inches thick, the container posed tremendous logistics problems.

To solve these, Henderson enlisted the aid and expertise of the Eichleay Corporation of Pittsburgh. The Eichleay Corporation, famous for heavy moving jobs, prepared a special 200-ton flatcar with recessed center. After making special arrangements with railroad officials, the transportation crew started Jumbo south from Barberton, Ohio, on a circuitous journey that led to New Orleans, across Mississippi and Texas, and on to Anthony, New Mexico, slightly south of Socorro.

When the train lumbered into view at Pope's Siding, the entourage revealed an engine and tender, a flatcar, and Jumbo, followed by another flatcar, and a caboose full of security personnel, guards, and railroad workmen. Jumbo, none the worse for wear, except for a rip in the tarpaulin covering it, then had to be carted thirty miles inland to its final desert destination one-half mile from Ground Zero. Assisted by the Eichleay experts, the engineers transferred the vessel onto a huge pneumatic-tired trailer with sixty-four wheels and a big flatbed on top. The idea was to use six caterpillar D-8 tractors, three pushing and three pulling, to move the vessel over the desert to Ground Zero. Once on location near the shot tower, Jumbo sat awaiting the decision that would determine its role at Trinity.[66]

By June 11, 1945, with more plutonium being produced, the prospects for a one-shot Trinity test were improving. Nevertheless, as Bainbridge wrote to Commander Norris E. Bradbury just one month before the test: "Jumbo is a silent partner in all our plans and is not yet dead."[67] Designed to withstand the detonation of 5,300 pounds of chemical high explosive, the container had several drawbacks.

First, the vessel, as constructed, would not be easy to use. Since the largest allowable opening in the vessel compatible with the pressures expected was significantly smaller than the bomb, the gadget would have to be assembled inside the vessel—a difficult task. Second, the scientists felt it would be hard to measure the actual energy released by the nuclear reaction when so much of it would be absorbed by the vaporization of Jumbo. They felt also that the phenomena associated with the ball of fire would be different from what they would have been if Jumbo were not used. In other words, scientists feared that their sophisticated instrumentation and test data would be compromised or greatly complicated by vaporization of the container. Furthermore, vaporization of Jumbo would contribute to radioactive fallout and atmospheric contamination.[68]

Moreover, the scientists had grown increasingly confident that the implosion device would be successful. Two major factors had contributed to their optimism, namely, the refinement of the implosion lens system, and improvements in the chemistry and quality control used in the manufacture of the explosive, Composition B.

Approximately two weeks before the Trinity shot, the decision was made. Jumbo would not be used. Henderson explained the general reaction: "We all breathed a sigh of relief when we didn't have to use it; but our tests had proved it would work."[69] Test director Kenneth Bainbridge expressed similar views: "Jumbo," he said, "represented ... the physical manifestation of the lowest point in the Laboratory's hopes for the success of an implosion bomb. It was a very weighty albatross around our necks."[70]

The story of Jumbo, however, did not end with the Trinity shot. Although the firing crumpled the seventy-foot steel support structure used to erect the vessel, it remained upright and intact. On August 27, 1945, Henderson checked the condition of Jumbo, found it undamaged by the July 16 blast and still sitting absolutely vertical on its foundations. Henderson then fabricated and installed a heavy weather-tight manhole cover for the opening. As added protection against corrosion damage, he covered it with a tarpaulin held in place by steel banding tape. The component parts of the vessel closure were cleaned, thoroughly greased, crated in four boxes, and stacked on six-by-six-foot sleepers in the field in back of the Fubar [fouled up beyond all recognition] warehouse at Trinity Site.

Later, afraid that he would be criticized for building this million-dollar monster without ever using it, General Groves ordered the Commanding Officer of Sandia Base to detonate something in it so that he could say that it had at least been used. The GIs, unaware that the explosives should be geometrically centered, placed six 100-pound bombs on the bottom of the upright vessel, slapped on some plastic explosives, and promptly blew a hole in it. The Babcock and Wilcox people, who had plans for a history of the vessel, were furious to learn of this irresponsible action.

After allowing the container to lie on its side in the desert for a period, General Groves decided that Jumbo should have a decent burial. This act of humiliation further irritated Babcock and Wilcox, who wanted to conduct tension and fabrication tests on the vessel. After sectioning the container and taking samples, they covered it up again.[71]

Selection of a Test Site

Along with decisions to field test the implosion device and to develop Jumbo, still another preliminary task along the road to Trinity was the selection of an appropriate test site. Physicist Kenneth Bainbridge, who was to direct all on-site operations,

Bob Henderson, explosives expert George B. Kistiakowsky, and Major William A. "Lex" Stevens flew over a good portion of the southwestern United States in a C-45 aircraft as part of the initial survey.

A relatively flat terrain was needed in order to gauge accurately blast effects and to be able to transport Jumbo. The area also had to be situated in a locale where weather conditions were generally good. And the site had to be isolated from population centers but close enough to Los Alamos to facilitate the transportation of men and equipment. Secretary of the Interior Harold Ickes imposed a final criterion: "Under no circumstances should any region be taken over which would involve the displacement of a single Indian."[72]

The preferred test site was a desert area on the banks of the Colorado River, just north of the small California town of Blythe; however, when Groves learned of the location, he subsequently declared it off limits because General George Patton was using it as a training ground. Second choice was the Malpais south of Grants, but access was difficult because of the hardened lava flows.[73] A total of eight different locations across the southwestern United States were considered.

Ground search for a suitable site began early in May 1944. Plowing their way through a late spring snow, Oppenheimer, Bainbridge, Lex Stevens—who would serve as construction engineer for the camp—and project intelligence officer, Major Peer De Silva, set out from Los Alamos for the headwaters of the Rio Puerco. Bob Henderson, and, occasionally, Roy Carlson, who were responsible for the logistics of transporting Jumbo, also accompanied Bainbridge on siting trips during the spring and summer of 1944.

The third and succeeding trips concentrated on the desert area known as the Jornada del Muerto near Socorro, New Mexico. Situated between the San Andreas Mountains on the east and the San Mateos on the west, the terrain is dotted with yucca, broomweed, and, closer to the Oscura Mountains, a smattering of piñon and juniper. As a part of the old Camino Real, the Jornada served as a highway for Spanish caravans en route from Mexico to settlements in the north. As these early travelers soon learned, the region was well named. Attacks from marauding Apaches and the lack of water often made the area a Route of the Dead Man.

By the 1940s, a few ranching families such as the David McDonalds had established themselves in the region, trying to eke out a living by grazing cattle and sheep. In 1942, however, the government leased a few hundred square miles of the area for the Alamogordo Bombing Range. The leased acreage would prove to be a source of contention between the United States government and the inhabitants for years to come.

After Groves approved the Jornada region September 7, 1944, arrangements were made with Alamogordo Air Base Commander, Colonel Roscoe Wriston, for the use of the northwest corner of the bombing range (Target B-4) for the Trinity test site. Inhabitants of the area were not happy about the prospect. Although the search team had the foresight to enlist the aid of the local U.S. Marshall Ruben Cole, who was described as "an authority on protocol in the rough and ready area," they were not met with open arms.[74]

On one occasion, the team took a car east toward Mockingbird Gap in the Oscura Mountains. There, between the Jornada and the broad valley in which Carrizozo is located, they came upon a small ranch house. When they got within "hailing distance," the Marshall told the passengers to get out and stand in front of the radiator. He then called out to the house and at first got no response. After the third call, a voice asked: "What do you want?" "We want to fill our canteens with water," the Marshall answered. Another pause, and the voice said, "Okay, come on up."

Still, there was no sign of life. The search team climbed back in the vehicle and proceeded toward the house. As they got closer, the men could see that they were looking right down the barrel of a rifle. When the door opened, a boy about fifteen years old stepped out holding the gun. "This was our first introduction to the McDonald clan," Henderson recalled. The main McDonald ranch building ultimately housed Jack Hubbard's weather group, the photography group under Julian Mack, and the assembly point for the bomb core.[75]

On a later trip to conduct surveys to lay out the observation stations and Ground Zero, the team noticed that they were working fairly close to one of the bombing range targets. Because arrangements had already been made with the commander to cease bombing activities temporarily, the crew wasn't worried about its position—that is, not until they looked up to see a flight of B-17 bombers approaching with bomb bay doors open. "It occurred to me that bombardiers are famous for their inaccuracy," Henderson said, "so the best place was to sit right on the target. We leaned right up against the canvas cover and watched the practice bombs come down. Our supposition was correct; not one hit the target; and we all walked away somewhat shaken."[76]

By November 1944, Captain Samuel P. Davalos and the Army Corps of Engineers had the construction of Trinity site well under way. Under the expert coordination of John Williams, leader of the Electrostatic Generator Group, miles of wire for telephones, radios, and public address systems soon provided a complete communications network in the desert. Two old windmills pumped an almost unusable water supply filled with gypsum and chemicals, requiring that Army trucks haul in water regularly. Diesel generators provided electricity; portable outbuildings were erected, as well as four barracks, a commissary, mess hall and kitchen, vehicle repair shop, and the storehouse nicknamed Fubar.

An Albuquerque contractor named Ted Brown, with a crew of 200 laborers, worked seven-day work weeks setting up hundreds of poles connecting Ground Zero to the shelter locations, constructing three earth and concrete bunkers, and the tower. Brown's crew also maintained the roads and scraped the twenty-five-mile path from Pope's Siding for the transport of Jumbo to its assigned location near Ground Zero.[77] The desert was being converted into a functional scientific laboratory.

Back on the "Hill," Groves ordered a freeze on implosion bomb design in February 1945. According to Groves' directive, all efforts were to be concentrated on the lens implosion method with a modulated nuclear initiator. To ride herd on the bomb program, Oppenheimer appointed a special "Cowpuncher" Committee consisting of Samuel K. Allison as chairman, Robert Bacher, George Kistiakowsky, Charles Lauritsen, William Parsons, and Hartley Rowe. Specialized committees helped capitalize on the advice of senior consultants such as Niels Bohr, I. I. Rabi, and Lauritsen. These individuals, according to Project Y historian Hawkins, "served in the capacity of Elder Statesmen to the Laboratory."[78]

March 1945 was a pivotal month at Los Alamos. In this month, in addition to the Cowpuncher Committee, a Special Weapons Committee was created with Ramsey as chairman. The Committee, which reported to Parsons, had as members A. F. Birch, Brode, Bradbury, Louis Fussell, and Glenn A. Fowler. Responsible for "all phases of work peculiar to combat delivery," the Weapons Committee soon became a part of Project Alberta, which, along with the Trinity Project, was formed in March 1945 with the status of divisions. Project "Alberta," headquartered at Wendover Field, Utah, had extended responsibility for movement overseas.[79] For the future of Sandia, the establishment of the Alberta Project would be especially significant since some of the same personnel would become part of the first contingent to be stationed at Sandia Base.

Pioneering Field Testing Techniques

As test director for Trinity, Bainbridge, assisted by William Penney of the British mission, made preparations for a trial test involving the detonation of 100 tons of conventional high explosives. For the trial test, crates of high explosives were transported to the site from Fort Wingate and stacked on a platform resting atop a twenty-foot tower. The crew then inserted tubes containing 1,000 curies of fission products from a Hanford "slug" to simulate effects of radioactive material from the actual test.

The trial test met its objectives of providing data for calibration of instruments for blast and shock velocity measurements. It also introduced the crew to the difficulties of field testing. On May 7, 1945, observers reported that the TNT detonation created a brilliant orange fireball that could be seen in the predawn darkness as far away as the Alamogordo air base sixty miles to the southeast.[80]

Meanwhile, last-minute procurement and communications problems for the "hot run" added to an increasingly tense situation. The Raytheon Company failed to meet its delivery schedule for X-units (high-voltage power supplies), and the development and production of full-scale lenses had been slowed because of a delay in the delivery of lens molds.

Snafus also occurred with maddening regularity. Although the loss of 10,000 feet of garden hose may not appear to be a crisis-making event, its impact was significant because the hose was used as protection from the weather to encase cables leading to sensitive instruments. Similarly, the teletype service between Trinity and Los Alamos was so bad, according to one participant, that "you never knew if the test site was asking for a tube or a lube job."[81]

Despite such frustrations, however, the scientists and engineers managed to set up their experiments and the necessary radar and searchlight operations to record the blast and its effects. A lead-lined Army tank carrying Herbert Anderson and Enrico Fermi prepared to move into the test area immediately after detonation to recover equipment and pick up earth samples. Photographic specialists Julian Mack and Berlyn Brixner constructed special shielding devices and viewfinders for the fifty-odd cameras strategically positioned to record the history-making event.[82]

Assisted by Ben Benjamin, who would later become supervisor of Sandia's Photo-Optical Group, the photographic crew prepared to track the pattern of the fireball. Their objective was to determine the yield of the Trinity device by observing the velocity of the fireball as photographed by super-high-speed movie cameras. This was only one of numerous field testing techniques to be pioneered at Trinity.[33]

In addition, a variation on the key-and-lockbox security system developed at Trinity for arming and firing the device is still in use. Other experiments to quantify yield included radiochemical analysis of the debris and the recording of blast pressure versus time and distance—all of which were important in predicting damage that could be caused by enemy nuclear attack. Group Leader John Manley, who had charge of the West 10,000 shelter, set up the instrumentation to measure pressure in the blast wave at varying distances from Ground Zero. "We even provided the G-2 Security people patrolling the area at different distances with instrumentation and measured earth shock with equipment used in oil exploration," Manley said.[84]

Radiochemical data confirmed the fireball yield method, and patterns of debris deposition were used to develop the first fallout model. Based upon these pioneering techniques, diagnostics would continually be improved.[35] In retrospect, the magnitude of the accomplishment signified by the success of the Trinity test, according to one participant, represented "a will to do that has long since disappeared."[86]

Indeed, the effects of the Trinity blast were destined to reach far beyond the isolated site of the Jornada del Muerto, although the impact of the actual test was at the time incomprehensible. Profoundly impressed by the cataclysmic unleashing of the atom, those present had only an inkling of the worldwide consequences of their work. Only a privileged few knew that across the ocean an equally significant operation was already under way—an operation that would not only end the war with Japan but would forever alter the defensive and political posture of the nations of the world.

GENESIS: THE LABORATORY ON THE HILL

Ben Benjamin was representative of the young military men who also contributed to establishment of the scientific community known as "the laboratory on the hill."

Tinian Island in the Pacific, 1945.

Chapter III.

DESTINATION:
Hiroshima and Nagasaki

> *There are those who considered that the atomic bomb should never have been used at all . . . that rather than throw this bomb we should have sacrificed a million Americans and a quarter of a million British in the desperate battles and massacres of an invasion of Japan.*
> *The bomb brought peace, but man alone can keep that peace.*
>
> Winston Churchill
> August 16, 1945

The history-making atomic blast in the New Mexico desert represented the successful culmination of Project Trinity, but it was only one facet in the grand design to win the war. Project Alberta, activated along with Trinity in March 1945, had been placed under the overall command of Captain William S. Parsons. The project extended the activities of Norman Ramsey's Delivery Group (O-2) to include the planning and establishment of the advance base where the bombs were assembled, modification of suitable aircraft, supervision of field tests on bombs without active material, the assembly and loading of bombs onto planes, and the testing and arming of the bombs in flight. Roger Warner and troubleshooter Hartley Rowe handled in masterful fashion the field logistics of moving the bomb from Los Alamos to Tinian. A few months later at a lush tropical setting in the Pacific, Project Alberta with its mission to deliver a practical airborne military weapon gained momentum.[1]

To the men of the 509th Composite Group of the Twentieth Air Force, the small coral island in the Marianas "looked like the Garden of Paradise."[2] Situated 1,450 miles from Tokyo, Tinian (or "Destination" as it was code-named), had been secured only one year earlier after a massive battle for neighboring Saipan.

Among those assigned to depart for Destination on July 1, 1945, was Army Air Corps Lieutenant Leon Smith. Trained as a weaponeer, Smith was flying in from Wendover, Utah, with Lieutenants Morris Jeppson and Phillip Barnes. Thus far, the trip had been long, but largely uneventful. The flight from Hamilton Field in San Francisco to Hawaii took sixteen hours, during which time the soldiers were moved back and forth to "trim" the aircraft. As the C-54 "Green Hornet" with its human ballast neared Johnston Island, landing was delayed by a downed aircraft on the runway. Flying on to Kwajalein, the soldiers noted that the climate had become more tropical with fluffy cumulus clouds and rain showers. The approach to Tinian showed high hills, heavy vegetation, and four long parallel runways. "It was the most magnificent field I had ever seen," Smith recalled, "and a delight after Wendover."[3]

New Yorkers arriving on the island felt especially at home. Not only was the island shaped roughly like Manhattan, but the network of good roads left by the Japanese had been given familiar names—42nd Street, Riverside Drive, Madison Avenue, and the Columbia University District, where the 509th took up residence. Under the technical direction of Norman Ramsey and Norris Bradbury, the scientific contingent from Los Alamos made themselves at home in the vicinity of Times Square.[4]

Selected by Commander Frederick L. Ashworth, an Annapolis graduate and aviation ordnance expert, the island was especially well suited to carry out the mission's objective in complete secrecy. There, Colonel E. E. Kirkpatrick of the Manhattan Engineer District directed a battalion of Seabees in the construction of the largest bomber base in the world. With three bomb assembly huts, four air-conditioned quonsets for lab and instrument work, two loading pits equipped with hydraulic lifts, a shop building, administration building, five warehouses, a

DESTINATION: HIROSHIMA AND NAGASAKI

Leon Smith stands at the corner of 42nd Street and Broadway, Tinian Island.

a wooden mess hall, an outdoor movie theater, and a tent city, Tinian was ready for the 509th when they first began arriving from Wendover Field, Utah, in late April 1945. Although the greater part of the Composite Group arrived at its overseas destination during the month of May, both individuals and equipment continued arriving until the first week in August 1945. Many of the scientists, civilians, and SEDs, including Robert Serber, Luis Alvarez, Arthur B. Machen, Harlow Russ, Theodore Perlman, Roger Warner, Jr., Vincent Caleca, and Eugene Nooker, arrived during June and July.[5]

In connection with the Project's mission to deliver a combat-ready weapon, it had been necessary to establish test ranges at far-flung locations, modify airplanes to carry the payloads, and train special crews. The three bomb assembly huts at Tinian, for example, were first tested at Inyokern, California, before being shipped to the Pacific site.

Inyokern

Inyokern was the field site for "Camel," a project conducted by the California Institute of Technology in collaboration with Los Alamos to deal with problems relating to implosion design and combat delivery of the bomb. Equipped with their own procurement, laboratory, and field test facilities, scientists of the Camel Project researched and engineered special components of the implosion assembly, detonators, lens mold design, impact and proximity fuzes, and high-explosive components. In addition, when implosion design work was frozen at Los Alamos in April 1945, Camel initiated its own implosion program as an alternative. The Camel Project produced implosion bomb mock-ups, known as "pumpkins" (also called "blockbusters"), which were prototypes to be used in practice bombing runs on enemy targets. Some of these were loaded with concrete, others with HMX cast explosive. Bright orange, equipped with special impact fuzes, and made to simulate the shell of the Fat Man, the pumpkins were drop-tested at Inyokern, Wendover, and at Salton Sea in California.[6]

Coordination between Los Alamos and Camel required considerable cooperation and understanding of responsibilities. To facilitate these arrangements, Oppenheimer appointed Edwin M. McMillan to act as a special liaison. Improved communications and regular flight schedules resulted, and freight and passengers were soon being transported much more efficiently between Los Angeles, Inyokern, Santa Fe, Sandy Beach on the Salton Sea, Wendover, and Sandia Base in Albuquerque.[7]

Wendover

Alternately referred to as "Kingman," or "W-47," Wendover was another isolated field testing spot where highly secret assembly and production shops had been established under direction of the Ordnance Division from the Hill. At this remote site, personnel loaded bombs into B-29s, checked electrical systems, and conducted other bomb preparation and monitoring work in the delivery craft up to "bombs away." These activities were carried out by the Army Air Corps and that facet of the Ordnance Group that would be designated by July of 1945 as the Z-Division of Los Alamos Laboratory. Los Alamos personnel determined the types of tests; the Air Corps conducted them, and the recorded results were then returned to Los Alamos for analysis. As Ramsey indicated, the objective was to test units that approached "more and more closely the final model."[8]

Visitors from Los Alamos arrived at Wendover in a disguised "Green Hornet" aircraft. Members of the Army Air Corps made similar trips to Los Alamos. All visits were authorized by military special orders such as those issued to G. C. Hollowwa:

> Fol named AC O, Sq A 216th AAFBU (S$_p$), WP on TDY to ************ o/a 13 Aug 45 by mil acft on matters pertaining to Project W-47 and upon compl this TDY will ret this sta.

Translated, the orders read:

> Following named Air Corps Officers, Squadron A 216 Army Air Force Base Unit (Special), will proceed on temporary duty to (classified) on or about 13 August 1945 by military aircraft on matters pertaining to Project W-47 and upon completion of temporary duty will return to this station.[9]

On these trips, personnel, although in uniform, were not allowed to wear Air Corps insignia. Hollowwa recalled that "we borrowed whatever we could, that is Signal Corps, Medical Corps, Ordnance Corps, etc. This led to interesting encounters with other military personnel during stopovers in Albuquerque. 'What service school did you attend?' 'What do you think of equipment Model XYZ?' "[10] Also at Wendover were members of the 509th who were training in preparation for Destination.

Located near the Bonneville Salt Flats in Utah, Wendover was a barren expanse of land that nudged the Nevada state line. Bright, treeless, and flat, the area rested on a salty marsh. Hollowwa, who came to Wendover as part of the backup squadron for the overseas contingent, noted that the terrain was devoid of vegetation and inhabited only by rattlesnakes and black crows. Closer toward the mountains lay the little town of Wendover, but the base itself was the typical tar-paper shack affair with a primitive officers' club, bowling alley, and the State Line Hotel with Western Bar and Casino. Adding variety to an otherwise nondescript complex of buildings were a library and a swimming pool fed by a hot spring. Uninviting as it might appear, Wendover suited demands for a highly secret base for big bomber operations. And, as Leon Smith soon learned, secrecy was all-important.

Upon Smith's arrival in November 1944, Robert Brode drove him out to a rocky hillside: "We sat on a rock," Leon said,

> and Brode told me a little about what we were to do. He described it as a fuzing job with six to seven cubic feet of available space. I remarked that that seemed like an awfully large volume for a fuzing system and innocently asked if this was a biological or atomic bomb that we were to work on. Later, I was told that the question got my name in a black book and resulted in an investigation. . . . I wasn't supposed to know so much.[11]

DESTINATION: HIROSHIMA AND NAGASAKI

A B-29 at Wendover Field, Utah.

Wendover, Utah.

A sign reading:

> What you hear here—
> What you see here—
> When you leave here—
> Let it stay here!

meant exactly that.[12]

Security problems were compounded by the presence at Wendover of a group of Russian "allies," which meant that extra care had to be taken to prevent them from getting access to information about the "Project." Assigned to the base as "liaison," this group of U.S.S.R. officers were politely kept at arm's length. However, as one engineer observed: "They, nonetheless, were with us every night at the officers' club."[13]

Hollowwa was also impressed with the sensitive nature of the work at Wendover. He noticed that

> The fuzing and firing activities were conducted in separate buildings, and we were strongly discouraged from visiting. (I was in the firing building only once during an interval of several months.) The firing unit in the nose of the Fat Man was connected to the fuzing unit, in the tail, by a two-wire cable. When the fuzing unit determined the proper altitude for detonation, it sent a signal down the cable. Thus the fuzing personnel did not have a need-to-know about details of the firing mechanism and vice versa.[14]

When Hollowwa first arrived at Wendover, he found the atmosphere charged not only with secrecy, but also with a sense of mission. Members of the 509th were packing and readying their planes for departure to Tinian. When Leon Smith transferred overseas, Hollowwa assumed his job and was assigned to continue the day-to-day field testing.[15]

In addition to field testing, there were other secret projects going on at Wendover, among them experimentation with television-guided bombs and captured German V-1 "buzz" bombs. The work week was a full six days, Monday through Saturday. Passes to Salt Lake City were highly prized. On those special weekends, five men or more would pile into one of the few available cars to make the trip and return by Sunday noon.[16]

DESTINATION: HIROSHIMA AND NAGASAKI

In the Tech Area at W–47, the men worked in metal assembly buildings connected by a narrow cinder road. As the sun beat down on the corrugated roof, the 1561 Model of the Fat Man slowly took shape. The unit, hung from overhead hoists, was manhandled into position by the crew. After assembly, the center of gravity was determined and each unit weighed—information that was needed by the bombardier of the plane. The men then hoisted the unit onto a wooden cradle atop a flatbed trailer called a "float," covered the device with a tarpaulin, and transferred it to the loading pit. For security purposes, pieces of lumber were inserted strategically under the tarpaulin to camouflage the shape of the unit underneath as it was taken to the loading pit.

At the pit, a C–2 automobile wrecker moved into position so that its hook could be attached to the lift lug of the unit. The unit was then carefully maneuvered over the center of the pit—a procedure that caused the front wheels of the wrecker to lift off the ground some fifteen to eighteen inches. Meanwhile, in a dangerous operation, two crewmen remained inside the pit to help position the cradle to receive the unit. The aircraft was then taxied toward the pit area where, in an exacting maneuver, the pilot positioned it with open bomb bay directly over the pit. A clevis and bar arrangement, later called a "linkage," was used to hoist the bomb into the plane. Later developments incorporated the use of a hydraulic piston to raise the bomb into the bomb bay, where it was latched into place. This arrangement, much like raising an automobile on a grease rack in a garage, was a much safer procedure than the method used to load the Little Boy. Procedures used for loading the Model 1791 Little Boy and for the pumpkins involved "pumping up" the shock absorber on the nose wheel, removing the bomb bay door, and rolling the bomb trailer under the airplane where the bomb was then hoisted into position.[17]

As the atomic energy program reached a climax, the training schedules and deadlines imposed upon W–47 personnel became more demanding. Roger Warner provided the schedules for the test program and acted as liaison between W–47 and Los

Alamos. By June 1945, the schedule called for nearly twice as many tests as required in May. To meet the increased demands for assembly, Captain James Les Rowe, the officer in charge of the Tech Area, set up a circus tent where boxed components were uncrated and assembled. In Rowe's opinion, "the speedup resulted in extreme physical hardship, and in the latter part of June we entered a period of high risk as explosives became a part of every assembly."[18]

Eight or nine miles west of the airfield, a bombing range was set up for the testing of ballistics and fuzes, although some of the tests were conducted at Salton Sea with a more elaborate instrumentation system. Another range, situated about sixty-five miles from W-47, was used for drop testing, but models fitted with high explosives were usually tested at Inyokern, which was better equipped and therefore preferable.[19]

Sandy Beach

Sandy Beach was a Naval Auxiliary Air Facility established on the southwest shore of the Salton Sea. The Salton Sink, a gourd-shaped depression thirty miles long and eight to fourteen miles wide, offered the advantage—for high-altitude test purposes—of being 250 feet below sea level. Separating the Salton Sink from the north end of the Gulf of California was the Imperial Valley. Originally a desert that rivaled the Sahara in aridity and high temperatures, it was converted into a fertile garden area through the introduction of irrigation. In addition, a profitable salt mining business was started by the New Liverpool Manufacturing Company, which shipped over 1,500 tons of salt to the San Francisco market each year.

In 1905 severe spring floods so eroded the protective banks of the large gap leading into the Imperial Valley that torrents of Colorado River water rapidly converted placid ditches into raging waterways. For sixteen months the river flooded the Valley, spilling northward into the Salton Sink, inundating the salt mine, and recreating the Salton Sea.[20]

It became apparent to William Parsons that a sea-level range with more suitable weather conditions would be essential for certain crucial tests. In October 1944 he proposed the Salton Sea site as an auxiliary to Wendover. General Groves granted permission; and in the following month, Admiral William R. Purnell directed Parsons to establish a bombing range in the vicinity of the Naval Auxiliary Air Station at the sea's Sandy Beach.[21]

The test drop area was a fifteen-mile-long stretch of brackish lake water. Here, Parsons made arrangements for facilities to accommodate a small observation party of not more than ten officers and civilians and ten enlisted Army personnel. Roads were cleared, observation points installed, and a bombing target constructed on pilings. Admiral Purnell specified that the facilities be ready by December 1. Among the personnel flying the 600 miles from Wendover to the Salton Sea range for test drops were weaponeers Leon Smith, Morris Jeppson, and Phillip Barnes.[22]

Training the 509th

In addition to establishment of facilities and test sites, the members of the 509th Composite Group had to be shaped into an effective combat element. In August 1943, Colonel R. C. Wilson and Colonel M. C. Demler visited Los Alamos and recommended that the Air Force begin immediate training of a combat unit for delivery of the atomic bomb. This assignment fell to Colonel Paul W. Tibbets, Jr., recognized as one of America's most distinguished flyers.[23]

After selecting a hand-picked group of pilots, enlisted men, and a highly skilled ground crew, Tibbets groomed the 509th through an intensive training program that extended through the winter of 1944 and into the spring of 1945. Under the code name "Centerboard," plans were made to drop the bomb as visually close to the center of the target as possible.[24]

By the first of the year, training for this specialized mission had intensified, and on January 7, 1945, an advance party left Wendover for temporary duty at Batista Field, Cuba. In Cuba, the Squadron simulated combat missions to Beriquen, Bermuda, the Virgin Islands, and up as far as Norfolk. This was the beginning of a period of extensive training that was to last until the end of February. In March 1945, the First Ordnance Squadron, a unit especially designed to carry out the technical aspects of the Group project, joined the 509th under the command of Captain Charles Begg. The advanced air echelon arrived on Tinian in May, with the first Technical Detachment, SEDs, military personnel, and writer William Laurence reporting in June.

Upon arrival in Tinian, all combat crew personnel underwent additional training relating directly to the combat operation ahead of them. Training flights included missions against the Truk and Marcus Islands. Finally on July 20, the 509th made their first individual strikes against the Japanese Empire.[25] Rather than joining the large fleets of bombers launched against Japan, the pilots of the 509th were sent to drop a single bomb, a training tactic that led the Japanese to believe that the United States was involved in some strange form of psychological warfare. According to Radio Tokyo, the raids were "sneak tactics aimed at confusing the (minds of the people)." Instructed to drop the bomb over the target visually, the American pilots were to release it, then immediately make a turn of about 150 degrees. They were cautioned about being over Ground Zero when the device exploded.[26]

Leon Smith recalled one run in particular: "We were about 8,000 feet over Iwo Jima practicing final arming of the Little Boy when Captain Parsons called: 'Jump down here and help me!' This didn't sound too exciting because to jump down, you would land on the bomb bay doors, and I knew of instances when they had popped open."[27] Smith jumped, nevertheless.

Project Silverplate

The B-29s flown by the 509th also had to be groomed to suit the mission. In June 1943, Norman F. Ramsey had first investigated aircraft that might be suitable to carry the gun model bomb. He found that among those available, only the B-29 could accommodate the lengthy Thin Man, and then only by joining the two bomb bays. He considered the British Lancaster as a possibility, but maintenance and operation of the plane from American bases would have been difficult. Furthermore, there seemed to be some moral objection to the idea. When General Groves told Commanding General Henry H. ("Hap") Arnold that the Lancaster might be used, Arnold replied that "there was no way they were going to deliver the American atomic bomb in a British plane; tell him [Arnold] what to do, and he would see that it was done."[28]

By the fall of 1943, it became apparent that in addition to the Thin Man, the external shape of the Fat Man bomb would also have to be considered in modification plans for the B-29. Shortly thereafter, Ramsey and General Groves met with Colonel R. C. Wilson of the Army Air Corps to implement the program code named "Project Silverplate."[29]

Modification of the first B-29 officially began November 29, 1943, under Colonel Wilson, who was the Army Air Corps Project Officer for all aspects of the program. Colonel Don L. Putt was placed in charge of the division at Wright Field under whose jurisdiction the modification took place. However, a large part of the mechanical-engineering design was done by civilian engineer Charles Speer of Boeing. The crew assigned to test the B-29 included Major C. S. Shields (pilot) and Captain David Semple (bombardier).

During December 1943 and January 1944, modifications on the prototype B-29 continued in earnest. Included were: installation of a carrier frame, changes in the bomb doors, sway bracing and release for the Fat Man bomb model, installation of release and sway bracing for the Thin Man, and installation of the special wiring circuits required by the Fuzing Group. Several Eyemo cameras were installed in the bomb bay for photographing the release and initial bomb flights. The release used was a modification of a standard glider tow release; two of these, tied together mechanically, were used. The bomb models were also fitted with suspension lugs.

With modifications completed, the prototype B-29 was flown to Muroc Army Air Base, California, to undergo the first series of tests, arriving there February 20, 1944. Two test drops were made on February 28 with standard inert bombs for camera practice by the ground crew. The first drop of the Thin Man model from a B-29 took place March 3, 1944. Aerodynamic problems with the bomb shapes as well as problems with the release mechanism of the bomb bay doors were solved during this testing period. The test program necessitated considerable interaction between Muroc personnel and representatives such as Sheldon Dike from Los Alamos.[30]

In August 1944, it was decided that additional aircraft should be modified for use. The decision to "freeze" bomb shapes and sizes in September 1944 eliminated problems associated with continually changing specifications. The Glenn L. Martin Nebraska Company, using the Muroc plane as a model, contracted to modify the first B-29s at their Fort Crook modification center. Charles Speer continued with supervision of engineering on the modifications required, although Sheldon Dike, as the Project Alberta representative, engineered the special wiring.[31]

In all, seventeen aircraft were modified and were soon set apart by their unique markings—a large, black circle bisected by a black arrow on the tail section. Adding to the distinctive appearance of the "Silverplates," as the B-29s came to be known, was the absence of the turrets, which were removed to provide maximum speed and altitude. The Martin-modified aircraft were delivered to Wendover Field, Utah, in September 1944, where testing continued under a Test Section headed by Major Charles Sweeney.

The Silverplates were eventually replaced by newer B-29s with the latest improvements, including fuel injection engines, new reversible-pitch propellers, pneumatic bomb doors, and a special bomb rail design supervised by Milo M. Bolstad. Again, Dike was responsible for the reengineering of Silverplate modifications at Omaha. Ray Brin, who had worked for Dike at Wright Field, helped develop the hoisting techniques and cradles used on the modified B-29s and was sent to Tinian as chief bomb loader. Among those planes sent to Tinian was Aircraft No. 44-27292, the *Enola Gay*.[32]

The Glory Boys

Because of the special treatment accorded the 509th and the secrecy surrounding its mission, members of the group were dubbed "the Glory Boys" and made the butt of a certain amount of razzing. A clerk in Base Operations composed a verse that expressed the feelings of many of those on Tinian who were not members of the 509th:

> Into the air the secret rose,
> Where they're going, nobody knows.
> . . . .
> Don't ask about results or such,
> Unless you want to get in Dutch.
> But take it from one who is sure of the score,
> The 509th is winning the war.[33]

While the atomic device was undergoing a full-scale field test at Trinity, the nuclear components for the bomb and the active materials were being sent piecemeal to Tinian. One part of the U-235 was sent to Tinian by cruiser; the other three were transported in three Air Transport Command C-54s.[34]

Even at this late date, however, events occurred that could have changed the course of history. On July 26, 1945, the cruiser *Indianapolis* unloaded her shipment of U-235 at Tinian and set out to sea. Four days later she was torpedoed off the shores of Leyte and went down with nearly all hands.[35]

Still another near-miss involved the Fat Man components being flown in by air. Three Fat Man assemblies, complete except for the nuclear material, were transported by truck from Los Alamos to Kirtland Field, Albuquerque. There, three B-29s of the 509th met the contingent and flew the bombs to Mather Field in Sacramento, where they arrived July 28 and 29.[36] However, one of the planes, the *Laggin' Dragon*, piloted by Captain Edward M. Costello, had a near-accident when the life raft door flew open, damaging the right control elevator in the tail. Fully expecting to crash-land, Costello managed to bring the plane in for a perfect landing. After servicing, the three B-29s took off for Tinian.[37]

Operation Centerboard

The Japanese rejection of Truman's Potsdam ultimatum meant that Operation Centerboard took effect. For two days, the superfortress, the *Enola Gay*, sat awaiting orders for takeoff, her mission still a mystery to many. Poised on the edge of the runway, and piloted by Colonel Tibbets, the *Enola Gay* was loaded with the world's first atomic bomb to be dropped in combat.[38]

The bomb, nicknamed the Little Boy, was a gun-assembly type weapon that bore strong resemblance to a whale. It weighed approximately 8,900 pounds, measured ten feet in length, and was twenty-eight inches in diameter. Slated to be armed en route, it was composed of three mechanical components; the gun and target case; the tail and panel assembly; and the structure on which the fuze and panels were mounted. The fuze consisted of a set of four Archies (tail warning radar), four battery assemblies, and the necessary operating and test cables. The three principal arming and firing devices (Archies, clock switches, and baros) were connected so that two units of the Archies or four of the baros could fail, or one of each could operate prematurely without affecting fuze operation. The Archies were, therefore, part of a relay logic system engineered so that any two of the four could provide a firing signal.

Back at Los Alamos, two young lieutenants, one of them George H. "Howie" Mauldin, had spent months redesigning and modifying existing fighter plane tail warning radars to specifications. Their major challenge had been to increase the frequency of the radar to prevent jamming. In flight, however, it was the responsibility of the weaponeer to see that the Archies were hooked up properly.[39]

The controls and monitoring for all functions were put together in a Flight Test Box, located just aft of the flight engineer in the front compartment of the B-29. Leon Smith and three other members of the Ordnance Squadron, Phillip Barnes, Morris Jeppson, and Bruce Corrigan, spent all day August 5 installing the black box in the *Enola Gay*. Various elements of the fuzing system had to be checked out in flight and prior to drop, particularly the radar fuzes and the lead acid batteries used as the power supply. "We were considerably interested in knowing if the *safety systems* were intact," Smith added in classic understatement. Smith, who had helped design the Flight Test Box while on special assignment to Project Y, flipped a coin with Jeppson to see who would be the weaponeer to support Parsons on the Hiroshima flight. Jeppson won, and Smith flew on to Iwo Jima, where he was to assist with the backup plane.[40]

On that same Sunday afternoon, August 5, the word came: Weather conditions over Japan were suitable for a daylight bombing raid. The crew received its final briefing, and the *Enola Gay* suddenly became the center of attention. Huge banks of floodlights gleamed and flashbulbs popped, as photographers recorded the event for posterity.

At 2:45 A.M. on August 6, shuddering under the strain of the bomb and a full load of gasoline, the bomber began its ascent. Bystanders held their breaths as the plane seemed to use up an inordinate amount of runway before gaining the necessary momentum for liftoff. The mission, however, proceeded uneventfully; and at 9:15 A.M., bombardier Major Thomas Ferebee released the first atomic bomb used in wartime on the Japanese city of Hiroshima on the island of Honshu.

A mile away, in the instrument plane, *The Great Artiste*, bombardier Kermit Beahan opened the bomb bay to release the blast gauges that would be floated to the ground by parachute. At the same time, only two miles away, *Marquardt's 91* moved into position to take photographs of the blast. A sharp diving turn and the *Enola Gay* was away as the bomb's firing set activated at 2,000 feet above the ground. The first shock wave, then another, hit the plane with a jarring thrust. On the ground below near the epicenter of the hit, there was only a wasteland. Fire storms extended the devastation. In the ultimate field test, the single atomic bomb, eighteen kilotons in yield, destroyed sixty percent of the city of Hiroshima.[41]

Unlike the Fat Man tested at Trinity, the Little Boy had never been tested in full scale. As R. W. Henderson recalled: "We had fired it five times as an inert system at Los Alamos and knew it would work, albeit horribly nuclearly inefficient as compared with the Fat Man."[42] In addition, three Little Boy inert units were drop-tested at Tinian. Nevertheless, only seventeen days after the Trinity test, the first combat bomb, although a different model, had been successfully deployed against the enemy.[43]

Operation Centerboard II followed three days after the mission against Hiroshima. The second strike, commanded by Major Charles W. Sweeney, did not proceed as uneventfully as the first. Originally planned for August 11, the flight had to be moved up to August 9 to take advantage of the weather. By dawn, the weather was already deteriorating, causing cancellation of plans for the bomb carrier to rendezvous with instrument and photography planes on Iwo Jima.

There was also a change in the aircraft slated to carry the bomb. Because of modifications to *The Great Artiste*, the decision was made to transfer Sweeney's crew to *Bock's Car*. Contrary to general misconception, perhaps created by the last-minute change, *The Great Artiste* would serve as the instrument plane rather than the bomb carrier.[44]

The weapon aboard *Bock's Car* was different as well. Rather than a gun-assembly bomb, the Fat Man, Model 1561 (a derivation of the 1222) was an implosion weapon weighing 10,800 pounds and measuring 128 inches in length and 61 inches in diameter. Inside the elliptical-shaped bomb casing rested the implosion sphere enclosed by a six-segment support casing. The plutonium and the initiator were at the center of the pit. Surrounding the sphere were H.E. lens charges of Barotol and Composition B overlying an inner charge of Comp B. Tail-warning radars with thin Yagi antennae projected outward from the waist. The electrical firing mechanism was crammed into the nose and connected to dual-routed cables that would carry the detonation signal.[45]

In the assembly hangar on Tinian, Art Machen, later of Sandia, and Bernard J. O'Keefe, who would become chairman of the board of EG&G, worked around the clock. Until the last minute, there had been considerable concern over the safety of the firing sets, which incorporated a spark gap dump switch—the only bridge between the high-voltage power supply capacitor bank and the detonators. As Bob Henderson explained: "We knew the repeatable characteristics of the detonators, but were worried about the spark gap switch. Our worry lay in nonstatistical knowledge of *that switch.*"[46] Consequently, the first test of a Fat Man with an active firing set did not occur until August 4, 1945, at Wendover, with the second test at Tinian August 8, 1945, just one day before its use in combat.

In the final checkout, it was U.S. Navy Ensign Bernard O'Keefe's job to connect the two ends of the cable between the firing sets and the radars. Alone in the assembly room with only an Army technician, O'Keefe proceeded with the final checkout and tried unsuccessfully to plug the cable into the firing set. "To my horror," O'Keefe recalled, "there was a female plug on the firing set and a female plug on the cable . . . . What had happened was obvious. In the rush to take advantage of good weather, someone had gotten careless and put the cable in backward."[47] The remedy was equally obvious to O'Keefe; but to disassemble and reassemble the weapon would take the better part of a day and delay the mission that should end the war.

HERITAGE

There was only one solution, O'Keefe decided: Break the rules. With the assistance of the technician, O'Keefe found two long extension cords, which they plugged into an outlet outside the fire door of the assembly room (a safety violation in itself since the door had to be propped open). He then carefully removed the backs of the firing set connectors, unsoldered their wires, reversed and resoldered the plugs onto the correct ends of the cable, and reassembled the connectors. "I spoke to no one about what I had done," O'Keefe admitted. The operation proceeded on schedule.[48]

Centerboard II continued to be bedeviled by problems even during flight. On August 9, 1945, at 3:45 A.M., *Bock's Car*, piloted by Major Sweeney, took off with an assembled nuclear system. Captain Kermit K. Beahan was bombardier; Commander Frederick L. Ashworth, bomb commander. Lt. Phillip Barnes assisted Ashworth as "electronics test officer." Ray Brin, as part of his duties, had checked out the aircraft, hovered around for the loading of the bomb, then "held his breath."[49]

Among other last-minute problems, an auxiliary gas tank had to be discarded because of leaking, which cut down on the fuel safety factor; Kokura, the primary target, was too overcast for visual bombing; and the second target, Nagasaki, was also shrouded in clouds. Finally, Beahan, a veteran of numerous European bombing missions, spotted a break in the clouds, and Sweeney decided to take it. At 11:50 A.M., August 9, 1945, Nagasaki became the second Japanese city to experience the devastating power of the atomic bomb. Forty-four percent of the city was destroyed.[50]

The tragic dimensions of the bombing of Hiroshima and Nagasaki were soon all too apparent—in the burned and naked survivors, the razed homes, and in the lives lost. Peter Wyden, in his book *Day One*, graphically and with sensitivity records the terrifying experience through the eyes of the Japanese survivors interviewed. Neither would the American bombing crews forget. In the throes of the war mood that prevailed, however, moral judgments would be a while in coming.

The return flight was not without its tense moments. The numerous sallies to look for alternate targets had used up a great amount of fuel; Sweeney decided to land in Okinawa. Without the proper codes, the crew fired flares to signal an emergency landing. As they taxied off the runway, one of the engines stopped. They barely made it.[51]

Yet, the mission was accomplished; the atomic bomb once again passed the test. But a greater test remained—how to perfect, manage, and control the awesome power unleashed by mankind.

Scientists and officials alike realized full well the responsibility inherent in their handiwork. Vannevar Bush, for example, in a statement to the National Policy Committee acknowledged that the field of "nucleonics" would bring with it changes "fully as drastic" as those resulting from the Industrial Revolution. "With this realization in mind," he continued, "it would be little short of criminal negligence not to take every conceivable step to prepare ourselves to make this social transition with the minimum of hardship." To meet the challenge of postwar control of nuclear power, he advocated that we consider its impact upon our own social structure, carefully analyze international agreements and commitments, develop a means of licensing and control, and above all, continue research in the field to protect our national interests.[52]

HERITAGE

Headquarters of the 509th on Tinian Island, 1945 (above and below).

DESTINATION: HIROSHIMA AND NAGASAKI

The *Green Hornet*.

HERITAGE

The *Enola Gay* and her crew return home after successfully completing their fateful mission.

Detonation of the Little Boy gun assembly bomb over Hiroshima, Japan, August 6, 1945, resulted in destruction of sixty percent of the city and a tragic loss of life.

HERITAGE

Major Sweeney piloted *Bock's Car*, carrier for the Fat Man bomb dropped over Nagasaki, Japan.

DESTINATION: HIROSHIMA AND NAGASAKI

Weaponeers (Above, left to right) Morris Jeppson, Phillip Barnes, and Leon Smith set up testing labs on Tinian Island to check out the fuzing, firing, and safety systems of the bombs. Jeppson was the weaponeer for the Hiroshima mission; Barnes for the Nagasaki flight. In addition to the military pup tents featured above, residential quarters on the island of Tinian included tents with screened "porches" such as the one pictured below. Seated, left to right: Norman Ramsey, William Parsons, E. E. Kirkpatrick, Edward Doll, and Dick Ashworth.

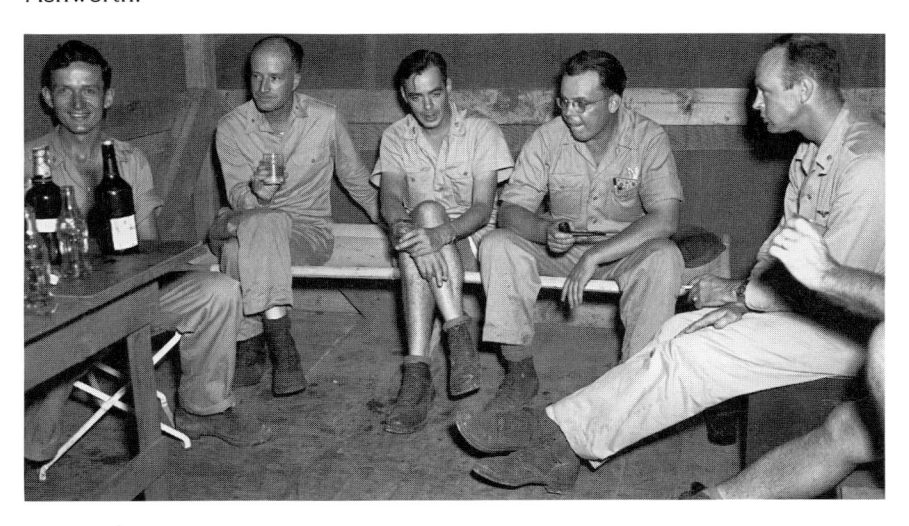

HERITAGE

Project A Organization at Tinian

Officer-in-Charge, Commodore W. S. Parsons, J.S.N.
Scientific and Technical Deputy to Officer-in-Charge, N. F. Ramsey
Operations Officer and Military Alternate to Officer-in-Charge,
   Commander F. L. Ashworth, U.S.N.
Fat-Man Assembly Team, R. S. Warner, Jr.*
Little-Boy Assembly Team, Commander F. A. Birch, U.S.N.R.
Fuzing Team, E. B. Doll
Electrical Detonator Team, Lieutenant Commander E. C. Stevenson, U.S.N.R.
Pit Team, P. Morrison and C. P. Baker
Observation Team, L. W. Alvarez and B. Waldman
Aircraft Ordnance Team, S. H. Dike*
Special Consultants, R. Serber, W. G. Penney,
   Captain J. F. Nolan, AUS

Project Technical Committee

Chairman: Norman F. Ramsey

Team Members:

Agnew, Harold	Linschitz, Henry
Anderson, David L., Ensign	Machen, Arthur B.*
Bederson, Ben B., T/5	Mastick, Donald, Ensign
Bolstad, Milo M.*	Matthews, Robert P., T/3*
Brin, Raymond, T/Sgt.*	Miller, Victor A., Lt.*
Caleca, Vincent, T/Sgt.*	Motechko, L. L., T/3*
Camac, Morton, T/Sgt.	Murphy, William L., T/Sgt.
Carlson, Edward G., T/Sgt.	Nooker, Eugene L., T/Sgt.
Collins, Arthur, T/4	Olmstead, T. H.
Dawson, Robert, T/Sgt.	O'Keefe, Bernard J., Ensign
Fortine, Frank J., T/Sgt.*	Perlman, Ted*
Goodman, Walter, T/3	Prohs, Wesley R., Ensign
Harms, Donald C., T/3	Reynolds, George, Ensign
Hopper, J. D., Lt.	Russ, Harlow W.*
Kupferberg, J., T/Sgt.	Schreiber, R. E.
Johnston, Lawrence H.	Thorton, Gunnar, T/Sgt.
Larkin, William J., T/Sgt.	Tucker, J. L., Ensign
	Zimmerli, Fred, T/4

---

*Indicates personnel who eventually came to Z Division of Los Alamos or Sandia Corporation.

Los Alamos Daily Log

<u>5 June 1945</u>

Staley returns from emergency furlough.
<u>Destination</u>--Advised Ramsey it will be necessary to ship material by air. Doll attended meeting in Jones office.

<u>6 June 1945</u>

<u>Destination</u>--Revised scheduled time of departure of O-3 crews as follows:

<u>Squadron Personnel</u>
<u>June 5</u>
1st Lt. George Koester
Cpl. Albert Lawrence

<u>July 1</u>
2d Lt. Richard Podolsky
2d Lt. Bruce Corrigan
Cpl. LeRoy Stradford
Cpl. William King
2d Lt. Leon Smith
2d Lt. Phillip Barnes
2d Lt. Morris Jeppson
Col. Richard Verts
Pfc. Carl Kasalek
Pfc. Leo Raub

Y Personnel
<u>First Departure -- July 1</u>
E. B. Doll
Ens. W. K. Prohs
T/3 Robert W. Dawson
T/5 Henry B. Silsbee
T/5 Frederick H. Zimmerli
T/3 William J. Larkin

Sandia Base, Albuquerque, New Mexico (February 1946).

Chapter IV.

THE POSTWAR YEARS:
Z Division

> *Never in history has society been confronted with a power so full of potential danger and at the same time so full of promise for the future of man and for the peace of the world. I think I express the faith of the American people when I say that we can use the knowledge we have won, not for the devastation of war, but for the future welfare of humanity.*
>
> *To accomplish that objective we must proceed along two fronts—the domestic and the international.*
>
> Harry S. Truman
October 4, 1945

In the spring of 1945, despite their concentration on Projects Trinity and Alberta, Oppenheimer and his technical adviser, Hartley Rowe, realized the need for postwar planning. According to Oppenheimer, "We wished to make provision for the continuation of weapons development, especially in its non-nuclear aspects, at a site convenient to Los Alamos—as Wendover was not—immediately accessible to aircraft and air strips, and not itself part of Los Alamos."[1]

Los Alamos—never a comfortable settlement at best—was by June of 1945, bursting at the seams. Local humorists, in fact, liked to claim that Los Alamos suffered from a "lack of women and water" throughout the duration.[2] As Oppenheimer explained:

The decision-making of Robert Oppenheimer, General Leslie Groves, Robert Sproul, President of the University of California, and Captain William S. Parsons (shown here during post-WWII ceremonies at Los Alamos) influenced the future of the nuclear ordnance and assembly facility known as Z Division.

THE POSTWAR YEARS: Z DIVISION

> We had not nearly enough housing, and no prospect of getting any soon. More important still, we were out of water. It was our belief that any major enlargement of activities there would not be practical; and that it was important to make the non-nuclear side of things easily accessible to members of the military services.[3]

General Groves concurred: "It was simply impossible to keep on increasing the activities at Los Alamos," he said. "Relief was essential." Groves also preferred to "separate, physically, production engineering and production from research and development."[4] Among other considerations were the costs, difficulties, and time delays involved in transporting materials and equipment up to Los Alamos and the finished product back down again.[5] Rowe and Oppenheimer also agreed that the need existed for a means to continue "to organize and direct engineering activities not immediately concerned with the Trinity test or with military operations in the Pacific."[6] A separate division, they felt, would fulfill these functions.

Hartley Rowe, who acted as a troubleshooter for General Eisenhower during the war in addition to serving in an advisory capacity to Oppenheimer, noted that "when the matter was called to my attention by Dr. Oppenheimer, I made several trips to Albuquerque to see what use could be made of the buildings at the Kirkland [sic] Air Base to accommodate the Z Division."[7]

In June of 1945, Captain Shelton A. Musser, later an executive of Zia Corporation, also came to Albuquerque to search for a suitable location. Musser reported that the old Oxnard Field was available. Located on the East Mesa at the base of the Manzano Mountains, the area conveniently adjoined Kirtland Air Field.[8]

Oxnard Field

Known originally as the Albuquerque Municipal Airport, the land in the desert-like southeastern sector on the outskirts of the city was first leased by airplane enthusiasts Frank G. Speakman and William Franklin in 1928. With the support of Major Clyde

Tingley, who provided them the city's road equipment during off-hours, Speakman and Franklin completed the airfield; and on May 15, 1928, Californian Ross Hadley landed his plane to become the first customer. Just the year before, Charles A. Lindbergh had made his solo flight across the Atlantic, signalling the growing popularity of aviation as an industry.

Among the crowd who welcomed Lindbergh at Le Bourget Field in Paris was James G. Oxnard of New York. Oxnard, described as a farsighted person with an extensive interest in aviation, later came to Albuquerque as part of his plan to build a series of airports around the country. In Albuquerque, Oxnard purchased the Franklin and Speakman interests, and with his brother Thornton, and an experienced administrator, Dayton E. Dalbey, formed Aircraft Holdings, Inc. Dalbey, who became Vice-President and General Manager for the Company, oversaw construction of an administration building and hangar, and contracted with F. B. Schufflebarger for placement of the Gilmore Plastic runways needed to bring commercial aviation to Albuquerque. The Oxnard team provided the leadership for initiation of numerous industry services, including a restaurant referred to as "The Nightclub on the Mesa." Recognized as the most luxurious restaurant in town, the building that housed it still exists as part of the Sandia Area Federal Credit Union.

The airfield hosted such aviation dignitaries as James Doolittle of Air Force fame, and powder-puff fliers Ruth Nichols and Jacqueline Cochran. Eventually, however, increases in engine power and landing speeds caused pilots to feel that the airport was located too close to the Sandia and Manzano Mountains. Use of the Oxnard Field declined as a result, and plans were made to construct a new airfield on the West Mesa. For a time, Oxnard continued to be used for charter operations and flight instruction, but even this business began to fade.

A country at war, however, meant that the old runways could once again be put to good use. On April 3, 1942, the Secretary of War (as Director of the Army Air Corps) appropriated 1,100 acres of land on Albuquerque's East Mesa, including Oxnard Field.

Officially acquired by the government for military purposes on May 12, the site was to be used as an Air Depot Training Station to prepare aircraft mechanics and repairmen for overseas duty. Civilian Conservation Corps–type hutments were built or hauled in from other locations. At the height of site activity, 150 officers and 2,100 enlisted men occupied the premises. According to former Sandia historian Frederick C. Alexander, Jr., "It was in this era that the unofficial term 'Sandia Base' began to be used by the construction engineers active in the erection of facilities."[9]

After the last group was transferred, the Base became an Army Air Field but saw little use until its conversion to the Army Air Corps' Convalescent Center. During this stage, the previously deserted buildings were converted into quarters for 800 men, and the shops were reactivated and used for occupational therapy. When the Convalescent Center was officially closed April 11, 1945, it once again became an Army Air Field, used by the Reconstruction Finance Corporation (later called the War Assets Administration) as a graveyard for battle–scarred and obsolete airplanes.[10]

After the closing of the Convalescent Center, the Secretary of War and the Defense Plant Corporation negotiated a new contract. For Los Alamos this contract was significant because it allowed the "licensee" to assign the reservation to "any other branch, Department, or agency of the United States Government."[11] Thus, by mid–July, the Defense Plant Corporation and the War Department agreed that jurisdiction over the Army Air Field should be transferred from the Army Air Forces to the Army Service Forces, Chief of Engineers, and thereafter assigned to the Manhattan Engineer District.

Representing the U.S. Engineers' Manhattan District at the official change of command July 21, 1945, were Colonel Lyle E. Seeman and Captain Sam A. Musser. The transfer included the Air Base and all leased facilities. Various hangars, warehouses, garages, a service club, and officers' mess and quarters were included as part of the transaction.[12]

HERITAGE

Tijeras Canyon is visible in the background of this aerial view (above) of the Albuquerque Airport, circa 1931. Below, a transcontinental air transport Ford trimotor, also known as the Tin Goose, stands ready on the ramp next to Hangar 1 at the Albuquerque Airport, circa 1929.

Los Alamos first became aware of the value of the general location of the Air Base during the process of converting the atomic bomb into a practical airborne weapon. To meet transporation needs associated with Projects Alberta and Trinity, Los Alamos had used Kirtland Field. Named for Colonel Roy S. Kirtland, one of the Army's pioneer military pilots, the field became an important staging ground for the ferrying of men and material to the various field sites and for use as a bomb-loading facility.[13]

The loading pit constructed at Kirtland, although a primitive, manually operated affair, served the purpose until a hydraulic lift was installed in December 1945. This loading pit and the proximity of Kirtland Air Field had a significant bearing on the decision by Los Alamos to acquire the nearby Oxnard Field as the future site of Z Division.[14]

Formation of Z Division

The postwar period was a time of transition and rapid change. At Los Alamos, the long months of intense effort had climaxed in victory, followed by a crashing letdown. Before the summer of 1945 ended, a mass exodus of students, scientists, and technicians took place as they attempted to resume the prewar pattern of their lives. The status of the Laboratory itself was faltering; there was a lack of national policy to determine the use and control of atomic energy; and little direction was being provided for the future course of the Laboratory on the Hill.

General Groves visited the Laboratory, ostensibly to persuade the staff to stay—at least, that was the understanding. According to Ray B. Powell: "There were loudspeakers set up both inside and outside the building, and a large crowd gathered to hear Groves offer us words of encouragement, but what we heard instead was a monologue of how great General Groves was; as a result the exodus from Los Alamos accelerated."[15]

Oppenheimer was also one of those eager to return to academe; however, before he resigned, he continued with the organizational planning necessary to activate the new division in Albuquerque and continue weapons development. To head the organization, Oppenheimer brought in Jerrold R. Zacharias, who had been active in the radar group at the Massachusetts Institute of Technology's Radiation Laboratory.

Formed in July 1945 (although not functional until late September 1945), and referred to as Z Division after the first letter of Zacharias' name, the new organization became the nucleus from which Sandia Corporation would evolve. It was soon recognized that establishment of the new division offered solutions to some of the major postwar problems facing Los Alamos.[16] The new division, for example, would not only provide an outlet for the overcrowded conditions on the Hill, facilitate transportation, and establish a separate production engineering branch, it would also provide a sense of direction during the turbulent transition period.

Those responsible for Z Division from inception to activation showed an almost visionary confidence in its future. On the same historic day that the atomic bomb was dropped on Hiroshima—August 6, 1945—Zacharias wrote a prophetic memorandum to Director Robert Oppenheimer. In reference to the newly organized group, he said: "The division is likely to become so large and take on so many duties of so many different divisions that a more vertical organization than is usual in Project Y will have to be sought."[17]

Slated to take over practically all the functions exercised in an advisory manner by the Weapons Committee, the majority of Committee members, including Albert F. Birch, Norris Bradbury, Robert B. Brode, Glenn A. Fowler, Lewis Fussell, Norman Ramsey, and Roger Warner, were to play important roles in establishing Z Division.[18]

On August 20, 1945, just one week after the Japanese surrender, Oppenheimer made the first formal public announcement of the organization of the new division. In a statement made to the Coordinating Council at Los Alamos, he explained that "Z Division had absorbed the functions of 'O' Division."[19]

Only three days earlier, as if in anticipation of Oppenheimer's announcement, Robert Brode had written a highly commendatory interoffice memorandum. "The abrupt termination of the World War is in no small measure due to the work of Group O-3 . . . ," he said. "For our field work at Muroc, Kingman, Sandy Beach, and Destination, we have been complimented in every case by the commanding officers for the outstanding way in which our work has been done." Continuing, Brode explained that ordnance work under the new program—which would include the production of a stockpile of fuzes and test equipment—"would require some reassignment of duties."[20]

This reassignment of duties would soon be more clearly specified. At a conference held August 21, 1945, Oppenheimer formally announced the twofold mission of Z Division—a mission that would indeed include responsibility for the assembly of a stockpile of the existing Fat Man model, including development and surveillance tests, and the development and testing of new models. Improvement of the present model, however, was listed as "the principal function of the 'Z' Division."[21] The staff also hoped to improve safety standards. They were especially concerned with reducing the possibility of an accidental nuclear reaction during servicing, loading in aircraft, on takeoff, or in flight to the target.[22]

This additional charge was based upon a real danger that had been graphically demonstrated the night before the Nagasaki mission when two B-29s crashed and burned on takeoff. The accident caused gasoline fires so intense that six fire engines working twenty minutes had not been able to stifle the blaze. In view of this hazard, it was felt that the redesign of the Fat Man should be carried out so as to make it possible to insert the active material after the plane had left the ground.[23] The mission gave Z Division a sense of direction and a focal point for the future.

After Oppenheimer's public announcement of August 20, the organization of the division began to crystallize rapidly. As it was then constituted, the division consisted of five major groups: Experimental Systems, Assembly Factory, Firing Circuits, Mechanical Engineering, and Electronic Engineering. Seeman and Bradbury assisted Zacharias as Associate Leaders. Seeman, in fact, served as the Executive Officer at Sandia and, with the assistance of Lieutenant Colonel Lockridge (Z-2A), he also handled production and procurement problems.

Separation of the project from headquarters at Los Alamos meant inevitable delays. Furthermore, strict security measures requiring the use of code terms and the movement of purchased material through several transfer points added to the difficulty of the operation. Z-2 took responsibility for receiving and storing, schedules and overseas shipments, assembly and instruction manuals, and mechanical engineering for factory, test, and handling equipment. Since the basic mission of the division was to assemble a stockpile of atomic bombs, personnel concerned with assembly activities had to make the move early in the organization's history. Thus, the Procurement and Storage Group (formerly O-7) under Lockridge transferred from Site Y to Sandia "on or about" September 27, 1945.[24]

Recognizing, too, that the newly improved bomb models would have to be tested, Oppenheimer soon implemented plans to establish a testing program with Glenn A. Fowler (Z-1C) slated to play a leading role. Under the initial direction of Norris Bradbury—head of Z-1 (Experimental Systems)—airborne testing, including ballistics and aerodynamics, radar, telemetry, and coordination with overseas divisions as well as with Seeman's Assembly Factory, would be carried out. Dale Corson (Z-1A) and Jerome B. Wiesner (Z-1B) managed Airborne Testing and Informer subdivisions. In October, a C-45 was brought in from Wendover for the use of the Z-1A Flight Test Group, which moved to Albuquerque on a permanent basis December 12, 1945.

Lewis Fussell, head of Z-3 (Firing Circuits), had charge of detonator design and development engineering, detonator production, firing circuits, research and development, developmental engineering and production, and production control of firing circuits. Z-4 (Mechanical Engineering) under Robert W. Henderson, although organized in August 1945, did not make the move to Sandia until early 1947, but coordinated from Los Alamos, where this group handled general engineering changes involved in the development of a streamlined Fat Man to become known as the Mk 4. Z-5 (Electronic Engineering) under Brode took responsibility for research and development of fuzes and fuze production.

Although Z-Division Groups 1-5 were the first to be organized, others soon were added. Zacharias, in fact, also had plans for the establishment of a Mechanical Engineering for Production Group, but as he explained: "I would prefer to put nothing on paper at this point except to say that Frank Oppenheimer is carrying the job of System Coordination for Streamlined Gadget without a suitable organizational plan to back him up."[25]

By October 1, however, the situation had been clarified and William T. Theis had been appointed to head Z-6 (Mechanical Engineering for Production). The function of this group, "to design, install, and operate a mechanical test laboratory," was the forerunner of environmental testing at Sandia. In this lab, set up in Building 828, the weapon and its components were subjected "to all conditions that . . . would be encountered from the time the weapon . . . was assembled until the time the drop . . . was completed."[26]

Personnel reporting to Theis included Grant Cline, who took charge of lab equipment and building design work; Harlow Russ, who was responsible for design checking; John Riede, in charge of design stress; and Arthur Machen, who handled procurement and procurement liaison. Lieutenant V. A. Miller assisted Theis as Alternate Group Leader.[27]

When Zacharias revealed plans to return to MIT, Norris E. Bradbury—selected to succeed Oppenheimer—officially notified all division and group leaders that Roger S. Warner, Jr., would take over as the acting Z-Division leader, effective October 17, 1945. Warner, who served at Los Alamos as the organizer of the Tinian operation and its field equipment needs, was

Z DIVISION

September 1945

Jerrold Rheinach Zacharias, Leader

Z-1—Experimental Systems, Norris E. Bradbury
    Z-1A—Airborne Testing, Dale R. Corson
        Glenn A. Fowler (Alternate)
    Z-1B—Informers (Telemetry), Jerome B. Wiesner
        B. Wright (Alternate)
    Z-1C—Coordination with Using Services,
        Glenn A. Fowler

Z-2—Assembly Factory, Colonel Lyle E. Seeman
    (originally Production)
    Roger S. Warner, Jr., Deputy
    Z-2A—Procurement, Storage, and Shipment,
        Colonel Robert W. Lockridge
        Major Parker (Alternate)
    Z-2B—Production Schedules, Manuals
        Roger S. Warner, Jr.

Z-3—Firing Circuits, Lewis Fussell, Jr.
    Commander Stevenson; Earl Thomas (Alternates)

Z-4—Mechanical Engineering, Robert W. Henderson
    Richard A. Bice (Alternate)
    Frank Oppenheimer (Coordinator for Redesign)

Z-5—Electronic Engineering, Robert B. Brode
    R. B. Doll (Alternate)

THE POSTWAR YEARS: Z DIVISION

Jerrold Reinach Zacharias (shown during a lab experiment), who had been active in the radar group at MIT, was brought in during the final days of the Manhattan Project to direct the newly formed division in Albuquerque. The group was named Z Division, after the first letter of Zacharias' name. Below: Roger S. Warner, Jr. assumed the reins of leadership as acting Z–Division leader in October 1945.

recognized as a superior engineer and a down-to-earth individualist. It wasn't at all unusual for Warner, who was a bachelor, to put on his scruffy, fur-lined mackinaw, pile into a drafty military aircraft, and fly all night to his hometown of Boston for a quick visit. A day or so later, he would fly back to Albuquerque, his pockets filled with live lobsters. Warner's pragmatic approach to problem-solving would be put to good use at the new ordnance division on Sandia Base.[28]

Meanwhile, reserve Navy Commander Norris Bradbury, appointed by General Groves and Oppenheimer to serve as the interim director, had the immediate task of forging a cohesive organization out of the general confusion. The job would extend into a twenty-five-year commitment.

The wisdom of General Groves and Oppenheimer in selecting Bradbury was quickly proven. To end the doldrums that seemed to hang like a pall over the Laboratory, he decisively delivered an ultimatum to Project Y personnel: "You've got three months to get off the Hill or go to work." "That shook up a few," Bob Henderson recalled, "but those that decided to stay, went to work, and the bickering and bitching stopped."[29]

To create a sense of mission, Bradbury—supported by Warner, Henderson, Bice, and Fowler—charged personnel with the reevaluation and incorporation of improved weapons concepts and systems that promised to double the effectiveness of the implosion bomb and to double the size of the stockpile. The focal point of the effort, therefore, was "to reengineer the Mk 3 into a device that could be easily assembled by the military and stored in the assembled form." The result of development work on the Mk 3 was to be the improved Mk 4. Toward this end, Group Z-7 (Assembly), under the leadership of Wilbur F. Shaffer, was officially formed by the first week in December 1945.[30]

The original contingent of this group, which included assembly personnel from Wendover, had arrived in Albuquerque by the end of November 1945. Upon advice of James Les Rowe, who had been asked to set up the assembly facilities, the mechanical assembly building (T-941) from Wendover was carefully dismantled and shipped to Sandia for reassembly. Soon

A-bomb material, fixtures and jigs, weapon components, and pumpkins began arriving from the Pacific and other parts of the country.[31]

During the Christmas holidays, Rowe's final-assembly crew started pouring slabs for the concrete floor of the first assembly building, located on the north end of the lot where B-840 now sits. Plans for the building called for the addition of a forty-foot machine shop and entrances on the east and west ends. On December 31, the crew poured the last slab for this adjacent shop area. To protect the concrete from the cold weather, the men covered it with tarpaulins and set up space heaters, which had to be watched throughout the night. After assigning two men to babysit the concrete, Rowe and the rest of the crew went to the New Year's party at the Club. They took turns making hourly checks on their buddies, and each check included drinks for the men serving as monitors. According to Rowe, "Nothing froze and both the sitters and the concrete were well cured."[32]

After his discharge from the service, in March 1946 Rowe succeeded Shaffer as group leader for Z-7 (Assembly), at which time Shaffer took over as head of Z-9 (Stockpiling). Shaffer, appointed chairman of the Stockpile Committee in February 1946, formulated the original plan for stockpiling the Fat Man and Little Boy components at Sandia. As leader of the Stockpiling Group, and with the assistance of Walt Treibel, he devised methods for packaging, inspection, storage, and cataloging.

Shaffer's Stockpiling Group, originally comprising nine enlisted men and nine University of California Sandia Branch employees, began the process of identifying and collecting useful components from wartime weapons. The salvage and assembly operation depended upon piece parts coming in from Los Alamos, storage facilities at Kirtland, World War II vendors, and the 1946 Crossroads full-scale test program in the Pacific.

As a security measure, remnants from bomb components had to be gathered from test sites and brought to Sandia for classification. G. C. Hollowwa and Arthur Machen were the resident experts in providing such identification. These early stockpiling efforts were a major focus that impacted different areas of the Laboratory. Mechanical engineer Raymond H. "Ray" Schultz, for example, remembers being borrowed from his engineering development assignment for a year to assist the Stockpiling Group.

Assembly and Production personnel worked to streamline and improve their operations. According to typical procedures, the Stockpiling Group performed their duties, then transferred the arming, fuzing, and firing components to R. L. Colby and O. L. Wright's Development Group for inspection, modification (if needed), and testing. Wright and Colby, in turn, sent the components to J. Leslie Rowe's Production Group for assembly into complete mechanical units—a task complicated by the lack of definitive blueprints or specifications.

Progress was made in solving the latter problem in 1947 when the drafting organization developed the first drafting standards manual, "LAMS-635." The document, commonly referred to as "The Little Red Wagon," represented one of the first attempts at standardization of specifications. This simple title took note of the fact that a child's red wagon was used to illustrate the desired breakdown of assembly and detail drawings.

Other groups established after the March 1946 reorganization included Z-8 (Informers) under T. J. Anderson, who was followed by William Caldes, appointed in December 1946; and Z-10 (Technical Area Supply). Henry Moeding became leader for this group (Z-10) in December, and in August 1946, a Special Weapons Group for the Little Boy (Z-11) was set up under Harlow Russ. Of the last two Z-Division groups, Z-12 was in existence for approximately seven months and Z-13 for only one month before the reorganization of March 1948. (See Z-Division Leadership Chart.)[33]

With servicemen both mustering out at Wendover in Utah and returning from duty in the Pacific, many of these personnel—as in the case of Rowe's Assembly Group—were used to activate the new division. Wendover, in fact, was terminated through organization of Z Division.[34] In September 1945, at a conference set up to decide administrative policies, the memo

"The Little Red Wagon"
LAMS-635

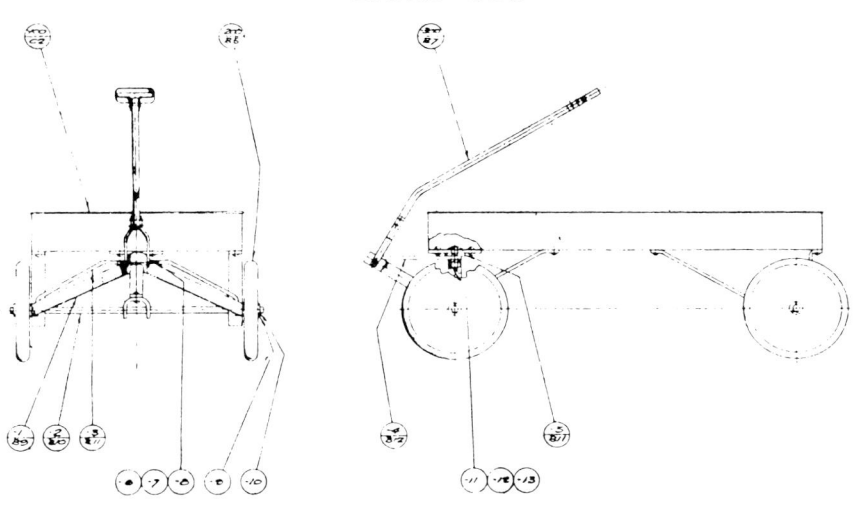

The Little Red Wagon was used to illustrate the desired breakdown of assembly and detail drawings in the first drafting standards manual, LAMS-635.

Z DIVISION LEADERSHIP (Ordnance Division)

Formed 7/45; functional by 9/45; Transferred to Albuquerque mainly between 3/46 and 7/46.

GROUP	LEADERSHIP
Z, Ordnance Division (7/45–4/48)	Jerrold R. Zacharias, Division Leader (7/45–10/45)
	Roger S. Warner, Division Leader (10/45–1947)
	Dale R. Corson, Actg. Div. Ldr. (10/45–7/46)
	Ellis E. Wilhoyt, Actg. Div. Ldr. (3/46–7/46); Alt. Div. Ldr. (7/46–4/48)
	Robert W. Henderson, Division Leader (1947–4/48)
Z–1, Experimental Systems (Group dissolved 1946)	Norris E. Bradbury, Group Leader (became Lab Director 10/45)
Z–1A, Airborne Testing (dissolved 3/46)	Dale Corson, Section Leader (8/45–10/45)
Z–1B, Informers (dissolved early 1946)	Jerome Wiesner, Section Leader (8/45–3/46; joined B–12)
Z–1C, Coordination with Using Services	Glenn Fowler, Section Leader (8/45–3/46)
Z–1, Field Test (Formed 3/46)	Nathan Eisen, Group Leader (3/46–12/46)
	Glenn Fowler, Group Leader (12/46–3/48)
Z–2, Assembly Factory at Sandia	Lyle E. Seeman, Group Leader (8/45–3/46)
Z–2A, Procurement, Storage, and Shipment	Robert W. Lockridge, Group Leader (8/45–3/48)
Z–2, Air Coordination (3/46–3/48)	Glenn Fowler, Group Leader (3/46–12/46)
	Richard A. Bice, Group Leader (12/46–3/48)

THE POSTWAR YEARS: Z DIVISION

Z-3, Firing Development	Mr. Thomas, Co-Group Leader (3/46-7/46)
	Donald Hornig, Co-Group Leader (3/46-7/46)
Z-3, Firing Circuits (absorbed into Z-5 3/46)	Lewis Fussell, Group Leader (8/45-3/46)
Z-3, Assembly Training (12/46-3/48)	Arthur Machen, Group Leader (12/46-3/48)
Z-4, Mechanical Engineering (name changed to Engineering 3/46; group dissolved by mid-47)	Robert W. Henderson, Group Leader (8/45-3/46)
Z-5, Electronic Engineering (absorbed into Z-5 3/46)	Robert B. Brode, Group Leader (8/45-3/46)
Z-5, Firing and Fuzing (moved to Albuquerque 12/46)	R. L. Colby, Group Leader (5/46-12/46)
	Otis L. Wright, Group Leader (12/46-3/48)
Z-6, Mechanical Engineering for Production (name changed to Mechanical Laboratory 3/46)	William T. Theis, Group Leader (10/45-3/46)
	Alan Ayers, Group Leader (3/46-3/48)
Z-7, Assembly	Wilbur F. Shaffer, Group Leader (12/45-3/46)
	James L. Rowe, Group Leader (3/46-3/48)
Z-8, Informers (3/46-3/48)	T. J. Anderson, Group Leader (3/46-12/46)
	William Caldes, Group Leader (12/46-3/48)
Z-9, Stockpiling	Wilbur F. Shaffer, Group Leader (3/46-3/48)
Z-10, Technical Area Supply	Henry Moeding, Group Leader (12/46-3/48)
Z-11, Special Weapons (Little Boy)	Harlow W. Russ, Group Leader (9/46-3/48)
Z-12, Name Unknown	Estel L. Cheeseman, Group Leader (9/47-3/48)
Z-13, Name Unknown	John T. Risley, Group Leader (3/48-3/48)

for record stated that "the primary mission of Sandia would be to take over the duties now being performed at W–47." Administrative work, however, would still be carried out at Los Alamos. Captain Roerkohl, as Commander of the Troops, reported to Colonel Seeman for Technical Area matters and Colonel Tyler for Post matters.[35]

SEDs transferred to Sandia found the place drab, dusty, and uninspiring. There was little "real work" to do. As one former SED explained: "The buildings were closed and there wasn't anyone here to do menial tasks." Yet, to muster out of the service, the soldiers had to accumulate a certain number of points; therefore, many of the men found themselves working with the motorpool, manning the switchboard, or assigned to duty in the mess hall.[36]

Sandia Base, Albuquerque

These first arrivals at Sandia Base found conditions far from ideal. Located east of the historic Rio Grande, the military complex sprawled on a mesa some six miles outside the city of Albuquerque. Toward the eastern horizon, the beauty of the Sandia and Manzano Mountains provided geographic contrast to the desert landscape. In the evenings the rays of the setting sun bathed the western face of the mountains in a rosy glow that gave the more imaginative observer the illusion of giant slices of watermelon, hence the Spanish name—*Sandia*. In the harsh glare of daylight, however, the scenic mountain backdrop did little to soften the reality of the primitive and highly secure complex developing at the end of Wyoming Boulevard in early 1946.

As the main entrance to the Base, Wyoming Boulevard led to a motley assortment of wood–frame and tar–papered structures. A guard station on Gibson Boulevard provided alternate access. Similar to Los Alamos, security at Sandia was of prime importance. A pass system for entrance to the facility was implemented, although the practices of mail censorship and telephone monitoring were discontinued. As an additional security measure, the entire area was enclosed by an inexpensive hogwire fence. The objective was to prevent Reconstruction Finance (RFC) personnel from entering the area and to keep Sandia personnel off RFC territory, "where a number of planes . . . were kept with very little guard."[37] At the outer entrance gates, the driver of a car was required to stop, show his Sandia identification card, and display a metal auto tag inside the windshield.

The scene inside the Base resembled a penal colony. Wooden guard towers stood like sentinels in strategic positions, and at night searchlights illuminated the fenced-in Technical Area, home of Z Division. A block to the east of Main Street lay a dusty parade ground, bordered by poplar trees and devoid of grass. A small white chapel situated between the stockpile storage area and the airplane graveyard added architectural relief to the monotony of Army-style barracks.

Security measures were stringent, and fully armed Military Police, assisted by members of the K-9 Patrol, roamed the Base. Early Sandians remember the MPs as being somewhat "overzealous." Many were transferred from caretaker assignments at Trinity Site—otherwise known as the "Siberia for troublemakers." The MPs were known to take their jobs very seriously. Richard A. Bice recalled that "anyone entering the Tech Area at night might find himself suddenly spread-eagled against a wall and searched."[38]

Z-Division employees working late were subject to preemptory challenge upon leaving the building. If challenged, an employee had to place his badge on the ground, do an about-face, and retreat ten paces while the MP, holding a forty-five-caliber automatic in readiness, examined the badge and identified the person.[39]

New construction and modification of existing buildings, started in September 1945, gradually gained momentum. Colonel Lyle E. Seeman, as the officer in charge, had responsibility for overseeing the necessary improvements, estimated to cost $100,000. This initial expenditure was allotted to cover housing and mess accommodations for military and civilian personnel, the security fence, and warehousing facilities.

HERITAGE

With vacant housing in the city of Albuquerque practically nonexistent, plans were made by October to obtain married quarters on Kirtland Field for civilians and the military. Since Kirtland Field was then being deactivated, the Manhattan Engineer District requested that the vacated facilities be reserved for the use of Z-Division personnel, and especially the advance Flight Test contingent commanded by Major Edward Jacquet.[40]

The facilities requested included 100 family units—which were given first to families with children—WAC and officers' quarters, aircraft hangars, and an ordnance area for explosive storage and assembly. To supplement, sixty prefabricated houses were moved to Sandia Base in April 1946 from the Hanford Works of the Manhattan Engineer District. The Hanford houses—square, one-story structures patterned after those used by the Tennessee Valley Authority—were placed just east of the parade ground.[41]

As the tempo of activities and the number of transients increased, it became apparent that additional bachelor officers' quarters (BOQs) were badly needed. General Groves, therefore, approved an additional $50,000 for new construction in January 1946.[42]

Ray B. Powell, who came to Sandia in the summer of 1947 to initiate a recruiting drive and establish a Personnel organization, testified to the primitive conditions of the BOQs. "They afforded only a marginal kind of existence," he said, "the high point of which was sweeping the building out after windstorms. The buildings we worked in were the same. They were so loosely constructed that the dust sifted in during the summer and snow drifted in during the winter. In the fall of 1947, we had to shut down some of the buildings to caulk the cracks."[43]

For a period, the scarcity of housing at Sandia limited the size of the new division. Los Alamos Director Norris Bradbury, meeting with division leaders, warned that "the project must be cut down to match the housing we can furnish."[44] Nevertheless, group leaders from both Z Division and the military attempted to make the transition as smooth as possible. Fuzing and Firing Group Leader R. L. Colby, complimented especially the efforts of of Base Commander Lieutenant Colonel A. J. Frolich and Dale R. Corson. Colby reported that

> Mr. Corson and [Lieutenant] Col. Frolich are both anxious to get this new organization underway and do everything possible to make the operation of it satisfactory. The attitude of these two men . . . is such that on the strength of the cooperative spirit shown there, it has been possible to talk several people into a move who would otherwise surely not have even considered moving.[45]

Most of those who did join the Division were former members of the Manhattan Project or its contractors, who felt a continuing sense of mission associated with the work they were doing.

About this time, members of the 509th also transferred in from Tinian in the Pacific to Roswell, New Mexico. Stationed conveniently close to Kirtland, the airmen and their B–29s would be available to provide air support for Glenn Fowler's testing group—one of the initial contingents to be established at Sandia.

During this transition period, military and civilian leaders were analyzing the position of Sandia Base in the overall weapons picture. At a conference held May 2, 1946, in Los Alamos, the decision was made to have Sandia Base function as an ordnance activity administered by the military. According to a Table of Organization submitted to the War Department, a military staff—part of a group to be called the First Engineer Special Battalion—was to be provided for Sandia. A combined work force composed of civil service and military personnel would handle the assembly, stockpile, surveillance, and field test operations. Colonel Gilbert M. Dorland, on July 29, 1946, was assigned as Commanding Officer for the Base and for the 2761st Special Engineer Battalion, which immediately took over Kirtland high–explosive work.[46]

Meanwhile, political maneuvering in Washington occurred that would soon reverse the trend toward military control of Sandia activities.

HERITAGE

A cadet in the U.S. Army Air Corps stands outside one of the military hutments that offered minimal living accommodations to early residents of the Base.

Interior of military hutment.

THE POSTWAR YEARS: Z DIVISION

In February 1946, the Z-Division technical area was situated amid a graveyard for battle-scarred and obsolete airplanes. The base was a maze of abandoned hangars, warehouses, garages, and temporary buildings, one of which had been transported from Wendover, Utah, to be used for assembly operations. The Administration Building 818 can be seen at center-top behind the base chapel. Early attempts at building the stockpile are visible in the rows of crates in the foreground. The crates in rows 1-2 (left to right) contained mounting cones for electronic components; rows 3-4, tail cones; rows 5-7, nose cones; rows 8-13, beta crates.

HERITAGE

A Graveyard for Surplus Airplanes

After the war, the Reconstruction Finance Corporation (RFC), an agency of the government, was assigned the task of disposing of numerous war surplus material from around the United States. Some items were sold; however, three-quarters of the various kinds of aircraft that were delivered to Sandia Base were dismantled and melted down. J. J. Miller, who supervised the war surplus aircraft center at Kirtland in 1946-1947, provided these photographs of the aircraft dismantling and disposal operation. While the center was open, the operation ran twenty-four hours a day, seven days a week. Below, a worker removes an engine from a surplus plane.

THE POSTWAR YEARS: Z DIVISION

Airplane guillotine operation (above, left): A piece of ship's armor plate, measuring twenty feet by five feet by three feet and weighing around five tons, was used as a huge "knife." A crane was used to hoist the piece of steel over the nose of a B–17. As the crane operator allowed its heavy cargo to drop, the steel would shear off the nose of the aircraft. Above, right: Remnants of the dismantling operation were left in piles about the base awaiting compacting and melting. Below, left: Pieces of raw material are being pushed into the furnace to be melted down. Thick black smoke, caused by elements such as burning insulation, oil, and grease, lasted only a few minutes. Below, right: A worker directs molten aluminum into iron molds for eventual resale.

Los Lunas Test Range, twenty-five miles southwest of Albuquerque.

Chapter V.

ORIGINS OF AN ETHOS:
Field Testing

"Our ethos is that of 'Can-Do'—the attitude that we are able to accept a challenge, produce on time, on budget—an attitude that originated from the Field Test ethic."

Ethos, Fall Management Conference Notebook, Sandia National Laboratories, September 24–25, 1985

On a warm spring day in April 1945, Glenn A. Fowler arrived at the Albuquerque Airport as instructed. Feeling somewhat like a government agent on a secret mission, he thought back over the events that brought him to the small pueblo-style airport on the arid mesa.

In 1941, after graduating from the University of California in Berkeley with a degree in electrical engineering, Fowler had signed on with the Radiation Laboratory at MIT. There he helped develop one of the first working models of a short wavelength radar. Referred to as "World War II's most famous example of a reverse lend-lease," the device was built around the British magnetron tube.[1] During the war the British used long wavelength radar to warn against Luftwaffe air strikes, but they were not in a position to develop a radar that could be used for finer resolution like that needed by antiaircraft gunners and fighter plane pilots. This led to MIT's involvement in the project and Fowler's participation.

With the new radar mounted in a lend-lease B-24, he spent the spring of 1942 flying sorties over the Irish Sea to evaluate the radar's efficiency against ship and submarine targets. The result of the MIT project was a production version of the short wavelength radar that was widely used by U.S. forces throughout the duration of the war.

From MIT, Fowler went to the Pentagon as a technical consultant to the Army Air Corps. While still in Washington, he heard through Norman Ramsey of a secret and important project going on in New Mexico and soon thereafter found himself being recruited for the mysterious Project Y (Manhattan Project). Like Dale Corson, also of MIT and also recruited by Ramsey, Fowler accepted the assignment largely on faith and good friendship.

In the lobby of the Albuquerque Airport, he waited for his contact. Before long, a WAC corporal approached and, mistakenly assuming that he was another scientist, addressed him as "Dr." before inviting him to go for a ride in a khaki-colored Army sedan. The driving skills of this particular WAC had made similar trips memorable for many of her passengers. Fowler was no exception.

While his chauffeur took the curves of the old road to Santa Fe with what appeared to be reckless abandon, Fowler wondered what he had gotten himself into. After a session at the badge office in Santa Fe, they proceeded through Española, down the west bank of the Rio Grande, then up the steep winding road to Los Alamos where the pine trees contrasted sharply with the drier desert-like environment around Albuquerque.

Within a week or two, Fowler was well integrated into the extraordinary community of scientists and engineers that made up the Los Alamos Scientific Laboratory. He was appointed secretary of the Weapons Committee and began an orientation that included seminars in applied nuclear physics and indoctrination into the engineering aspects of weaponry.

In July 1945, he returned to Washington, D. C., to bring his wife, Mary Alice, to Los Alamos. That trip also had its memorable aspects. To make the move, he bought a surplus Army truck, and with furniture piled high, headed west, "marking progress by blown tires—one a day." In November 1945, the Fowlers would make still another move from the Hill to Albuquerque.[2]

ORIGINS OF AN ETHOS: FIELD TESTING

Glenn Fowler, who directed field test operations, stands with wife Mary Alice in front of the surplus Army truck that brought them to New Mexico in July 1945.

Z-1A – Airborne Testing	Dale Corson and Glenn A. Fowler
Permanent Deputies at Sandia	Captain Nathan Eisen and Ensign D. P. Irons
Flight Section	
Communications	Coye Vincent
Field Team	Ensign C. H. DeSelm
Fuzing	Captain C. H. Staley
Firing	Lieutenant W. O. McCord
Informers	Lieutenant T. J. Anderson
Photo Lab & Cameras	H. C. Barr
Analysis	J. J. Miller
Test Project Office	T. B. Lanahan

ORIGINS OF AN ETHOS: FIELD TESTING

Los Lunas Test Range

By the end of the summer, plans for establishment of Z Division were well under way. Fowler was assigned to help set up a flight and lab test capability at the Sandia Base location. This Field Test Group (Z-1A), organized in mid-September under direction of Fowler and Dale Corson, was further divided into sections responsible for the various phases of the tests. Captain Nathan Eisen and Ensign D. P. Irons served as permanent deputies.

As a base of operations, Field Test personnel set up offices in one of the old wooden sectional Civilian Conservation Corps (CCC) buildings at Sandia.[3] One of their first tasks was to investigate possible sites for drop testing of developmental bomb shapes.

On September 19, 1945, representatives of Z Division from Los Alamos, including Fowler and Clinton "Hilt" DeSelm made an aerial survey of the practice bombing range previously used by aircraft from Kirtland. As a result of this survey, Range S-1, approximately twenty-five miles southwest of Albuquerque, was selected. The Corps of Engineers negotiated new use agreements and leases, with custody and accountability for the range being placed with the Army at Sandia Base. One of the leases specified a payment of the "huge" sum of $3.25 per annum for the use of about 198 acres owned by Fred D. Huning of Los Lunas.[4]

Located west of Los Lunas, east of the Rio Puerco, and bounded roughly by today's Interstates I-25, I-40, and New Mexico Highway 6, a part of the range rested on land belonging to the Isleta Indian Pueblo. Fowler made arrangements with an official of the tribe for permission to use this area, and on November 21 and 22, 1945, supervised a ground survey of the site. By December 1945, the Field Test group began setting up equipment. Soon the Range was functional.[5]

Each day the test group loaded personnel and equipment, hauled both to the test site, performed their tests, loaded up

Above: A temporary base of operations: Each day the testing group loaded personnel and equipment into trailers and Army trucks and hauled them out to the Los Lunas Test Range to conduct testing activities. At day's end, the group packed up their vehicles and returned home. Right: A field tester stands next to a crater made by a nonexplosive impact test unit dropped at the test range.

again, and returned to Sandia Base. During 1946, all bomb instrumentation was completed at Los Alamos and the units convoyed to Sandia Base for final checkout by Z-Division personnel. Since this was before the days of AEC security guards or couriers, Z-Division personnel controlled the units even to the extent of driving the vehicles.

With red lights flashing, the convoy passed through the streets of Santa Fe en route to Albuquerque. Citizens "in the know" gave the vehicles plenty of room. After final instrument checkout at Sandia, the unit was loaded in a B-29 aircraft at Kirtland Air Base. The Z-Division convoy, escorted by a military jeep in front and one in back, would then proceed down U.S. 85 to the Los Lunas Range and set up for testing.[6]

In many respects, the operation at Los Lunas exemplified the problems and primitive conditions that would continue to characterize mobile field test operations. Trailers served as buildings; there were no roads; and Post Supply on Sandia Base provided a cook and rations for mess facilities on the range. The eating and cooking accommodations, however, left something to be desired. Lieutenant B. H. Schaffer, in a memo to the Commanding Officer at Sandia, complained that the utensils were "unsatisfactory" and requested that corrective action be taken. Shaffer also requested that civilian personnel using the mess shack do their share of the work and that cots be furnished to the military police on duty at night. The sentries, it seems, had been forced to sleep on the dining room tables, which, as Schaffer explained, "was neither sanitary nor desired."[7]

The original range instrumentation at Los Lunas consisted of three K-24 camera stations, one telemetering receiver station, two Mitchell turret cameras, and three transit stations for locating the impact point. Early in the program, a German-made Askania phototheodolite was used on a trial basis. Later, three more Askanias were set up and "debugged" prior to installation at Salton Sea.

Testing proceeded at the rate of two drops per week. The objective was to determine the reliability of the various Fat Man models as indicated by sequence timing on fuzing and firing equipment and the vibration and general flight behavior (pitch and yaw). Other drops tested components and fuzing and firing units, as well as completely new Fat Man models. Before proceeding with the Fat Man program, however, Corson and Fowler ordered a number of preliminary drops with inert standard bombs fitted with informers to ensure that the data could be recorded and transmitted.[8] ("Informer" was the World War II name for an airborne instrumentation subsystem that transmitted information from the bomb in-flight to a ground receiver station via radio frequency link—today called "telemetry.")

Tom Pace, then a member of the Communications group and responsible for timing and control for the tests, recalled that "we had a couple of old military radio trucks dragging a power unit behind it and used that to communicate with the drop aircraft which had been loaded at Kirtland and Sandia."[9] "The bombing accuracy," according to Howard Austin, "was not too good, and it didn't improve a lot through the years."[10] In addition to informers, the Field Test group used ground-based and airborne cameras.

The telemetering program provided information on the performance of each individual component to the extent that actual flight conditions experienced by the bomb could be duplicated in the laboratory under controlled standards. The Mechanical Test Laboratory, designed and set up by Theis' Z-6 group, provided support by responding to requests for specific tests.[11]

Selection of Salton Sea Site

It soon became apparent that a larger and better instrumented range was needed. "Almost simultaneously with the decision to go to Los Lunas," Fowler said, "we started thinking about a more permanent place."[12] The Los Lunas Range, at an altitude of 5,000 feet, simply did not permit evaluation of the effect of dense, sea-level air conditions on ballistic performance. Conversely, the Salton Sea Sandy Beach location, used during the war for test drops originating out of Wendover, was a logical consideration. In November 1945, Fowler, accompanied by C. Hilt DeSelm and Captain David Semple, made an aerial and ground inspection of the site. The men also surveyed the Palm Springs Army Air Field and El Centro Naval Air Station; however, the Salton Sea range had certain distinct advantages.

As Fowler explained:

> At 200 and some odd feet below sea level, it erased all questions about density of the atmosphere, and it had a security aspect as well—When the weapons impacted, they disintegrated and went to the bottom of this murky sea . . . and into the soft mud at the bottom. In addition, . . . the shore rose gradually from the sea and you were looking down with an unimpaired view right down to impact.[13]

Commodore Parsons was the first to advocate that the Salton Sea Test Base be retained. "This below sea-level target," he said, "has some of the best all-season, high-altitude weather in the U.S.A., and is otherwise ideal for our kind of tests." Parsons, therefore, recommended that the site be acquired, and on December 10, 1945, formally requested use of the Salton Sea facilities for the Manhattan Engineer District.[14]

The transfer involved several sections of land in Imperial County, California, two of which were within a national wildlife refuge belonging to the Department of the Interior. This proved to be no problem, although lease agreements had to be negotiated for property not already owned by the federal government. These agreements covered land completely under water, including 920 acres rented from the state of California for $27.91 per year. The other contract, with the Imperial Valley Irrigation Company, provided for the use of 16,024 acres at a lump-sum rental of $100.00 for fifty years, payable $99.00 at the beginning and $1.00 at the termination of the agreement. In June 1946 the Navy's buildings and property were transferred on "Memorandum Receipt" from the Commander, Eleventh Naval District, to the Army Office of the Area Engineer, Los Angeles Area, for use as a bombing range by Z Division.[15]

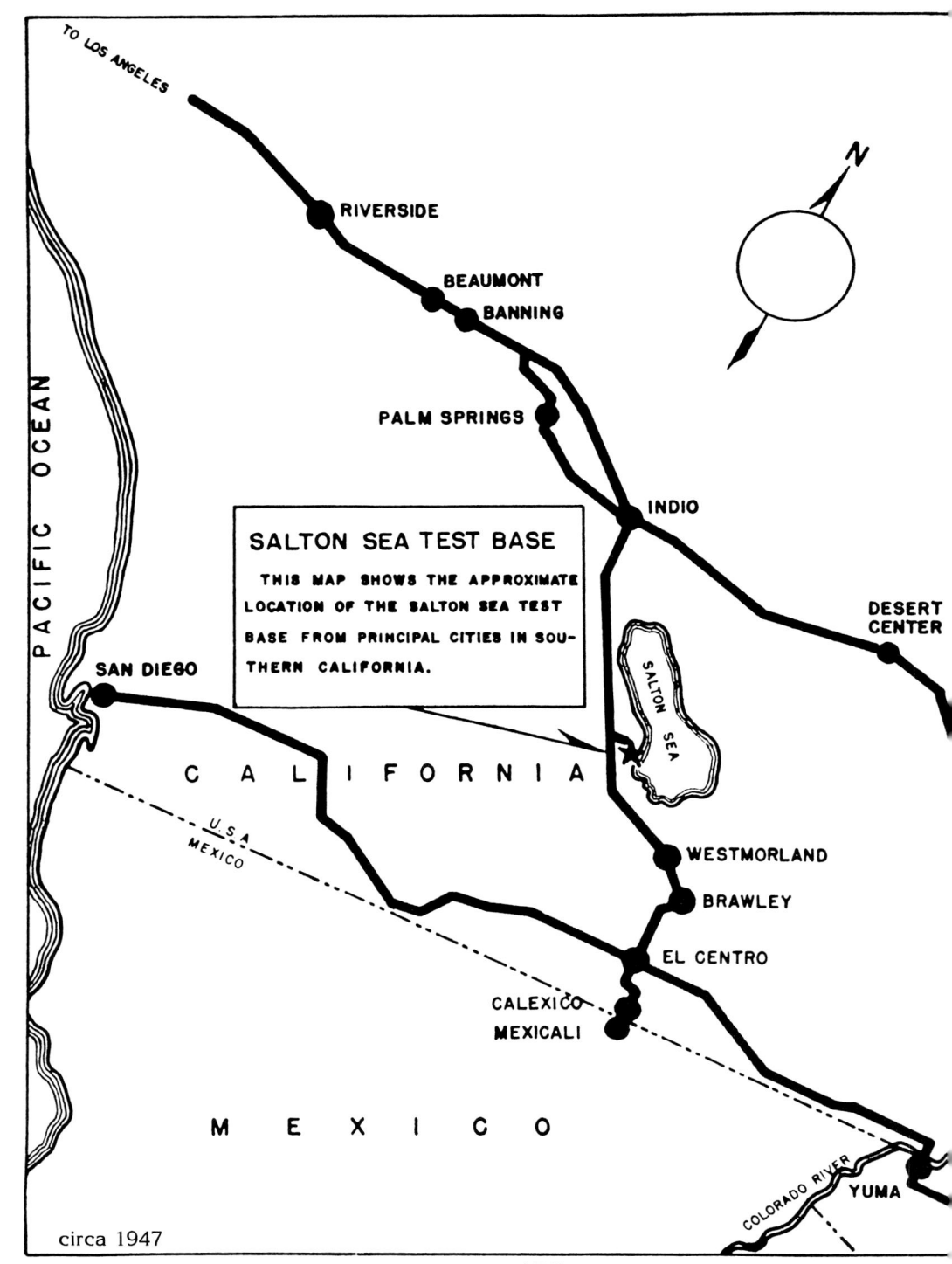

HERITAGE

SALTON SEA TEST BASE

THIS MAP SHOWS THE APPROXIMATE LOCATION OF THE SALTON SEA TEST BASE FROM PRINCIPAL CITIES IN SOUTHERN CALIFORNIA.

circa 1947

In that same month, a small crew of Army personnel and one civilian employee from Z-1 at Sandia started work on installation of limited range facilities at Salton Sea. This advance group, under the supervision of Howard Austin, rebuilt the old Navy target and anchored it in position in the water, installed camera pads, bulldozed dirt roads, strung power and communication lines, repaired motor vehicles, and taped up field wire where the rabbits had eaten through the insulation. "The Field Test business," as Austin noted, "was always plagued with problems you didn't think about."[16]

In addition to rabbits, the area also had a plentiful supply of sidewinders. One of the Base photographers, Jim Karo, had a real interest in snakes and would stash them in the refrigerator with his film in order to photograph them in a hibernating state. Austin recalled that "if you ever walked into the lab and saw Jim on his hands and knees, you'd turn around and walk back out. Chances are one of his snakes had gotten loose."[17]

In November 1946, ownership of real property at the Base had been vested with the Manhattan Engineer District of the Army Corps of Engineers at Sandia Base. Instructions specified that the military would operate the Test Base, but Z Division would have control over technical activities.[18] Although relations between the military and the civilian contingent were generally amiable, the field test engineers enjoyed observing the difference between the structured "go-by-the-rulebook" operating style of the military officer in charge and their own rather freewheeling "get-the-job-done" attitude.

This difference became particularly apparent to two observant Z-Division members. On this particular day the wind was blowing a gale, sending sand in swirls across the old concrete slabs scattered about the Base. To their amusement the men noticed a group of GIs with push brooms trying to sweep the concrete slabs during the windstorm. The officer in charge drove up in his truck, rolled down the window, and yelled at the GIs:

"How are you boys doing?"

"Think we ought to quit, Sir. It's awful windy out here," one of them responded.

> "You're right," the Captain said. "It ain't fit out for man nor beast, but GIs are gonna work."[19]

And with that he rolled up his window and drove off.

Although establishment of the Atomic Energy Commission transferred jurisdiction over the Salton Sea Test Base to the AEC in December 1947, the military, under the Armed Forces Special Weapons Project, continued to operate the Base until the following July. Meanwhile, Sandians continued efforts to make the Base a functional test site.

Facilities, including several Askania phototheodolites used to photograph bomb trajectories, were transferred from the Los Lunas Test Range to Salton Sea. An old airport control tower was converted into a test control room, and telemetry ground-station equipment was installed in the sand dunes to the northwest of the tower. The Navy Operations building housed the Field Test office. The crew used to advantage the experience gained at Los Lunas in modifying equipment for improved operation at Salton Sea. Since there was little field test equipment available commercially, they ingeniously designed and built much of their own.[20]

During this early period many of the photographers and telemetry people commuted all the way from Albuquerque. Sandia commuters generally made one or two round trips a week in single-engine Bonanzas, or D-18 Twin Beeches. The aircraft were owned and operated by Carco, predecessor of Ross Aviation. Gordon Miller, however, arrived at Salton Sea for the first time in his own single-engine Culver prop-type airplane that he piloted himself. There, he went to work for Howard Austin, supervisor of Field Operations, and with the assistance of Tom Pace and Greg Abeyta, set up the timing control system around the range. Another employee, Curly Saxton, simply rode up on his motorcycle and was hired. "Back then," as one engineer observed, "the important thing was getting the job done."[21]

Other field testers, such as Guy Willis and buddy Don Brandeburg, drove in from Albuquerque on U.S. 66. Around Flagstaff, Willis and friend saw a "Watch For Snow Plows" sign, which they filched and placed on a post alongside the access road at Salton Sea. In the southern California desert where temperatures frequently exceeded 110 degrees, it left a lot of people shaking their heads.

Another sign that caused a bit of headshaking was one that read: "Dips Next 17 Miles." It was also filched from an Arizona highway and relocated at the north end of the range where the graded roads stopped and numerous trails for four-wheel drive vehicles began.

The young bachelors assigned to Salton Sea were housed in an old Navy BOQ. It was a one-story U-shaped building with rooms and a screened porch along the legs of the U. The building had been abandoned for several years and was in a bad state of repair. Sand scorpions and sidewinders sometimes crawled through the windows and doors. "It became routine procedure," Willis said, "to dump your boots before putting them on."[22]

The social center in the early days was the medical building or sick bay. It, too, was an old Navy building with a screened porch on three sides. Nurse Mildred Whitten, who ran the facility, lived there with her mother and her young daughter. These three were the only females living on Base at the time, so their front porch became the gathering place for the evening social hour.

Well known for ingenuity, Sandians found various ways to entertain themselves—even at Salton Sea. With the help of Base carpenter Bob Sparks, for example, one young man made a pair of water skis. "Don't tell Glenn Fowler," Willis confided, "but it wasn't unusual to ski back and forth to North and South Islands."[23]

Testing at Salton Sea

A B-29 bomber from Kirtland Air Force Base made the first test drop at Salton Sea on March 12, 1947. This was followed by additional test drops of production units, as well as experimental models of a new design. The aerodynamicists, it seems, had so many bomb designs that the shops couldn't keep up with the demands for different shapes. Therefore, the Field Test crew was asked to drop bomb shapes filled with concrete. This led to some humorous situations. One time, according to Howard Austin, the bomb was wrapped with white tape that sent streamers unfurling as it came down. Moreover, the units were classified, which meant that the chips and pieces of the model had to be dredged from the murky lake bottom. "To do this," he added, "we used a vehicle that looked like a bathtub with wheels. We called it 'The Duck'."[24]

Ground crews provided telemetry support. Through the use of telemetry, the time and sequence of various operations within the bomb could be recorded, as well as other pertinent data such as pressure, temperature, vibration, acceleration—all of which were necessary to evaluate weapon designs. Using cinetheodolites captured from the Germans, the men were able to triangulate the position of the free-falling bomb. "We set up the first good optical measurement system with those captured cameras," Miller recalled. Orville T. Howard, who joined the Laboratory's telemetry group in 1951, noted that the instrumentation assemblies for ballistic-type weapon replicas were built at headquarters in Albuquerque for testing at the Salton Sea Base.[25]

Early bomb releases took place without the benefit of control by ground radar, which sometimes resulted in a high degree of inaccuracy. On occasion, test drops became exciting affairs. Field Test Group Leader Glenn Fowler, accompanied by his wife, who was helping out in Personnel, would sometimes fly out to the Salton Sea Base to observe the test shots. The setup included a tower for communications with the aircraft and a test conductor stand.

On one memorable test, Fowler, acting as test conductor, was sitting on the stand watching the trajectory of the bomb with his field glasses. Mary Alice stood directly behind him. Those present heard the tone indicating that the Mk 4 had been released successfully and looked up to watch its descent. Leaning back in his chair, Fowler kept his field glasses trained on the bomb, which seemed to be aimed directly at the console. Soon all anyone could see was the nose of the bomb. Mary Alice grabbed her husband's shoulder: "Let's go! Let's go!" she urged. Glenn looked back at her and said pointedly: "*Where?!* "

"Glenn expressed all our feelings," Austin said. "We all felt we should *Go*, but the question was *Where?!* " In a few seconds, the bomb hit just behind secretary Gladys Keller's office and the control tower. Gladys came storming out: "What are you SOBs trying to do?" she demanded. "Kill us all!"[26]

There were also other times when test bombs would release accidentally during flight to the target, to fall harmlessly in the sea or the surrounding isolated area. These accidents were usually caused by a malfunction in the aircraft wiring or release mechanism, both of which were frequently modified between tests.

An accident of this nature—although one with wider repercussions—occurred on May 25, 1949. Tom Pace happened to be in charge of communications when the aircraft from Davis-Monthan Air Force Base in Tucson, Arizona, set out on a training mission. As the aircraft approached a little town called Niland on the southeast side of the Salton Sea, Tom suddenly heard the bombardier say: "The damn thing came out!" In answer to his call, the pilot confirmed that they had indeed "lost one."[27] The bomb, which had been dropped in a carrot patch a scant 100 feet from the Southern Pacific railroad tracks and five miles from Niland, created quite a stir.

The local rancher who saw the plane and heard the unit hit the ground described it as some sort of "missile," leading to the speculation that it carried a nuclear warhead that was on the verge of exploding. The result was the "Niland Bomb Scare," written up in local and Los Angeles newspapers. Reporters persisted in trying to get information from embarrassed Air Force personnel, and souvenir seekers created security problems at the crater site. Two days later, the Air Force released a statement admitting that a "practice bomb" had been dropped accidentally, but assured the public that it contained no explosives of any kind.[28]

Still another accident occurred nearly a year later, in March 1950, when a group of Navy planes from nearby El Centro Naval Air Station mistook the Salton Sea water target for one of their own and lobbed practice bombs at it. Two Sandia employees who were repairing the target at the time failed to appreciate the Navy's skill.[29]

Robert Hepplewhite, who later became manager of the Salton Sea operation, recalled still another accident that occurred December 10, 1954, in the middle of a dust storm. The flight path to the target at that time paralleled the shoreline; therefore, if a test bomb fell short or beyond the target, it landed harmlessly in the sea. On this particular test run, however, the aircraft—because of strong crosswinds—was permitted to approach from a heading that was approximately twenty degrees off course. This change let the plane's flight line extend over the Base, which in itself was not serious, but it happened that the test bomb was to be parachute-retarded. When the chute failed to open, the weapon fell far beyond its intended point of impact.

These two errors combined to place the unit right on the concrete tennis court, a scant forty feet from the San Felipe Lodge. The impact threw chunks of concrete into the air, and the shock wave lifted people in nearby offices from their chairs. No one was injured, but tragedy had narrowly been averted; children on the Base had been using the tennis court as a play area a few days before. Seven days later, an ad appeared in the *Lab News* in the "For Sale" column:

> TENNIS RACKET. Sprink, Salton Sea Test Base.

T. A. Sprink was Department Manager of Salton Sea operations.

Precautions were taken to assure that such accidents would not be repeated. Future test operations placed control over release of the bomb with the Sandia crew on the ground, rather than with the aircraft commander or bombardier. In this manner, ground stations tracked the plane's approach by radar and denied permission to release the bomb if the course to the target was not within specified limits. To provide added safety, the target was moved from 4,000 feet to 7,000 feet offshore.[30]

ORIGINS OF AN ETHOS: FIELD TESTING

The Salton Sea tennis court was a source of recreation until a practice bomb hit the court and turned it into a "swimming pool." A cleanup operation is in progress in the photo below. The "Salton Sea dog" (left-hand corner) earned his nickname by always being on the spot for any excitement.

HERITAGE

Tommy Taylor operates a 35-mm Mitchell camera. The Mitchell camera, developed for Hollywood use, was not designed for instrumentation work, but it was the best camera available to field testers at the time.

Sandia field testers at an Askania station include Henry Sweeney on the left and Whitey Hollenbeck on the right.

ORIGINS OF AN ETHOS: FIELD TESTING

Glenn Fowler oversees a Salton Sea test from the observation deck of the control tower. His low-key management style, adventuresome spirit, and get-the-job-done attitude made him a well-liked and respected manager. Fowler, with binoculars, is sitting at a desk built for him by his crew.

Sandia field testers at work in the control room include (left to right) Billy Mitchum, Tom Pace, Howard Austin, and Ed Stout.

HERITAGE

Roads to and from Salton Sea were marginal or nonexistent. The photo at bottom-left shows a new road under construction leading to one of the outlying instrument stations at Salton Sea. The concrete mixer was used in the construction of concrete pads for photooptical and instrumentation stations. Above, Howard Austin (at right front wheel), supervisor of field operations at Salton Sea, works to free a vehicle stuck in the sand.

ORIGINS OF AN ETHOS: FIELD TESTING

The blustery gusts of a typical daily windstorm make the journey back to the photo lab difficult for camera operator Jimmy Allen.

Frequent windstorms damaged communication lines.

167

HERITAGE

Whitey Hollenbeck, a field testing technician, works in a Salton Sea shop.

Base photographer Jim Karo works on his equipment in the camera shop at Salton Sea. During leisure hours, Karo's interests extended to photographing snakes, which he kept in the refrigerator of the photo lab.

ORIGINS OF AN ETHOS: FIELD TESTING

Jack Merillat, Don Brandenburg, and Guy Willis (left to right) were among the many ingenious and adventuresome field testers who lived and worked at Salton Sea. The "Watch for Snow Plow" sign, "borrowed" by Brandenburg and Willis and posted on the access road to Salton Sea where temperatures frequently exceed 110 degrees, caused passing motorists to stop and stare.

Bombing targets at Salton Sea (above) had to be rebuilt every few months because of the frequency of windstorms and the destructiveness of the waves. The coastline at Salton Sea could be calm one minute and turbulent the next. Windstorms came without warning and frequently exceeded forty miles per hour.

ORIGINS OF AN ETHOS: FIELD TESTING

Housing at Salton Sea included a trailer park, . . .

three-bedroom houses, . . .

and an apartment complex.

Improvement of Facilities

The transfer of responsibility for the Base from the military to the Atomic Energy Commission made it necessary to evaluate the future expense of operating the Salton Sea range. It was apparent that facilities would have to be improved if the test base was to be used to maximum efficiency. The tests, if not the personnel, required it.

Several hundred miles of wiring, on poles aboveground, were urgently needed, as well as more adequate camera stations to protect equipment from dust and heat. Wind and sand made maintenance and repair of delicate instruments difficult and added to the discomfort of operating personnel. There was very little dust proofing or air conditioning of the temporary buildings, and at temperatures ranging from 30 degrees above zero in the winter to 128 degrees in the summer, conditions were far from ideal. Progress reports, for example, frequently mentioned that "considerable time was spent during the past month cleaning sand out of the equipment."[31] Roads to many of the test facilities were marginal or nonexistent, requiring use of four-wheel-drive vehicles for travel on the range. The nearest town, Westmorland, was situated some thirty miles south across the desert toward Brawley and El Centro.

Consequently, Robert W. Henderson, acting leader of Z Division in 1947, prepared a report of technical range requirements and forwarded it to Los Alamos in July. The possibility of replacing the Salton Sea Base with an entirely different test range was considered. A site team investigated other military installations, including the Muroc Dry Lake range. Geographical features at Muroc were found to be favorable, but many changes would have been required to provide adequate operational facilities. Furthermore, Sandia would be required to share the range with the existing operator, the Aberdeen Bombing Mission; this raised fears that Sandia test programs might be delayed. Therefore, the decision was made to retain the Salton Sea Range and upgrade the facilities.

To provide the new facilities, the Atomic Energy Commission awarded a contract to the Trepte Construction Company of Los Angeles. Costs for the project came to slightly over $2.1 million upon completion. Additions to the base included four Askania stations; three Mitchell motion-picture camera stations with cameras mounted on modified gun turrets to facilitate tracking; two radar stations to provide ground control for aircraft; telephone and instrumentation lines strung on poles; and radio transmission and receiving stations. A meteorological building including weather-balloon inflation equipment was also built. New housing for personnel included a trailer park, eight apartments, and two three-bedroom houses. Modern air-conditioned buildings were constructed for offices and shops. There was also a fire station, medical building, standby power plant, and water-treatment plant.

Additions such as these made the range one of the most modern in the country. Personnel especially appreciated the air conditioning, which increased efficiency and protected delicate equipment from desert heat and sand.[32]

Other changes at the Salton Sea Base included establishment of two island stations. With the increase in speed of carrying aircraft such as the B-47 jet bomber, the release points over the water target had to be altered to accommodate the greater distance the bomb would travel after release. The longer distance made it difficult for shore camera stations to track the bomb from the moment of release to impact; therefore, the decision was made to set up two island stations closer to the release point. The islands, designated South Island (S-1) and North Island (S-2), permitted better camera coverage of the test drops.

Construction of these stations proved to be no small task. The shallow water surrounding the islands meant that channels had to be dredged to allow passage of the boats. Diesel generators, portable concrete mixers, oil storage tanks, and tracking cameras were transported to the islands by means of a manmade barge constructed of two fourteen-foot motorboats bridged by a wooden platform. Another boat towed the barge. To meet the need for manpower, Field Test engineers doubled as riggers to get the equipment set up. A concrete platform and a small building were installed on each island to support the cameras and house the instruments.

The crew built a small, flat-bottomed boat, powered by an outboard motor and officially christened the SS *Fowler*, to transport personnel and equipment to the island stations. Later, a large diesel-powered barge was brought into use for construction projects.

Several factors led to the eventual phaseout of the island stations. The use of irrigation in the Imperial and Coachella Valleys increased drainage into the Salton Sea; and rising water, coupled with wind and wave erosion, began to wash away the soil under the island installations. Rehabilitation work became necessary. It also became necessary to rotate the flight path to the target in a north-south direction to provide air space for a Navy gunnery range that was reactivated just beyond the Chocolate Mountains east of the Sea. This change in flight path eliminated most of the advantages of the island stations, and during 1952 they were abandoned—new stations being set up on land west of the target.[33]

GAFCO

Although many valuable tests were conducted at the Salton Sea Range, Field Test personnel used only inert nonexplosive test units because of limited space and the possible danger to personnel and equipment. The test drops provided training for the military crews that flew the aircraft and for the Strategic Air Command, which used the water targets for practice bombing on a noninterference basis. Sandia provided communications and scoring assistance for the drops. The Navy also used the Salton Sea as an emergency landing area for its seaplanes operating along the Pacific coast.

The number of test drops per year averaged about 150, with a peak of 223 being reached in 1952. In 1949, just after completion of the new facilities, Fowler, as part of a grand tour, arranged for visiting dignitaries to view one of these test drops.

ORIGINS OF AN ETHOS: FIELD TESTING

Meanwhile, the Field Test crew went to the dump where they found an old kitchen sink, put a bomb rack on it, and hung it in the bomb bay of the aircraft. Later, visitors watched with amazement as the porcelain "bomb," filled with rolls of toilet paper, dropped from the aircraft. All Fowler could say was: "Well, I'll be damned." As Austin explained, "Fowler was fond of saying that we had dropped everything out there but the kitchen sink, so that time we did!"[34]

GAFCO, otherwise known as "The Glenn A. Fowler Company," couldn't resist the opportunity to pull this good-natured prank on the boss. The test drop of the kitchen sink exemplified the camaraderie that existed among field testers. It also reflected their liking and respect for Fowler, whose low-key management style seemed particularly appropriate for the work involved. Newcomers to the Field Test cadre, for example, soon learned the meaning of "Fowler's Law." The message was quite simple: "Follow the rules and regulations—they're there for your guidance. But don't let the rules make you do anything stupid."[35]

The kitchen sink incident demonstrates the good-natured camaraderie that existed among field testers.

As former Vice-President Richard A. Bice, who once worked for Fowler, observed: "He was one of the best at getting the most out of his people in creative, innovative programs." He was also astute in selecting people with promise.[36] In later years, the meaning of GAFCO would take on an added dimension as Field Test personnel proved their "can-do" adaptability by branching out and helping to develop many other viable technological initiatives within the Laboratory.

An adventuresome type, who at one time held the official New Mexico high-altitude record for soaring, Fowler led a Field Test group that attracted men of a similar mold. According to Fowler:

> "Field Test people have to . . . rely on themselves, be willing to make judgements, and take the consequences. . . . Being placed in such a situation tends to develop a certain type of leadership, whereas in an organizational set up, where there are multiple judgements on anything anybody does, . . . a person is not put on the spot quite as clearly as the person who is out in the field. . . . On the other hand, you don't want someone in the middle of a very complex design problem making independent decisions that might affect other people and blow the whole project. Fortunately, Sandia is big enough that we have a place for both kinds of people."[37]

Wallace T. "Watertight" Smith, in slouch hat and comfortable khakis, represented still another variation of the Field Test breed. "Smitty," as he was affectionately referred to, became a legend and a source of inspiration to young Field Testers such as William C. Myre and Lyle Hake who, after a stint with the Road Department, transferred to Field Test. Myre, recognized as one of the best Field Testers in the business, observed that "Fowler and Smith had a vision of what testing ought to consist of. All of the rest of us followed their lead. Bob Henderson was the same kind of person in the weapons design process in that time frame."[38]

With emphasis during the early period on drop testing of new weapons, preparation of telemetry packages for bomb drops at Salton Sea was a significant aspect. It was in this area that W. T. Smith became involved when he joined the Labs in 1949.

Although Smith was not the long-range visionary that Fowler was, according to Myre, his leadership style made an indelible imprint on the lives of those who worked for him and with him. "People either worshipped Smitty," Myre said, "or thought he was crazy and didn't want to have anything to do with him." He instilled the attitude in people to work hard and to try. "When you succeeded," Myre recalled, "that was fine, and when you failed; well, all he asked was that you not make the same stupid mistake next time."[39] Under the competent leadership of such individuals, field test activities at Salton Sea would continue until 1960.

In the interim, an increasing number of full-time resident employees, supplemented by those who commuted from nearby towns, would handle all technical and maintenance matters at the site. Living conditions, as well as technical capabilities, would also improve.[40] The tests themselves, from "pumpkins" to guided missiles, reflected the technological advancements being made. Yet, some things remained unchanged—the men still worked long hours; the desert environment was still hot and uncomfortable; but field testing, for many Sandians, became a chosen way of life and the matrix from which other major activities would evolve.

At headquarters in Albuquerque, conditions were not much better. And in Los Alamos, plans had already been activated for a major test event in the Pacific that threatened to diminish the operational capabilities of the Division before it was fully functional.

Fowler had a liking for boats, so it wasn't surprising when he designed and had built a flat-bottomed boat used to transport personnel and equipment to the island stations at Salton Sea. Field testers officially christened the boat, which was powered by outboard motor, the *SS Fowler*. In the photo above, Gordon Miller (at outboard motor) and Tom Pace (standing) were among the early field testers who made frequent use of the *SS Fowler*.

ORIGINS OF AN ETHOS: FIELD TESTING

Field testers not only worked hard, they played hard. Here the group takes a break during a raft trip down the Rio Grande in the Santa Elena Canyon of the Big Bend area of Texas. Not shown is photographer Ben Benjamin, who snapped the picture. Back row (left to right): H. H. Patterson, Jack Hinde, Hilda Patterson, Helen Steck, W. T. Smith, George Steck, and W. C. Myre. Front row (left to right): Rush Robinett, Glenn Fowler, Mary Alice Fowler, Gene Harling, and Jim Scott.

Test Baker, Operation Crossroads, July 24, 1946.

Chapter VI.

CROSSROADS:
Impact and Legacy

"A Pinpoint in the Sea of Human Affairs."

David Bradley, No Place to Hide

It was a bright tropical day—that March 17, 1946. The brilliant sun glanced off the wing of the aircraft, momentarily blinding the men attempting to get a glimpse of the land below. Looking down, the team could see the irregular chain of islands of the Bikini Atoll, a part of the Marshall Islands. The advance inspection party, composed of Edward B. Doll, John Williams, Richard A. Bice, Herbert M. Lehr, and Harlow Russ, was flying in to examine the Bikini-Kwajalein area in preparation for Operation Crossroads. Shimmering in the heat, like emerald and topaz jewels on a sea of cobalt, one of these small Pacific islands was soon to become the staging ground for the world's fourth and fifth nuclear detonations.

On closer view, the men saw a small, oval island protected by a barrier reef of red coral. Where the land met the ocean, breakers rushed in on small islets and sandspits. Farther inland a dense thatch of palms and a cluster of dwellings showed proof of human habitation. Located approximately 200 miles from Kwajalein to the southeast, the atoll below them was ten miles wide and twenty miles long. "Bikini"—hitherto insignificant and unknown—was destined to become "a pinpoint in the sea of human affairs, truly a crossroads."[1]

The world at large stood at a crossroads as well—peace or atomic destruction; secrecy and control or nuclear proliferation. The advent of the atomic age not only fueled scientific curiosity, it also raised the consciousness of the military. In the aftershock of Trinity, military strategists posed questions to which there were no answers. Former instruments of warfare were outmoded. A weapons revolution was in the making.[2]

An internal revolution also brewed among the branches of the armed forces. At the end of World War II, the Air Corps actively began to seek separation from its parent service, the Army, and attempted to spread a possessive wingspan over "virtually all military aviation." From the Navy's viewpoint, this bid for autonomy and expansion posed a personal threat, namely, the possibility that it might lose its new capital ship, the aircraft carrier. Furthermore, speculation that "the atomic bomb had made warships obsolete" presented an even greater threat.

While Hiroshima and Nagasaki revealed the devastating effects of the atomic bomb on land targets, the effects of such nuclear force upon other environments, including ships, had yet to be determined. In the weeks following the war, attention focussed increasingly on the possible effects of atomic bombs used against naval vessels.

In a speech presented August 25, 1945, Senator Brien McMahon, who would become chairman of the Senate Special Committee on Atomic Energy, made the proposal official. He suggested that the bomb be tested against naval vessels at sea and that the captured Japanese fleet be used as a target. In his opinion, these ships could be put to no better use.[3]

The Joint Chiefs of Staff took the recommendations seriously and proceeded by appointing a committee under the direction of General Curtis LeMay to determine the nature and scope of the tests. The Joint Chiefs of Staff then acted upon the recommendations of the LeMay Committee. The President gave his full support to the proposed Bikini tests, saying: "I am in complete agreement with the joint chiefs of staff . . . in their view that these tests are of vital importance in obtaining information for the national defense." Continuing, Truman said, "These tests, which

B DIVISION
Operation Crossroads

Symbol	Group Name	Office	Office Holder
B-DO	Division Office	Division Leader	Marshall Holloway
		Assoc. Div. Ldr.	Roger Warner*
B-1	Firing	Group Leader	William O. McCord*
B-2	Fuzing	Group Leader	George A. Koester*
B-3	Assembly	Group Leader	Arthur Machen*
B-4	Engineering	Group Leader	Robert W. Henderson*
B-5	Pit Assembly	Group Leader	Raemer E. Schreiber
B-6	Logistics/Supply	Group Leader	Harlow W. Russ*
B-7	Air Coordination	Group Leader	Glenn A. Fowler*
B-8	Air Collection	Group Leader	Thomas V. Davis*
B-9	Photography	Group Leader	Berlyn Brixner
B-10	Fast Neutron	Group Leader	Gustave A. Linenburger
B-11	Gamma Timing	Group Leader	Norris Nereson
B-12	Electronics (Timing & Firing)	Group Leader	Jerome Wiesner
B-13	Airborne Blast Damage	Group Leader	James Wieboldt
B-14	Radiochemistry	Group Leader	William Rubinson
B-15	Phenomenology	Group Leader	Joseph O. Hirschfelder
B-16	Radiography	Group Leader	James L. Tuck

* Denotes members of Z Division.

are in the nature of a laboratory experiment, should give us the information which is essential to intelligent planning in the future and an evaluation of the effect of atomic energy on our defense establishment."[4]

A Joint Task Force of Army, Navy, and Air Force personnel would be responsible for the operation. To head the Task Force, the Joint Chiefs on January 11, 1946, appointed Vice Admiral William Henry Parnell Blandy, who had been serving as Deputy Chief of Naval Operations, Special Weapons. The following day, Blandy, soon dubbed "the Atomic Admiral," christened the Operation "Crossroads."[5]

As their official mission, Joint Task Force 1 (JTF-1) was charged with determining "the effects of the atomic bomb upon naval vessels in order to gain information of value to the national defense." However, JTF-1 was also interested in studying the effects of the bomb upon aircraft; consequently, target planes from the Bureau of Aeronautics were to be used in support of the mission. Surplus military aircraft placed on the ships' decks would simulate normal disposition of shipborne aircraft on combat vessels, thereby providing the means to ascertain the effects of the bomb upon planes as a function of distance from the center of burst.[6]

Z-Division Personnel to Overseas Duty

At preliminary meetings held in December 1945 and January of 1946, Los Alamos learned that its laboratory was to support JTF-1 by providing technical direction of the tests and by supplying the atomic weapons. Dr. John Williams, who had so successfully organized and conducted the Trinity Test, was asked by Bradbury to prepare a plan for the Crossroads event. The Williams Plan, as it was known, proposed that the Los Alamos Laboratory and the Manhattan Engineer District assume responsibility for the coordination of all instrumentation, assembly, and delivery of the atomic bombs, fuzing and firing equipment, remote control radio timing and firing signals, measurements to determine the efficiency of the nuclear reaction and total radiation, preparation of the radiological safety plan, and for making available technical data for other participating activities. Overall test direction would come from the military task force, and authority to assemble the weapons would come from the President.

To meet the demands of such a colossal test operation, Dr. Norris E. Bradbury, on February 1, 1946, officially organized a special "B" Division. Dr. Ralph A. Sawyer of the Navy would serve as Technical Director. Reporting to him was Marshall Holloway, head of the experimental group. Assisting as Associate Division Leader was Roger Warner, who left the newly established Z Division at Sandia Base in the capable hands of Dale Corson. Corson, however, soon found that he had fewer and fewer people to manage. As the stronghold for Los Alamos Assembly operations, Z Division at Sandia Base became a major source of personnel for the overseas effort.

The Informer and Fuzing and Firing Groups also suffered severe depletions of manpower. As Corson wrote to Bradbury:

> The Crossroads operation has sapped the strength of the Informer Group to such an extent that the program has been unduly delayed and there is some question about being able to remove all the "bugs" from the system. Crossroads also took several of our Fuzing and Firing personnel and there is some question about being able to complete the Stock Pile before the remaining personnel capable of testing and inspecting the equipment going into it leave the project.[7]

Assignment of most of the Informer Group to the Bikini operation meant that vibration studies actually had to be suspended until after July 1946; however, assignment of the Assembly Group made the greatest impact. As Corson said,

> The most serious implication of this personnel loss is the fact that we will be unable to assemble any more atomic bombs. Furthermore, if the personnel situation continues to deteriorate at the present rate there will not only be no personnel trained in bomb assembly and testing, but there will be no one capable of teaching the art to new personnel.[8]

The gravity of the situation from a national security standpoint encouraged Corson to support the recommendation that assembly operations ultimately be turned over to a purely military organization—the U. S. Army Special Battalion—a recommendation that fueled the smoldering controversy over military versus civilian control.

Meanwhile, Los Alamos and Z Division were once again operating in crisis mode with an original deadline of May 15, 1946—"D" Day for test *Able*, the airborne detonation over an array of ships. Originally, the second test, *Baker*, was to have been a surface burst, and the third, *Detector*, an underwater burst. Robert Henderson, responsible to JTF-1 for preparations for *Detector*, carried out plans for the underwater burst concurrently with *Baker* until the decision ultimately was made to substitute *Detector* for the surface burst. In this manner *Detector* became the shallow underwater test *Baker*. In addition to *Baker*, the Task Force made tentative plans for a third test, *Charlie*, to be detonated in deep water at a later date.[9]

Preparations for Operation Crossroads proceeded, with coordination taking place between Los Alamos, Sandia Base, and the Bikini-Kwajalein area in the Pacific. A special "Crossroads Airline," operated by the 508th wing stationed at Roswell, New Mexico, transported personnel between Santa Fe and Albuquerque, New Mexico, to Washington, D.C., Hamilton Field, California, and overseas to Kwajalein and Bikini.

In the interim, the advance inspection team had flown from Kwajalein to Roi, to Enewetak*, and on to Bikini. There, Edward B. Doll of Los Alamos, with John Williams as consultant, took charge of general administrative detail. This included transportation, living accommodations, and initial plans for establishment of Glenn Fowler's overall air-coordination and communications

---

*Formerly Eniwetok. "A better understanding of the Marshall Islands language has permitted a more accurate transliteration of Marshall Islands names into English spelling." See DNA 6033F, *Operation Sandstone 1948*, p. 1.

network. Harlow Russ was responsible for installation of the bomb loading pit on Kwajalein and facilities for bomb handling, while Bice and Lehr coordinated technical requirements for the various groups. When Williams and, later, Doll reported back to Los Alamos at the end of March, Bice, Lehr, and Russ remained to continue overseeing the establishment of communications systems, laboratories, and camera towers.

Construction on the islands was carried out largely by Seabees personnel under the competent direction of Major Dubose at Kwajalein and Commander Lovell at Bikini. Dr. Williams, in his March 29 report of the inspection trip, complimented these men for their diligence and expressed his confidence in the ability of Richard Bice to lead the group "in a competent fashion."[10]

Two months earlier, at the Terminal Island Naval Shipyard, Long Beach, California, a party consisting of Major General Leslie R. Groves, Richard Tolman, Colonel H. C. Gee, William S. Parsons, Norris Bradbury, Ralph Sawyer, Bob Henderson, Roger Warner, and other B-Division representatives had inspected the ships to be used in the operation. Subsequently, they wrote specifications for alterations to adapt a seaplane tender, the U.S.S. *Albemarle*, for bomb assembly. Modifications included the construction of an elevator to transport the bombs from magazines in the bottom of the ship to the assembly hangar on the aft deck. The *Cumberland Sound* also had to be modified for laboratory and measurement functions and the Tank Landing Craft, LCT 1359, for bomb firing purposes.

The *Albemarle* and *Cumberland Sound*, used principally by the Los Alamos group, were thus transformed into well-equipped, floating laboratories. The *Burleson*, known as "the animal ship," became a palatial Noah's Ark for test animals. The *Kenneth Whiting*, *Wharton*, *Avery Island*, and the WWII hospital ship *Haven* also had special laboratories aboard, while the Landing Ship Mechanized, LSM-60, remodeled to accommodate lowering of the bomb in a watertight caisson through a centralized well, was designated the "zeropoint ship" for the *Baker* test. The *Saidor* became the operation's photographic lab.

Henderson and Warner, along with General Groves, Richard Tolman, Colonel H. C. Gee, Admiral Parsons, Dr. Sawyer, and Dr. Bradbury, made a final inspection of the laboratory ships at Terminal Island, California, on March 22, 1946, just three days before the *Albemarle* was scheduled to sail with the Los Alamos Bomb Preparation Group. During this inspection, Admiral Blandy announced that President Truman had postponed the date of the tests by six weeks so Congressional observers could complete their sessions and still see the tests.[11] Crossroads, it seems, was to be a highly publicized event.

Some Congressmen, however, felt the tests should not take place at all. Supported by members of the recently formed American Federation of Scientists, Senators Scott Lucas of Illinois, James Huffman of Ohio, and J. William Fulbright of Arkansas, among others, questioned the wisdom of conducting tests while the United States was involved in negotiations for the international control of atomic energy. The tests, they felt, would cast doubt on the sincerity of American intentions and could reflect a disparity between stated intent and action. Nevertheless, while Wall Street millionaire Bernard Baruch presented the American plan to activate the international machinery to prevent atomic warfare, preparations for the Bikini operation proceeded without interruption.[12]

As a gesture of international goodwill, President Truman approved the invitation of twenty-one foreign visitors to view the tests. Over one hundred members of the press were also invited, and a special ship, the *Appalachian*, was designated for their use.[13]

Staging the Crossroads Colossus

The operations plan and technological offensive for Crossroads overshadowed Hollywood's largest extravaganza. Roger S. Warner, trying to rally support from the Los Alamos Personnel Office during the initial planning stages, humorously expressed the magnitude and complexity of the problems, both major and minor, when he wrote:

> It is high time that all personnel offices recognize that they must function as a mass mother and deal with all those petty problems of living which confront each individual but are common to all. The greater the group, the harder it is to organize each individual to put his rubbers on at the proper time without last minute crises. All such details must be planned, organized, and handled.[14]

In response, E. J. Demson, speaking for Personnel, pointed out that although they thought it was the duty of the "American Red Cross" to "mass mother," they would shoulder the burden, "recognizing that the job had to be done."[15]

The "job," as Norris Bradbury acknowledged, was a big one. Los Alamos and Z-Division personnel were responsible for furnishing the bomb components, assembling them, and measuring practice drops to provide ballistics data. To measure its efficiency, they used radio-chemical techniques, calculated blast pressures, predicted bomb effects to plan instrumentation better, measured neutron flux, and provided timing signals.

Logistics for the Crossroads operation, under the direction of the joint services, included staging rehearsals for both tests *Able* and *Baker*, the mobilization of 242 ships, 156 airplanes, and an impressive cast of nuclear physicists, chemists, mathematicians, spectroscopists, roentgenologists, veterinarians, engineers, seismologists, and varied types of scientists, all working at the frontiers of experimental science. Numerous Army, Navy, and Air Force personnel increased the total number participating to 42,000.[16]

Coordination of such a massive operation required the finest in orchestrational abilities; yet, basic to the success of the overall mission was the technical competence of the people involved. Among those people were numerous well-qualified military personnel with technical responsibilities who had to be hurriedly discharged from the service so they could participate in special capacities. Leon Smith, for example, recalled that after some initial delay, a call from General "Hap" Arnold directly to the base

Assembly Section of B Division

Roger S. Warner, Jr.
Lt. Col. Ellis E. Wilhoyt, Jr., Glenn A. Fowler, Robert W. Henderson,
Thomas B. Lanahan, H. G. Greening, Donald Mastick

B-1 FIRING

Group Leader:
William O. McCord, Jr.

Ensign David L. Anderson
Frank Fortine
Donald C. Harms
Arthur Collins
Robert M. Mainhardt
Walter J. Rohlfing
E. C. Hermann

DETS:
George P. Tilley
Vincent Caleca

B-2 FUZING

Group Leader:
George A. Koester

William Hempker
Leon D. Smith
Philip M. Barnes
Leo G. Raub
Joseph Klein
William Larkin
Leroy M. Stradford

B-3 ASSEMBLY

Group Leader:
Arthur Machen

Bryan E. Arthur
Alvin D. Van Vessem
William R. Stewart
Philip H. Dailey
Ira D. Hamilton
Kenneth O. Roebuck

HE:
Eugene H. Eyster
John H. Russell

B-4 ENGINEERING

Group Leader:
Robert W. Henderson

Richard A. Bice

B-5 PIT

Group Leader:
Raemer E. Schreiber

Louis S. Slotin
Harold T. Hammel
Neil Davis
LeRoy B. Thompson
Ted Perlman

B-6 LOGISTICS/SUPPLY

Group Leader:
Harlow W. Russ

John R. Riede
Raymond Schultz
Avery Bond
H. G. Schwaner

Technical Stockroom
Arthur Newell
Theodore Montgomery

B-7 AIR COORDINATION

Group Leader:
Glenn A. Fowler

Asst. Group Leader:
C. Hilt DeSelm

Wright Field Liaison:
Maj. Robert L. Roark

Air Coordination:
Capt. W. F. Hartshorn
Robert A. Knapp

Armament:
Ray Brin

Air Collection:
Thomas V. Davis
W. A. Jamieson

Communication:
Howard B. Austin
Martin Warren

Photography Supply:
Thomas R. Logan

Design:
Virgil Harris
Arthur M. Barrett
John Miles
Michael Clancey

commander at Lowry Field, Denver, got him mustered out of the Army Air Corps in a record thirty minutes.[17]

By mid-April 1946, approximately 100 scientists, technicians, and administrators from Los Alamos and Z Division were Bikini-bound. They made an intermediate stop at Pearl Harbor, where support equipment and thousands of instruments were installed before proceeding on toward the Marshall Islands. JTF-1 Technical Director R. A. Sawyer arrived at Bikini on May 29, 1946, to coordinate the offensive from his headquarters on the *Kenneth Whiting*.[18]

As *Able* Day approached, activity increased and tension mounted. In the air, C-54s, B-29s, Navy Hellcats, and torpedo planes went through various test maneuvers. Guided by mother ships in the air, pilotless drone planes sought out their predetermined altitudes at which they would experience the force of the atomic whirlwind. On the sea below, remote-controlled boats were strategically placed to test the effects of the blast and retrieve water samples within the radioactive area.

Fuzing and Firing personnel, under George A. Koester and W. O. McCord, Jr., tested and retested equipment, while Arthur Machen's crew of Bryan E. Arthur, Alvin D. Van Vessem, William R. Stewart, Philip H. Dailey, Ira Hamilton, and Kenneth O. Roebuck, practiced assembly techniques to rote perfection. Scientists and civilian specialists put the final touches on setups for experiments to determine the intensity of the blast wave, the kind of radiation produced, the effects of blast and radioactivity on test animals, fish, plant life, and the structure of the atoll, as well as the extent of the dissemination of radioactive fallout carried by air and water currents.[19]

The military photography brigade and Berlyn Brixner of Los Alamos readied cameras numbering in the hundreds, some perched atop steel towers and others installed in airplanes. More than 50,000 stills and 1.5 million feet of motion picture film would be taken. Communications specialists Howard Austin and Martin Warren checked and rechecked linkups that would relay information and provide communication between the island and headquarters. From the *Cumberland Sound*, which functioned as the nerve center of the Task Force, timing signals would coordinate activities for the entire operation.[20]

On June 24, 1946, after an initial postponement caused by bad weather, the dress rehearsal for Test *Able*, designated "Queen Day," took place successfully with Frank Fortine of the B-1 Firing Group and Philip Barnes of B-2 Fuzing aboard the bomb-carrying plane. Six days later, based upon the weather expert's prediction of twenty-four hours of good weather, June 30, 1946, was set as D-Day. The news flashed to all ships and, indeed, the world.[21] Scientists fine-tuned their instruments; animals were placed at their assigned positions; support ships and personnel were evacuated. By dusk the target vessels floated like gray silhouettes on the calm waters of the lagoon.

Tests Able and Baker

Able Day began inauspiciously. The sky was completely overcast; rain came down in torrents, and with it the hopes of Z-Division personnel and other members of JTF-1. However, by 6:45 A.M., blue sky could be seen on the horizon. The first of the Bikini atomic tests was still the order of the day.[22]

At Kwajalein, the day before, the special B-29 selected to make the airdrop had been readied. *Dave's Dream*, named after bombing expert David Semple, who was killed in a crash a few months earlier near the Los Lunas Test Range, rolled into place over the loading pit. An Army truck marked with the insignia "X-Roads" transported the bomb to the pit after the Los Alamos crew had assembled it in the *Albemarle's* hangar. They off-loaded the transport trailer onto the Kwajalein dock using the seaplane crane. To make sure that New Mexico personnel received just recognition for their substantial technical contribution, Senator Carl Hatch, a member of President Truman's Evaluation Commission, chalked "Made in New Mexico" on the side of the bomb's canvas security cover.[23]

CROSSROADS: IMPACT AND LEGACY

Weaponeer Leon Smith uses the Flight Test Box to make last-minute tests on the bomb prior to release for the *Able* test drop at Operation Crossroads.

The B-29 selected to make the *Able* airdrop was named *Dave's Dream*.

193

By 5:34 A.M. *Able* Day, the plane stood ready and waiting. The crew that boarded had won their spot after serious competition with other flight crews in training exercises at bombing ranges near Albuquerque and Alamogordo, New Mexico, and Erik Island in the Pacific. Commanded by Major W. P. "Woody" Swancutt, the crew and passengers included General R. M. Ramey, Task Group 1.5 commander; Colonel W. J. "Butch" Blanchard, air attack commander; Captain W. C. Harrison, Jr. (Army) copilot; Major H. H. Wood, bombardier; Colonel J. R. Sutherland, bomb commander; weaponeers Ensign D. L. Anderson and Mr. L. D. Smith; Captain P. Chenchar, Jr. (Army), flight engineer; scanners Corporal R. M. Modlin and Corporal H. B. Lyons; and Technical Sergeant J. W. Cothran, radio operator.[24]

The final practice run began at 8:20 A.M. The bomb commander and weaponeers made last-minute adjustments. "In the process of checking out the bomb," Smith recalled, "a series of amber, green, and red lights flashed on the Flight Test Box. To the pilot, red meant emergency; to us, it meant that certain events had occurred that should have; but the pilot turned around to see the red lights flashing and his face became almost ashen."[25] Reassured that all was proceeding according to plan, he continued the mission. Visibility was good.

On the lagoon below, one could see the aircraft carrier *Saratoga*, queen of the U.S. fleet, bearing the scars of World War II; the *New York*, battlewagon of 1912, floating in cumbersome contrast to the streamlined cruisers, *Salt Lake City* and *Pensacola*; the Japanese battleship, *Nagato*, pride of the Japanese Navy; the handsome German cruiser, *Prince Eugen*, and the U.S. submarine, *Skate*, among others.[26]

Aboard the electronic ship, *Avery Island*, electronics assistant, Ensign George W. Anderson, Jr., and other crew members located their goggles and binoculars and donned protective clothing. Over the loudspeaker came the warning: "All hands. Remember, protect your eyes."[27]

For months, inspection teams had worked to repair, catalog, and calibrate the electronic gear installed on the target ships. After the test they would return to conduct a systematic analysis of the gear for comparison with pretest conditions. Now, as "Mike" hour approached, aboard the *Avery Island*, the flagship *McKinley*, and other ships, the electrical instrumentation in the forefront of attention was the television. Just before the evacuation of Bikini, a member of the island staff activated the system linking television transmitters on the two towers to shipboard installations outside the lagoon. One camera focussed on the beach, one on the fleet.[28]

All hands not standing engine watch or monitoring a channel found an observation point on deck. Over the speaker a voice blared: "Live bomb run. Live bomb run. Ten minutes until release time. Ten minutes until release time. Stand byyyy."[29]

Bombardier Harold Wood hailed from Bordentown, New Jersey, but for some reason, over the intercom, he sounded like "a rebel from the deep south." Others recognized the drawl as "a Harvard accent." On the horizon, Enyu Island could be seen clearly. Target ships floated just to its right. The *Nevada*, brightly painted in "International Orange," stood out like an inflamed bull's-eye on the blue lagoon, signifying its purpose in the target array.

Minutes passed, and the "southern" voice came on again. "Coming up on release," he warned. Within seconds, at 9:34 A.M.: "Bomb away!" *Dave's Dream* quickly executed a breakaway turn.[30]

In this maneuver, the aircraft dropped into a steep bank, during which time, according to the weaponeer: "We exceeded the red line on the air-speed indicator." When the air blast struck the plane, Commander Jack Sutherland was looking downward through the drift meter. "The impact forced the soft rubber eye piece of the drift meter into his eye socket." Smith said. "He ended up with a cut all the way around from the impact."[31]

At age twenty-six, with more than 200 flight hours on test runs, engineer Leon Smith considered the Bikini enterprise "just another flight." "We were so busy making the breakaway turn after drop—to get as far away as fast as possible—that I didn't see the actual explosion," he explained. "The first look I had at the thing was about thirty seconds later."[32]

HERITAGE

Glenn Fowler, aboard General White's command aircraft, witnessed the detonation from the air. Shipboard observers had a closer view. With the initial burst, an entire hemisphere of air seemed suddenly to catch fire. "The incandescent sphere snuffs out and the column starts. Boiling white, pink, peach-colored steam," a crewman wrote. In the air, Smith saw the thin spiral of smoke, topped by a mushroom cloud. "We were flying high—but the cloud still went above us," he noted.[33]

The loudspeaker picked up the Stateside broadcast, spewing out adjectives. "Too damn many adjectives," one Task Force representative observed. "It's terrific and beautiful, but not that terrific."[34]

There were fires on the *Nevada* and *Saratoga*, but the bomb had missed its target. Few ships were immediately sunk. Although the drop occurred within the area designated for probable error, the explosion took place several hundred yards west of the target ship and, therefore, west of the closely spaced array of capital ships.[35] Thus, test *Able* sank only five ships and craft of the total number of ninety-two. Three miles away, Bikini showed no wave or blast damage, but its sands, like that of neighboring islands, bore traces of radioactivity and the refuse of civilization—a panoply of broken boxes, torn mattresses, tires, and tools spread across the beaches.

Scientists set about retrieving instruments, and with them, records of the blast pressures and neutron intensities. Thermal radiation had seriously burned exposed test animals; the shock wave twisted and bent masts and destroyed equipment, but the superstructures and hulls of those ships still afloat had suffered little serious structural damage.[36]

By July 19, "William Day," amid a tropical rainstorm, Task Force One rehearsed its second atomic test. The underwater shot *Baker* followed six days later. The crew assembled the bomb in the modified hangar aboard the *Albemarle*, then maneuvered it, minus tail fins, to rails aboard the emplacement ship

CROSSROADS: IMPACT AND LEGACY

LSM-60. Whereas the four previous atomic test detonations had included one tower shot and three airdrops, this one was to be suspended beneath the LSM and detonated electrically by radio linkage.[37]

On July 25 on a frigate outside the lagoon, Fowler observed that weather conditions were perfect. The day started with a beautiful sunrise and a gentle breeze. The explosion itself occurred as a spectacular climax. At 8:35 A.M., the device detonated underwater, creating a gas bubble with an illuminated dome. Pushed out of the water by force, the ship momentarily rode the crest of the dome, then blew into fragments as the mushroom cap became a giant fireball, followed by a towering waterspout that reached a height of 5,500 feet. A base surge of spray, mist, and air rolled outward, leaving an invisible blanket of highly radioactive contamination on many of the target vessels. The underwater shock waves radiating out from Zeropoint did the greatest damage. On the surface, water waves of unprecedented magnitude and speed swept toward Bikini, tearing out palm trees and washing back into the lagoon some 50,000 tons of beach sand. Nine ships and craft went to watery graves, including the 26,000-ton battleship, *Arkansas*, and the aircraft carrier, *Saratoga*. Nonetheless, it was the extent of radiological hazard that far exceeded scientific expectations.[38]

Radio-controlled drone boats darted in and out of the target area, collecting samples of contaminated water. Geiger counters buzzed and crackled as the coral, floating oil slicks, and target ships broadcast their contamination. Four days after the *Baker* detonation, it was still unsafe for inspection parties to board the target vessels. Later, crew members would spend futile days trying to decontaminate their ships with water, lye, foamite, and soap. Liberal amounts of Navy profanity failed to help. They even tried sandblasting, which removed the radioactivity—and the paint. This procedure was discarded when someone realized it would be impossible to sandblast an entire ship under battle conditions.[39]

Fallout

To some observers, the Bikini tests were a disappointment. After Hiroshima and Nagasaki, the public (because of the Press) had developed a healthy respect for the cosmic force of the atom. At Bikini, casual observers apparently expected to see massive tidal waves or immediate obliteration of the entire fleet. When this did not occur, their sense of awe was replaced by a more rational viewpoint.[40] But scientists and government leaders in informed positions better understood the grim realities.

Statistics supported their concern. When the final score was tallied for both tests *Able* and *Baker*, all but nine of the ninety-two target ships were either sunk, suffered damage, or reflected the seriously lingering effects of dangerous radioactive contamination. Yet, paradoxically, from the perspective of the Navy, the tests were a success. Crossroads demonstrated to political leaders and the Navy Command that the U.S. fleet was far from obsolete. Although it was determined that ships should undergo certain design changes and have washdown systems installed to control radioactive contamination, the overall conclusion was that "the atomic bomb is not at present, nor will it be for some time to come, a practical weapon for use against a fleet at sea."[41]

From the data gathered, scientists also felt they had gleaned enough information to predict accurately the effects of a deep-water explosion. This factor, in addition to the high costs of the test operation, plus technical and personnel problems, resulted in a decision to eliminate the third atomic test, *Charlie*.

While the Bikini operation took place in the Pacific, a political game of "watching and waiting" continued on the atomic energy control front. In Congress there was a delay on the McMahon Bill, designed to set up the machinery for control of atomic energy within the United States. Internationally, the United States proposed creation of an atomic development authority under the jurisdiction of the United Nations, but Crossroads cast doubt upon our true motives and provided Soviet leaders the opportunity to accuse American leaders of militarism. "If the atomic bomb did not explode anything wonderful," *Pravda* charged, "it fundamentally undermined the belief in the seriousness of American talk about disarmament."[42]

Contradictory as it may seem, the United States' attempts at peacekeeping and control on the one hand and nuclear testing on the other reflected an ambivalence imposed by circumstance. While esoteric philosophical and political debates surrounding the use and control of atomic energy continued at national and international levels, the President's Test Evaluation Board presented their concluding observations on the Crossroads operation. They expressed, as well, what they believed to be our only recourse as a nation. "One enduring principle of war has not been altered by the advent of the atomic weapon," they maintained, adding:

> Offensive strength will remain the best defense. Therefore, so long as atomic bombs could conceivably be used against this country, the Board urges the continued production of atomic material and research and development in all fields related to atomic warfare.[43]

Thus, Crossroads had ramifications that went beyond accomplishment of its professed scientific and military objectives. At the domestic level, the operation served as a mandate for continuation of postwar reconstruction and production efforts at the national laboratories—a *modus operandi* considered to be essential if the United States was to retain its leadership role in the atomic energy arena.

The Advisory Board to Operation Crossroads coordinated activities of the War Department and the Department of the Navy with the activities of the Manhattan Engineer District. Left to right: Roger S. Warner, Bradley Dewey, K. T. Compton, Brigadier General T. F. Farrell, General Joseph Stilwell, Brigadier General K. D. Nichols, Lieutenant General Louis Brereton, Vice Admiral Hoover, Rear Admiral Ralph Ofstie, and Lieutenant Colonel J. H. Derry.

Roger Warner, Associate Director of Z Division, discusses a point with William S. Parsons, a distinguished Naval officer. Parsons' assignments and ranking changed often during the postwar years. He held the rank of Commodore, or Captain, during his Manhattan Project days and served as weaponeer and bomb commander aboard the *Enola Gay* when the first atomic bomb was dropped on Hiroshima August 6, 1945. In December of that year, Parsons was appointed Assistant Chief of Naval Operations for Special Weapons. In January 1946, in addition to achieving the rank of Rear Admiral, he was also appointed by the Joint Chiefs of Staff to Joint Task Force One as Deputy Commander for Technical Direction at Crossroads.

Members of the Z-Division Assembly Group take a break during testing activities at Operation Crossroads. Front row, left to right: Phil Dailey, Kenneth O. Roebuck, Arthur Machen, Ira "Tiny" Hamilton, and Bryan Arthur. Back row, left to right: Roger S. Warner, Major Robert L. Roark, Colonel Jack Sutherland (seated behind Roark), Glenn Fowler (kneeling), Alvin Van Vessem, William O. McCord, and Gene Eyster (seated).

Test *Able*, conducted June 30, 1946, involved the airborne detonation of an atomic device over an array of ships. Only five of ninety-two ships actually were sunk during the test.

HERITAGE

Test *Baker* was conducted on behalf of the Navy to determine the effect of an underwater detonation on ships at sea. The picture above shows the swell of water as the device was detonated. A brilliant glow of orange and peach illuminated the crown-shaped formation. Below, water is forced skyward in a column as steam vapor escapes from the developing cloud above.

CROSSROADS: IMPACT AND LEGACY

When the explosion reaches maximum altitude, the familiar mushroom cloud becomes visible. Sand, pieces of coral, and other residue from the ocean floor can be seen in the mushroom formation following the underwater detonation.

On December 31, 1946, President Harry Truman signed the executive order that ended the wartime Manhattan Project and formally transferred control of the nation's atomic energy program from military to civilian authority. On hand for the signing were (seated, left to right) Carroll L. Wilson, AEC General Manager; President Truman; David E. Lilienthal, AEC Chairman; (standing, left to right) Sumner T. Pike, AEC member; Colonel Kenneth D. Nichols, Deputy Chief of the Army's Manhattan District; Robert Patterson, Secretary of War; Major General Leslie Groves, head of the Manhattan Project; and Lewis L. Strauss and William W. Waymack, AEC members. Truman had signed the Bill creating the Atomic Energy Commission on August 1, 1946.

Chapter VII.

MANAGING THE NUCLEAR ARSENAL:

The Atomic Energy Act

Will military minds prevail?
Will . . . other councils fail?
Will nations go to war again?
Will atomic horror reign?
Will the world be steeped in sorrow?
*Will the sun arise tomorrow?**

Richard West

During the summer of 1946, aboard the U.S.S. *Cumberland Sound* floating off the island of Bikini, Richard Bice, Norris Bradbury, and Robert Henderson sat up late one night making plans for the move of additional personnel from Los Alamos to Z Division at Sandia Base. It was ultimately decided that the last group, Z-4 Engineering, would move to Albuquerque by January 1947. The emphasis on "readiness" as an increasingly important part of postwar defense strategy had sanctioned the decision to establish the engineering and production facility.[1]

Six months later in the nation's capital, a more formal meeting took place. The ten men in attendance sat around a conference table in a small office on the sixth floor of the War Department building. Outside, cloudy skies cast a grey pall over

---
*Used with the permission of Richard West, author of *Three Songs and Other Poems*.

the city and transformed the New Year's snow into mud and slush. The dismal atmosphere on that January 2, 1947, created an appropriate stage for the serious nature of the discussions taking place. The occasion was the first meeting that year for the newly formed Atomic Energy Commission (AEC).

Those present included Chairman of the Commission, David E. Lilienthal, a "courageous lawyer" who had earned recognition as head of the Tennessee Valley Authority, and General Manager Carroll L. Wilson, a thirty-six-year-old engineer, who had assisted Vannevar Bush in the establishment of the National Defense Research Committee during the war. Other members were New England businessman Sumner T. Pike; gentleman farmer and newspaperman from Iowa, William W. Waymack; and investment banker Lewis Strauss, who was accompanied by his administrative assistant, William T. Golden. Herbert S. Marks, the Commission's General Counsel, was there as well as his assistant, George Fox Trowbridge. A young Army officer, Colonel Kenneth E. Fields, attended the meeting, along with Richard C. Tolman of the California Institute of Technology. Scientist Robert F. Bacher was away on an inspection tour of the nation's atomic stockpile.

Upon these men rested the mantle of responsibility for the vast complex known until its transfer to the AEC at midnight December 31, 1946, as the Manhattan Engineer District.[2] Their decisions would determine to large degree the future course of atomic energy policy in the United States and would affect, either directly or indirectly, the operation of the small organization that would become Sandia Corporation.

The Quest for Control

In the realm of atomic energy, the summer of 1946 through the winter of 1947 was a time for decision-making and debate. Around the world, statesmen, scientists, and engineers pondered the problem of establishing controls for atomic energy while maintaining national security. Driven by humanistic ideals to find a means of either sharing, abolishing, or harnessing the awesome power they had unleashed, world and national leaders realized that the United States' monopoly of the atomic bomb could not endure.

These were not new concerns. Even before the end of the war, the nation's scientific and political leaders had given serious consideration to postwar atomic energy activities and the establishment of a commission to provide government control and administration. As Secretary of War Robert T. Patterson publicly acknowledged on the date of the actual transfer: "We will have carried out the long-range plans of President Roosevelt, President Truman, Secretary Stimson, and General Marshall, who months before Hiroshima clearly recognized that Congress should create an independent agency of the government to carry on this vital work."[3]

Nonetheless, as in a master's game of chess, both national and international attempts to gain political and managerial control over the power unleashed by the atom had met with a series of checkmates. Not long after the end of the war, Secretary of State Dean Acheson, as head of the government committee on atomic energy, initiated the first move by gathering together a group of consultants to formulate policy. David Lilienthal, chairman of the Tennessee Valley Authority, was selected to head the board. Subsequently, Acheson and Lilienthal, with the assistance of Robert Oppenheimer, drafted a report for presentation to the United Nations. Known as the Acheson–Lilienthal Report, the document was poorly received by the press and in Congress.[4]

In an effort to attract more support from conservatives, President Truman appointed Wall Street millionaire Bernard Baruch to head the United States delegation to the United Nations. Baruch modified the Acheson–Lilienthal Report to encompass total disarmament, abolition of the veto, and unrestricted inspection of all military facilities.

On June 14, 1946, the silver-haired statesman stood before the United Nations General Assembly to offer his plan for the international control of atomic energy. "We are here to make a choice between the quick and the dead," he solemnly intoned. "That is our business. Behind the black portent of the new atomic age lies a hope, which seized upon with faith can work our salvation. . . . Let us not deceive ourselves. We must elect world peace or world destruction."[5] The oratorical dramatics, however, failed to sway the Russians, who responded to the Baruch Plan, first with silence, then with a comprehensive counterproposal.

The Soviet representative, thirty-six-year-old Andrei Gromyko, proposed that an international convention be established to prohibit the possession, production, or use of atomic bombs. "The Gromyko Plan," as it was called, included safeguards and scientific interchange of information but did not provide for inspections or sanctions against violations. The strongest point of contention was the veto, which Gromyko felt must be retained "under any circumstances."

On July 5, Baruch rejected the Soviet proposal. The Russians were not surprised. Only four days earlier, the first postwar tests of an atomic bomb had taken place in the Pacific. The Bikini Operation, according to some political analysts, was "a more forthright exposition of U.S. policy than anything said at the United Nations."[6] With this turn of events, as Prime Minister Winston Churchill had accurately predicted, the Iron Curtain descended with reverberations felt around the world.[7]

The Great Debate

On the domestic front, the battle for control of atomic energy raged for nine long months. Originated by Senator Edwin C. Johnson, Democrat of Colorado, S. 1463 was a companion measure to HR. 4280, introduced into the House of Representatives October 3, 1945, by Andrew J. May, Democrat of Kentucky and Chairman of the House Military Affairs Committee.[8] Said to represent the Army viewpoint, the May–Johnson Bill—with its proposals for strict military control and drastic penalties for disclosure of atomic secrets—provoked a legislative debate of epic proportions. Of particular concern was the Bill's provision

MANAGING THE NUCLEAR ARSENAL: THE ATOMIC ENERGY ACT

for a part-time commission with membership that could be composed of officers of the Army or Navy. This last proposal raised the hackles of American scientists, who quickly threw themselves into the fray with passionate fervor and unexpected organizational talent.

In that same month of October, the Senate set up a special committee under the chairmanship of Brien McMahon, freshman senator from Connecticut, to study the problem of control of atomic energy. After extensive hearings, including testimony from leading scientists, the committee discarded all previous bills and wrote its own. The McMahon Bill, S. 1717, introduced by McMahon in the Senate and by Congresswoman Helen Gahagan Douglas of California in the House, proposed to vest control over the development of atomic energy in a full-time commission of five civilian members, appointed by the President and confirmed by the Senate.

Senator Brien McMahon spearheaded the Bill through Congress that established the AEC and placed atomic energy under civilian control.

HERITAGE

David E. Lilienthal

Gordon Dean

Lewis L. Strauss

Carroll Wilson

Following the advice of the scientific contingent, the Bill placed strong emphasis on the greatest possible freedom of atomic research for peacetime, as well as military, uses. It also reversed the power structure proposed by the May–Johnson Bill by giving little voice to the military services.[9]

On January 31, 1946, President Truman at his regular press conference voiced his preference for civilian over military control. Having gained presidential support, McMahon prepared to do legislative battle. Soon, however, the War Department, under direction of Secretary of War Patterson, dealt the McMahon forces a severe blow by strongly urging greater participation and control by the military as well as closer liaison between the Commission and the War and Navy Departments.

On February 16, 1946, the McMahon camp received a second severe setback when news of the Canadian spy scandal made headlines. Washington was stunned to learn that British physicist Alan Nunn May, who had been working on the Canadian atomic project, had disclosed information about American efforts to Russian agents.

The psychological impact of the Canadian spy scandal was revealed in a public clamor for secrecy and in diminished support for the McMahon Bill. Conservatives, such as General Leslie Groves, used the spy case as an effective rationale for the continuation of strict military security. In Senate testimony, for example, he skillfully capitalized on this state of affairs by declaring that the security of the nation "depended upon the Army and the Navy having a voice in developing 'the most powerful military weapon in existence.' "[10]

In an attempt to resolve the civilian–military control issue, Michigan Senator Arthur H. Vandenberg introduced his own amendment. Considered by McMahon as a real threat to civilian control, the Vandenberg Amendment essentially gave the military veto power over the Commission. The Federation of American Scientists also viewed the amendment as a threat, and their efforts, combined with public pressure, induced Vandenberg to accept substantial modifications. As AEC historians Richard G. Hewlett and Oscar E. Anderson, Jr., observed, the amendment ultimately proved to be a "blessing in disguise" because it captured the interest and imagination of the American public, which, in turn, provided the McMahon Bill the necessary popular support.

After overcoming opposition in the House Military Affairs Committee, the Bill passed as amended. On August 1, 1946, President Truman signed into being the Atomic Energy Act, clearly establishing civilian control of atomic energy.[11] The "keeper of the keys" had been selected, but custody of the stockpile would continue to be a source of controversy.

The Political Machinery

The Atomic Energy Act of 1946—marked as it was by political maneuvering—remains a remarkable and effective document, in some respects as revolutionary as the scientific discovery that caused its creation. Therein lies an interesting paradox—the 79th Congress of the United States, predominately conservative in character and preoccupied with security, had enacted "a thoroughly radical piece of legislation." As effective as it remains, even in its amended format, the Act offers no ultimate solution to the problem of military versus civilian control. Because of the flexibility of the formula, interpretations of the Act continue to be shaped by political pressures and public opinion.

As an administrative agency, the Atomic Energy Commission was vested with sweeping authority and responsibility, the magnitude of which constituted a government monopoly of the sources of atomic energy. Although operated under the control of civilians, military involvement and oversight functions had to be provided. With these objectives in mind, three major advisory committees and a Division of Military Application were established as part of the Vandenburg Amendment to the McMahon Bill—which together made up the Atomic Energy Act.

Organization of the AEC, 1946

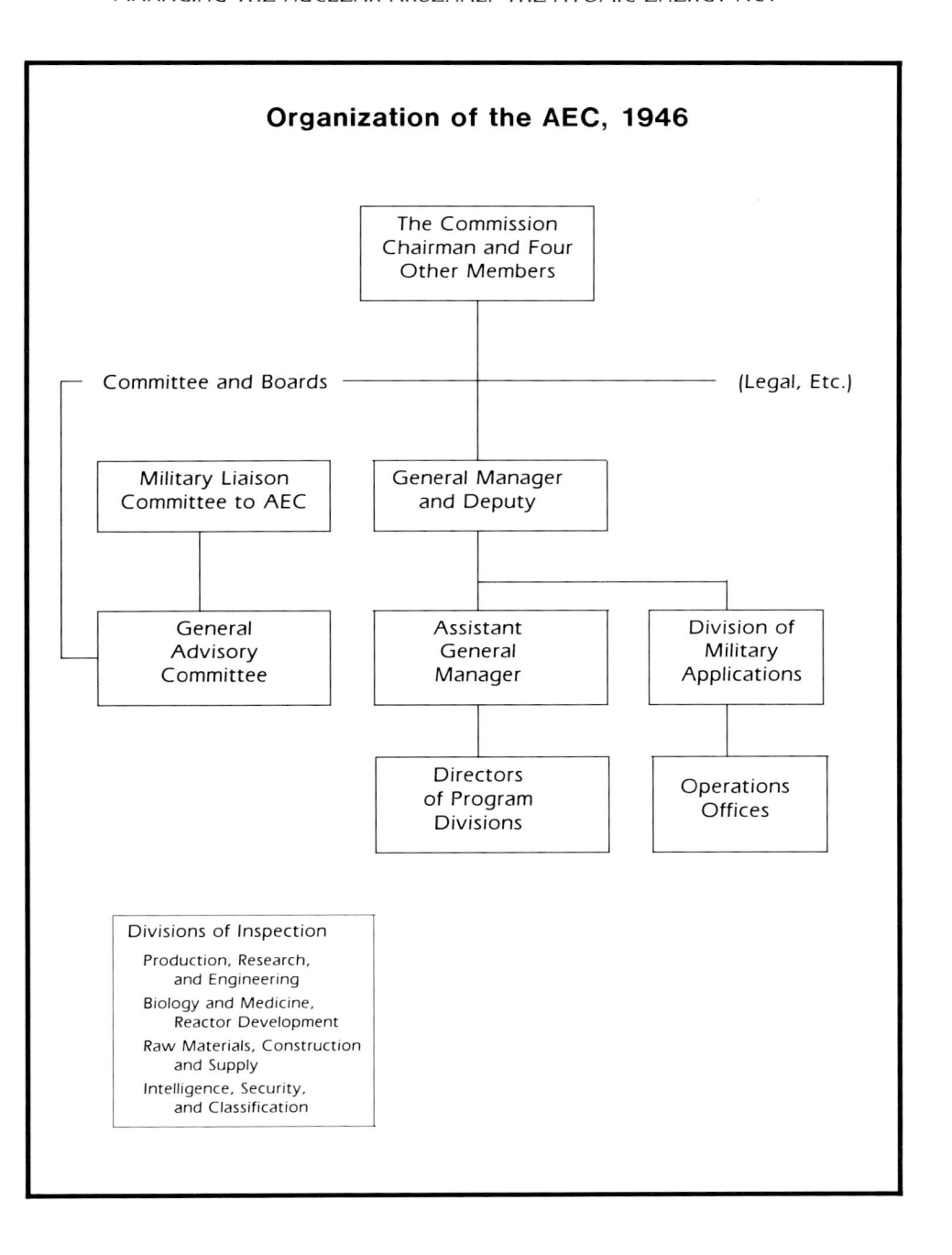

- The Commission: Chairman and Four Other Members
  - Committee and Boards
  - (Legal, Etc.)
  - Military Liaison Committee to AEC
  - General Manager and Deputy
    - General Advisory Committee
    - Assistant General Manager
      - Directors of Program Divisions
    - Division of Military Applications
      - Operations Offices

Divisions of Inspection
- Production, Research, and Engineering
- Biology and Medicine, Reactor Development
- Raw Materials, Construction and Supply
- Intelligence, Security, and Classification

Of these groups, the Military Liaison Committee (MLC) became the formal channel between the military services and the AEC. The MLC dealt with weapons requirements and other atomic energy interests of the military, although by statute, the onus was placed on the AEC to

> advise and consult with the Department of Defense, through the Committee, on all atomic energy matters which the Department of Defense deems to relate to military applications of atomic weapons or atomic energy including the development, manufacture, use, and storage of atomic weapons, the allocation of special nuclear material for military research, and the control of information relating to the manufacture or utilization of atomic weapons.[12]

In essence, the MLC would act as the interface between the Commission and the Department of Defense, which had come into being with passage of the National Security Act of 1947.[13]

Initially, the statutes of the Atomic Energy Act allowed the Secretaries of War and Navy to assign as many members to the Committee as they desired. In 1946, for example, the first chairman was veteran Army Air Corps officer Lieutenant General Lewis H. Brereton. There were two Army members, Major General Leslie R. Groves and Colonel John H. Hinds, as well as three Navy members, Rear Admiral Thorvald A. Solberg, Rear Admiral Ralph A. Ofstie, and Rear Admiral William S. Parsons.

Under the extensive provisions of the National Security Act of 1947, and in an effort to quell interservice rivalry and concerns about military control, it was decided that the chairman of the MLC would be a civilian designated by the President, with two members appointed by each service with concurrence of the Secretary of Defense.

Whereas the MLC functioned as the AEC–military interfacing agency at the *policy* level, the Armed Forces Special Weapons Project (AFSWP) functioned in a similar manner at the *operational* level. AFSWP, established effective midnight December 31, 1946, would, therefore, play a major role in the development of the military applications of atomic energy.

The Commander of AFSWP, General Leslie Groves, had the difficult task of operating without a specific charter, which, in turn, created operational problems at Sandia during the first half of 1947. By midsummer, however, a revised charter, approved July 8, 1947, clarified the role of AFSWP, although not to Groves' satisfaction.

Limited to noncommand functions, Groves did retain command of nonnuclear areas of ordnance work and the technical training of military personnel at Sandia Base. In his autobiography, the general recalled:

> We . . . organized a unit at Sandia Base near Albuquerque. Here groups of carefully selected young officers and senior noncommissioned officers were trained in the details of atomic bomb assembly and organized into teams thoroughly capable of doing the job under the kind of field conditions that had existed at Tinian. The unit was responsible for procuring and developing the equipment the assembly teams would need in the field. Throughout, the aim was to give each man as much technical information as he could absorb. The whole purpose of the operation was to make absolutely certain that in case of war, or even the threat of war, the Defense Department would have at its instant disposal teams ready and trained to assemble atomic weapons.[14]

The training of Groves' select cadre would be conducted by Z-Division personnel.

From Groves' point of view, the AFSWP charter was constraining; yet, as Commander, he was in a position to respond to the needs of the military with the production of nuclear weapons. Moreover, as a member of MLC, he could influence the Commission to produce fissionable materials. When Groves retired from the Army to enter private business in February 1948, he was succeeded by his assistant from the early days of the Manhattan Project, Major General Kenneth D. Nichols.[15]

A second committee established under the auspices of the McMahon Bill was the important Joint Committee on Atomic Energy (JCAE), which acted as the Congressional board of directors for the atomic energy program. Referred to as "the most powerful committee in the history of the nation," the JCAE was the only permanent joint committee ever created to enjoy "continuing legislative responsibility."[16]

Required to maintain a bipartisan staff, the JCAE retained as consultants various authorities on nuclear science and technology, as well as former AEC Commissioners. Composed of nine members from each House, with no more than five from the same party, the JCAE urged "a vigorous, imaginative, and aggressive atomic-energy program," characterized by "boldness and risk-taking rather than caution and economy."[17]

Although the Committee generally acted as a strong protagonist for the AEC, there would be times (as during the tenure of Lewis Strauss) when the JCAE felt that its prerogatives had been denied and reacted accordingly. As Senator Clinton Anderson of New Mexico recalled, the Staff Director of the Joint Committee, James T. Ramey, who had been appointed in 1955, was told to "watch the AEC." "If we found they'd swept out the office and hadn't told us about it," he continued, "we had somebody down there to find out why."[18]

As part of its "overseer" function through the years, the Committee expected scientists, including directors from the various laboratories, to provide direct and candid testimony relating to atomic energy matters. Because of its independent position and continuing influence, the JCAE was able to effect important decisions relating to the nuclear weapons program, including persuading President Truman to overrule the AEC Chairman and the AEC General Advisory Committee in 1950 to develop the hydrogen bomb. To the Committee's credit, it also encouraged expansion of plants necessary for production of fissionable materials; urged development of lightweight nuclear bombs to be carried by fighter aircraft; advocated the installation of electromechanical devices on nuclear weapons to improve U.S. custody and control (particularly for NATO deployments); and supported construction of nuclear-powered submarines and aircraft.[19]

A third group established under the Atomic Energy Act, the General Advisory Committee (GAC), numbered among its membership many well-known scientists and engineers. The GAC, although appointed by the President, reported directly to the Commission and, therefore, exerted considerable influence on the nation's emerging nuclear policies.

Those serving on the first GAC included: Robert Oppenheimer, Chairman; James Conant, Lee DuBridge, Enrico Fermi, Isidor Rabi, Hartley Rowe, Glenn Seaborg, Cyril Smith, and Hood Worthington. Philosophically, the Committee recognized the need to rebuild the morale of the AEC laboratories, which had been depleted by the exodus of leading scientists during the postwar period. On the other hand, the GAC—perhaps owing to the background of its membership—had a vested interest in encouraging basic research at the universities. They maintained, too, that development projects should be carried out under contractual arrangements with industry and that quality should be emphasized over quantity or size—attitudes that would affect the operational style of the various laboratories.

Throughout the early fifties, the GAC would enjoy an impressive amount of influence until its opposition stance against the development of the hydrogen bomb led to Truman's decision to overrule them. Subsequently, the influence of the GAC on nuclear programs experienced an ebb and flow, dependent upon the makeup of the group and the rapport—or lack thereof—between the President and the Commission. Basically, the GAC was to serve largely in a technical advisory capacity, although during the earlier years, it did venture advice on national and international nuclear policy.[20]

To maintain responsibility for the nuclear weapons program, the Atomic Energy Act provided for a Division of Military Application (DMA), which had as its director a member of the Armed Forces. The Commission appointed the Director, but the DMA established its own management practices with guidance from the Commission. In addition to responsibility for weapons production programs, the DMA had contract administration of production facilities and the weapon design laboratories at Los Alamos and Sandia.[21]

AEC Commissioners
Postwar Decade

	From	To
Sumner T. Pike	10/31/46	12/15/51
David E. Lilienthal	11/01/46	02/15/50
Robert F. Bacher	11/01/46	05/10/49
William W. Waymack	11/05/46	12/21/48
Lewis L. Strauss	11/12/46	04/15/50
Gordon Dean	05/24/49	06/30/53
Henry DeWolf Smyth	05/30/49	09/30/54
Thomas E. Murray	05/09/50	06/30/57
Thomas Keith Glennan	10/02/50	11/01/52
Eugene M. Zuckert	02/25/52	06/30/54
Joseph Campbell	07/27/53	11/30/54
Willard F. Libby	10/05/54	06/30/59
John Von Neumann	03/15/55	02/08/57
Harold S. Vance	10/31/55	08/31/59

AEC Chairmen

	From	To
David E. Lilienthal	11/01/46	02/15/50
Gordon Dean	07/11/50	06/30/53
Lewis L. Strauss	07/02/53	06/30/58

General Managers

	From	To
Carroll L. Wilson	12/31/46	08/15/50
Marion Boyer	11/01/50	10/31/53
Kenneth D. Nichols	11/01/53	04/30/55
Kenneth F. Fields	05/01/55	06/30/58

Joint Committee on Atomic Energy

Chairmen	Dates of Service
Brien McMahon	1946 –
Burke B. Hickenlooper	1947 – 1948
Brien McMahon	1949 – 1952 (d. 07/28/52)
Carl T. Durham (Acting)	1952 –
W. Sterling Cole	1953 – 1954
Clinton P. Anderson	1954 – 1956
Carl T. Durham	1956 – 1958
Clinton P. Anderson	1959 –

Military Liaison Committee

Chairmen	Dates of Service
Lt. Gen. Lewis H. Brereton, USAF	1946 – 1948
Donald F. Carpenter	1948 –
William Webster	1948 – 1949
Robert F. LeBaron	1949 – 1954
Herbert B. Loper	1954 – 1960

General Advisory Committee

Chairmen	Dates of Service
J. Robert Oppenheimer	1946 – 1952
Isidor I. Rabi	1952 – 1956
Warren C. Johnson	1956 – 1959

As the first director of the DMA, Brigadier General James M. McCormack, U.S.A.F., and the new manager of the Commission's western domain, Carroll L. Tyler, were faced with the enormous task of rebuilding an effective nuclear weapons complex from the shambles of the war project. Under the DMA organizational structure, Santa Fe Directed Operations (SFO), as it was initially called, was set up at Los Alamos in 1947. By 1956, headquarters for the atomic empire had been relocated in Albuquerque.

Albuquerque Operations (ALO), situated for a time on the campus of a former girls' school, moved into four surplus barracks buildings on Sandia Base early in 1958. San Francisco Operations, started in 1952, originally functioned under the Division of Reactor Development, but was transferred to DMA jurisdiction.[22] Ten years later, a Nevada Operations Office (NVOO) was established at Las Vegas and given responsibility for the nuclear testing programs in Nevada, the Pacific, and elsewhere. These field offices, then, under programmatic guidance from the DMA and administrative and production guidance from ALO, were to serve as important cogs in the political machinery established to provide domestic control over atomic energy.

Revitalization

When Carroll Tyler first arrived in Los Alamos in July 1947 to assume his duties as manager of SFO, the crumbling temporary buildings and conglomeration of prefabricated plywood homes, hutments, and trailers hardly resembled "one of the world's most famous scientific laboratories."[23] The physical facilities at Z Division on Sandia Base were equally unimpressive, but reconstruction was in progress. Tar-papered temporary buildings (T-Buildings), dating from Air Depot training days, housed the nucleus of early weapons production and engineering-research activities, although some new semipermanent buildings were erected. Design work—largely instrumentation design and fabrication—was conducted in T-824, mechanical in T-828, systems development in T-838, and electrical componentry in T-839.

With no natural gas and no steam heat, many of these early buildings were heated by coal stoves. To accommodate the occupants, Post Engineer Captain Luther J. "Luke" Heilman, whose duties began in the fall of 1946, set up a "do-it-yourself" distribution system. A dump truck was sent to Madrid, New Mexico, for a load of coal, which was dumped in a stockpile on the east side of the base. It then became the responsibility of designated individuals to pick up the coal, fill wooden bins at their buildings, and stoke the stoves. Between the period when coal was used and gas introduced, Kresge oil heaters were installed in many buildings.[24]

An old Civilian Conservation Corps barracks building, 818, located in the northwest corner of the Tech Area, was the first occupied and served as the first AEC/Z-Division headquarters. Of unusual design, the building had three corridors with a connecting hall and managerial offices that were paneled in an attractive knotty pine. T-818 was not only the first headquarters building, but also had the distinction of being the only building at the Laboratory to have floor space eventually condemned as unsafe for occupancy. The AEC manager occupied the northwest corner with his staff located down the west wing, and Z-Division management occupied the northeast corner with their staff in the east wing. Plant Services occupied neutral territory in the center.

One of the first improvements to rehabilitated buildings was the addition of transite (cement/asbestos) shingles as siding over tarpaper, which decreased heat loss. Although air conditioning came from open windows, complaints about environmental controls were virtually nonexistent. Six of the original temporary buildings, considerably modified for present-day use, are still occupied some forty years later, giving one a feel for the atmosphere of earlier times.[25]

When Tyler took over the revitalization effort in 1947, he had fewer than 400 Commission employees—a considerable decline from the workforce of 5,000 Army troops and civilians in residence the year before. A local contractor, Zia Company, handled general maintenance of Los Alamos schools, homes, and the electrical power system. At Sandia Base, however, the Army Corps of Engineers, under the supervision of Heilman, performed the housekeeping and maintenance activities.

Building 828, the first building occupied by Z–Division personnel, was designed by Virgil Harris at Los Alamos during the summer of 1945. Plans were then sent to Heilman for construction. Originally planned for bomb assembly, this building had a monorail hoist in the center. Other buildings constructed during this period and classified as "semi–permanent," were designed largely by Assistant Post Engineer Dave Tarbox. At that time, the first permanent structure, an 8,000–square–foot Plant Services Shop (P500), was added.

The construction of a complex of fourteen permanent buildings on Sandia Base soon followed. This work was not "bid," rather it was "negotiated" with local contractor Robert E. McKee. McKee, who had cleared workmen and subcontractors, appointed Guy McCullough as the resident manager for the Division's work.[26]

The new construction took place within Technical Area I—an area set aside for personnel with security clearances who were conducting classified activities. Bounded by F, H, Fifth, and Eighth streets and fenced in the shape of two rectangles connected by a narrow corridor, the unique conformation of the Tech Area permitted the Base Motor Pool to be kept outside the fence. During the 1946–1947 period, the Tech Area covered approximately 100,000 square feet of space.

Administration, research, and development of the Laboratory remained under the auspices of the University of California, with 1200 employees on the Hill under the direction of Norris Bradbury and 320 scientists, engineers, and technicians at Z Division on Sandia Base. Major administrative headaches involved procurement of supplies and personnel problems. In keeping with the postwar mood of reaction, many individuals questioned the propriety of engaging in armaments work.[27] The low salary structure and the lack of adequate housing in Albuquerque and on the Base further complicated recruitment efforts. With Air Force reactivation of Kirtland Field, for example, FHA housing had to be vacated to accommodate the military. The AEC then agreed to construct a separate civilian housing area in the northwest corner of Sandia Base.[28]

Acting Z-Division leader during the 1946 Crossroads operation, Dale Corson, called attention to other problems in a memo to Bradbury: "It is difficult to find adequately trained personnel who are willing to join this project, and some of those who *have* been found, accept other employment during the long delays attendant with security clearances and job offers." Corson concluded by noting that he felt it would be "worthwhile" to look for a new division leader since he and Warner both planned to leave in the near future.[29]

As he had indicated, Corson left during the summer of 1946, and Roger Warner, who had served as the head of Z Division since October 17, 1945, went to Washington to assume a post as Director of Engineering for the AEC in November 1946. After Warner left, Bradbury asked Robert W. Henderson if he would be willing to take over as head of Z Division on a temporary basis. Henderson, who had planned to leave, agreed, and ran Z Division from the Hill until February 1947 when the move of his Z-4 Engineering Group consolidated all units at Sandia Base in Albuquerque. Henderson's "temporary" arrangement was to extend throughout a thirty-year career.[30]

At the national level, activation of the political machinery to establish domestic control of atomic energy represented a major accomplishment prerequisite to maintaining our national security. But there was much more to be done at the operational level. The more mundane but equally essential tasks associated with building of the weapons complex were still in the implementation stages. For Z Division the final years of the postwar decade were to be characterized by positive change and accomplishment, despite adversity.

Of unusual design, E-shaped Building 818 had three corridors with a connecting hall. The AEC manager and staff occupied the northwest corner and west wing; Z-Division management and staff were in the northeast corner and east wing. Plant Services was located in the center.

Nearly three decades after Sandia was first established, "temporary" Building 818 was still in use.

MANAGING THE NUCLEAR ARSENAL: THE ATOMIC ENERGY ACT

Building 828 (above) and temporary Building T-933 (below) housed some of the earliest Assembly operations. The church steeple is visible on the left of Building 828.

HERITAGE

Building 835 is under construction in the foreground (above). In the background of the photo (left to right) are Buildings 855, 860, and 840. The first ambulance belonging to Medical is shown parked outside the old Safety Stores building (middle-right). Bernice Beeson, a registered nurse (at back of ambulance in photo below), was one of the first members of the Medical staff.

228

Bob Hopper's home on Sandia Base was representative of many of the homes in which Sandians lived in 1948. The cost for housing on base was as follows:

GENERAL INFORMATION ON AEC HOUSING

The rent and utility rates for the various AEC Housing Units are listed below:

Type of Housing	Rent	Utilities	Total
3 Bedroom House with Fireplace	$71.00	$8.43	$79.43
3 Bedroom House without Fireplace	67.50	8.43	75.93
2 Bedroom House with Fireplace	63.00	7.10	70.10
2 Bedroom House without Fireplace	59.50	7.10	66.60
2 Bedroom Apartment	46.00	6.10	52.10
1 Bedroom Efficiency Apartment	39.50	3.96	43.46
Private Dormitory Room	27.50	2.50	30.00
Semi-Private Dormitory Room	22.50	2.50	25.00

Members of the General Advisory Committee outside airplane at Santa Fe airport, April 3, 1947. Left to right: James B. Conant, Robert Oppenheimer, General James McCormack, Hartley Rowe (Los Alamos consultant and division director at NDRC), John H. Manley (Executive Secretary of the GAC), Isidor I. Rabi, and Roger S. Warner.

Chapter VIII.

PRODUCT OF POSTWAR READINESS

"The substantial stockpile of atom bombs we, and the top military assumed was there in readiness did not exist."

Chairman of the AEC, David E. Lilienthal

In early November 1947, Z-Division personnel were busy preparing for a visit by physicist John H. Manley. Manley, who was on an investigative tour for the General Advisory Committee (GAC) of the AEC, planned to meet with Dr. Norris Bradbury of Los Alamos, Robert A. Montague, Commanding General of Sandia Base, Colonel Gilbert M. Dorland, and other members of the Armed Forces Special Weapons Project, Z-Division Acting Director Bob Henderson, and other personnel. Manley's mission was to determine the status of technical activities at the Lab and report his findings.[1]

During 1946 and 1947, the atomic energy laboratories, including Z Division of Los Alamos, had undergone close scrutiny by the AEC. This latest visit represented the most recent of a series of fact-finding tours that had caused Washington officials increasing concern over the state of the nation's stockpile—a concern that would eventually result in the transformation of Z Division into an integral unit of the nuclear weapons complex.

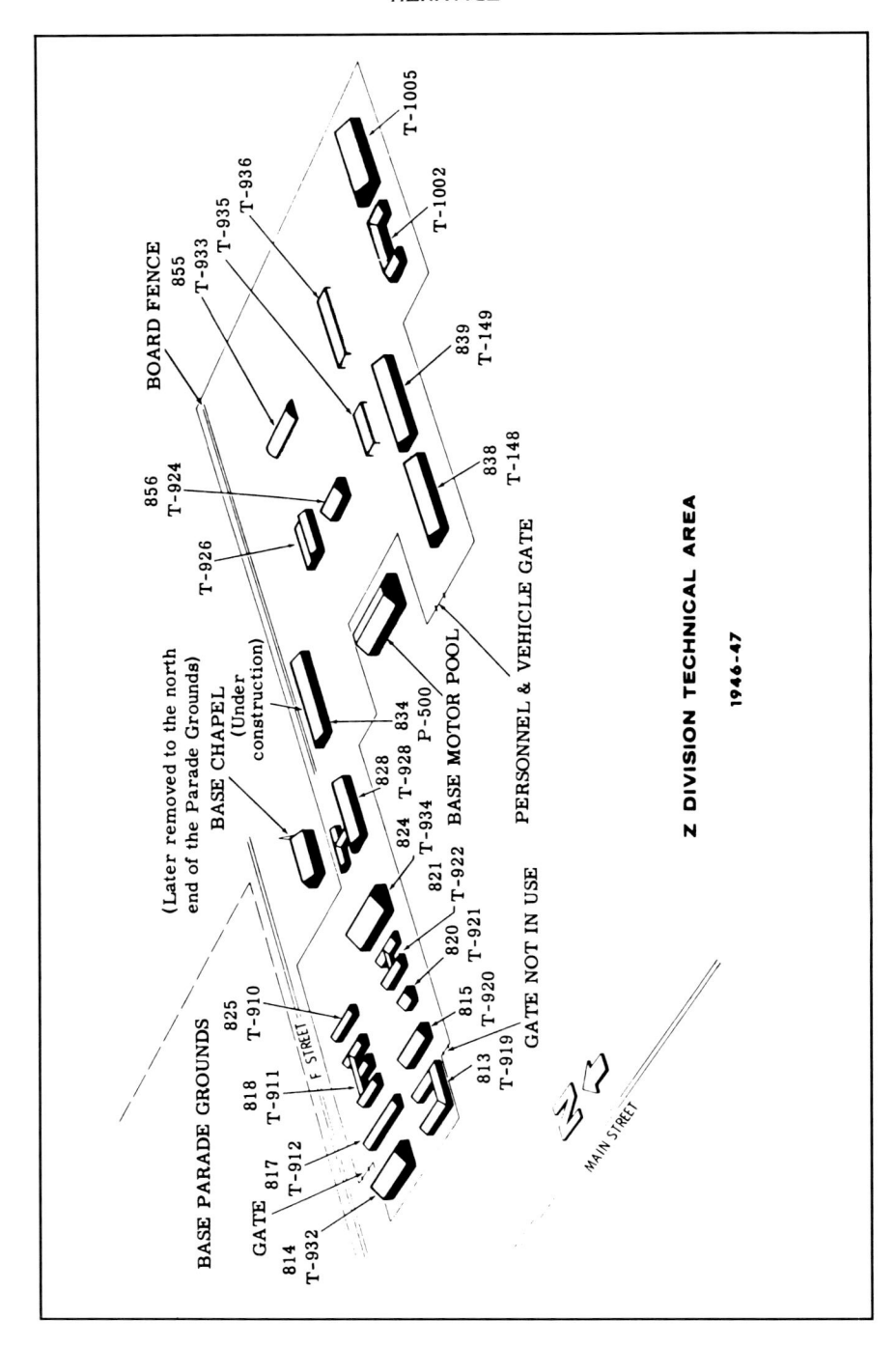

HERITAGE

Z DIVISION TECHNICAL AREA
1946-47

PRODUCT OF POSTWAR READINESS
The AEC Visits Sandia Base

The initial visitors to the laboratories were the new AEC commissioners, who felt that their first order of business after taking office should be to inspect the collection of plants and laboratories inherited from the military. On November 16, 1946, they began an all-day flight from Washington to Los Alamos. As the plane neared its destination, John Manley, who had walked up to the cockpit, noticed that the pilot was going by an outdated map, which would have put them down at the old Santa Fe airport at the corner of Cerrillos Road. Manley pointed out the error, and the pilot landed the plane safely some five miles further at the new airfield. Following the all-day flight, the commissioners received a briefing on weapons research and production, then drove to Albuquerque. Chairman of the AEC, David E. Lilienthal, commented upon the "rather somber but highly intelligent scientists" he had met at Los Alamos and "the alert and handsome young West Pointers" at Sandia Base who seemed "eager to learn the art of putting things together." He noted that he "learned quite a lot" during his visit, particularly about "what had *not* been done in the way of planning, coordination, and the like."[2]

Six weeks later, in early January 1947, AEC commissioner Robert F. Bacher also visited Los Alamos with the objective of assessing the nation's stockpile. His report, presented to a meeting of the General Advisory Committee (GAC), three commissioners, and several members of the Military Liaison Committee (MLC) on February 1, 1947, gave those present a better—if more depressing—view of the state of the nation's nuclear arsenal.

Among the topics of discussion at the GAC meeting was the future of the weapons laboratory at Los Alamos, including Z Division at Sandia Base. There was talk of moving Los Alamos from its isolated location, but before the meeting adjourned, committee members agreed that Los Alamos should be revitalized and weapons research accelerated, particularly in view of the failure of various disarmament plans in United Nations' negotiations.

On the weekend of March 28, at the next meeting of the GAC, the Commission's general manager, Carroll Wilson, announced that he had extended the Laboratory's management contract with the University of California to July 1948. The committee agreed that the number-one priority should be placed on weapons development and testing at Los Alamos, although ordnance and production, they decided, should be transferred to the Sandia Base facility under conditions acceptable to both the AEC and the military. Further decisions on a "new course for weapon production" would be delayed until after the committee got a "firsthand view of the situation."[3]

To obtain this firsthand view, Chairman of the GAC Robert Oppenheimer and the Weapons Subcommittee, made their own tour to assess the nation's weapons production complex. At Los Alamos, Oppenheimer's successor, Director Norris Bradbury, anticipated their concerns with a comprehensive report. He commented on the implosion and gun-type weapons, and on recent improvements made by the Los Alamos scientists. Bradbury also stressed that it would be essential to test stockpile models as well as new models under development. The Weapons Subcommittee came away convinced of the need for another scientific test of advanced designs of the atomic bomb in the spring of 1948.[4]

On April 4, before returning to Washington, Oppenheimer's group stopped at Sandia Base where Z-Division personnel were operating largely upon their own initiative and with little direction from the Hill. There, technicians sorted weapon components from the wartime project, tested new ones, and transferred them to the ordnance section at Kirtland where high-explosive charges would be added when available. Completed weapons were to be stored in igloos located in an arroyo south of the runways. Richard A. Bice's Engineering Group supervised production of mechanical mockups of standard weapon stockpile models; Military Liaison under the direction of Arthur B. Machen trained officers in assembly, developed standardized handling and test equipment, and documented new technology in military manuals. Groups under O. L. Wright and Alan N. Ayers subjected electronic and electromechanical components to various tests, while Glenn Fowler's Field Test personnel conducted tests on completed weapon models at Salton Sea in California. Despite the activity, the ordnance branch was still a long way from production-line status.

As AEC historians Richard G. Hewlett and Francis Duncan wrote: "To realize that the nation's vaunted power to wage nuclear war rested on this slender reed must have been a sobering experience." Although the visitors admitted that they saw "clear signs of initiative, enterprise, and even enthusiasm," this was not enough to dispel their concern about weapon production and the state of the atomic arsenal.[5] Overall, according to Lilienthal:

> The result of these inspections was a shock. The substantial stockpile of atom bombs we and the top military assumed was there, in readiness, did not exist. Furthermore, the production facilities that might enable us to produce quantities of atomic weapons and weapons so engineered that they would not continue to require a Ph.D. in physics to handle them in the field, likewise did not exist. No quantity production of weapons was possible under the existing handicraft setup.[6]

On April 16, 1947, Lilienthal briefed Truman on the actual number of bombs in stockpile. The President was visibly concerned. A particular problem was the lack of high-explosive castings and initiators. As Bradbury had indicated in his report to the commissioners: "The rate of stockpile attainment," he said, "is determined by the rate of metal fabrication at Los Alamos, by the rate of H.E. charge production at NOTS [Naval Ordnance Test Station], Inyokern, California, and by the rate of initiator production currently being carried out at Los Alamos." But Truman was not the only one concerned. The quantity of bombs in stockpile, or the lack thereof, was also at the top of General Groves' list. Speaking of the stockpile, Bradbury recalled: "I knew that number every hour on the hour! Because if you think Truman was concerned, . . . everytime that number would change (and believe me, for technical reasons that number would sometimes go down), I'd get a call from the good General saying, 'What the hell happened now?' I knew that number every hour on the hour."[7]

The Manley Evaluation

The ability of Z Division to carry its load in building the stockpile also rested heavily on the minds of the Commissioners. Subsequently, in November 1947—almost a year since the first inspection tour—John H. Manley of the GAC's Weapons Subcommittee, arrived at Sandia Base to conduct a special evaluation. Manley supported Bradbury's contention regarding the need for strong leadership. From the study of various documents, conversations with Los Alamos officials and members of the Armed Forces Special Weapons Project (AFSWP), he reported that the problems of Z Division were very much a reflection of difficulties experienced at all other AEC installations—these difficulties arising from the loss of key personnel, lack of overall policy since the close of the war, security restrictions, inadequacy of physical plant, and cumbersome personnel and procurement policies.

The flux of personnel, Manley felt, was a more serious problem than one might think. In his opinion, the result of having "different people with different concepts introducing different perturbations that might last beyond a person's tenure would be a lack of standardization of components and manuals and the eventual retardation of development programs."[8]

Manley also suggested a more careful criterion for selection of AFSWP military assembly personnel. Although impressed by their intelligence, he observed that they lacked the necessary manual skills or "electrical-mechanical sense." To illustrate, he cited instances where AFSWP personnel used a hacksaw for minor repairs without regard for cuttings falling into electrical contacts, the passing of cables with inadequate insulation, and the use of incorrect test equipment.

John Manley

In the technical areas, Manley found serious problems in the Fuzing and Firing and Stockpile groups. "Fuzing," according to Manley, had not "appreciably progressed since the war." He noted that the fuzing unit in use rarely met specifications "in toto" and consequently advocated strenuous development work on a replacement. Baroswitches failed to meet requirements "due to an ambiguity in the wording of the specifications"; pressure transmitters for X-units were showing deterioration; and the Mk 2 X-unit inverter, provided by Bendix, was also unsatisfactory. Although Manley considered the fuzing problems to be "the most serious in the whole Z Division," he advised that the entire stockpile should be gone over in detail, adding that it "suffered from a lack of systematization."[9]

Not all groups came under criticism, however. Manley lauded Alan N. Ayers' Mechanical Test Group for "discharging its present responsibilities in a well-functioning manner and . . . planning for the future in an enthusiastic, sound fashion." He also had a "favorable impression" of the direction and activities of Glenn Fowler's Field Test Group and Arthur B. Machen's Assembly and Training. Manley commended Machen's people for their "high degree of technical competence," despite the fact that they had "far too great a load per person" and were the first to feel criticism from the military.[10]

In colorful terms, Machen expressed his own opinion of the difficulties he encountered, expounding on the "pseudo importance placed on the manuals as a panacea for bomb builders." The manuals produced by Machen's group were originally designed for use as field handbooks before the Model 1561 Fat Man bomb design was frozen. This meant they contained information that was "highly controversial." Yet, these manuals were being indiscriminantly distributed by "another agency" before changes could be implemented. But Machen's group took the gaff.[11] Manley concluded his report by recommending a careful evaluation of the organizational structure at Z Division, which he still considered to be basically "a shoestring operation."

In Washington, the MLC and the AEC reviewed the Manley Evaluation at their meeting of December 17, 1947. The document, which General Brereton classified as "a very excellent and searching report," raised significant questions about the stockpile and about the relationship between Z Division and the AFSWP. Those present considered the situation serious enough for immediate action and responded by announcing the appointment of a new Associate Director and by informing the Congressional Joint Committee on Atomic Energy of "measures taken or contemplated" for improvement of the Z-Division operation.[12]

Interaction with the Military

Meanwhile at the ordnance lab in Albuquerque, the brunt of responsibility for trying to rebuild during this difficult transition period fell on Z-Division's Acting Director, Robert Henderson. The problems he faced were both operational and political in nature. Henderson's task involved not only weapons-related demands, such as the design of new weapon devices for the Sandstone test series and the hiring of personnel under less than desirable circumstances, but also the oftentimes sensitive interaction with the military.

Referred to as "Mr. Weapons," Robert W. Henderson came to Z Division from the Manhattan Project at Los Alamos. An award-winning engineer, Henderson co-designed the containment vessel called Jumbo and the Trinity shot tower. In 1947 he served as the acting director of Z Division and eventually assumed the position of Vice-President for Weapons Programs, the first executive Vice-President-level position at Sandia. Henderson retired early from Sandia in 1974 with a history of capable leadership.

At the operational level, the director had the problem of training and working with military personnel who felt that "the shoe was on the wrong foot," in Henderson's terminology. The bomb, they felt, should belong to and be controlled by the military, with civilian input from the AEC being strictly in a monitoring and advisory vein. Moreover, Z Division was in the undesirable situation of being a civilian tenant on a military installation, an arrangement that inevitably led to confrontations.

On one occasion, a representative from General Montague's staff showed up at Henderson's office in Building 818. "We're going to restructure you according to Civil Service regulations," the visitor announced.

"The hell you are!" Henderson responded. A phone call to Washington took care of the problem.

Sometime later, Major General Montague himself marched into Henderson's office and flopped a paper down on the desk. "This is an eviction notice," declared the bearer of good tidings. "You occupy a lot of housing on the east side of the base, and I want it all vacated in two months. This is your legal notification."

Henderson just looked at Montague and picked up the phone again. "The military had problems in accepting the fact that they were expected to provide housekeeping duties for us," Henderson observed.[13]

The military did receive preferential treatment in housing and facilities, and Z-Division personnel did have to depend upon them for their day-to-day existence, but the arrangement was mutually difficult. General Montague and his military personnel were unaccustomed to adhering to power in the hands of a civilian AEC.

At Los Alamos, Bradbury viewed the Z-Division-military interaction difficulties with a broader perspective. He attributed a part of the problem, particularly those of an operational nature, to the Laboratory's lack of status and reputation, but felt that such clashes—although inevitable—did not preclude cooperation in the future. He had reached the conclusion, however, that the division needed to be headed by a "very senior man with considerable prestige" to stabilize the Laboratory's position and prevent "continual sniping from both the military and the Commission." In retrospect, Bradbury explained that this decision did not reflect on Henderson, "who is very capable."[14] In addition, Bradbury had plans for Henderson to serve as Assistant Technical Director for the Sandstone tests being planned for the spring of 1948.[15]

Los Alamos Director Norris E. Bradbury (center, with plans in lap) confers with civilian and military leaders in the center office of the first AEC/Z-Division headquarters Building 818. Thomas Lanahan stands on the far left and Major General Nathan Eisen stands on the far right.

The Custody Issue

The division's problems with the military were compounded by an undercurrent of competition that still existed between the military and the AEC over the custody of nuclear weapons—a problem exacerbated by General Groves' strong advocacy and promises to his officers at Sandia Base "that Commission fumbling would soon put weapons activities back in the hands of the military where they belonged."[16]

The argument championed by Groves seemed to be that the military could not rely on nuclear weapons unless military personnel had had experience in handling, storing, and maintaining them. The civilian argument, espoused by Lilienthal as the chairman of the AEC, held that the intent of the Atomic Energy Act of 1946 and subsequent Executive Orders made custody a policy decision for the President.

Bradbury, as Director of the weapons laboratory at Los Alamos, voiced the technical argument, maintaining that the weapons in stockpile were more laboratory devices than production models. The weapon design, therefore, required scientists with laboratory instruments rather than military technicians with check lists. Bradbury openly doubted that the armed forces had the kind of talent needed to maintain the stockpile in a state of readiness. Further, he disagreed with the military argument that preparedness required custody. Training with dummy components, he felt, should be adequate. He reportedly "bridled at the assumption that the Commission was merely a service and procurement organization for an operation the military intended to control."[17]

In early 1948, the custody issue arose again—this time with a slightly different emphasis. Concerned over worsening relations with the Soviet Union, the Military Liaison Committee in a report presented to the GAC on February 6, recommended that the AEC transfer the stockpile back to the military. The question that loomed large in their minds was whether or not emergency transfer from the AEC to the military could be facilitated expeditiously in times of crisis.

Certain technical difficulties did exist. The Military Liaison Committee cited the "divided responsibility between the Commission and the military in the existing situation at Sandia [Base]" as a source of concern and the possible cause of confusion in an emergency.[18] At the time, however, it was decided that the Commission should retain custody. The question would be reconsidered.

In February 1948, General Groves retired as head of AFSWP—but not without making one final bid to have the custody of nuclear weapons turned over to the military. As the retiring chairman of AFSWP, he drafted letters presenting the position of the military establishment in the development of atomic weapons to General Omar N. Bradley, Chief of Staff, U.S. Army; Admiral Louis E. Denfeld, Chief of Naval Operations; and General Carl Spaatz, U.S. Air Force.

In each case, Groves explained that he was vacating his post with "mixed pride and regret" and proceeded to reiterate his long-held position on custody: "The present system, whereby the military assists an extramilitary service only when and if invited to do so constitutes a serious hazard to the preparedness of the Nation," he declared, adding,

> It is axiomatic that the Armed Forces must have control over the weapons upon which our National Security so largely rests. It is essential that completed atom bomb parts be given over to the custody of the Armed Forces for stockpiling. I cannot urge this point too strongly.

Elaborating further, Groves said:

> A situation in which the Armed Forces are denied the use of the Nation's most powerful weapon until an Agency, which is in no wise charged with National Security, and possibly is not even really concerned, deems fit to deliver it, will lead to confusion and delay at a time when prompt retaliatory action will be vital to our country.[19]

Despite the urgings of Groves and the MLC, it was decided that the Commission should retain custody, at least for the present. It was felt that the appointment of a civilian chairman to the MLC, as well as the restructuring of the AFSWP, would help alleviate the interservice rivalry and help to quell the competition for custody.

On March 11, 1948, the President himself stepped into the dispute to grimly demand constructive cooperation between General Nichols, Groves' successor as head of AFSWP, and Lilienthal of the AEC. Truman bluntly informed the two men that he would not tolerate "the kind of squabbling that had prevailed in 1947."[20]

In the interim, civilian Donald F. Carpenter, formerly Vice-President of the Remington Arms Corporation, replaced General Lewis H. Brereton as chairman of the MLC. In April, Carroll Wilson and James McCormack personally visited Sandia Base where they saw to their satisfaction that there would be "absolutely no delay" in the emergency transfer of weapons from the civilian AEC to the military.

Impressed with Wilson's report on the improved situation, Carpenter supported continued AEC control of the stockpile. Despite a personal bent toward control by the military, he felt that too many technical difficulties still existed for the change to be made at this time. At a meeting held April 12, 1948, the new chairman suggested that the Commission and the MLC also visit Sandia Base to judge for themselves the problems of transfer—if any.[21] One week later, the MLC followed up by inviting representatives of the AEC to meet with them to discuss aspects of the transfer problem.

The proposed conference on weapons custody took place on May 24, 1948, in Albuquerque, New Mexico, as planned. After a nonstop flight from Washington National Airport to Kirtland Field, Carpenter, members of the newly constituted MLC, McCormack, and the Commissioners (except for Lilienthal) gathered in a classroom at nearby Sandia Base. Z-Division personnel presented a briefing on storage facilities and took the VIPs on a guided tour of a temporary storage igloo. The next day, military personnel demonstrated assembly operations and showed how technicians were trained in surveillance activities.

The issues of military custody and transfer were not discussed until the group went on to Los Alamos, where they met

Bradbury and staff. It was at this meeting that Bradbury diplomatically presented the technical arguments against transfer to the military. As the director pointed out, even if the military did control the weapons, they would have to be returned to the Sandia Base facility for modification and updating; therefore, transfer in reality, "would only be temporary."[22] Furthermore, Z Division would be required to perform surveillance activities to determine the need for modification and reliability—still another rationale for the continuation of custody by the AEC.

General Nichols presented the standard military arguments for custody, using the construction of storage sites at remote locations under military control as support for his position. Since these sites were under the protection and operation of the military, surveillance—according to Nichols—should be a military responsibility.

The MLC and AEC representatives acknowledged that the concerns of the military were not without some basis in fact. However, the major concern relating to the efficacy of emergency transfer proved to be unfounded. After listening to both sides, Carpenter felt that he had the basis for compromise, but he had not reckoned with the strength of the Commissioners' feelings on the matter. Commissioner Lewis Strauss cited the poor condition of the new storage sites as evidence of the military's lack of technical capability. But the major issue remained one of policy and vested power; only the President had the right to decide the custody issue.[23]

The arguments and recommendations seesawed back and forth until July 21, 1948, when the AEC released their Fourth Semiannual Report. Using this document as his forum, President Truman made public his decision regarding the transfer of custody:

> As President of the United States, I regard the continued control of all aspects of the atomic energy program, including research, development and the custody of atomic weapons, as the proper functions of civil authorities. . . . There must, of course, be very close cooperation between the civilian Commission and the Military Establishment.[24]

It is significant that during the peak of the Soviet crisis, Truman decreed that custody of bombs in the nuclear arsenal remain in civilian hands. The President's stance was strongly supported by certain members of his Cabinet. Budget Director James Webb, for example, expressed his sentiments in plain language: "(T)he idea of competing, jealous, insubordinate services fighting for position with each other is a terrible prospect."[25] Easier to verbalize than to enforce in all its subtleties, the issue of custody would continue to be raised by the national military establishment.

In the summer of 1950, the outbreak of the Korean War prompted the development of a plan for transfer between the AEC and the Pentagon. Truman then released a number of nonnuclear components to the military—a move that paved the way for eventual transfer of custody. By this time, it was felt that the military had reached technical maturity in the area of nuclear weapons maintenance and operation.[26]

In June of 1952, a further arrangement allowed the President to authorize the AEC chairman or Commissioners to use nuclear weapons. In the event these individuals were killed in a surprise attack on Washington, the AEC's manager of operations in Albuquerque would transfer custody to the military after notification by the Commanding General, AFSWP, that Presidential sanction had been preauthorized.[27] By 1957, the military and the Commission were essentially sharing control of the stockpile.

The Readiness Problem Defined

On the world scene, tensions between the United States and the Soviet Union during these years were increasing. By 1947, General Markos and his Communist guerillas had entrenched themselves in the Greek hills, and Klement Gottwald had established a Communist dictatorship in Czechoslovakia. In January, a blizzard—the worst in the century—paralyzed England, contributing to the nation's economic plight and inability to maintain its anti-Communist commitments in Greece and Turkey.

PRODUCT OF POSTWAR READINESS

President Truman on March 12, 1947, launched a counterattack. In a two-pronged strategy, first with the Truman Doctrine and later, the Marshall Plan, he provided both the military and economic means to resist Communist subjugation.

Positive action came none too soon. On May 31, the U.S.S.R. ordered inspection of all military trains moving from West Germany to Berlin. The following March 1948, the alleged suicide of Jan Masaryk, the Czechoslovakian foreign minister, marked the capitulation of that country to Communism. By June the U.S.S.R. had blockaded Berlin. Then, as General Douglas MacArthur phrased it, during the last week in June 1950: "North Korea struck like a cobra."[28] Such aggressive actions focussed the attention of U.S. policy planners on the readiness problem.

One of the most basic steps in meeting that problem, as Lilienthal and the Commissioners acknowledged, was to redesign the bomb into a genuine field weapon. To do this, they agreed, "would require industrial experience and scientific experts with experience in dealing with industrial problems." What was required, according to Lilienthal, was "not something that could be done in a laboratory alone, but in a production center, with . . . factory management." They realized, too, the need for technical experience in dealing with weapon systems and weapon development; what they envisioned was "a team working together as a unit." "To go out and create such an organization," Lilienthal acknowledged, "was out of the question. There was not time."[29] The AEC would have to build on what it had.

At Sandia Base, the Commission had the semblance of a team, in fact, the only team with assembly expertise. Z-Division personnel had the technical experience, but they lacked an effective organizational structure and a functioning production center. To meet these deficiencies, the AEC took a series of major steps involving the interaction of government and industry, and the development of cooperative research and engineering.

Part II.

NUCLEAR ORDNANCE ENGINEER FOR THE NATION

Paul J. Larsen, Director, Sandia Branch of Los Alamos Scientific Laboratory.

Chapter IX.

MEETING THE CHALLENGE:

Sandia Laboratory— An Independent Branch

"Responsibilities of Z Division have grown to such an extent that divisional status became completely inappropriate."

*Norris E. Bradbury to
University of California President R. G. Sproul*

As part of the nation's campaign to meet readiness needs during the postwar period, the AEC made a concerted effort to improve the ordnance facility at Sandia Base and build up the nuclear stockpile. The Commission and Norris Bradbury first attacked the leadership and organizational aspects of the problem. The production issue would follow.

Paul Larsen Appointed Director

The AEC's solution to the leadership problem arrived December 4, 1947, in the person of Paul J. Larsen. Handsome and silver-haired, Larsen exuded determination and a certain distinctive flamboyance that would stand him in good stead in his new position as Director. A native of Denmark, Larsen first came to the United States at the age of ten. During his early career he worked on the first transatlantic radio broadcasting station for the Marconi Wireless Company and managed research laboratories for the Radio Corporation of America in New York. His professional background also included Bell Telephone Laboratories, Warner Brothers, Baird Television, and the Department of Terrestrial Magnetism of the Carnegie Institute. At the Applied Physics Laboratory at Johns Hopkins, he had just completed a commendable job of production–engineering the proximity fuze, a talent that would be put to good use at the Sandia Base facility.[1]

Larsen addressed the job of building up Z Division with determination and quickly established a reputation for getting things done. He seldom permitted defeat of his aims and carried his battles to Washington if he met opposition at the local level. His predecessor, Bob Henderson, considered him "to be the perfect man for the time," adding, "Larsen could talk himself into a situation, then talk himself out before he got into trouble."[2] Admittedly, the international situation lent force to Larsen's efforts. By 1948, the military and civilians alike realized the need for national readiness.

A memorandum from Bradbury outlined the new director's duties, including the development of new and improved components for weapons. He was directed to "devote immediate and personal attention to the electronic problems of the Sandia Laboratory including fuzing, firing, test equipment, and telemetering" as Manley had recommended in his evaluation.[3] A primary objective was the incorporation of the Mk 3 firing unit, the next milestone being to establish the proper "form" and electrical components for the Mk 4 weapon.

The Laboratory's mission, therefore, would remain oriented toward "weaponization" . . . the design of fuzing and firing systems to detonate a weapon at a particular time and altitude, assuring optimum destruction of a special target. In addition, the Sandia Laboratory would continue to design casings to house the explosive systems produced at Los Alamos.

One of Z Division's first crash programs, called "Chicken Pox," involved preparation of a Mk 3 bomb for delivery, essentially during flight from the storage site to the point of bomber takeoff. A Boeing C-97 cargo plane, modified to accommodate the flying assembly operation, was used as a carrier.

To facilitate such weapon activities, Bradbury in his memorandum to Larsen urged that the Sandia Lab assume even more of an independent posture, which would require organizational changes and expansion. Larsen, whose title was abbreviated in July 1948 from "Associate Director, Sandia Laboratory, Branch of Los Alamos Scientific Laboratory" to the less cumbersome "Director, Sandia Laboratory," responded by delegating certain assignments to Associate Director Bob Henderson. "The activities and volume of work of the Sandia Laboratory have expanded so much in the past nine months," Larsen explained, "that it is now necessary to split operational responsibilities."[4] Larsen continued to formulate policy and handle matters related to production, but all research, development, engineering design, and field activities involving weapons work were placed under Henderson's jurisdiction.

The responsibilities of the Lab remained essentially the same with the exception of added emphasis on support for still another full-scale atomic test in the Pacific. Scientists at the Los Alamos Laboratory had realized since late 1946 that it would take field tests and actual detonation with proper instrumentation to prove the experimental designs of atomic weapons and verify theoretical calculations.

In connection with the test series, Bradbury planned to call upon the full-time services of Henderson "in his capacity as Assistant Technical [sic] Director of the Pacific Proving Ground Operation."* Bradbury stressed that the Sandstone operation, designed to verify new weapons design principles, should have highest priority and that "he [Henderson] must have first call upon the facilities and personnel of . . . Z Division in bringing about its successful accomplishment."[5]

---

*Henderson's official title at Sandstone was "First Assistant Scientific Director."

Early Mockup of the Chicken Pox Flying Assembly Lab
(1946-1947)

These photos of an early mockup of the Chicken Pox concept illustrate the feasibility of disassembling and reassembling the Model 1561 unit within the C-97 aircraft. The sphere case shown here was probably just the bomb casing, as the aircraft cargo hoist would not support the full weight at that time. A wooden mockup of the later hoist is shown in the final photo. The handling equipment shown (chocks, tong, and dolly) reflect the original equipment, which was later replaced by more sophisticated, lightweight items using aluminum rather than steel.

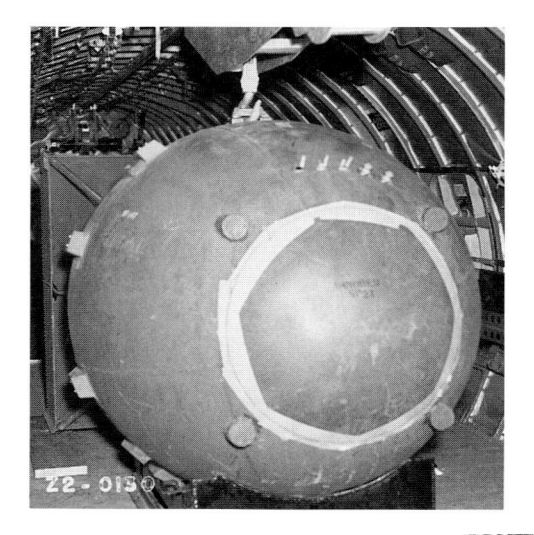

The fully assembled unit was loaded into the aircraft.

The nose cap and forward components were stored at the front of the aircraft.

MEETING THE CHALLENGE: SANDIA LABORATORY, INDEPENDENT BRANCH

The sphere was lowered on chocks in the center of the aircraft.

The tail section was stored at the back of the aircraft behind the aft case.

Aft components were stored behind the sphere in this manner.

Wooden mockup of a later hoist supports a casing of the Model 1561 bomb.

255

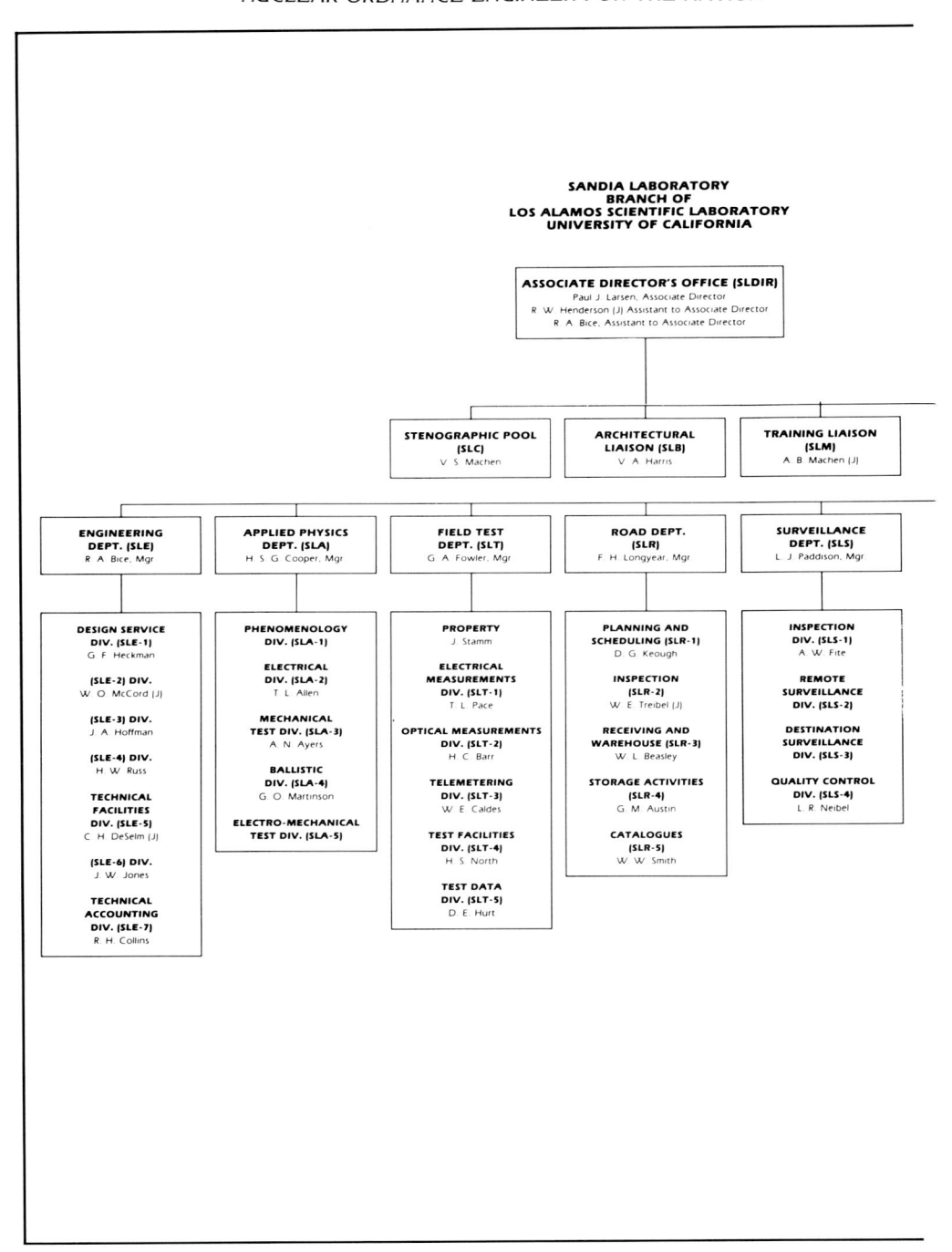

SANDIA LABORATORY
BRANCH OF
LOS ALAMOS SCIENTIFIC LABORATORY
UNIVERSITY OF CALIFORNIA

ASSOCIATE DIRECTOR'S OFFICE (SLDIR)
Paul J. Larsen, Associate Director
R. W. Henderson (J) Assistant to Associate Director
R. A. Bice, Assistant to Associate Director

STENOGRAPHIC POOL (SLC)
V. S. Machen

ARCHITECTURAL LIAISON (SLB)
V. A. Harris

TRAINING LIAISON (SLM)
A. B. Machen (J)

ENGINEERING DEPT. (SLE)
R. A. Bice, Mgr.

- DESIGN SERVICE DIV. (SLE-1) — G. F. Heckman
- (SLE-2) DIV. — W. O. McCord (J)
- (SLE-3) DIV. — J. A. Hoffman
- (SLE-4) DIV. — H. W. Russ
- TECHNICAL FACILITIES DIV. (SLE-5) — C. H. DeSelm (J)
- (SLE-6) DIV. — J. W. Jones
- TECHNICAL ACCOUNTING DIV. (SLE-7) — R. H. Collins

APPLIED PHYSICS DEPT. (SLA)
H. S. G. Cooper, Mgr.

- PHENOMENOLOGY DIV. (SLA-1)
- ELECTRICAL DIV. (SLA-2) — T. L. Allen
- MECHANICAL TEST DIV. (SLA-3) — A. N. Ayers
- BALLISTIC DIV. (SLA-4) — G. O. Martinson
- ELECTRO-MECHANICAL TEST DIV. (SLA-5)

FIELD TEST DEPT. (SLT)
G. A. Fowler, Mgr.

- PROPERTY — J. Stamm
- ELECTRICAL MEASUREMENTS DIV. (SLT-1) — T. L. Pace
- OPTICAL MEASUREMENTS DIV. (SLT-2) — H. C. Barr
- TELEMETERING DIV. (SLT-3) — W. E. Caldes
- TEST FACILITIES DIV. (SLT-4) — H. S. North
- TEST DATA DIV. (SLT-5) — D. E. Hurt

ROAD DEPT. (SLR)
F. H. Longyear, Mgr.

- PLANNING AND SCHEDULING (SLR-1) — D. G. Keough
- INSPECTION (SLR-2) — W. E. Treibel (J)
- RECEIVING AND WAREHOUSE (SLR-3) — W. L. Beasley
- STORAGE ACTIVITIES (SLR-4) — G. M. Austin
- CATALOGUES (SLR-5) — W. W. Smith

SURVEILLANCE DEPT. (SLS)
L. J. Paddison, Mgr.

- INSPECTION DIV. (SLS-1) — A. W. Fite
- REMOTE SURVEILLANCE DIV. (SLS-2)
- DESTINATION SURVEILLANCE DIV. (SLS-3)
- QUALITY CONTROL DIV. (SLS-4) — L. R. Neibel

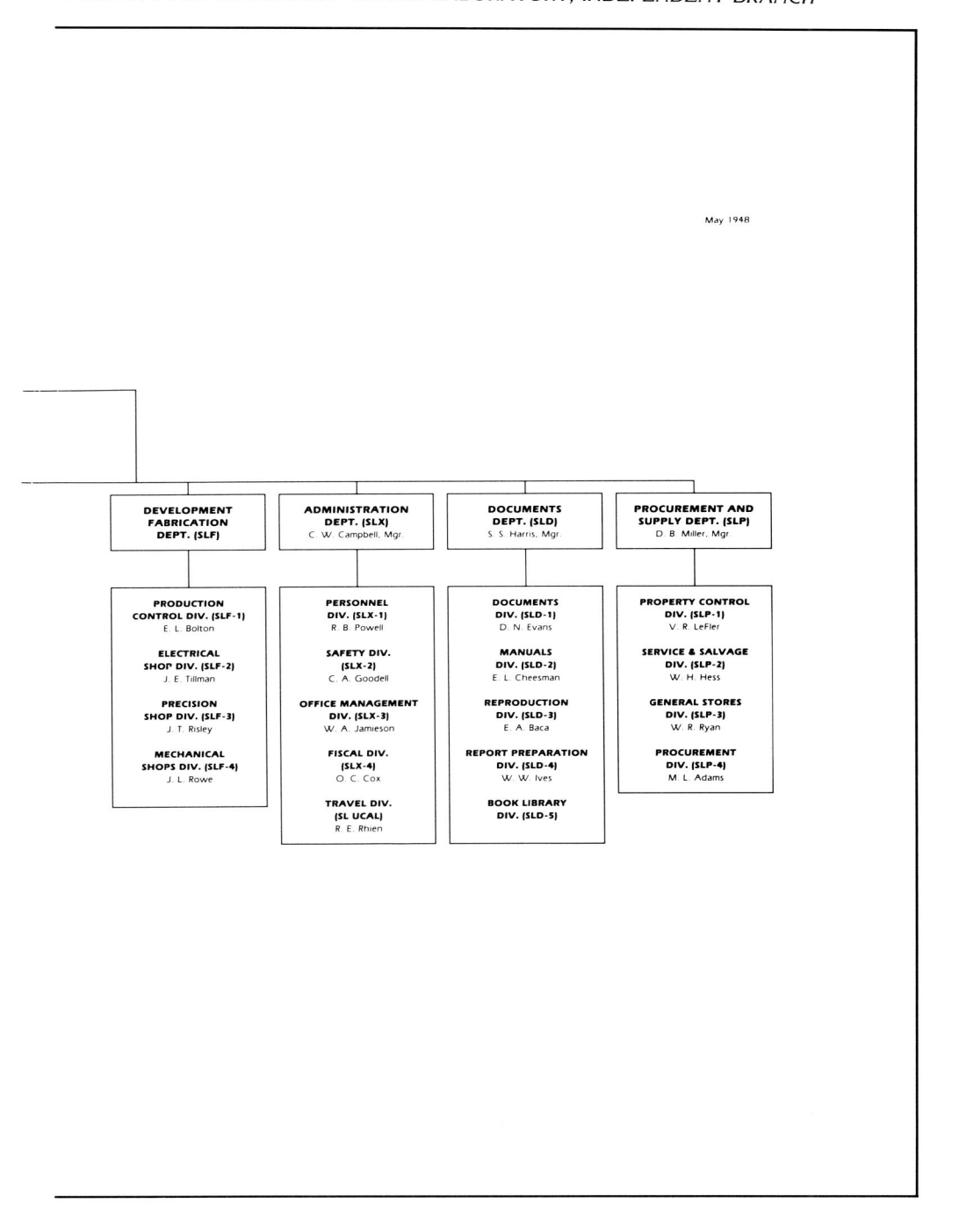

MEETING THE CHALLENGE: SANDIA LABORATORY, INDEPENDENT BRANCH

May 1948

DEVELOPMENT FABRICATION DEPT. (SLF)	ADMINISTRATION DEPT. (SLX) C. W. Campbell, Mgr.	DOCUMENTS DEPT. (SLD) S. S. Harris, Mgr.	PROCUREMENT AND SUPPLY DEPT. (SLP) D. B. Miller, Mgr.
PRODUCTION CONTROL DIV. (SLF-1) E. L. Bolton	PERSONNEL DIV. (SLX-1) R. B. Powell	DOCUMENTS DIV. (SLD-1) D. N. Evans	PROPERTY CONTROL DIV. (SLP-1) V. R. LeFler
ELECTRICAL SHOP DIV. (SLF-2) J. E. Tillman	SAFETY DIV. (SLX-2) C. A. Goodell	MANUALS DIV. (SLD-2) E. L. Cheesman	SERVICE & SALVAGE DIV. (SLP-2) W. H. Hess
PRECISION SHOP DIV. (SLF-3) J. T. Risley	OFFICE MANAGEMENT DIV. (SLX-3) W. A. Jamieson	REPRODUCTION DIV. (SLD-3) E. A. Baca	GENERAL STORES DIV. (SLP-3) W. R. Ryan
MECHANICAL SHOPS DIV. (SLF-4) J. L. Rowe	FISCAL DIV. (SLX-4) O. C. Cox	REPORT PREPARATION DIV. (SLD-4) W. W. Ives	PROCUREMENT DIV. (SLP-4) M. L. Adams
	TRAVEL DIV. (SL UCAL) R. E. Rhien	BOOK LIBRARY DIV. (SLD-5)	

With the test operation planned for April–May 1948, Henderson made it quite clear to weapons development personnel that the Sandia Laboratory had "a tremendous responsibility" to have the bomb ready. "Either the Mk 4 is ready by that time," he said, "or all work must cease, and a new project along different lines begun."[6] As Larsen soon learned, the work environment at Sandia Base was one of crash programs and crisis management.

By April 1, 1948, in an attempt to restructure the Laboratory more efficiently, Z Division was officially declared a separate branch of Los Alamos Scientific Laboratory. This action was supported by University of California personnel who were beginning to have increasing qualms about involvement of an academic institution in ordnance activities during peacetime.

Five days later, Larsen reorganized the Laboratory into nine departments. Three support groups—Training Liaison (SLM), Secretarial Training and Coordination (SLC), and Architectural Liaison (no symbol) were attached as staff to the Director, with no supervisors named. As Bradbury explained to University of California President R. G. Sproul: "Responsibilities of Z Division have grown to such an extent that divisional status became completely inappropriate."[7]

Advances on the Administrative Front

Energetic and convincing, Larsen launched a series of successful campaigns that included a major building program and increases in personnel and funding. He was assisted in these efforts by the establishment of a full-fledged Administrative Department (SLX) under Charles W. Campbell.

Campbell, who spent the war years in the Navy, received a telegram from A. W. "Cy" Betts in 1946 asking him to join the Laboratory at Los Alamos, New Mexico. Campbell wrote and thanked him, saying that he had just completed two and one-half years of "foreign duty" and right then didn't want anymore. A year later Betts called again and offered Campbell a job in

Albuquerque. This time he accepted, and with the April 1948 reorganization set up a Personnel Division under Ray Powell, a Safety Division under C. A. Goddell, Office Management under W. A. Jamieson, Fiscal under Otis Cox, and Travel under R. E. Rhien.[8]

As head of Personnel, Powell spearheaded a recruitment campaign that brought in new employees at the rate of twenty-five per week. With the assistance of the Washington AEC office, which placed job queries across the country, he was successful in locating the talent necessary to get the job done. As a result, the personnel makeup of the new branch lab began to represent a cross section of industry.

There was considerable internal movement of personnel as well. Managers proselytized shamelessly to obtain the manpower they needed. Training programs had to be accelerated and efforts made to reduce the thirty to sixty days necessary to obtain security clearances.

Establishment of the Office of Business Manager at the Sandia Laboratory was a step in the right direction. Management took this action because the payroll had grown rapidly from a total of approximately $96,000 in December 1947 to $176,000 in June 1948, with plans for further expansion to $300,000. The rapid growth of the payroll made it inadvisable to continue operating out of the Los Alamos Business Office. The new arrangement saved time because it meant that there was no traffic of paperwork to the Los Alamos Accounting Office by way of Los Alamos. A. E. Dyhre, Project Business Manager at Los Alamos, observed that operations of the Sandia Business Office were based upon "a literal interpretation of the payroll, personnel, and travel policies; approvals were not relaxed." He added in what was undoubtedly an understatement: "Some verbal comment was received from certain Laboratory personnel during this period relative to the apparent inflexibility of the office."[9]

In April 1948, the first real purchasing activity at the Sandia Base facility got under way, although the initial effort was small and selective. Previously, under the University of California, purchasing was handled by Z-10 (Supply), which screened requisitions for completeness, and then forwarded them to either Los Alamos or the University of California purchasing office for performance of buy. The Laboratory's first buyer, Henry Moeding, had been hired in 1946. Items purchased directly by the Lab, and largely from Albuquerque sources, were mostly bulky heavy items such as printing, plumbing supplies, and lumber or cement.

The logistics and particulars of the transfer of administrative functions to the Branch Lab did not take place without considerable debate with Los Alamos. It was regarded as "a signal accomplishment," for example, when the Sandia Laboratory was allowed to have its own letterhead stationary.[10]

Other negotiations focussed on whether the Branch Lab should have its own newspaper. Sylvan Harris, Manager of the Document Department (SLD), led the campaign. He first broached the subject with L. G. Hawkins, the University of California's Business Manager for the Sandia Laboratory. Hawkins seemed to think that a publication similar to the Los Alamos *Bulletin* should be more than acceptable to personnel, but Harris had higher expectations. "In my opinion," he said,

> this "Bulletin" contains nothing more than notices such as will doubtless be fastened by thumb tacks to bulletin boards in the Tech Building when completed. This sort of thing, after once looked at, finds its way quickly into the waste basket, exactly as do the copies of the daily Army Bulletin that cross my desk everyday.
>
> The idea of the "Sandia Laboratory News" goes far beyond this.[11]

Harris laid out his plans for the proposed publication, which he felt would have an important influence on the morale of the Laboratory. Both Paul Larsen and the Personnel Department, he reported, were "highly" in favor of it. Harris estimated the cost of the paper to be approximately seventy-five dollars per month "or less" for 1,200 copies.[12]

Hawkins questioned whether the University of California would approve expenditures for other than "research and development" items. Harris responded by pointing out that administration, housing, security, etc., were not "research and development"; yet, they were also important. "Building and upholding the morale of the employees is just as essential," he maintained.[13]

Hawkins took the problem to Business Manager A. E. Dyhre. Hawkins admitted that he had tried to discourage the idea, but, he said, "As you can tell . . . I was not very successful." He agreed with Harris that a paper "could very well have a positive effect on the morale, and . . . loyalty of the employees at large." He added that "the sudden and startling growth of the Sandia Laboratory has resulted in the throwing together of a large group of people who do not know each other or the Lab's connections with Los Alamos and the University of California. . . . Therefore," he said, "the more I think of it, the more I like the idea of a good, informational publication coming out each month about the Laboratory at large."[14]

By mid-November 1948, the Branch Laboratory had its own *Sandia Lab Weekly Bulletin*. It did not meet Sylvan Harris' expectations for a full-fledged newspaper, but it was a good beginning. Published each Friday by the Employee Services Section and edited by Jackie Downing and Elizabeth Wallick, its gossipy style introduced employees to each other, posted ads, and welcomed in new transfers.[15]

During this same period, employees, with management support, made an organized effort to establish what would eventually be known as the Coronado Club. At the time, the Branch Lab was an isolated facility located some six miles from Albuquerque. Many employees of the Sandia Lab and the AEC lived on base, so there was definite need for a social and recreational center.

In early 1949, a three-man task force composed of Les Rowe, Virgil Harris, and Hal Gunn studied the basic problem and recommended a larger committee effort. AEC Field Manager George Kraker and Sandia Laboratory Director Paul Larsen added another nine members to the original task group to make a committee comprising six Laboratory and six AEC representatives. Construction on the building itself, originally the AEC Community Center, began in early 1949.

Within a year's time, the Community Club Committee and the Sandia Laboratory/AEC management agreed that the Lab would operate the dining room and that the remaining facilities would be operated as a social club by a separate organization known as the Coronado Club. The Club's first Board of Governors was appointed, and in May 1950, the Coronado Club of Sandia Base, Inc., became a reality. Officers Harold Gunn, President; Jack Hansen, Vice-President; and Ted Sherwin, Secretary, carried out the details of incorporation. Shortly thereafter, Gunn recalled: "We held a membership drive, incorporated the Club, wrote proposed by-laws, house rules, organized the bingo parties, started bowling leagues—even purchased furniture and kitchen equipment and supplies, all before the Club actually opened."[16]

The original club, which cost the AEC an estimated $442,000, was finished in late spring of 1950. Located at 3210 West Sandia Drive, the facility included a single swimming pool and a small patio. The basement held a four-lane bowling alley, game rooms for Ping Pong and pool, and a small party room. A second pool would be added in 1956, as well as other additions and improvements through the years. Approximately 2,500 people attended the dinner dance celebrating the Club's grand opening on June 9, 1950.

Ed Brawley of the AEC was the Club's first elected President, and he was succeeded in 1951 by Dave Tarbox. By the end of the year, the Club had 1,400 members and was still growing, justifying the addition of the La Caña room. Charles Campbell became President in 1952; Al Gruer in 1953; and Charles O'Keefe in 1954. Under O'Keefe's tenure, the Board improved recreation programs; modified accounting systems; and began construction on a new swimming pool.

Above: Construction of the Coronado Club, the center for social activity on the base, was completed in late spring of 1950. Below: Board members of the Coronado Club meet to discuss Club business in this 1954 photo.

Bowling was a popular activity for many early Coronado Club members. Shown here are the bowling alleys located in the basement of the Coronado Club.

The banquet staff prepares to serve Club members buffet style.

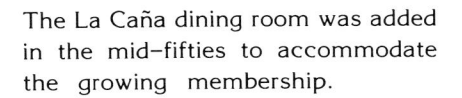

The La Caña dining room was added in the mid-fifties to accommodate the growing membership.

During these early years, the Club experienced a number of controversies and crises, ranging from opposition to liquor sales on Sundays to misunderstandings over the use of Base Military Police for protection at the Club. Because the land that the Coronado Club sits on is an island within Sandia Base belonging to the AEC, Base MPs had no legal law enforcement status on Club grounds. Only the U.S. Marshal could exercise jurisdiction on the premises. However, Sandia did have an agreement with the Base Provost Marshal to allow the Club to call on the Military Police when needed. Such a need, in 1958, led to one of the biggest, and in retrospect somewhat humorous, disturbances ever to occur.

That summer, as in previous years, the Coronado Club held its annual Beachcomber's Ball. Traditionally, the Beachcomber's Ball was a very gay "summer New Year's Eve" party at which many of the members got drunk and went swimming. In 1958, however, the Manager decreed that swimming would not be allowed and made several announcements to that effect. As the evening progressed, many members decided that this was an unreasonable request and went in swimming anyway. The Manager panicked and called the MPs. Angry Club members in and around the pool retaliated by attacking the police, one of whom ended up taking an unplanned swim.

Such shenanigans led to a massive investigation, a visit by a New Mexico Congressional delegation, and extensive reports to the AEC. Several people were arrested by the MPs, tried by the U.S. Marshal, and given light fines for drunkenness. Needless to say, there have been no more Beachcomber's Balls. Over the years, however, the Coronado Club has weathered these and other crises to remain a much appreciated employee benefit.[17]

In the realm of employee morale and support services, it was recognized that a banking institution was also needed to assist personnel in meeting financial obligations. With this in mind, fifteen Sandia employees applied to the Federal Government for a charter to start a credit union. On October 18, 1948, the government formally granted the charter, and two days later, at a meeting convened by a Federal Credit Union Examiner, Virgil

The Sandia Laboratory Federal Credit Union, established in 1948, was first housed in a temporary building.

Harris was elected President. J. Leslie Rowe became Vice-President, and Adrian J. Verburg, Secretary. During 1949-1951, employee members performed all bookkeeping and routine clerical work. The Union's original assets of $510 in two years' time increased to $25,000. In 1955, the Sandia Federal Credit Union passed its first million-dollar mark.

Under Larsen's direction, the Branch Lab continued taking steps toward organizational independence. On July 1, 1948, Sandia Laboratory, as it was more frequently referred to, began to assemble cost records and reports to serve as budget and planning guides.[18]

The reorganization meant changes in operating style for both the administrative and technical sides of the house. Whereas each department had maintained its own support services such as secretarial, drafting, and purchasing, technical departments henceforth drew upon centralized service units.

"A Rube Goldberg Affair"

In the technical area, Larsen faced challenges of long standing. Roger Warner, before his departure, summarized part of the problem graphically:

> As we reach the end of 1561 stockpiling for mechanical and electrical components, we are finding more and more subassemblies which are marginal in functional reproducibility or reliability. In certain cases, where we have taken steps to bolster one weakness, another develops, and so on *ad nauseum* until we reach the stage where by physical handling and reworking, the component is worn out even before it reaches the stockpile.[19]

When Larsen arrived in December 1947, the 1561 Mk 3, essentially the Fat Man model used at Nagasaki, was the only type of nuclear weapon in the U.S. stockpile. Described as "a Rube Goldberg affair that took an assembly team of scientific experts a week's worth of effort to assemble," the Mk 3 was a far cry from a production-type bomb.[20]

The assembly process for this bomb involved placing individual high-explosive lenses on an aluminum polar cap to form about a half sphere. Next, assembly personnel placed a cored metal sphere in the vacant center of the H.E. "hemisphere." Assembly of the explosive segments continued, culminating in the insertion of a "trapdoor" set of segments at the top. During the procedure, the H.E. sphere was held together by pieces of an aluminum belly band loosely joined with assembly bolts, which were replaced upon final tightening with structurally strong bolts.

The resultant configuration, referred to as a nuclear subassembly, then had to be rotated ninety degrees so that the trapdoor could be directed forward in the completed bomb. Sheet aluminum mounting cones were bolted onto the polar caps to allow mounting of the electrical and mechanical components of the arming, fuzing, and firing subsystem. As a final step, assembly personnel bolted the tail assembly onto the rear ballistic case ellipsoid. Initially, there was no special handling equipment, which meant that personnel became very inventive at devising imaginative improvisations.[21]

The power supply for this first stockpiled weapon was a primitive one utilizing the NT-6 lead-acid battery, a minor modification of the standard automobile battery manufactured by the Willard Storage Battery Company. Several batteries connected by pigtail cable leads made up the bomb's power supply. Since the batteries were somewhat marginal in capacity, the main capacitor bank was charged using aircraft power prior to release of the bomb from the aircraft. Hence the batteries only needed to power the fuze and to replace the power lost via the main capacitor bank bleeder resistor. These lead-acid batteries had numerous drawbacks. They had to be cycled through three sixteen-hour charging and discharge periods; therefore, it took three days' preparation time to get a weapon ready for drop. If the bomb was not used immediately, on the third and sixth days, the batteries had to be recharged, and on the seventh they had to be replaced, which entailed partial disassembly of the weapon.

As military crews were trained and the bomb became more and more a field item rather than a laboratory "gadget," the need for a suitable power supply became urgent. Subsequently, Sandia placed a contract with the Willard Company for development of a more advanced lead-acid battery. The result was the improved Willard ER 12-10 battery, which consisted of five two-volt cells assembled with a molded plastic case. These batteries, which charged for twenty hours, had a usable shelf life of about three weeks. If a five-hour booster charge was used, the life of the batteries could be extended an additional two weeks. But the ER 12-10 still had limitations.

The use of sulfuric acid as the electrolyte, for example, resulted in some acid spillage during preparation. Environmental control was also a problem. Heaters were required to protect the batteries at temperatures below minus four degrees Fahrenheit. They could not withstand temperatures above 130 degrees Fahrenheit; nor did they stand up well under shock and vibration environments. Despite these deficiencies, the ER 12-10 batteries were used until 1953, when more advanced power supplies became available.[22]

As Manley had recommended in his evaluation, and as Bradbury dictated, the electronic problems of the Lab had to be solved. The Laboratory, therefore, became increasingly involved in the development of fuzing and firing technology.

A priority item for the Sandia Lab was the development of an improved X-unit—the electrical device used to fire the weapon's detonators. The purpose of the X-unit was to store and release electrical energy. It provided the required high-voltage pulse to the detonators with a high degree of precision, but it was a slow charging device. Power could be supplied either by the batteries and inverter in the bomb, or by auxiliary power from the aircraft. As the weapon fell toward the target, the fuzing system supplied first arming and then firing signals to the X-unit.

The first X-unit had been developed in a very short period of time by Los Alamos personnel in conjunction with Raytheon Manufacturing Company of Boston, Massachusetts. It was used successfully over Nagasaki, Japan, on August 9, 1945. But there was a problem. Since its bank of electrical capacitors was charged to several thousand volts while the bomb was still inside the delivery aircraft, the situation created a hazard to the aircraft crew and also made it necessary to pressurize the X-unit to prevent failure by arcing at high altitudes.

Consequently, Laboratory personnel worked to develop an improved rapid-charging X-unit that could be charged during drop. The new X-unit featured a large inductor that responded to twenty-eight volts provided by the new type of lead-acid storage battery. The new X-unit and battery, along with improved weapon detonators developed by Los Alamos, became a package available for production by late 1948. Bombs produced with these improvements carried the nomenclature Mod 1.

Among the components under study were spring-wound clock-timers that served to isolate the radars electrically from the X-unit for the first fifteen seconds after the bomb was dropped. The clocks were started by the withdrawal of arming wires attached to the aircraft, pulled as the bomb dropped away. After fifteen seconds—time enough to minimize possible jamming of the radar fuzes—a switch would close, completing triggering circuitry to the X-unit.

Barometric switches were used as additional safety devices to keep the radars from transmitting until just before reaching the desired burst height. The barometric pressure outside the bomb was fed from inlet ports in the bomb case through tubing to the baroswitches. At a preset altitude, the baroswitches would close electrical contacts and start the Archie radars transmitting. This design allowed only a short time for possible jamming of the radars by enemy countermeasures. The probability of jamming, however, would continue to be a matter of serious concern to Sandia Lab engineers and the focus of a major research effort.[23]

Component modifications were also directed at reducing weight as well as improving aerodynamic characteristics. The weight of the Mk 3 X-unit, for example, was reduced to 335 pounds from 700 pounds, and the ellipsoidal configuration of the weapon was changed to a blunt-nosed, somewhat stubby shape in the advanced design, the Mk 4. The new design of the case simplified final assembly and improved ballistics as well.

Production of the Mk 4 for stockpile, initiated by Henderson's weapons group at Los Alamos shortly after V–J Day, superseded demands for older models and incorporated the improved fuzing and firing components developed for the Mk 3, Mod 1. These components were arranged in a single cartridge for ease of assembly and testing, and included the ER 12-10 batteries.[24] The firing set for the Mk 4 Mod 0 was engineered by Leon Smith, Philip Barnes, Joe Dawson, and the firm of Edgerton, Germeshausen, and Grier (EG&G).[25]

Among other improvements, a new fuzing-antenna system for the Archie radars, mounted flush with the envelope of the bomb, was designed for the Mk 4 to replace the exterior Yagi antenna used on the earlier weapons. The Yagi antenna had been susceptible to breakage and added to the aerodynamic drag. Another redesign on the radars, designated AR-10A, allowed for selection of desired burst height and improved reliability. Despite the advances being made in solving problems of the early Mks, however, there was much to be accomplished.[26]

Sandia Laboratory Adds an Applied Physics Department

To help the Lab solve its immediate problems and to improve the design and effectiveness of atomic weapons in the long term, the decision was made to establish an Applied Physics organization at the time of reorganization. The Applied Physics Department (SLA), formally constituted at the same level as the Engineering Department (SLE) under Richard A. Bice, added a research emphasis to the engineering, production, stockpiling, surveillance, and training aspects of the Laboratory's mission. As head of the new group, Harold S. G. Cooper was recruited from the Applied Physics Laboratory of Johns Hopkins. Cooper, however, was soon succeeded by R. P. Petersen.[27]

The Applied Physics Department established headquarters in a complex of buildings that originally housed the Sandia School for Girls. Located just northeast of the Carlisle entrance to Kirtland Air Force Base, the buildings during World War II served as a hospital for convalescing pilots and crews. In 1948–1949, however, the facilities were being used for proximity fuze development by the New Mexico School of Mines in Socorro, which had a contract to conduct ordnance tests in an area near Coyote Canyon several miles south of the cluster of buildings that constituted Sandia Base. Future Sandians Louis Paddison, Lawrence Neibel, and George Hansche (who would later join Sandia as the Laboratory's first Ph.D.), all worked for the School of Mines at this time. When the Applied Physics Department took over the girls' school in early 1949, the complex was renamed Sandia West Lab.

As part of its assignment, Applied Physics personnel were to "establish contact and liaison with all possible sources of basic scientific research data . . . both domestic and foreign." They were to evaluate the data for possible application to problems of the Laboratory.[28]

The newly established Applied Physics group did not lack for problems to solve. Soon after reorganization, Headquarters, AFSWP requested that Physics personnel conduct a feasibility study to provide in–flight adjustment of the baro operation on the Mk 4. The objective was to make possible the assignment of primary and alternate targets at substantially different altitudes. Second, they wanted an inflight adjustment for height–of–burst in the fuzing system being developed. Such an adjustment would allow the military to vary blast intensity and area to suit the particular target attacked. They hoped to be able to use this type of setting on the Abee radar under development. Still another technical goal was to accomplish nuclear disarming of the Mk 4 bomb in flight.

As an initial test, Laboratory personnel conducted a demonstration for AFSWP personnel on an inert Mk 4 bomb inside a grounded B-29. Without the benefit of special handling tools, three technicians worked through the pressure bulkhead of the forward end of the bomb bay to open the bomb. In fifteen minutes, they had withdrawn the capsule. And, in an additional fifteen minutes, they had reinserted the capsule and replaced all the components. The demonstration showed that the nuclear capsule could be removed in flight as an emergency measure, and the bomb case jettisoned after the capsule had been removed.

In the process, however, the technicians identified certain problems. Although there was sufficient space for clearance of components, there was too little space for any extensive amount of handling equipment. AFSWP planned to request that the problem be investigated fully.[29]

Assignments such as these meant that research personnel were soon involved in long-range studies and research in the fields of electronics, telemetering, metallurgy, chemistry, aerodynamics, ballistics, mechanics, heat, low temperatures, sound, and physical effects of nuclear explosions. The procedure followed was for Applied Physics personnel to take ideas that needed to be proven, then work them into an experimental configuration and weapon development mode. The components, equipment, or devices had to be proven functional and tested. From these efforts, a formalized environmental testing activity and component development program would gradually evolve. The testing activities, especially, required additional space for test and handling equipment. Establishment of the Applied Physics Department, therefore, contributed to the Laboratory's expansion.

As a final step, the working models were transferred to the Engineering Department for application development. Engineer G. C. "Corry" McDonald, who joined the Laboratory on the same day as the new director, observed that "Applied Physics worked well because it tended to give our engineering activities a solid basis for proceeding."[30]

Organizations for Interaction

Actual development of a specific nuclear weapon system began only after the military had requested through AEC channels that such a weapon be added to the stockpile and the research organization had determined its feasibility. Consequently, the AEC, the Laboratory, and the military all recognized that an effective working relationship between the Laboratory and AFSWP personnel was essential. The Sandia Research and Development Planning Board, proposed by Larsen in December 1947, and approved by Bradbury the following month, provided one such forum for interaction.[31] Officially, the Board reported to Bradbury, who planned to incorporate the Board's recommendations in his submission on weapons components to the AEC. Bradbury advised the AEC, however, that the Laboratory would not be bound by "any majority decision of the Board" and questioned how Board and Laboratory recommendations would be received in Washington. Despite apparent reservations regarding the Board's effectiveness, Bradbury urged Larsen to appoint "his most senior personnel" as members.[32]

On March 2, 1948, the Research and Development Planning Board met for the first time in the conference room at Post headquarters on Sandia Base. Larsen, Richard Bice, and Arthur Machen, who was standing in for H. C. Cooper, represented Sandia; and Colonel William M. Canterbury, Commander J. T. Hooper, and Major A. W. Carney represented the Armed Forces Special Weapons Project. Bice was chosen as chairman for the first six months period; Bob Henderson was to serve as alternate for Larsen when necessary.

At this first meeting, a primary issue to be resolved concerned the scope of the Board. Some members felt that Board responsibility should go so far as to establish concrete requirements for numbers of pieces of equipment and parts for use in the assembly and storage of atomic weapons. But the majority view seemed to be that the Board should function in an advisory capacity. Larsen recommended that the Board first consider problems unquestionably within its purview and postpone decisions on other items until they had surveyed members of their respective organizations. Of immediate concern to all members were the technical problems being encountered in Production.

Larsen took the lead in establishing the Research and Development Planning Board as a forum for communication on the different aspects of atomic weapons development. He got the group off to a good start by initiating discussion on the progress of current weapon projects of mutual interest. He announced that he expected drops and laboratory tests on the Mk 3 X-unit to be completed by May. And if use of the unit was approved, he predicted that it would be ready for stockpile with newer bomb models starting September 1, 1948. Drop tests on the Mk 4 package were turning out satisfactorily.

Larsen reported also that the Laboratory expected initial engineering models of the Mk 4 X-unit to arrive the following month from RCA. Lab personnel planned to test these models, and if the tests were satisfactory, contracts would be placed for development units. Production models would be ready on January 1, 1949, if the design could be settled on by October. In essence, the Lab was pressing ahead, although there had been a delay in the delivery of "clocks" contracted with Bendix in Kansas City, Missouri.[33]

On March 12, 1948, the name of the group was shortened to the Sandia Research and Development Board. The committee would undergo two more name changes, first, to become the Sandia Weapons Development Board in May 1950, with Robert Poole as chairman, and by May of the following year, to the Special Weapons Development Board (SWDB). There was some question as to whether the Board should be continued, but Chairman of the MLC Robert LeBaron went on record in support of what he considered to be "its very important function."[34]

Whereas SWDB represented Laboratory interests, a TX-5 Steering Committee presented the design interests of Los Alamos (as well as the design interests of the Sandia Laboratory) and provided further coordination on development of a weapon to

NUCLEAR ORDNANCE ENGINEER FOR THE NATION

The Special Weapons Development Board was organized in March 1948 to evaluate and make recommendations to the AEC concerning nuclear weapons development programs. The Board reached an important milestone on April 25, 1956, when members—past and present—gathered for their 100th meeting. Members on hand for the occasion included—top row, left to right: First Lieutenant A. K. Roesener; Max K. Linn; Major L. D. Walker; L. M. Jercinovic; First Lieutenant E. T. Naden, Jr.; H. M. Willis, Jr.; James W. McRae; L. J. Paddison; Hoyt Westcott; Colonel G. C. Darby; Rear Admiral F. O'Beirne. Middle row: Brigadier General W. M. Canterbury; Captain D. B. Young; Commander D. K. Ela; R. E. Schreiber; K. W. Erickson; M. J. Kelly; Captain R. S. Mandelkorn; K. F. Hertford; A. B. Machen; S. C. Hight; A. E. Uehlinger. Front row: Rear Admiral L. R. Daspit; L. A. Hopkins, Jr.; Brigadier General L. K. Tarrant; H. M. Agnew; R. W. Henderson; R. E. Poole; M. D. Martin; Colonel P. J. Long; Colonel G. B. Webster, Jr.; Commander P. G. Schulz.

follow the Mk 4. On October 31, 1947, the MLC had expressed interest to the AEC in a bomb both smaller and lighter than the Mk 3 and Mk 4. On December 10, 1947, the AEC replied, pledging a vigorous program to develop a bomb more flexible in operational utility. The Mk 5 program, activated at Los Alamos, received ordnance support from the Sandia Lab. A joint coordinating committee was formed to facilitate interaction on the project; and on November 3, 1948, the TX-5 Committee, which logically took its name from the weapon under development, held its first meeting.[35]

During this period, the Sandia and Los Alamos Laboratories engaged in some technological territoriality, as might be expected from institutions experiencing the throes of organizational change. Basically, however, Los Alamos continued to concentrate on the development of the explosive systems to meet military requirements, while the Sandia Lab "packaged the bang."

In addition to design and development of the casing, or "package," for the bomb, the internal arming, fuzing, and firing components remained a primary responsibility of the Sandia Laboratory. Because of this, and the need to integrate the technology of both Laboratories, the opportunity to interface through committees such as the TX-5 was essential. In the 1948-1949 time period, this committee met once a month to coordinate development of the Mk 5 and to discuss manufacturing problems in the developing integrated contractor system. Later simply referred to as the "TX Committee," the group continues to provide interface between the two Labs to the present day.[36]

Other groups organized to facilitate interaction between the military and the Sandia Laboratory included the Road Materiel Committee. This committee, composed of AFSWP and Sandia personnel, developed catalogs of weapon components and other equipment required for surveillance and training; recommended quantities of handling, assembly, and test equipment; established packaging procedures for components; and set up a technical program for carrying out surveillance. Board members presented status reports on the various weapons and made efforts to coordinate activities between the military and the AEC.

A Technical Facilities Committee dealt with problems associated in handling the weapon; a Manual Committee coordinated the preparation of manuals related to atomic weapons; and a Sandia Base Functions Committee addressed problems pertaining to utilities, medical services, and recreational areas. Max Roy served as the Los Alamos representative on these boards, but participated only as a nonvoting member.[37] Establishment of groups such as these improved relations among the Labs, the AEC, and the military and helped the Sandia Branch Lab meet the higher production schedules being levied.

Weapons Development, A Project Group Orientation

Sandia Laboratory personnel had their work cut out for them. The new stockpile goals and production schedules often meant twelve- to sixteen-hour days. Engineering and development work on the Mk 3 had to be carried on concurrently with work on the Mk 4 implosion bomb, the gun-assembled Little Boy, and eventually, the Mk 5.

The project group orientation of the Engineering Department proved to be effective in managing the increasingly heavy workload. The work on the Mk 3 was separated organizationally as SLE-2 under W. O. McCord, who hailed from General Electric; while J. A. Hoffman (SLE-3) handled the Mk 4, and Harlow Russ (SLE-4) worked on the Little Boy. These divisions as well as Technical Facilities (SLE-5) under C. H. DeSelm; Test Equipment (SLE-6) under J. W. Jones; Technical Accounting (SLE-7) under R. H. Collins; and Design Services (SLE-1) under G. F. Heckman in 1948 made up the Branch Lab's Engineering Department.

Manager of the Department, Richard A. Bice, had joined Los Alamos Scientific Laboratory in 1945 as Alternate Group Leader, at which time he was active in the design of explosives and testing techniques for the first atomic bomb. He moved to Albuquerque as a member of Z Division in 1946. By 1948, he was serving as Assistant to Associate Director, R. W. Henderson, in addition to managing the Engineering Department. Bice remembers this period as being characterized "by drive and energy, and an overriding and intense awareness of the job to be done."[38]

To meet these goals, early design engineers became involved in technological pioneering. Since many commercial parts were either unavailable or unsuitable, Laboratory engineers often designed their own. As the first step in actual development, the systems engineers would lay out the block design of the weapon system, with each block representing a particular function. They would then investigate each link in the system to determine what device or combination of devices could perform this function. If the desired device did not exist, as was often the case, the project group leader would ask the component development engineers to design one to meet desired specifications.

These Sandia Laboratory–designed components carried the label MC (Major Component). For the first decade or so, the weapon development project group retained responsibility for development of the major components of the weapon's arming and firing subsystems but relied upon a separate organization for development of major components of the weapon's fuzing system. This split in responsibility eventually led to the establishment of a separate discipline known as Component Development.[39]

In the design of the weapon system, simplicity, ruggedness, and reliability received prime consideration. It was essential then, as now, that a weapon fire at the proper time and place. An unreliable system might cause the weapon to detonate in the delivery aircraft before the target was reached, or result in a dud. In an actual military operation, second chances to bomb a target might be few, and the failure could be costly, if not catastrophic. Therefore, the design and manufacture of weapon components had to meet tighter specifications and closer tolerances than are normally required in commercial products.

If, for example, an electronic timing device was to be incorporated in the arming circuit of a new weapon system, the component development engineer would design the timer, then order a prototype model of it fabricated in the Laboratory's model shops. To make sure that the model would perform according to specifications under the most adverse conditions, the device would undergo numerous tests, and perhaps full-scale tests, before it could be considered acceptable. When testing was completed, the device would be sent to a selected manufacturer for production. Throughout the process, the project leader held "cradle to grave" responsibility for the device, a responsibility that is uncommon in U.S. industrial, governmental, or academic fields.[40]

The success of early component development programs depended to a large degree upon the competence of the project leader. Lee Schulz, who worked on the Mk 3 program, characterized a good project engineer as an individual with a certain "method of operation." "These people welcomed challenges," he said:

> They wanted to see a job done right. They had persistence. They usually did not take 'no' for an answer . . . ; they tried to figure out a way to get the job done even if someone posed an obstacle in front of them. They were not always highly talented or educated, but they had to be able to pull it all together into a successful end result.[41]

In the beginning, the project leader personally assumed many of the responsibilities associated with design and development. Such direct interaction between the engineer and technician had still another positive aspect—the technician felt very much a part of a team effort. "This was good for morale and high motivation," Schulz recalled.[42]

On occasion, it was necessary to search around the country for certain items. Leon Smith and Philip Barnes, for example, made a round-robin trip to various manufacturers looking for an improved capacitor for the Mk 4 fuze, only to conclude that the one they were using was the best available.

The timer for the Mk 4 would continue to be a trial to Sandia engineers. In the intervening months between the April 1 reorganization and the summer of 1948, work on this and other technological problems would be slowed due to the demands of another full-scale nuclear weapons test. Progress reports provided by the Sandia Laboratory made the situation clear: "Project Sandstone," the report read " was proceeding under heavy pressure" and made the "entire Mk 4 program unrealistic as to its established dates."[43]

X-Ray Shot, Operation Sandstone, April 14, 1948.

Chapter X.

SANDSTONE:

A New Realm in Weapons Development

"Nowhere did the anticipation and achievements of Sandstone have greater effect than in weapons activities at Sandia."

AEC Historians, Richard G. Hewlett and Francis Duncan

<u>The Atomic Shield</u>

The world's third nuclear test series was quietly announced in an innocuous sentence hidden away in the Commission's 1947 semiannual report to Congress. "The Atomic Energy Commission is establishing a proving grounds in the Pacific for routine experiments and tests of atomic weapons," the report read.[1] Unlike Crossroads, the Sandstone test series would be held with little fanfare and much secrecy.

Following the technical recommendations of Los Alamos scientists, AEC Chairman Lilienthal officially proposed the test series in a letter to the MLC in the latter part of April 1947. There was agreement on the need to test the new models; yet there was concern that there might be international repercussions. Important negotiations, both in the Security Council of the United Nations and in the Four-Power Foreign Minister's Conference, were in progress in London. Public release of information that the United States planned to conduct firing tests of new atomic weapons could have deleterious effects.[2]

Despite the international situation, on June 27, 1947, in a meeting at the White House, President Truman, members of the Joint Chiefs of Staff, Secretary of State George Marshall, and Lilienthal made the decision to proceed with the operation. On August 11, 1947, in support of the Joint Task Force, the Los Alamos Scientific Laboratory, including Z Division, agreed to accept direction of the technical phases of the operation.[3]

"Sandstone," also called "Operation Milestone," was to follow the pattern successfully established by Crossroads. Military personnel of all three services and civilian scientific personnel would be participating in a large-scale scientific expedition at a considerable distance from the United States. However, the Sandstone series marked a departure from Crossroads in that coordination involved the joint efforts of two major agencies of government: the National Military Establishment and the Civilian Atomic Energy Commission. The two operations also differed in that Crossroads had primarily provided nuclear weapons effects on ships, using atomic bombs of the so-called standard Nagasaki design, whereas the objective of the Sandstone series was to prooftest the effectiveness of new and previously untested experimental designs. Like Crossroads, however, the Sandstone operation was to be a massive undertaking requiring approximately 10,366 people and hundreds of ships, motor vehicles, and aircraft.[4]

Enewetak Selected as Test Site

The search for a test site began in September 1947, although the number of shots and the design of the devices had been decided upon several months earlier in meetings held at Los Alamos between members of the Scientific Lab and Sandia. Both Lilienthal and General Dwight Eisenhower believed that a remote location in the Pacific was preferable to a site in the continental United States. One of the specifications was that the location selected would be available as a permanent proving ground.

SANDSTONE: A NEW REALM IN WEAPONS DEVELOPMENT

Preliminary investigations convinced Lilienthal that the only site that would permit full realization of the objectives of the test series was an atoll comprising forty small islets in the western Pacific.[5]

During the month of October, JTF Scientific Director Darol K. Froman and Deputy Scientific Director Alvin Graves of Los Alamos, accompanied by military personnel, undertook a reconnaissance mission to the proposed forward areas. After a careful study, on October 11, 1947, they decided that Enewetak, the major island of the atoll, had more advantages as the test site than Kwajalein.

A typical atoll, Enewetak is a coral cap set on truncated volcanic peaks that rise to considerable heights from the ocean floor. There, a string of narrow islands, composed largely of coral and sand, surround a sheltered lagoon. Remote and isolated, Enewetak proved to be large enough to accommodate the three shots planned. The reconnaissance team also determined that the steady trade winds would carry fallout from the shots away from the island over the open ocean to the west. The fact that the atoll had a WWII airstrip was also an advantage.

Some weeks later, military authorities met with the island chiefs, the "dri-Enewetak" and the "dri-Enjebi," to explain the need for relocation. Approximately 136 islanders and their belongings would have to be moved to Ujelang, located 124 nautical miles southwest of Enewetak.[6]

On October 18, 1947, the Joint Chiefs of Staff officially established Joint Task Force Seven (JTF-7) and assigned the operation the code name "Switchman." The number seven was selected to denote the operation simply because it was a distinctive number. "JTF-2" was considered undesirable from a security standpoint because it would be reminiscent of JTF-1.[7]

The Joint Chiefs nominated Lieutenant General John E. Hull, Major General William E. Kepner, and Admiral William S. Parsons to serve as the Joint Proof-Test Committee, pending organization of the Task Force. Shortly thereafter, Captain James S. Russell, U.S.N., was designated by the AEC as its representative on the Committee. Russell would also serve as

NUCLEAR ORDNANCE ENGINEER FOR THE NATION

Firing and Engineering Branch

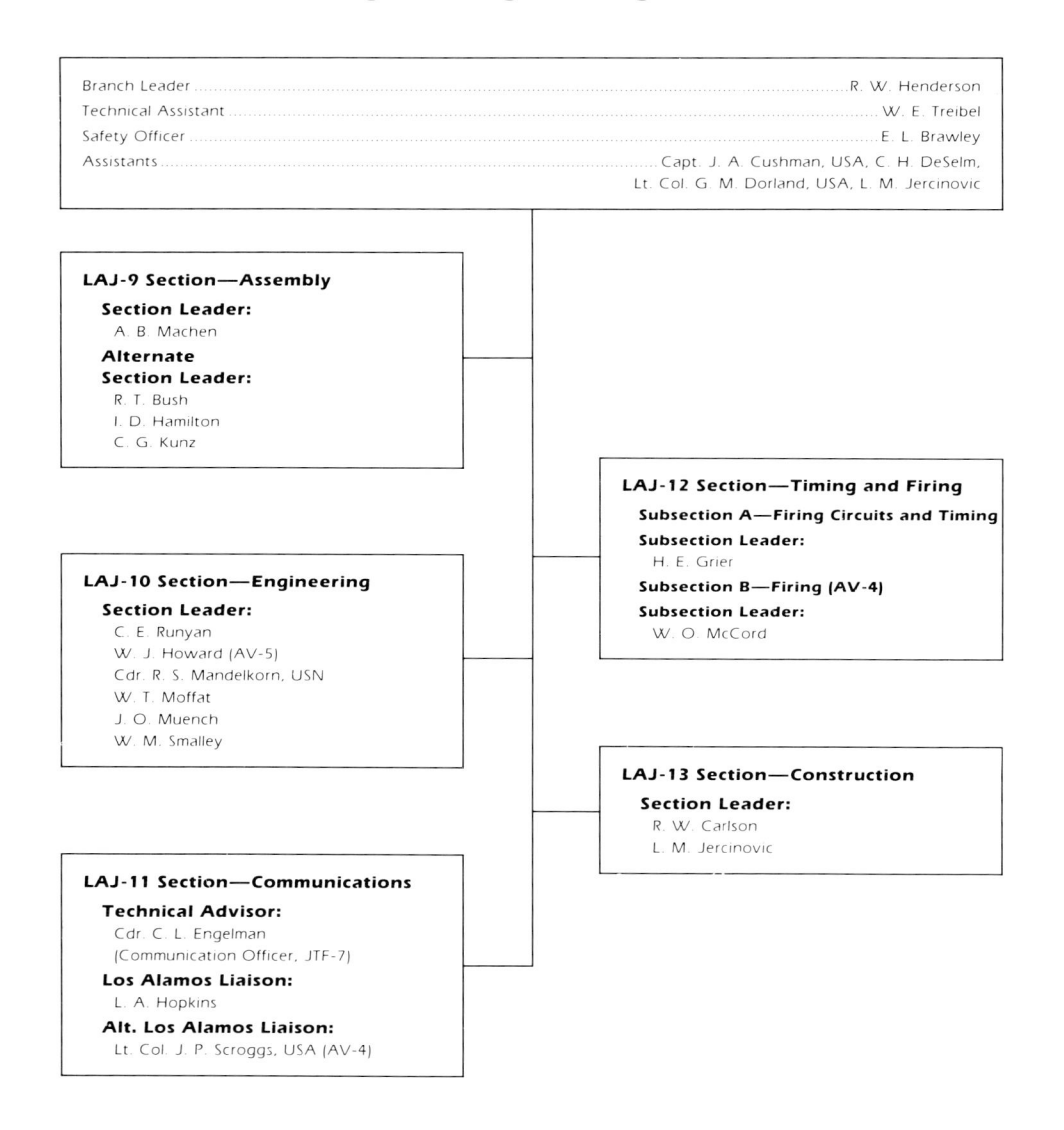

Branch Leader	R. W. Henderson
Technical Assistant	W. E. Treibel
Safety Officer	E. L. Brawley
Assistants	Capt. J. A. Cushman, USA, C. H. DeSelm, Lt. Col. G. M. Dorland, USA, L. M. Jercinovic

LAJ-9 Section—Assembly
  Section Leader:
    A. B. Machen
  Alternate Section Leader:
    R. T. Bush
    I. D. Hamilton
    C. G. Kunz

LAJ-10 Section—Engineering
  Section Leader:
    C. E. Runyan
    W. J. Howard (AV-5)
    Cdr. R. S. Mandelkorn, USN
    W. T. Moffat
    J. O. Muench
    W. M. Smalley

LAJ-11 Section—Communications
  Technical Advisor:
    Cdr. C. L. Engelman
    (Communication Officer, JTF-7)
  Los Alamos Liaison:
    L. A. Hopkins
  Alt. Los Alamos Liaison:
    Lt. Col. J. P. Scroggs, USA (AV-4)

LAJ-12 Section—Timing and Firing
  Subsection A—Firing Circuits and Timing
  Subsection Leader:
    H. E. Grier
  Subsection B—Firing (AV-4)
  Subsection Leader:
    W. O. McCord

LAJ-13 Section—Construction
  Section Leader:
    R. W. Carlson
    L. M. Jercinovic

test director. Colonel John H. Hinds, U.S.A., was chosen as the MLC representative. Lieutenant General Hull, ultimately selected to head the Joint Task Force of Air Force, Army, and Navy personnel, would be responsible for the overall direction of the JTF-7 mission, which was "to construct an atomic proving ground at Enewetak Atoll . . . and . . . to support the AEC in the conduct of the initial test operations."[8]

Under Hull's direction, the coral soil of the atoll would soon be transformed into a gigantic tabletop. In this open-air laboratory, scientists, engineers, and technicians from Los Alamos and Sandia Branch, assisted by Department of Defense (DoD) civilians and contractors, were to play a major role. The Armed Forces Special Weapons Project, the triservice military organization formed to manage the receipt, storage, safety, security, and inspection of atomic weapons, provided 116 personnel to be integrated into the AEC Proving Ground Group.[9]

Sandia Branch Personnel Arrive at Sandstone

To support the Sandstone operation and manage the nuclear testing program for JTF-7, Bradbury had organized Z-Division and Los Alamos personnel into a special J Division. On September 18, 1947, he appointed Froman and Graves, along with First Assistant Scientific Director Robert W. Henderson, to assume overall responsibility for technical support of the operation. Since Los Alamos was interested principally in scientific tests and measurements, Bradbury delegated responsibility for engineering and related activities to Henderson.

Henderson, as head of the Firing and Engineering Branch, carefully selected Z-Division participants for the operation. Staff members included First Technical Assistants Walt Treibel, C. H. "Hilt" DeSelm, and Leo M. Jercinovic. Jercinovic, as the site liaison officer, was to accompany Second Scientific Director, John Clark, and Roy Carlson to Enewetak as part of an advance group charged with coordinating with Task Force engineers and construction troops.[10]

On New Year's Day, 1948, Jercinovic left Albuquerque by military transport. A number of high-ranking officers were also making the trip. From California to Hawaii to Johnston Island and on to Kwajalein, the old C-54 plane, minus air conditioning and insulation, made its noisy way to the Sandstone site. As the uncomfortable ride neared an end, the crew chief got on the public address system to announce that the DOD was making a documentary film of their arrival. Everyone was instructed to deplane in order of descending rank.

The airplane taxied to a big review stand by the runway where a sizable welcoming committee stood ready to greet them. The crew chief advised the passengers that they should step out and proceed down the receiving line as their names were called. Jercinovic relaxed, knowing that there would be quite a few "brass" in front of him. However, the sound of the first name read from the list, "Dr. L. M. Jercinovic," jolted him to attention.

To the strains of a military band playing in the background, the young engineer stepped out of the plane into the brilliant sunlight reflecting off the white sands of the airstrip. Waiting at the bottom of the steps to extend an official welcome to the island was General David A. D. "Dad" Ogden. Jercinovic couldn't help but hear someone muttering in the crowd: "How could that young kid be a doctor!" Jercinovic, who was only in his twenties, agreed. Obviously, there had been a mixup. The name should have been Dr. John Clark. After the welcoming ceremonies, the party enjoyed a formal luncheon, which included hearts of palm salad and the General's very best Scotch. A tour of the island followed.

Enewetak in January was a tropical paradise. A stiff breeze blew in off the ocean, cooling the temperature to a comfortable seventy-five degrees. Refuse from World War II—broken equipment, abandoned vehicles, and exploded ordnance—still blighted the natural beauty of the sandy beaches, but the Army Corps of Engineers was making headway. Jungle growth and palm trees were being removed and landing strips enlarged and lengthened. Footings for a photography tower had to be securely positioned in the live coral head that nestled at the bottom of the big blue lagoon in the center of the atoll. Natives from a Hawaiian-based construction firm handled the underwater work masterfully.[11]

Carlson and Jercinovic were soon involved in the preparation process. Carlson, as the expert on concrete structures, supervised construction of special limonite radiation-resistant bunkers and test facilities, while Jercinovic devoted his energies toward setting up the 200-foot and 75-foot towers that would be used for the test shots and photographic cameras. Generators had to be installed at each firing tower, timing station, photo tower, and at the control station on Parry Island. Henderson, Froman, and military staff members arrived at Enewetak during the first week in February to survey the island and check on progress made. They found preparations well under way.

In the ensuing weeks, Hilt de Selm, Lou A. Hopkins, and W. J. "Jack" Howard assisted Charlie Runyan in liaison work with Terminal Island Naval Shipyard near San Francisco. Specifications for alterations to the U.S.S. *Curtiss* had to be made in order to convert the seaplane tender to a weapon assembly ship. Special ramps for the landing craft tanks (LCTs) had to be built, and emergency hoists and slings required testing.

Walt Treibel assisted Henderson by carrying out technical liaison work at Terminal Island, Los Alamos, Washington, and Boston. Constant surveillance and coordination were necessary to correlate the detailed requirements of all participating agencies, such as contracting engineers, EG&G, and Jackson and Moreland, who designed the instrument buildings. EG&G had responsibility for the timing signal and remote control firing circuits necessary to fire the bomb and switch on experimental recording equipment. This group was to work with the Sandia Branch Lab and, later, Sandia Corporation on many a test. The Naval Ordnance Lab conducted blast experiments and measured blast overpressures, and the Naval Research Laboratory conducted radiation measurement experiments. Argonne Laboratory participated mostly in gamma ray experiments.[12]

On February 24, the remainder of personnel from Sandia Base arrived at Terminal Island where they spent a week aboard ship for orientation before proceeding on to Enewetak. L. A. Hopkins and W. T. Moffat went on ahead to the forward area to check out communications and engineering liaison matters.

When the major part of the Sandia Branch contingent arrived, they were shuttled to quarters on the particular island where they were to work. At Enjebi and at Enewetak, they found themselves housed in quonset huts, each of which accommodated about sixteen personnel. At Aomon and Runit, scientific personnel had quarters in regular Army tents, four persons per tent. By the usual stateside standards, living conditions were crude, but the food provided by the Army was quite satisfactory. Good Scotch could be purchased for eighty-five cents per quart, and there was plenty of cold beer. Ice cream was served at two meals each day.[13]

By the end of March, miles of underwater cable connected timing and firing systems; concrete bunkers buried in sand protected instrumentation; and airplane drones based on Kwajalein 200 miles distant prepared to sample the radiation-filled cloud after the blast. Cameras were positioned on distant islands to prevent them from being sprayed by salt water of the lagoon. The *Mount McKinley*, flagship for the operation, lay at anchor off Parry Island, while the U.S.S. *Comstock*, mother ship for the boat pool, waited to provide water transportation.

In addition to bomb assembly and handling, construction, engineering, ship modifications, and firing circuitry liaison, Sandia Branch personnel took part in the Radiological Safety Program. The objective of the Safety Program, in keeping with AEC directives, was to keep the radiation exposure level of personnel at the lowest possible dose and avoid inadvertent contamination of populated islands or transient shipping.

The radiation standard established by the AEC was 0.1 roentgen per twenty-four-hour period, with a maximum exposure of three roentgens for certain approved missions. By AEC directive, radiation dosimetry badges were issued, and an extensive weather forecasting group predicted wind directions and areas of potential fallout. The safety of personnel was an important part of pretest preparations.[14]

A Dry Run

As zero hour for the first shot approached, test personnel made ready for a dry run. Personnel from the Sandia Laboratory Branch assembled the dummy unit on board the U.S.S. *Curtiss*, brought it out on the fantail of the ship, and positioned it for pickup by crane. Alongside the ship, a landing craft about twelve feet wide and fifty feet long bobbed up and down in the ocean. The large crane was carefully attached to the special transporting trailer. With infinite skill, the Navy crane operator maneuvered the transport trailer to ferrying landing craft mechanized (LCM), while a bosun's mate rode the unit over the side, yelling directions during the thirty-foot descent to the well-deck of the landing craft.

The maneuver was accomplished "without a perceptible bump," according to Jercinovic, who was to serve as the main tower-hoist operator on both the *X-ray* and *Yoke* shots.[15] The landing craft drove right onto the beach and roadway. A truck hooked up to the transport trailer and delivered the unit to the foot of the tower.

The next step in the practice exercise was to attach the tower-hoist mechanism to the device, which was then lifted carefully into position through the center of the small twelve-foot-square metal shed at the top of the test tower. Art Machen completed the wiring and installation of the detonators, using dummy sugar-loaded components made up to match the real item. "Are you sure that's a dummy?" someone asked. "You'll be among the first to know if it isn't," Machen quipped.[16] The team from EG&G checked out the firing system and signalled the people in the bunker to press the button activating countdown.

Jercinovic, Machen, and other members of the crew relaxed, leaning against the unit, as the countdown began: minus thirty seconds . . . , minus twenty seconds . . . , minus ten seconds . . . .

Precisely as the radio voice reached zero, an enormous explosion shook the tower. After the initial shock, and realizing that they were still intact, the men stuck their heads out of the tower cab to see what had caused the perfectly timed detonation. It turned out that Navy frogmen had selected that inauspicious moment to blow up three coral formations in the lagoon next to the tower. "It made the trip down a bit shaky," Jercinovic recalled, "Fortunately, the tower had an elevator that ran up and down the outside. I don't think we would have made it if we had had to climb down hand over hand."[17]

Zero Hour

The day before the actual shot, which was planned for April 14, 1948, all personnel except for the firing crew were evacuated from the island of Enjebi to ships in the lagoon. An eerie sense of anticipation surrounded the firing party as they proceeded to the tower in the black silence of the early morning hours. The first test in the series, *X-ray*, was scheduled for thirty minutes before sunrise to have the darkness needed for cameras on the ground and the daylight at 30,000 feet that would allow cloud-tracking aircraft to navigate visually. Edwin C. Udey and Harold G. Sweeney, transferred from the Photo Lab at Sandia Branch, prepared to document on film the final assembly and handling of the atomic weapons to be used. Udey and Sweeney had to operate separately from the Task Force photo crews because it was felt that it would be inadvisable to allow the temporary photographers hired by Air Force General P. T. Cullen to have access to top secret assembly operations.[18]

At the tower, the firing party went through the final arming procedure. The twenty-knot trade winds caused the platform atop the 200-foot tower to sway alarmingly. And on the sands below, armed guards positioned at strategic points peered out toward the black ocean. There was fear that the Russians might somehow penetrate the strict security to come ashore and steal the unit or wipe out the entire scientific contingent. At the height of the Cold War, the Russian threat seemed all too real. However, Bernard O'Keefe of the EG&G firing team admitted that he "worried more about trigger-happy Marine guards" than the Russians.[19]

After plugging in the final cables and throwing the connecting switch, the crew made their way down to the base of the tower. According to security procedures, they were to dismantle the elevator while the guards made a final search of the area. Only then could they leave the premises to take a twenty-mile boat ride back to the control station on the island of Parry.

At zero hour, a flash of blinding light reflected off the control panels, signalling the successful detonation and the ascent of the fireball. The now familiar multihued mushroom cloud was followed by a shock wave that hit with a sharp crack. On the islands of Aomon-Bijiri and Runit, the scene would be repeated twice more.

The largest shot, *Yoke*, staged on May 1 at Aoman, registered a forty-nine-kiloton yield.* *Zebra*, the last test in the series, was detonated May 15, 1948, on Runit Island. Tests set up by Los Alamos personnel measured output of the nuclear devices—specifically radiation, blast, and yield. The results garnered from the tests added to the knowledge base of a relatively new science. The fact that the new models had detonated at all indicated what was termed "a stunning success." The series not only verified new designs but also suggested other possibilities for breakthroughs in nuclear technology. Basically, Sandstone proved that, through use of a fission igniter, a weapon with thermonuclear fuel could be designed.[20]

---

*One kiloton equals the approximate energy release of a one-thousand-ton TNT explosion.

Group leaders and advisory personnel discuss requirements for the Sandstone test series. Pictured are (standing, left to right) Captain John King, U. S. Army (G-2), Phillip Barnes, Raemer Schreiber, Harlow Russ, W. O. McCord, and Art Machen, (sitting, left to right) Glenn A. Fowler, Roger S. Warner, and Robert W. Henderson.

SANDSTONE: A NEW REALM IN WEAPONS DEVELOPMENT

Above, Hilt DeSelm, Robert W. Henderson, and Walt Treibel study a map of Enewetak Atoll to determine the best possible locations for unloading equipment and supplies. Below, left, Holmes and Narver contract personnel unload surveying equipment from an LCM. Below, right, some LCMs had to be unloaded short of the coastline because of their size.

An emplacement operation in progress. Before testing at Sandstone, the device was assembled onboard the U. S. S. *Curtiss*, placed on a weapons carrier, and transported aboard the Landing Craft Tank (LCT) to a good debarkation point on one of the sandy beaches. The test device on its transport trailer or truck was then taken to the base of the tower where it was hoisted to the top.

SANDSTONE: A NEW REALM IN WEAPONS DEVELOPMENT

The tower at Aoman can be seen through the palms in this April 1948 photo. Below, Hilt DeSelm (left) talks on the telephone at the base of Engebi tower. In a safety practice that began at Trinity, G. I. mattresses were brought in and stacked at the base of Engebi tower—in case the device should fall.

NUCLEAR ORDNANCE ENGINEER FOR THE NATION

The Logistics Group (above) was led by Charlie Runyan (at far left) and included (left to right) Joe Muench, Jack Howard, William Smalley, and Bill Moffatt. Below, Art Machen's Assembly Group included personnel from the Sandia Branch Laboratory and the military. Machen is seated at the far left of the middle row. Robert A. Knapp and Tiny Hamilton are found standing in the back row at the far left and third from the left, respectively. Bob Bush is on the floor at left.

SANDSTONE: A NEW REALM IN WEAPONS DEVELOPMENT

The group responsible for arming, fuzing, and firing included (standing, left to right) Leon Smith, Sylvester Zelenski, Group Leaders W. O. McCord, Phillip Barnes, (seated, left to right) Joe Dawson, and Bob O'Connor. Below, photographers Henry Sweeney and Ed Udey pose with the tools of their trade.

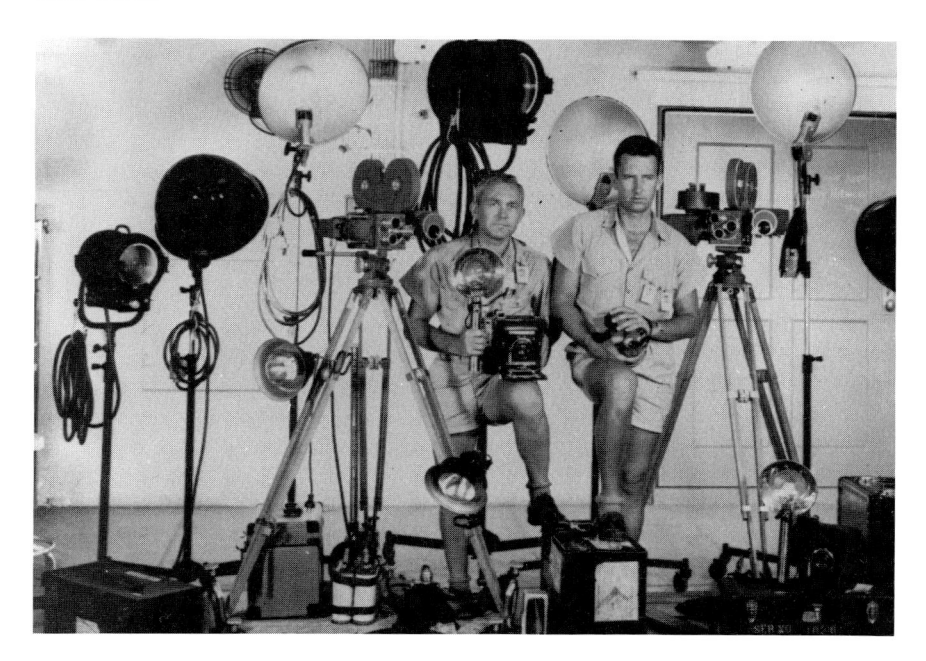

The determination that the new bomb models were more efficient than older ones was highly significant. In June 1947, for example, the U.S. had thirteen nuclear weapons in stockpile. One year later, despite heavy emphasis on increased production of fissionable material, the number was only fifty, far short of the quantity that military planners determined would be needed in a war with the U.S.S.R. However, the efficiency of the new models, coupled with the higher production rates of fissionable material, provided for expansion of the U.S. stockpile that would be evident by 1949. Thus, Sandstone—truly a milestone operation—opened a new realm in weapons development.[21]

Impact on the Sandia Branch Laboratory

In the opinion of AEC historians Hewlett and Duncan: "Nowhere did the anticipation and achievements of Sandstone have greater effect than in weapons activities at Sandia."[22] Interaction with the military during the operation had established common goals and helped to work out problems and procedures. The Joint Research and Development Board, formed just one month before the operation began, proved to be an effective vehicle for the planning and coordinating of activities related to weapons development and production. In addition, the skilled diplomacy of George P. Kraker, a retired Rear Admiral who was the Commission's representative at Sandia Laboratory, helped to "mesh" Sandia Base activities and other facets of the production complex.[23]

From the technological aspect, results of the Sandstone operation suggested even newer and more advanced weapon designs than those tested. Scientists envisioned smaller, lighter weapons—perhaps a variation of the Mk 5 weapon already under study. Bradbury had in mind an integrated weapon system that would join the expertise of his scientists with that of the aircraft industry. The result might be a bomber designed and adapted to its nuclear payload. It was hoped that the TX–5 could be tested at Enewetak in 1951. The Sandia Laboratory, as the nation's nuclear ordnance facility, was expected to respond. In view of these new demands, it was fortunate that reorganization and expansion at the Sandia Branch were well under way; yet, there were disadvantages associated with the almost simultaneous timing of the test operation and the restructuring of the Sandia Branch Laboratory.[24]

The job of implementing the reorganization plan at its inception in March of 1948 fell largely on Richard Bice, who ran the Laboratory for three to four months while Larsen was still involved part-time at Johns Hopkins and Henderson was overseas. During this time, teletypes kept the lines busy between Albuquerque, New Mexico; the Pacific test site; and Larsen in the East. The transition to branch status did affect the morale of personnel assigned to the Sandstone operation. As duly noted by Henderson, Treibel, and DeSelm in their official report of the Sandia Branch Lab's participation in the test series,

> It was extremely unfortunate from a psychological viewpoint for the reorganization and expansion of Sandia Base Branch of the Los Alamos Scientific Laboratory to occur when it did. . . . While individuals are not adverse to such opportunities in field work under normal conditions, the thought of being lost in the shuffle due to proposed changes in organization gave indications of hesitation and insecurity.[25]

Henderson had tried to reassure prospective members of J Division when he interviewed them for the overseas assignment, but the reorganization still made an impact. According to Henderson, employees naturally questioned "whether or not the best opportunities, in the long run, remained with the home office or in the field."[26] To the credit of Acting Director Dick Bice—and despite the problems of communication—he tried to take into consideration the interests of all facets of the organization, including those personnel in absentia. At the same time, he realized the overriding need to get the division operational under the new guidelines as quickly as possible to meet stockpile schedules.[27]

Morale problems relating to placement of employees returning after extended tours of duty away from the Albuquerque headquarters—whether to Amchitka, Alaska, or Washington, D. C.—would continue to surface throughout Sandia's history. The analysis presented to the administration by authors of the Sandstone Report in 1948 bears a timeless wisdom:

> Each employee is awaiting with keen interest his placement in the enlarged organization. This fact is significant— the success of future operations has a direct bearing on the manner in which personnel are absorbed by the organization.[28]

Concerns over reorganization did not dampen the esprit de corps that existed among Branch Lab personnel. Since many employees lived in base housing separated from the main community of Albuquerque by what was then a distance of some miles, they oftentimes shared recreational and pastime activities as well as work-related time. The camaraderie of these early days is reflected in the folksy, team-spirited style of a November 19, 1948, Sandia *Lab Bulletin*, in which the first notice read

> Glenn Fowler has made an appointment for a haircut. All of SLT is looking forward to seeing him again.

Another entry announced the wedding of Mary Lou Maloney and the gregarious and talented Mike Michnovicz, who came to Sandia Base shortly after the original Z-Division group.

> Mary Lou is Mrs. Michnovitz [sic] now. They had a beautiful wedding Saturday morning at Saint Charles.

Under "Sports," the reporter wrote:

> Sparked by Leo Gutierrez and Ted Allen, SLA's hardwood crew almost managed to whip the younger, sprighter [sic] Headquarter Detachment five points at the Post Gym last Monday night. Though they played hard and fast till the closing minutes, the Physics boys finally pooped out to the tune of 21-19. Besides Leo and Ted, the SLA players included Bob Buckner, Barney Geelan, Jack Bryson, Bill Hereford, Chet Clyde, and Lloyd Sheppard, a stray from SLT. Double-dribble McNabb, SLA's fiery manager, laid the loss to the lack of substitutes and swears he'll have ample reserve strength on the court next Tuesday nite when his charges tangle with the MP's.[29]

SANDSTONE: A NEW REALM IN WEAPONS DEVELOPMENT

Despite the pressures of crash programs and hardship tours in the Pacific, Laboratory personnel generally showed a strong sense of friendship and mission that stood them in good stead. However, the technological breakthroughs at Sandstone would add to the enormity of the engineering job facing the young Laboratory. In May 1948, even before the third Sandstone shot had been fired, orders from Bradbury completely revamped production schedules.[30] On the positive side, the Sandstone operation gave the Laboratory direction and a sense of mission. Laboratory personnel would have to pull together to meet the newly established quotas for the Mk 4; completion of stockpile items no longer held priority. With the Mk 4, the era of handcrafted weapons had passed; mass production of components and assembly line techniques became the order of the day.[31]

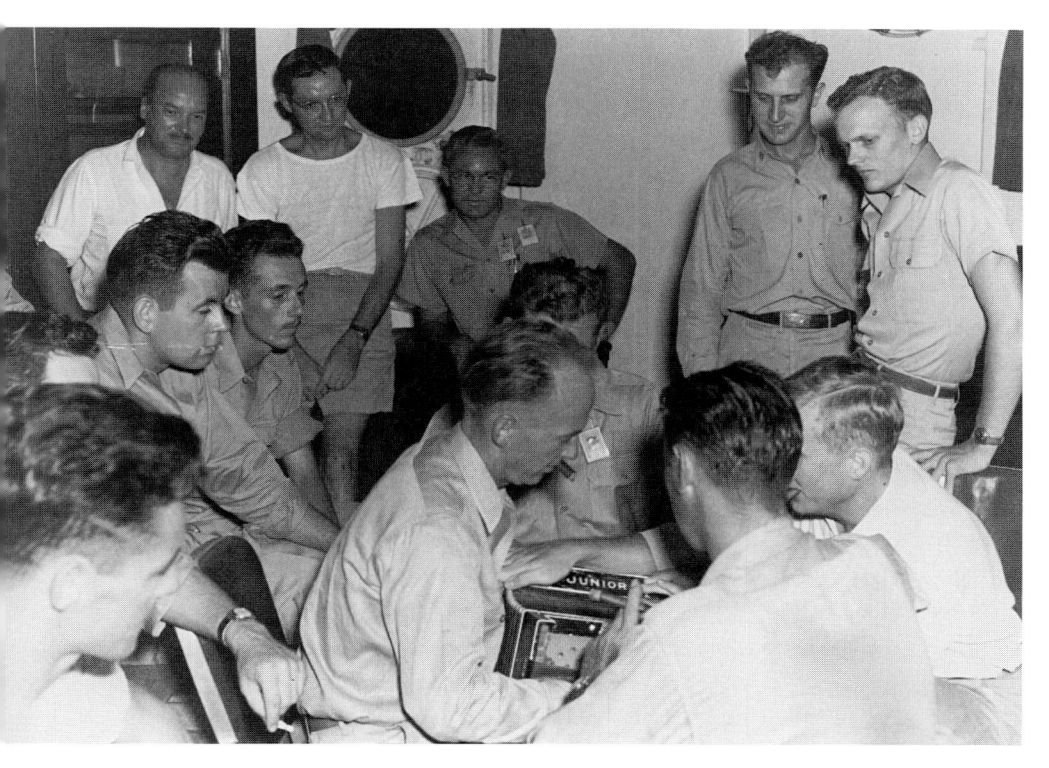

Art Machen plays the "squeeze box," or accordion, for a relaxed group during leisure hours on the U.S.S. *Curtiss*.

Personnel (above and below) relax during a cookout on one of the sandy beaches.

SANDSTONE: A NEW REALM IN WEAPONS DEVELOPMENT

The Sandstone test series completed, Sandia Branch Lab and military personnel on the deck of the U. S. S. *Curtiss* are welcomed home by this view of the Golden Gate Bridge. By ship, the trip from the Enewetak atoll to the San Francisco harbor took two weeks.

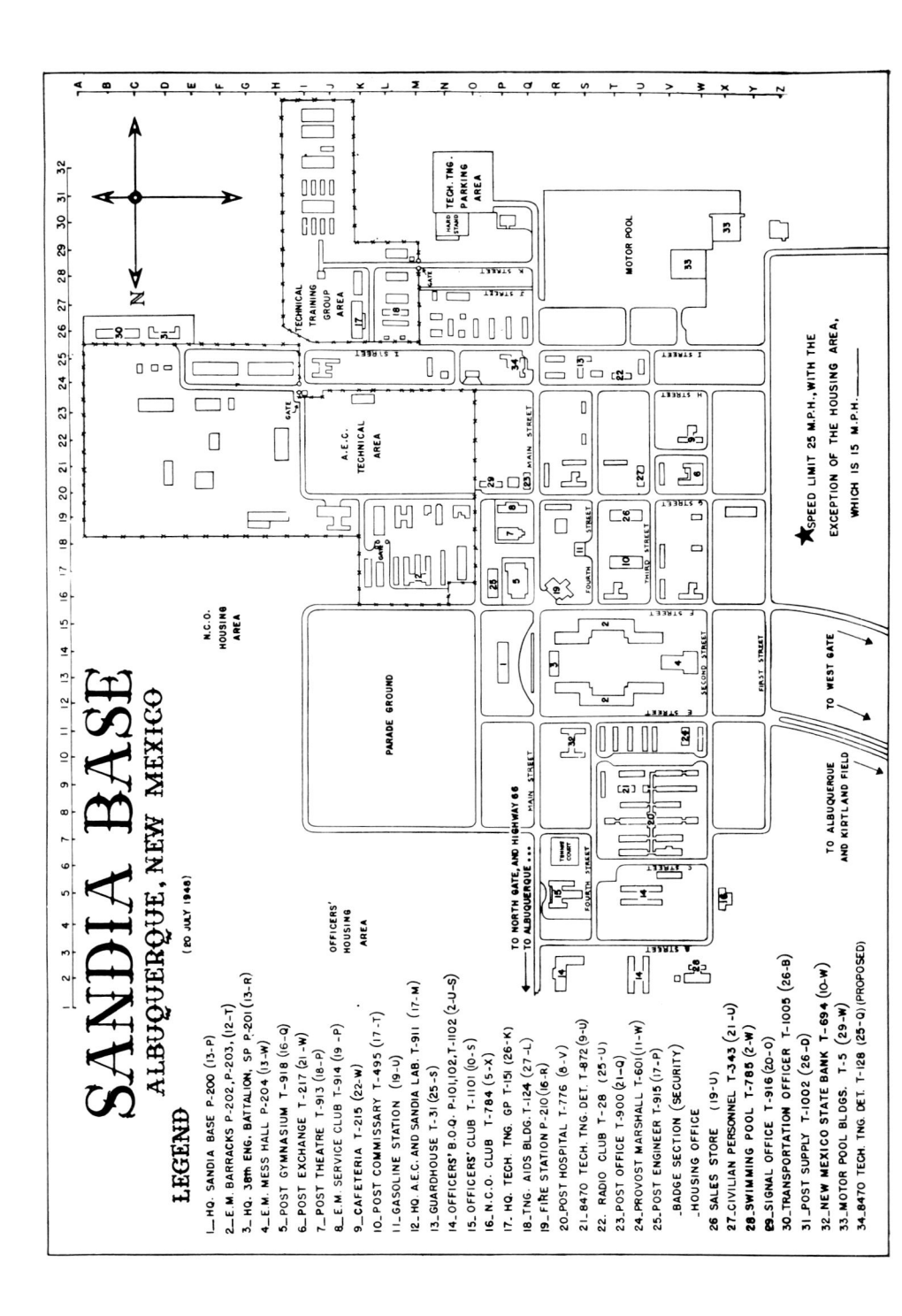

SANDIA BASE
ALBUQUERQUE, NEW MEXICO
(20 JULY 1948)

LEGEND

1 — HQ. SANDIA BASE P-200 (13-P)
2 — E.M. BARRACKS P-202, P-203, (12-T)
3 — HQ. 38th ENG. BATTALION, SP P-201 (13-R)
4 — E.M. MESS HALL P-204 (13-W)
5 — POST GYMNASIUM T-918 (16-Q)
6 — POST EXCHANGE T-217 (21-W)
7 — POST THEATRE T-913 (18-P)
8 — E.M. SERVICE CLUB T-914 (9-P)
9 — CAFETERIA T-215 (22-W)
10 — POST COMMISSARY T-495 (17-T)
11 — GASOLINE STATION (19-U)
12 — HQ. A.E.C. AND SANDIA LAB. T-911 (17-M)
13 — GUARDHOUSE T-31 (25-S)
14 — OFFICERS' B.O.Q. P-101, 102, T-1102 (2-U-S)
15 — OFFICERS' CLUB T-1101 (10-S)
16 — N.C.O. CLUB T-784 (5-X)
17 — HQ. TECH. TNG. GP T-151 (26-K)
18 — TNG. AIDS BLDG. T-124 (27-L)
19 — FIRE STATION P-210 (16-R)
20 — POST HOSPITAL T-776 (8-V)
21 — 8470 TECH. TNG. DET. T-872 (9-U)
22 — RADIO CLUB T-28 (25-U)
23 — POST OFFICE T-900 (21-Q)
24 — PROVOST MARSHALL T-601 (11-W)
25 — POST ENGINEER T-915 (17-P)
   — BADGE SECTION (SECURITY)
   — HOUSING OFFICE
26 — SALES STORE (19-U)
27 — CIVILIAN PERSONNEL T-343 (21-U)
28 — SWIMMING POOL T-785 (2-W)
29 — SIGNAL OFFICE T-916 (20-O)
30 — TRANSPORTATION OFFICER T-1005 (26-B)
31 — POST SUPPLY T-1002 (26-D)
32 — NEW MEXICO STATE BANK T-694 (10-W)
33 — MOTOR POOL BLDGS. T-5 (29-W)
34 — 8470 TECH TNG. DET. T-128 (25-Q) (PROPOSED)

Chapter XI.

"Let's Get the Show on the Road"

The pressure to "get the show on the Road" was relentless. There was the inevitable, constant reminder that this might indeed be the one, only and last opportunity.

Anonymous Report

By the summer of 1948, the Sandia Branch Lab was functioning under strong leadership and a new organizational structure. Yet management still had to address the pressing production issue—how to convert from a handcraft job shop arrangement to a more efficient factory production mode.

Following the successes achieved at Sandstone, the Research and Development Division of AFSWP, in recommendations to DMA, made their position clear. The "number one priority program," they said, "should be to get the Mk 4 weapon into production as soon as possible." Next on the list of priorities was the development of a subsurface weapon and the scaling down of the overall size of the implosion bomb. They also recommended development of a smaller gun-type airburst weapon and initiation of a thermonuclear program. At the time, however, members agreed that the thermonuclear project "was not of immediate importance,"[1] International events would soon change priorities.

News of the Berlin blockade in August, for example, placed even more pressure on Sandia to refine its production and assembly procedures. In response, the AEC informed the President that it would be able to produce more than had been

307

SLR—Road Department

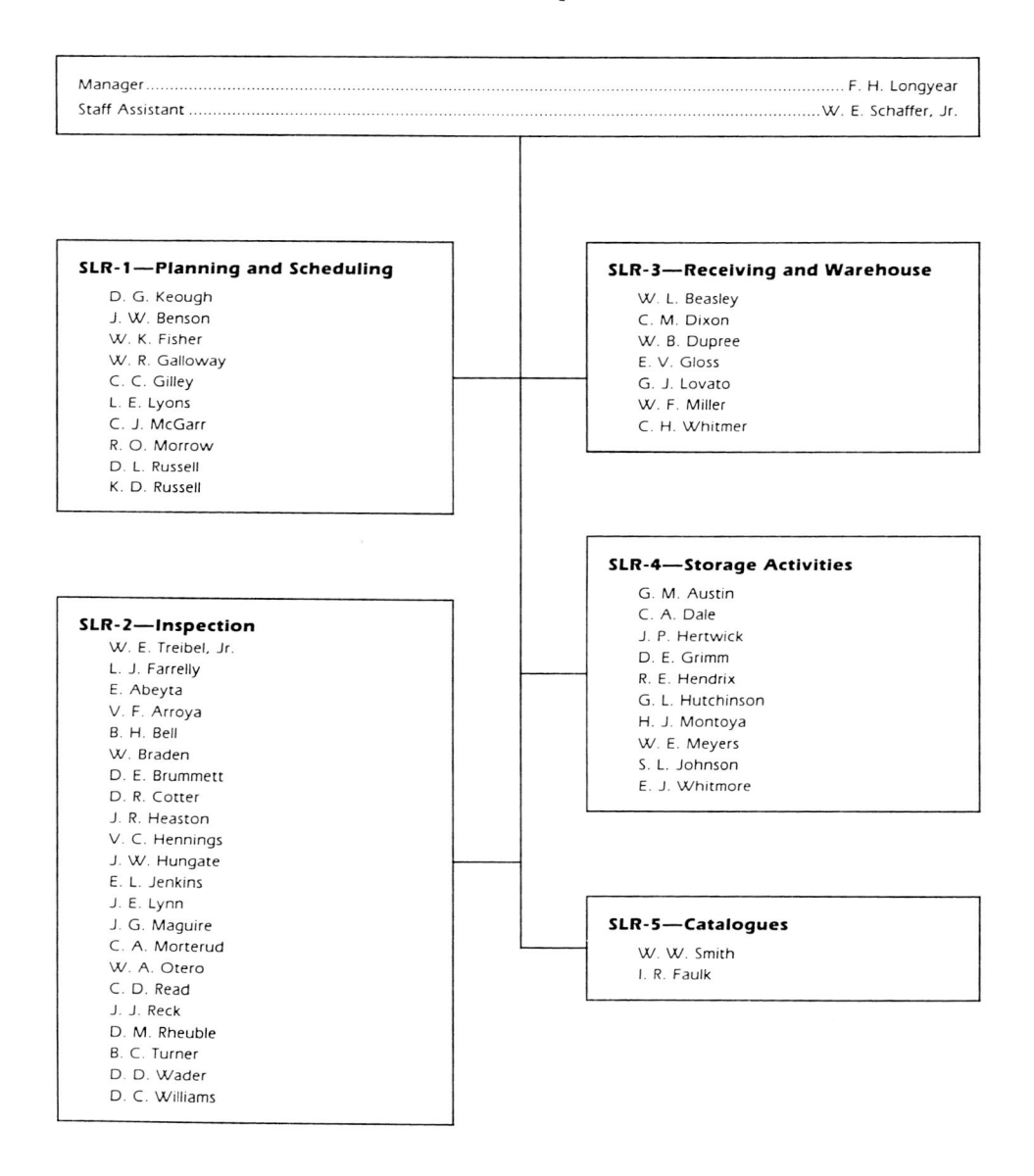

Manager .... F. H. Longyear
Staff Assistant .... W. E. Schaffer, Jr.

SLR-1—Planning and Scheduling
- D. G. Keough
- J. W. Benson
- W. K. Fisher
- W. R. Galloway
- C. C. Gilley
- L. E. Lyons
- C. J. McGarr
- R. O. Morrow
- D. L. Russell
- K. D. Russell

SLR-2—Inspection
- W. E. Treibel, Jr.
- L. J. Farrelly
- E. Abeyta
- V. F. Arroya
- B. H. Bell
- W. Braden
- D. E. Brummett
- D. R. Cotter
- J. R. Heaston
- V. C. Hennings
- J. W. Hungate
- E. L. Jenkins
- J. E. Lynn
- J. G. Maguire
- C. A. Morterud
- W. A. Otero
- C. D. Read
- J. J. Reck
- D. M. Rheuble
- B. C. Turner
- D. D. Wader
- D. C. Williams

SLR-3—Receiving and Warehouse
- W. L. Beasley
- C. M. Dixon
- W. B. Dupree
- E. V. Gloss
- G. J. Lovato
- W. F. Miller
- C. H. Whitmer

SLR-4—Storage Activities
- G. M. Austin
- C. A. Dale
- J. P. Hertwick
- D. E. Grimm
- R. E. Hendrix
- G. L. Hutchinson
- H. J. Montoya
- W. E. Meyers
- S. L. Johnson
- E. J. Whitmore

SLR-5—Catalogues
- W. W. Smith
- I. R. Faulk

specified in the Joint Chiefs of Staff schedule of late 1947. The MLC, in coordination with the AEC, strongly supported the effort to "substantially" increase and expand production efforts. This turn of events indicated that the nuclear weapons stockpile might ultimately involve thousands of weapons rather than the hundreds envisioned in 1947.[2]

The "Road" Department

The extensive reorganization of April 1948 proved to be the opportune time to convert the Stockpiling Group into a full-fledged department carrying the curious code name of "Road." The fact that Sandia assembled bombs was classified; therefore, a code name was essential because of the secrecy surrounding this part of the Laboratory's mission. Since most of the weapons had to be transported via Road to storage sites, the title was appropriate. There was also the connotation of "Let's get the show on the road," a phrase that indicated the concerted efforts being made to meet readiness needs.[3]

Wilbur F. Shaffer, Jr., who had been instrumental in formulating the original plan for stockpiling the Fat Man, recalled that "when Paul Larsen became Lab Director, we started a crash production program" that led to rapid expansion of the group.[4] Since no formalized production organization had existed at Los Alamos, there were few guidelines available for the establishment of a Road Department (SLR) at Sandia. It was realized, however, that extensive teamwork would be required to build the facilities, acquire the tools, and train and educate personnel for the task.

To head the group, Larsen brought in Frank H. Longyear, a man with many years of industrial experience. "SLR," as the group was called, planned the procurement of special weapon components and related equipment based on AEC directives, performed receiving and inspection functions on all electrical and mechanical materials and components, and processed components and related equipment for storage. The

Road Department also established and maintained accurate reports, records, and catalogs on all shipments of Road materiel to and from the Laboratory.[5]

Road Department personnel worked closely with Engineering, which furnished approved drawings and specifications. As the unit responsible for modifying existing weapons and redesigning weapons for tactical use, the Engineering Department had the right to waive specifications and decide on technical issues.

The Lab's cooperative goal was to put into Road a weapon capable of being manufactured by industrial production techniques for use in combat by regular military personnel. Consequently, the AEC, using the Manhattan Engineer District as a model, began organizing a complex of integrated contractors that would supplement and eventually replace Sandia's in-house production activities.[6]

In this manner, private industry would become increasingly involved in managing such government-owned, contractor-operated ("Go-Co") plants. These facilities, which also produced the high-explosive castings and pit materials needed for nuclear weapons, proved to be a highly effective means for the channeling of government funds into nuclear research and development.

Origins of the Integrated Contractor Complex

The process of setting up production facilities began in 1947. The first facility, the Y-12 Plant at Oak Ridge, Tennessee, was located some twenty miles from Knoxville in territory once under control of the Tennessee Valley Authority. As a part of the Manhattan Project, the plant was built to separate and concentrate the uranium 235 isotope by the electromagnetic process. After demonstration of the more efficient gaseous diffusion process, however, the Y-12 Plant was adapted to other nuclear weapons-related activities. The facility would conduct research in uranium and fabricate, test, and supply stockpile assemblies to the weapons complex.

In that same year, the AEC made the decision to adapt the Army Ammunition Plant in Burlington, Iowa, to meet new requirements for weapons production. In Sandia's Area III, a preproduction prototyping facility was set up to prepare for opening production at Project "Sugar," as Burlington was known. Robert Duff, who was then an AEC-AFSWP inspector on active duty with the Air Force and assigned to participate in joint AEC-AFSWP acceptance of assembled weapons, recalled that "both Los Alamos scientists and Sandia engineers worked hand-in-glove with the production people, managers, and assembly crew from IOP [Iowa Ordnance Plant]."[7] The contract for the operation of the plant was awarded in August 1947 to the Silas Mason Company (now Mason and Hangar—Silas Mason Company, Inc.). Eventually known as the Burlington Plant, the facility by 1949 was fabricating and assembling H.E. components used to form the inner and outer sphere for the implosion bombs. Used initially for the manufacture of munitions, the plant also assisted with H.E. development and configurations for the gun weapon.

By 1948, the Mound Laboratory at Miamisburg, Ohio, near Dayton was also in operation under management of the Monsanto Corporation. Situated on a 170-acre tract of farmland on a hilltop overlooking the city of Miamisburg, the Laboratory processed polonium and fabricated polonium initiators until the mid-fifties, when these were phased out in favor of initiating devices using other longer-lasting radioisotopes. Originally, all detonator production took place at Los Alamos, but arrangements were made in 1949 with the Ordnance Corps, U.S. Army to produce standard detonators at the Picatinny Arsenal. Production of initiators was moved to the Mound Lab in 1949, which left Los Alamos fabricating special components for specific research. The Mound Laboratory also produced small H.E. units, explosive components, and pyrotechnic devices.

To meet the need for ever higher production quantities and to prepare for the eventual divorce of production responsibilities from Sandia proper, it became apparent that a separate facility for manufacturing electrical and some mechanical components was needed. Investigation showed that the old Pratt and Whitney airplane–engine manufacturing plant in Kansas City, Missouri, would be suitable for production activities. In November 1948, the AEC selected the plant to serve as a dual facility to Sandia for the production of electrical and mechanical components. The plant's location in the midcontinent area, and inaccessibility to enemy attack, were distinct advantages.[8]

The Bendix Aviation Corporation, in a project known as "Royal," was chosen to manage the facility. In April 1949, Bendix officially subleased building space from Westinghouse Electric Company and assumed responsibility for nonnuclear ordnance production. Sandia's Road facilities could then be dedicated to making prototype models for development purposes.[9]

For the future of Sandia, establishment of the production facility at Bendix, Kansas City, had an additional significance. Operation of the plant relieved the University of California of production responsibility under its contract—a move that prompted Bradbury and the regents simultaneously to request release from administrative duties as well. The regents proposed that Bendix extend its area of operation to include the entire Sandia Branch, but Bradbury vetoed the suggestion, explaining that the arrangement might seriously damage the "absolutely essential close technical liaison" between Sandia and Los Alamos.[10]

The Department of Military Applications (DMA) also considered having Bendix take over Sandia production operations; however, this would have required the AEC to choose still another "research–minded" contractor to manage other aspects of the operation. The DMA, therefore, decided to avoid the added complexity of having two major AEC contractors at the Sandia location. Careful analysis of the situation seemed to point to an industrial contractor as being the best solution.[11] Nevertheless, establishment of the Bendix facility and the production issue moved Sandia one step closer to industrial management. In the interim, additional production facilities were planned.

"LET'S GET THE SHOW ON THE ROAD"
"Road" Becomes Production Engineering

Although the University of California regents wanted to divorce the institution from ordnance activities, they hoped to retain control over research and development. Sandia management balked at the idea. As Larsen explained:

> Close technical coordination must exist between weapon research and development, the ensuing production engineering phase, and the production and final acceptance of the end products, to insure that they meet the original required and planned specifications.[12]

Citing the AEC's desire to centralize continuous responsibility for quality and performance of weapons in the organization that originated development, Larsen strongly rejected separation of research and development from production. Bradbury agreed, maintaining that "research and development to the prototype model stage belonged with Sandia."[13]

The push for additional weapons production created some difficulty in engineering, particularly in coordination with the new Bendix organization. "The development engineering people," according to Bill Shaffer, "continually issued drawing changes, which made the job of stockpiling and components fabrication almost impossible and yet assure interchangeability of components."[14]

Not surprisingly, extensive dialogue was required. Certain questions inevitably arose, such as: "Who is responsible for design changes after transfer of components to the manufacturing area; how do we approve them, and how does the design agency maintain control?" Procedures had to be developed, and in the interim, Sandia employees would have the opportunity to practice their human relations skills.

Furthermore, there were justifiable complaints from the military, who had to train with the stockpiled weapons. These problems prompted Bill Shaffer to propose to Longyear and

Larsen that there was need for a "production engineering" department. The department would serve as a "middleman" or link between development engineering, the Road group, and Project Royal at Bendix.[15]

Larsen and Longyear bought the idea; and in May 1949, the two groups were combined—first into the "Road Engineering Department" under Shaffer, and by October 1949 into "Production Engineering" under L. J. Biskner, with seven subdivisions.[16]

Quality control and stockpile reliability would continue to be of major concern to the Lab. Improvements in the efficiency of production activities and increases in the number of weapons in stockpile by 1950 led to increased pressure to crystallize surveillance operations and to construct additional storage sites.

Stockpile Surveillance and Training Liaison

Stockpile reliability was not a new concern. "Reliability," as one early document quite simply read, "is of utmost importance."[17] Statistical requirements did not exist. The weapon simply should not fail. Inspection and testing responsibilities had been a part of the Sandia mission since Z–Division days. Science and engineering personnel accompanied materiel into the field, where they supervised the final assembly, testing, and use of weapons. No government inspection procedures existed, although, beginning in late 1947, military personnel of the AFSWP performed "over-the-shoulder" inspections at Sandia and Kirtland. Eventually military inspectors would join AEC inspectors in a joint AEC/AFSWP formal certification and acceptance program at Sandia, Sugar, and Royal. The inspectors indicated acceptance by stamping the weapons with large red stars.[18]

By the late 1940s, designs were becoming increasingly sophisticated. Production methods had replaced hand operations, and manufacturing and supporting responsibilities for atomic materiel were being dispersed throughout the developing

American industrial complex. It was evident that a surveillance system was needed to guarantee the quality of the stockpile.

At Sandia, Larsen in 1948 took steps to formalize surveillance activities.[19] Concurrent with the reorganization, he elevated the function to the status of a full department (SLS). To head the group, he brought in under emergency clearance a bright young man he had known at the Applied Physics Laboratory (APL).

The new department manager, Louis Paddison, like other early Sandians, had a varied professional background that included exploration work in the West Texas oil industry. But as Paddison put it, "West Texas made me hungry for greenery."[20] Consequently, with his brother's encouragement, he joined the staff of the Physics Lab at Johns Hopkins, where he became involved in a contract to test the proximity fuze. The APL contract took him back to the Southwest and to the Coyote Canyon Test Site near Albuquerque.

A stint with the New Mexico School of Mines Research and Development Division followed. In this capacity, Paddison first became involved in the technical aspects of construction of storage sites. By this time, he had placed his application with the Sandia Branch Lab. "I went to work in December of '47," Paddison recalled, "and the first thing Paul told me to do was to organize a surveillance function to take care of the weapons in stockpile and be sure they were in readiness."[21]

The first step was to find office and work space. Paddison and another new hire, G. C. "Corry" McDonald, searched the tech area for a vacant facility. Just south of the present engineering facility, they located a nondescript building used during the war for munitions storage. Once inside, however, they found that it was still filled with remnants of WWII ordnance. Undaunted, Paddison and McDonald began making space for their offices. "It was in that building," McDonald remembers, "that we first started thinking about Quality Assurance."[22]

Larsen recognized that a monitoring agency independent of production (Road) was needed. The term *independent* of production would become a significant watchword for Surveillance personnel. Implicit in the concept from the outset was the

Some staff members of the Physics Lab of Johns Hopkins, such as Louis J. Paddison (inset), came to Albuquerque to test the proximity fuze at the Coyote Canyon Test Site and later hired on at Sandia.

understanding that freedom from outside influence and production pressures was essential to the integrity and satisfactory performance of what would eventually become a Quality Assurance organization.

The concept of auditing quality from the customer's point of view was first developed by Bell Telephone Laboratories in the late 1920s and early 1930s. Thus, the pioneering work of Bell Labs staff members would influence the functions and method of operation of Sandia's Quality Assurance organization, as well as those of other industrial concerns.[23]

Sandia's Surveillance personnel were to act as auditors to improve quality and reliability of components, subassemblies, and completed bombs by monitoring fabrication, stockpile maintenance, and modification of weapons. They would perform tests on end products to determine if functional criteria had been met and, when required, modify weapons in the field. For reference, Paddison relied heavily on the military assembly manuals being prepared and used by A. B. Machen and G. C. Hollowwa of Training Liaison (SLM). SLM personnel, involved in the training of numerous officer/NCO teams for the AFSWP, also observed and critiqued service maneuvers.

Information from surveillance of weapons in the field was fed back to designers. As military organizations took over the maintenance of weapons, the Surveillance group would take on the function of auditing military procedures.

The year 1949 saw the move of SLM to a larger, specialized building that offered improved training facilities and allowed for establishment of weapon and component displays. From its inception, this training room has been a Laboratories' focal point for all military liaison functions. It continues to provide a window to the world of nuclear weapon hardware and research and development for countless military and civilian VIPs, including President John F. Kennedy. In the 1948–1949 time frame, Sandia Surveillance, Liaison, and military personnel worked closely together to meet reliability and readiness goals.[24]

Quality Assurance

During this period, AEC headquarters in Washington, D. C., began to take more of a direct interest in surveillance activities. In 1948, AEC Chairman David Lilienthal requested a formal report detailing particulars of Sandia's new stockpiling surveillance program. With the assistance of William C. Kraft and William Caldes, Paddison assembled the report, three-fourths of which consisted of a list of all the spare parts they could locate. The report did include a brief outline that became the basis for a quality control program.

Paddison sent the report on to AEC headquarters, where Lilienthal read it with interest. The chairman still had questions. This meant a trip to Washington for Paddison and a face-to-face question-and-answer session. At the end of the meeting, Lilienthal was satisfied with the plans to initiate a stockpile surveillance program. As the size of the stockpile grew, the need for a monitoring system of checks and balances became more and more clear.[25]

Among the factors leading to the growth of the stockpile and the need for quality assurance was the fear of Soviet aggression. As early as 1946, the Joint Chiefs of Staff had anticipated a Russian attack in Western Europe. The American response was formulated in terms of Operation Pincher, termed "the first theoretical plan for war with Russia." Not only did the Pentagon believe that the Russians would provoke a major conflict, they believed that it could occur with only a few months' advance warning. Historian Gregg Herken in his analysis of the plan explains that "most remarkable about Pincher was not its assumption of American military strength because of the bomb, but its tacit admission of relative weakness."[26] By 1948, the Joint Chiefs of Staff also envisioned a Russian attack on the continental U.S., a fear further exacerbated by the Berlin crisis. In this eventuality, they knew that the then-current goal of ten nuclear assembly teams would be inadequate. Reevaluation of the situation led them to establish as an objective a final assembly rate of

"LET'S GET THE SHOW ON THE ROAD"

QUALITY ASSURANCE
1950

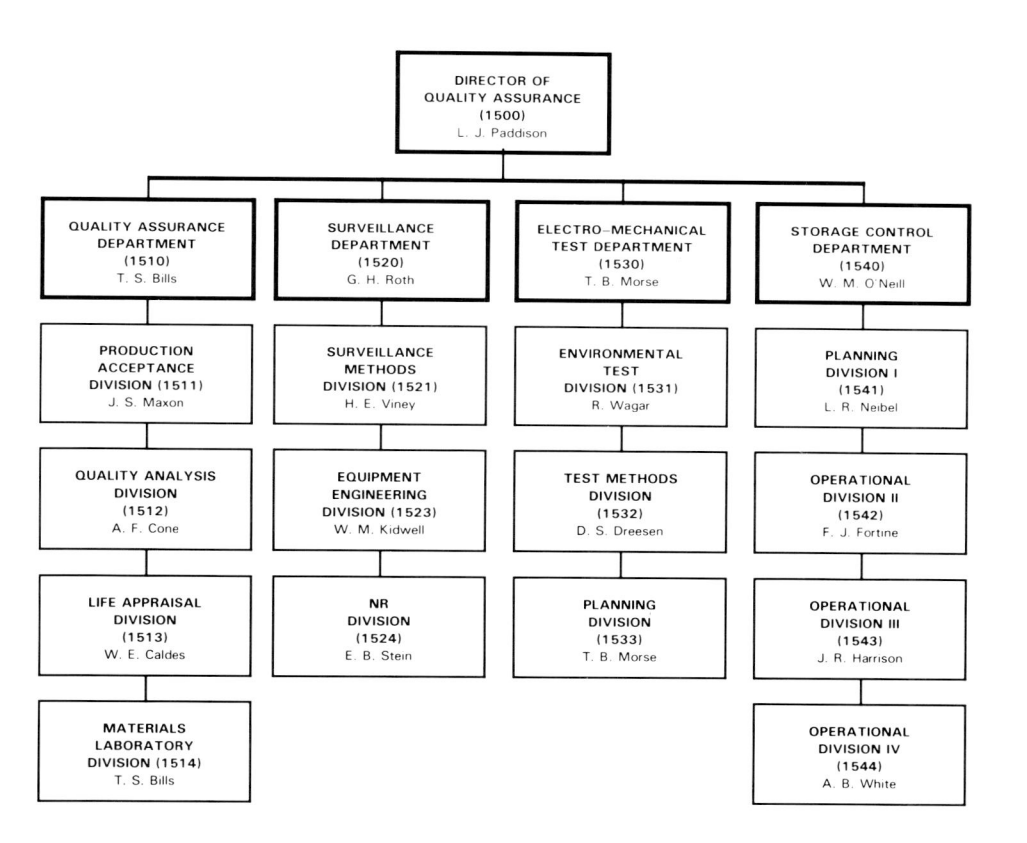

- **DIRECTOR OF QUALITY ASSURANCE (1500)** — L. J. Paddison
  - **QUALITY ASSURANCE DEPARTMENT (1510)** — T. S. Bills
    - PRODUCTION ACCEPTANCE DIVISION (1511) — J. S. Maxon
    - QUALITY ANALYSIS DIVISION (1512) — A. F. Cone
    - LIFE APPRAISAL DIVISION (1513) — W. E. Caldes
    - MATERIALS LABORATORY DIVISION (1514) — T. S. Bills
  - **SURVEILLANCE DEPARTMENT (1520)** — G. H. Roth
    - SURVEILLANCE METHODS DIVISION (1521) — H. E. Viney
    - EQUIPMENT ENGINEERING DIVISION (1523) — W. M. Kidwell
    - NR DIVISION (1524) — E. B. Stein
  - **ELECTRO–MECHANICAL TEST DEPARTMENT (1530)** — T. B. Morse
    - ENVIRONMENTAL TEST DIVISION (1531) — R. Wagar
    - TEST METHODS DIVISION (1532) — D. S. Dreesen
    - PLANNING DIVISION (1533) — T. B. Morse
  - **STORAGE CONTROL DEPARTMENT (1540)** — W. M. O'Neill
    - PLANNING DIVISION I (1541) — L. R. Neibel
    - OPERATIONAL DIVISION II (1542) — F. J. Fortine
    - OPERATIONAL DIVISION III (1543) — J. R. Harrison
    - OPERATIONAL DIVISION IV (1544) — A. B. White

Source: *Sandia Corporation Telephone Directory*, October 1950

twenty-five units per day, a correlating increase in the number of teams, and expansion of training programs. Plans for the construction and use of storage sites were also accelerated.[27]

At Sandia Base, procedures for emergency transfer of weapons were being refined. According to plan, the President, in an operations order, would notify the AEC chairman that one or more atomic bombs were to be released from AEC custodial control and be turned over to specified military units for preparation for operational use. All pertinent Washington political and military personnel, including the chief of AFSWP, would be alerted. The chief of AFSWP would notify the field organization. In turn, the Commanding General, Sandia Base, would inform the appropriate Sandia representatives, who used special code words to initiate the Sandia plan of action. Spelled out in a special "Operation Memo," the plan set in motion the chain of events resulting in delivery of atomic weapons. Shortly after notification of transfer, the AEC custodian was expected to arrive at the storage igloo where transfer would take place.[28] Such formalized procedures for transfer, inspection, and testing were being worked out not only for Sandia, but also for application at the various production centers and at other storage sites.

Alert Status, Site "Easy"

At the operational storage sites, certain cleared visitors were allowed to observe the alert, assembly, and surveillance exercise. Representatives from the integrated contractors, for example, might accompany Sandia Surveillance personnel to see firsthand how their components held up under actual assembly and test conditions. On one such occasion, Lou Paddison made arrangements for a group of visitors from Sandia and Bendix to fly to Site "Easy."

As the plane approached, the passengers saw below a cold-looking terrain with rolls and swales like the nearby sea. Patches of potato plants covered vast areas of rolling hills cleared for that purpose. Islands of spruce and poplar formed dark green shadows against a white cover of melting snow. The Air Force had first arrived here in 1947 to convert farmland and forest into a Strategic Air Command Base.[29]

Site Easy, the first of the "Operational Sites," like others, was located near SAC bases and operated by the military. Smaller than those known as "National Sites," Operational Sites usually contained only one assembly plant, fewer storage igloos, and fewer personnel. These sites were designed to be forward distribution points as contrasted to the major storage depots, or National Sites, which were operated by the AEC. The exact locations of the sites would remain highly secret.[30]

Despite some communication problems with a ground controller who at first refused to allow them to land, Paddison and observers arrived in time for the alert. The exchange of paperwork between the AEC and the military had taken place, initiating the transfer and ascertaining that all command channels concurred in release of the weapon. The AEC would release custody only upon receipt of an authenticated emergency action release message to the military. "In those days," Paddison noted, "the nuclear capsule or initiator was handled separately. Sandia had overall surveillance responsibility, but a separate AEC custodial group watched over the core, which was separately stockpiled and released."[31]

A phone call in the middle of the night rousted assembly team members out of bed. A rush trip to the site got them there within the specified period of time. Surrounded by two fences, one of which was electrified and illuminated by bright lights, the site looked secretive and forbidding. A patrol road manned by the resident Air Provost Security Squadron wound around the perimeter of the area. Security squadrons provided protection at all sites; however, an additional guard force was assigned to this particular site because of proximity to the sea and the Canadian border.

Team members and observers were informed of the specific mission, then entered the main tunnel, which extended some 200 to 300 feet before branching out into various rooms. On one side they could see the power plant where two large diesel motor generators stood ready as an emergency power supply. There were also a mechanical bay, an electrical bay, and the readiness or breakdown bay where team members stripped down the weapon for reassembly. The facility included workshops, machine shops, the nuclear inspection area, and a nuclear storage area, located at the end of another tunnel away from the main plant. Observers noticed that the tunnels and passageways were constructed so as to attenuate shockwaves in the event of an accident.

Team members charged up the batteries and began assembly of the training weapon. (Live weapons were never used.) Components were removed from the assembly line for testing at various stages. Assembly personnel took the completed unit, covered to protect its classified shape, to a waiting convoy surrounded by uniformed guards. The convoy then transported the unit to the aircraft, where the bomb was loaded on the plane. This was a difficult task because the early Marks were so large that the B-29 afforded little work room. With the bomb hoisted in the bomb bay, the plane prepared for takeoff. After the aircraft returned, the bomb was disassembled, tested, packaged, checked by the custodian, and replaced in stockpile.[32]

Although surveillance and assembly personnel regarded their jobs as serious business, the assignments sometimes had humorous sidelights. Sam Johnson, who succeeded Sam Egger as Sandia's technical representative at Site Easy, recalled how alarmed he became to learn that one of the bomb casings was missing. After a frantic search of the area, Johnson found the casing down by the lake. Apparently, a couple of GIs had borrowed it with the idea of using it as a fishing boat.[33]

Project "Water Supply"

Before establishment of the Easy Storage Site in 1950, three National Storage Sites "Able," "Baker," and "Charlie" had been constructed at a cost of about $28 million each. Code-named

Project "Water Supply," the plan for the design and construction of Storage Sites was originated by the DOD; however, Sandia personnel played an important role in their establishment and operation.

Structures expert Roy Carlson, who with Bob Henderson designed the containment vessel for the Trinity shot, coordinated contract work with the DOD to evolve plans for the construction of the various sites. Funding from the DOD was funnelled into the Corps of Engineers, which appointed as Project Manager, Colonel Benjamin Rogers.

Sandian Jerry Jercinovic, who served as technical liaison on Project Water Supply, described Rogers as "a crusty old war horse" who would typically blast into town by military transport and pick up Jercinovic to go check out the site construction. Not only was it important that the sites be built in strategically secure locations, they also had to be highly invulnerable to attack. For this reason, project engineers initially favored underground locations—also because they afforded greater protection for the surrounding populace in the event of an accident.

In keeping with the concept of strict security, code names based upon alphabetical sequencing were assigned to the sites. Later, with the establishment of foreign sites, the sequencing was intentionally skipped to emphasize the discontinuity of locations. For security purposes, not more than one-third of the total nuclear or nonnuclear stockpile was stored at any one site.[34]

Information regarding the sites was given solely on a need-to-know basis. Joe Heaston, for example, after learning that he had been assigned to Baker Site, asked Larry Neibel the location. Neibel, only half-joking, replied: "It is so secret, Joe, they'll probably blindfold you until you get there." Disgusted, Joe walked off. However, when he returned from his trip, he approached Neibel again: "Larry," he said, "I hate to tell you this, but we had a tragedy on this last trip—The engineer peeked and we had to shoot him."[35]

Baker was the first site finished. Movement of components into the site's igloos began on March 23, 1948. Complete Fat Man weapons less capsules, as well as urchins, detonators, and fabricated parts of the Little Boy weapon were moved into stockpile at this time. Atmospheric humidity posed some problems at the Baker repository. Los Alamos continued to provide transitory storage for nuclear components because of requirements for fabrication and repair, but it was not considered a National Storage Site.

The second National Storage facility, Charlie, housed complete weapon assemblies, including active materials. Completed by August 1, 1949, the AEC in January of the following year approved construction of additional aboveground igloos and warehousing facilities for all three original sites. The third site, Able, also opened for operation in 1949. Frank Fortine is remembered as the capable individual assigned the task of opening many of these early sites.[36]

Site construction teams sometimes encountered unique problems. While drilling the tunnels for one particular site, miners ran into an underground stream. The decision was made to reroute the stream and proceed. There were also problems with rats who enjoyed munching on the wiring. And on August 1, 1950, a flood hit the area, leaving a foot of mud blocking the entrance to Igloo Number One. Fortunately, inspection failed to show damage to the pits and tuballoy rings stored in boxes sitting in two inches of mud and water inside the igloo. At this same site there was more excitement when miners supposedly uncovered a promising vein of gold ore. The decision was made to seal it up and proceed.[37]

Sandians on Site

William O'Neill, who was hired in December 1948 to take charge of the Sandia group that manned the sites, brought with him valuable experience with the military. O'Neill recalled that Sandia project engineers were given a free hand in hiring people and in establishing logistic channels because these sites required tools and supplies that had to be forwarded from Sandia. "As each site opened," according to O'Neill, "it became an enormous task to coordinate all these shipments."[38]

News that a certain modification had been ordered meant that Sandians onsite must procure and arrange for shipment of all parts needed to replace or add to those already on the weapon. The parts had to be shipped to the sites under AEC vigilance.

It became imperative to establish an efficient logistics, records, and planning section at headquarters, Albuquerque. Since the early Marks were often modified in the field, errors in weapon status could be catastrophic. Jim Porter, remembered as "the prime record keeper" of those days, therefore, held an important and demanding position. Each site had a Base spare parts warehouse that stocked all the nuts, bolts, and screws as well as major components. Stock control was essential.

If the components needed could not be obtained through channels, Sandia project engineers on site purchased special supplies locally. To pay for such expenses, project engineers were given a bank account of $1,000, which afforded them a freedom and latitude uncommon in industry.

In the opinion of Larry Neibel, who was eventually placed in charge of Sandia activities at the Operational Sites: "One of the most important things we provided was a continuity in personnel, knowledge, and records." Noting that military personnel were transferred every three years or so, Neibel said: "We were that thread of communication back to the plants when weapons problems came up."[39]

A decision to drastically increase the number of bombs in stockpile led to a concerted effort starting in 1950 to increase the number of repositories for the national atomic stockpile. By 1956, a total of thirteen Storage Sites were in use. In the interim, Sandia's involvement in surveillance activities grew to the extent that in 1950 a formal Quality Assurance Directorate was established under L. J. Paddison.[40] Meanwhile, the University of California, the government, and industry had just completed negotiations that would chart the future course of the Sandia Laboratory.

SITE RESPONSIBILITIES
June 1950

Org.	Agency Head	Responsibility for All Sites	ABLE In Charge at Sites	BAKER In Charge at Sites	CHARLIE In Charge at Sites	Duties
AEC	Tyler	P. Agar	H. Ream plus 8 men (Mays)	G. Jelsen plus 8 men (Haley)	R. Stauffer plus 8 men (Sutton)	Custody of all components and kits in storage
AFSWP	Brig. Gen. Montague	Lt. Col. Free	Lt. Col. Brown Site Comdr. Comdr. Vannoy Deputy Site Comdr. 111th & 122nd S.W.G. plus 144th & 133rd S.W.G.	Lt. Col. Gandy Site Comdr. Lt. Col. Hikel Deputy Site Comdr. plus 510th Aviation Sqdn. plus 551st Aviation Sqdn.	Lt. Col. Cline Site Comdr. Lt. Col. Sheppard Deputy Site Comdr. plus 580th Aviation Sqdn.	Operate and maintain Site; provide security, handling & transportation. Also, provide man power & equipment to assist AEC and surveillance activity.
Chief of Engrs.	Lt. Col. Rogers, Dist. Engr.		Klenche*	Beasley*	Bordelon*	
S.C.	G.A. Landry	L.J. Paddison (O'Neill)	Fortine plus 3 men	Harrison plus 10 men	White plus 8 men	Maintenance and surveillance of stockpile
SAC	Gen. Lemay	SAC Hdqtrs.	**Gen. Bunker 15th Air Force	**Maj. Lucas 8th Air Force	**Lt. Col. Woolwine 2nd Air Force	Support to Site. Perform loading and armament function maintenance of air strip.

* - These men have dual functions:
1. Resident representative for Area Engineer in charge of construction. (Reporting to Lt. Col. Rogers).
2. Resident project post engineer reporting to site commander.

"LET'S GET THE SHOW ON THE ROAD"

Signs reminded all who entered the Kirtland Ordnance area that strict rules regarding security procedures would be enforced.

Mervin J. Kelly, President of Bell Telephone Laboratories, 1951–1959.

Chapter XII.

SANDIA CORPORATION:

From University to Industrial Management

"I am informed that the Atomic Energy Commission intends to ask that the Bell Telephone Laboratories accept under contract the direction of . . . Sandia Laboratory . . . . In my opinion you have here an opportunity to render an exceptional service in the national interest."

President Harry Truman to
Leroy A. Wilson, President, AT&T

The man at the head of the conference table sat in heavy-lidded silence. The rimmed glasses gave him a scholarly look, like that of a college professor. But those present knew Mervin J. Kelly as a hard-driving scientist and engineer with expertise in electronics and telecommunications. He was also one of the world's foremost scientific administrators. Bradbury wondered if he had fallen asleep. "Are you tired?" he whispered. Kelly opened his eyes. "No, I always listen with my eyes closed," he said.[1]

The meeting, held in Bradbury's office in one of the green wooden buildings at the Los Alamos Lab, proceeded. Conversation centered on Kelly's forthcoming evaluation of the technical programs at Los Alamos and at the ordnance branch in Albuquerque. Those who participated in this and other technical reviews conducted by Kelly through the years remember him as "tireless, demanding, and inspiring." His ability to absorb hour after hour of briefings on technical and administrative matters never ceased to amaze those who talked with him.[2]

Kelly, accompanied by James B. Fisk, Director of Research for the AEC, and James McCormack of the DMA, had arrived on the *Chief* on Sunday, March 6. This was the first of thirty-three trips to Sandia and Los Alamos that Kelly would make between the years 1948 to 1958.

His schedule on this initial visit began in late afternoon of the day of his arrival with a preliminary session with Carroll Tyler and Norris Bradbury. Monday evening, McCormack hosted a party for the visitors. He thought that the informal atmosphere might provide Kelly with a bit of relaxation and the opportunity for a more candid insight into the true status of affairs at the Labs. As McCormack explained to Bradbury, he felt that Kelly might "benefit from the slightly different form of wisdom that ... usually flows from your people after dark."[3]

In contrast, the daytime session held in the director's office a couple of days later was a forthright information-gathering session and only one of a series of consultations with members of the AEC, the military, Sandia, and Los Alamos personnel.

"This Whole Sandia Matter . . . Seems to Have Gotten Out of Hand"

When Kelly arrived in 1949, the Branch Lab at Sandia Base in Albuquerque, although modest in size, was recognized by the weapons community as the nuclear ordnance engineer for the nation. While this recognition offered some feeling of security to national leaders, it had created increasing concern on the part of the University of California Regents and Los Alamos Director Norris Bradbury. It was inappropriate, they decided, for a university to be associated with the engineering aspects and the actual production of weapons, particularly in peacetime. Yet, the Branch Lab at Sandia was doing just that. There was also an awareness of the need to strengthen the Laboratory's technical capabilities.

The concerns of the Regents were not new. The University had never been pleased with the de facto, but informal extension of its contract to manage the ordnance facility. As early as June 1947, Secretary and Treasurer of the Board of Regents, Robert Underhill, registered his disapproval to Bradbury. "This whole Sandia matter," Underhill said, "is one that seems to have gotten out of hand."[4] In his opinion, "the University could not protect its or the government's interest at branch stations." Consequently, he advised that the Lab should be given over "to the AFSWP" where he always thought it belonged.[5]

During the next two years, the University staged reruns of the same scenario. The plot was similar. The Board could sanction basic research in the field of nuclear science, such as that being done at Los Alamos; but research and development of nuclear ordnance, as well as manufacturing (even though performed by outside contractors), was not a proper academic pursuit.

Bradbury carefully fostered these attitudes. In later years when asked if he had been "aware at the time that this separation was in the planning stages of the concern of the Regents," he stated quite bluntly: "No. I invented the concern." Underhill elaborated: "The committee was down there [Sandia], and he had a little demonstration one day." (Underhill was referring to a speech Bradbury made to the Regents during their visit.)

Bradbury explained further: "Within a year [Z Division] was almost as big as Los Alamos, and I had lost control of it. . . . It didn't seem to be the Laboratory's business; it was ordnance engineering. And I sort of invented, with all respect, the idea that it wasn't any of the University's business either. . . . So I made this speech . . . , and it sold."

"It didn't take them long to figure [it] out," he added. "They saw airplanes buzzing around."

Underhill recalled the reaction of one Regent in particular: "When the word came that this was all research and development, and they [Sandians] were packaging things up and stockpiling them, I remember McFadden turned around to Norris and says: 'This is research?' "

Bradbury explained the difference in the Lab's function:

> Except for actual fission materials, things intimately concerned with the actual nuclear detonation itself, they [Sandians] were the in-between, but in the production empire of the AEC, and the stockpile, they did the ordering; they did the checking; they did the testing. . . . Sandia had the responsibility for the cases and the detonator, the fuzing, the firing, the ballistic shapes, the umbilical cords, the hooks and lifts and the handling equipment . . . ; which is very important, but it was of zero interest to me.[6]

Other factors also impinged upon the decision. For example, University officials felt that they were not in a position to provide the Branch Lab the type of administrative and technical support needed. DMA Director James McCormack agreed. He was concerned about the University's responsibility for the $100 million budget and felt that the magnitude of the financial accountability needed a businessman's rather than an academician's approach. He went so far as to contact various industrial companies to see if they would consider the possibility of taking over the Sandia operation, but the responses he received were indifferent.[7]

Roger Warner also made inquiries. In the fall of 1948, he talked to Oliver E. Buckley, President of Bell Telephone Laboratories (BTL), about the prospect of a takeover. At the time, however, Buckley indicated that they would not consider it. Internal policy, he said, restricted BTL to holding its defense business to fifteen percent of its total activity.[8]

On December 31, 1948, the Regents made their request official: They wanted the University of California disassociated from the production facility in Albuquerque and removed from their contract by July 1, 1949.[9] The onus was on the AEC.

In support of the move, Bradbury observed that, from the technical viewpoint, bringing in a contractor with an "extensive military engineering background" would improve the strength of the Branch Lab. And, too, a separate contractor would leave personnel at Los Alamos free to concentrate on warhead research and development without corresponding administrative worries.

From the personal perspective, as Bradbury candidly iterated: "I was running a division of the laboratory by remote control and by a charter they had set up." "Furthermore," he added, "Larsen was a real live wire, an operator. This made *me* nervous and the University of California nervous. I was running the place only nominally."[10]

The validity of Bradbury's observations on Larsen's operating style and personality could not be denied, but neither could one deny his accomplishments. By 1949, the major building program at headquarters in Albuquerque and at the Salton Sea site was thirty-five percent complete. The number of personnel on roll had grown from 370 to 1,720, and total disbursements per year had risen from $18 million to $61 million.

The amount of design work being turned out was equally impressive. The Little Boy weapon (previously in development) was released for production. Improved versions of the Fat Man offered enhanced reliability and safety, and a completely new model had been designed and placed in Road. Meanwhile, the Laboratory was continuing efforts to improve components, expand its training program for military personnel, plan for weapon storage sites, and organize a stockpile surveillance program. A program for the development of guided missiles was in the offing.[11]

Larsen had not tallied such an impressive array of accomplishments without personal sacrifice. The job took its toll. In December 1948, the forty-nine-year-old director collapsed in his garage—apparently from overwork. The incident concerned members of the military at Kirtland, who used the occasion of Larsen's illness to comment on the director's "growing ego and irascibility," felt to be "traceable to his lack of rest."[12]

The observations were not unwarranted. Despite the positive strides made under Larsen's leadership, the increasing demands upon the new Laboratory exceeded its ability to carry out all the complex procurement, production, and development tasks assigned to it. Relations with representatives of the armed forces had improved, but Sandia still came under fire from military brass who complained that the Laboratory displayed "consistently inferior capabilities to say nothing of third rate engineering qualifications."[13]

Underneath such outbursts smoldered the issue of civilian versus military control. Not surprisingly, the military establishment joined University Regents and Bradbury in voicing support for new leadership and transfer of the Laboratory's contract to an industrial concern.[14]

Larsen himself favored the idea of a transfer of the reins of management, and on January 11, 1949, proposed to the AEC that the Laboratory be operated as a New Mexico state corporation to be known as Sandia Laboratory, Incorporated. Rather than accept Larsen's suggestion, however, James Fisk suggested that it would be wise to select an impartial and qualified individual to look at the Los Alamos operations. Fisk, who still held close ties to BTL, felt that the Laboratory's Executive Vice-President, Mervin J. Kelly, would be a good choice.[15]

Kelly, who held a Master of Science degree from the University of Kentucky, earned a Ph.D. in physics from the University of Chicago, where he had been an assistant to the well-known physicist Robert Millikan. In 1918 he joined Western Electric as an engineer. As part of a small research group—the forerunner to Bell Telephone Laboratories—Kelly became part of a project to improve the then-new vacuum tube. Under his direction, the group successfully developed a reliable water-cooled tube that insured the success of transoceanic radiotelephony. Kelly went on to obtain seven patents and move rapidly up the Bell Laboratories' management ladder. During World War II, he had responsibility for BTL's 1,200 military research and development programs; and in the final months of the war*, he organized the team effort leading to invention of the transistor.[16]

Carroll Tyler, Manager, AEC's Santa Fe Operations Office, agreed with Fisk that Kelly would be a good choice. Subsequently, at McCormack's suggestion, Lilienthal approached Oliver Buckley to see if Kelly would be available to conduct the review.[17]

---

*President of Bell Labs from 1951 to 1959, Kelly was praised by Nobel recipient William Shockley for "the managerial skills that created the atmosphere of innovation at Bell Labs."

Buckley, known for his pioneering work on high-speed telegraph cables, also served as a member of the GAC. As such, he was well aware of the importance of the request. After coordination with American Telephone and Telegraph management, Buckley advised Lilienthal that Kelly's time would be made available. But there were contingencies: Kelly would act as an individual, rather than as a representative of Bell Laboratories. All arrangements would be made personally through Kelly, who would conduct the survey, give an evaluation, and make recommendations.[18]

There were other conditions: First, no written report would be made, and only one oral report. Secondly, if Kelly thought a change in management should be made, he would make no specific recommendation as to who the new management should be. Third, the AEC would back the change if one were suggested, since it was possible that "neither Bradbury nor Larsen would be enthusiastic about replacing the current management at Sandia Base."[19]

Mervin Kelly Recommends Changeover to Industrial Management

To understand the Sandia problem thoroughly, Kelly had to comprehend the physics and engineering of the whole weapons job. This meant spending time at Salton Sea as well as at Los Alamos and Sandia. There were three major factors under consideration: the departure of the University of California as contractor, the expanding program of the Sandia Branch Laboratory, and the increasing proportion of activities devoted to production rather than research and development. From the perspective of the AEC, there was also concern that the Commission's long-range goals for Sandia, including stockpile quotas, might not be taken into account.

Kelly conducted his review during the months of February, March, and April. By May 4, 1949, he was in Washington to make his oral report to the AEC. Bradbury of Los Alamos, Tyler, and

Admiral Kraker of the Albuquerque Operations Office were present, having flown in from New Mexico to hear the report. A stenographer sat ready to transcribe the presentation.[20]

Kelly began by commending Commission personnel in Washington, in Los Alamos, and at Sandia for their frankness and cooperation. General Montague's people in the Armed Forces Special Weapons Project, he said, "were equally frank, and in some cases quite critical."[21] Kelly continued by addressing the technical aspects of each operation, relations of the Labs with the military, and administrative concerns.

He reviewed the Los Alamos operation first. In looking at the work being done there, he found that the technical strength of personnel had been increased greatly since the Laboratory low in 1946. He was pleased to find pay scales and recruiting procedures comparable to those at Bell Laboratories. Top technical personnel, he said, showed maturity and experience with "a high degree of imaginativeness."

His only real criticism of the Los Alamos Lab related to the lack of emphasis on long-range research. To solve the problem, he advised that a nucleus of personnel be set apart to devote themselves to fundamental research. Accordingly, he recommended that the AEC consider the Los Alamos Lab as having "indefinite life."

As for interaction with the military, Kelly observed that the relationship between General McCormack (who supervised the Laboratory for the AEC, as well as its director) and Bradbury appeared to be cordial. He noted somewhat pointedly, that it was fortunate from his perspective "that the University of California did not intrude in the philosophy and functional operation of the weapons program." "The Los Alamos Scientific Laboratory," he concluded, "was the best looking from the points of view of scientific capacity, spirit, and sense of the job to be done of any Government lab he knew of."[22]

In contrast, Kelly described the picture at Sandia as "less rosy." Sandia—despite the progress made over the last three years, especially in the instrumentation of the Mk 4, Mod I and its external details—failed to show professional competence in other phases of work. The largest deficiency appeared to be in the applied physics area. Some six years later, while testifying before the Personnel Security Board in Washington during the Oppenheimer hearings, Kelly's assessment had not changed. The deficiency that remained in his mind "as the only blemish on the program" was the "inadequacy of the applied technology having to do with aerodynamics, electronics, and so on."[23]

In 1949, however, he was more specific. Leadership in the applied physics area, he said, until R. P. Petersen took over, did not have the necessary strength, and "despite the courage, clear thinking, and leadership of Dr. Petersen, there was still need for improvement."

Another pressing need, Kelly determined, was for professional and mature "design for manufacture" leadership, a person with experience in designing for endurance and reliability. The Road job and production engineering, he said, should have "the best professional handling the country could provide," especially since a manufacturer (Bendix) was coming into production at Kansas City.

Surveillance and quality control could also be improved upon, according to Kelly, although he viewed L. J. Paddison, as "a young man, quite capable . . . with vision and leadership." Kelly complimented other Sandians as well. He considered R. A. Bice, for example, "a good design engineer," but qualified his observations on the Laboratory overall by stressing repeatedly the need for more professionalism.[24]

In addressing the AEC-military relationship at Sandia, Kelly explained that this was an issue of operational and political importance. As the user for Sandia products, the military had to be considered in the design of a product.

Lilienthal interjected with a question. Since much of this weapons work was unattractive to civilians, why couldn't the whole job be turned over to the military?

"The two-headed management of the weapons development program would be wrong at the present time," Kelly responded. "Either the Commission or the Military Establishment should have the whole research and development job," he said, "and for the immediate future, the Commission is better placed for total control." Furthermore, if Sandia were turned over to the military, in Kelly's opinion, "the intimacy and freedom" of the relationship between Sandia and Los Alamos would be endangered. It was important, that the work at Sandia and Los Alamos continue to be considered "a unit job."[25]

The establishment of boards by the AFSWP, the AEC, and the University of California as formal meeting grounds for exchange of information and points of view, Kelly felt, was a positive step. He discussed ways of improving their functioning with Generals Montague and McCormack, and the Military Liaison Committee. Of the various committees, the Research and Development Board appeared to need the most structural change.

In the area of administration, Kelly acknowledged that "while . . . Larsen at Sandia reports to Dr. Bradbury, he had been detailed . . . a great deal more responsibility for operation than his equals at Los Alamos." Kelly complimented Larsen for his hard and conscientious work, but noted that it was impossible for one man to have the knowledge sufficient to meet "all functions of the job at Sandia." The only way these multiple requirements could be met and provide what the Commission needs, Kelly determined, was "to bring in an industrial contractor with experience, professional know-how, and a sense of public responsibility." Such a contractor, he said, could supply from within its own organization "a general professional direction of operations" beyond the wisdom of one individual. "A mature and public-spirited contractor," he concluded, "would be able to work out smooth relationships with Dr. Bradbury on the one hand and the Military Establishment on the other."[26] In accordance with the agreement, Kelly refrained from suggesting any given company.

The AEC Selects AT&T

Who should the contractor be? McCormack suggested a number of concerns, among them North American Aviation,

General Electric, Bendix, and Borg Warner. But the prime candidate was AT&T.[27]

Considering Kelly's professional affiliation with BTL, the choice of the American Telephone and Telegraph industrial complex may have raised questions about the selection process—but only to those unfamiliar with the Laboratories' outstanding reputation. As author John Brooks has written, along with being an extremely wealthy company, AT&T "is one of the most widely experienced."[28]

Experience, from the viewpoint of the AEC, was an important consideration. AT&T's range of experience dated back to 1875 when Alexander Graham Bell's invention of the telephone launched a communications revolution. Ten years later, AT&T's predecessor company, American Bell Telephone, became American Telephone and Telegraph.[29]

In view of the vital nature of the defense-oriented missions at Los Alamos and Sandia Laboratories, AT&T's track record of success and its broad experience base were decided assets. In addition, a company offering a wide range of capabilities was essential. Kelly's report stressed not only the importance of improving production capabilities at the Sandia Lab, but also the need to provide support for long-term research efforts at Los Alamos. The Commission, therefore, was searching for a contractor that could meet both the direct management needs of the Sandia operation, and indirectly, because of close technical liaison, the needs of Los Alamos. The industrial complex selected should be capable of starting with research, carrying through development, and continuing with prototype production.[30]

AT&T offered this combination. On the one hand, it had Bell Telephone Laboratories, skilled in research and development related both to pure and applied science, and on the other, the Western Electric Company, the Bell System's manufacturing arm.

The genesis of Bell Telephone Laboratories can be traced to 1907. In that year a small nucleus of personnel from the engineering departments of AT&T and Western set up shop on the outskirts of New York City's Greenwich Village. By January 1, 1925, the Western Electric Research Labs and part of the engineering department of AT&T were joined to form Bell Telephone Laboratories, Inc. In this year BTL came under joint ownership of AT&T and Western Electric. BTL's function as design–innovator for the equipment that Western Electric manufactured and the Bell System used, soon made the Laboratory an integral cog in the Bell System wheel. The Laboratory's systems approach (which would duly influence the Sandia operation) involved studies to determine, systematically, research and development projects that should be pursued.[31]

Operating as a regulated monopoly in a competitive society, AT&T held a unique and paradoxical position. Its history of complex government relations and contributions to defense efforts, and its acquisition of Western Electric, one of the nation's largest manufacturing companies, made AT&T well qualified for the task for which it was being considered.[32]

Like BTL, Western Electric had a long roster of accomplishments. Its predecessor, Gray and Barton, began operations in a Cleveland loft in 1869. In the 1870s, the firm moved to Chicago and changed its name to Western Electric Manufacturing. The company became a haven for inventors who produced, among other items, the first commercial typewriters and Edison's electric pen, a device that led to development of the mimeograph. The Bell System's purchase of a major interest in the manufacturing company in 1881 heralded its entrance into the production of telephones and related equipment. By World War II, the company could add to its credits the high–vacuum electronic amplifying tube that paved the way for coast–to–coast telephone calls and sound movies. The company also made important contributions to radio broadcasting, television, and radar for the armed forces, the first of many defense–related systems.[33]

As the primary supplier for the Bell System, Western Electric inevitably came under fire as the epitome of vertical integration. The Bell System's first confrontation with the federal government occurred in 1913, when the threat of antitrust action forced the Corporation to accept the Kingsbury Commitment, by which it agreed to refrain from attempting to monopolize U.S. telecommunications. But the move to separate Western Electric failed.

A similar confrontation was in the offing in 1949 when AT&T was being considered as contractor for Sandia. Ironically— and perhaps fortuitously for the ordnance lab—one arm of the government, the Justice Department, threatened AT&T with an antitrust suit, while another, the Atomic Energy Commission, beckoned it into the fold with the intent of asking a large favor. One action was destined to influence the other.[34]

President Truman Intercedes

On May 10, 1949, the Commission agreed that the Bell System was "outstandingly qualified as the industrial contractor (for Sandia)."[35] With the strong support of the GAC and the concurrence of the national military establishment, it was decided that Lilienthal should enlist the aid of President Harry Truman. "After careful consideration," Lilienthal wrote, "we believe it would be most valuable in approaching that organization (Bell Telephone Laboratories) and its parent organization, the American Telephone and Telegraph Company, if you would indicate a personal interest in the matter."[36] Attached to the memo were drafts of letters for the President to consider. All were similar in content, but the one Truman selected was well-phrased, succinct, and to the point.

The President acted quickly. In letters dated May 13, 1949, he informed AT&T President Leroy Wilson and BTL President Oliver E. Buckley that the AEC planned to ask AT&T to accept direction of the Sandia Laboratory at Albuquerque, New Mexico. At stake were not only the successful operation of a small nuclear ordnance and assembly facility located in the southwestern outback, but also the growth and reliability of the nation's nuclear stockpile.

Truman made a strong case for the government position: "This operation," he said,

> which is a vital segment of the Atomic Weapons program, is of extreme importance and urgency in the national defense, and should have the best technical direction.

> *I hope that after you have heard more in detail from the Atomic Energy Commission, your organization will find it possible to undertake this task.*

He then added a sentence that could not help but appeal to AT&T's patriotism:

> *"In my opinion,"* he said, *"you have here an opportunity to render an exceptional service in the national interest."*[37]

Four days later, Wilson replied. He promised to give the matter "prompt and sympathetic consideration," although he had not yet heard from the Atomic Energy Commission, nor did he know the "details of the problem." Buckley also responded positively and promptly. As a member of the GAC, he was well aware of the circumstances leading to the request for change-over. He promised to give "full consideration and weight to the importance and urgency of the project in light of the situation as I have come to understand it."[38]

On Memorial Day, May 30, 1949, Lilienthal, Carroll Wilson, and DMA Director General James McCormack met the AT&T President at his home to discuss the matter further. At this meeting, Wilson expressed concern about the pending antitrust suit, which had been filed January 14, 1949, under the Sherman Antitrust Act. How could one reconcile being asked to do the job suggested when the Justice Department was trying to break up the very combination that made it so suitable for the task? There was a logic in the AT&T position. Wilson wanted assurance that the Bell System's acceptance of the charge would not add fuel to the fire and be held against them. His point was well taken.[39] The matter would have to be resolved.

In the discussion that followed, the question of fee was also an issue. There were overtones of the "Merchants of Death" stigma. Like DuPont, AT&T *could* suffer from adverse publicity if the Bell System reaped a profit from a weapons operation. The situation, after all, was reminiscent of the circumstances that led to charges against DuPont in the 1920s and motivated passage of the War Powers Act of December 1941.[40]

AT&T, as Wilson pointed out, was already heavily committed to defense work and really did not relish another great load. The

Leroy Wilson, President of AT&T, agreed to accept responsibility for managing Sandia Laboratory, in the national interest. However, Wilson insisted that management of the Sandia operation should be on a no-profit, no-fee basis. On July 1, 1949, he formally accepted the contract on behalf of AT&T for operation of Sandia.

Oliver E. Buckley, President of Bell Telephone Laboratory and a member of the General Advisory Committee, was well aware of the circumstances leading to Truman's request that AT&T assume the management of Sandia Laboratory.

Corporation would accept the assignment, in the national interest, but management of the Sandia operation, at Wilson's insistence, would be on a no-profit, no-fee basis.

The following day, McCormack reported the results of the meeting to the Santa Fe Operations Office. "I visited with Wilson and Buckley of AT&T yesterday," he wrote, "and they are prepared to take the Sandia job, providing the Commission can establish the position relative to pending antitrust action by the Department of Justice." He indicated that he was preparing a written outline of the formal proposal and a description of scope of work as a first step in clarifying "this governmental issue."[41]

On the afternoon of June 6, 1949, Joseph Volpe, Jr., and Bennett Boskey, General Counsel for the AEC, met with Peyton Ford, Assistant to Attorney General Tom Clark, and Herbert Bergson, Assistant Attorney General in charge of the Antitrust Division of the Justice Department. Among the problems related to the changeover, the matter of the pending antitrust proceedings was high on the agenda. The discussion ended with Ford and Bergson in agreement: "There was no reason, from an antitrust standpoint, why the Commission should not go ahead promptly with its plans to obtain as the Sandia Laboratory contractor, a unit or combination of units within the Bell System."[42]

One week later, McCormack met with Wilson and Buckley at headquarters in New York. Wilson appeared to be pleased when he heard of the actions taken regarding the antitrust suit, but he had other concerns. He wanted the record to show that *AT&T* had not sought the Sandia job, rather that the *government* solicited AT&T for the job. He also requested a statement for stockholders explaining to them the reasons for the undertaking. He was especially concerned that AT&T might be accused of devising the Sandia contract as "a wedge" in the pending antitrust action. "He is most anxious . . . ," McCormack reported, "that the AT&T case . . . stand on its overall merits, which he believes to be conclusive without regard to the particular application of the integrated resources of AT&T to the operation of the Sandia Laboratory."[43]

Wilson added that he would like the issue carried one step further than the formal record. President Truman, he felt, should be made *personally* aware of Wilson's concerns and of the Commission's reasons for its proposal to AT&T. This would prevent the President from being surprised by a question or a comment from a newspaper reporter. If uninformed, the President's response could prove detrimental to AT&T, particularly in view of judicial scrutiny.[44]

As Wilson requested, Lilienthal, representing the AEC, met with President Truman to discuss AT&T's management of the Sandia operation. A memo for record in Wilson's files indicates that the AT&T chief was indeed given certain assurances. Lilienthal, according to Wilson,

> pointed out my personal concern . . . and also told the President of his discussions with the Attorney General [Tom Clark]. The President stated, according to Mr. Lilienthal, that he appreciated why we had been concerned but felt that both we and the atomic group should have no concern about the problem. Mr. Lilienthal did not say, but he indicated in an indirect way, that *there is no program by the Department of Justice to press the Western suit.* Again, in inference and not without words, there was a feeling that the Bell System could carry on just as we were without further attack. While these were only inferentially indicated, Lilienthal closed by being extremely complimentary about the people in the Bell System with whom he had talked on this subject and about our prompt and sympathetic response, stating that the nation would be far stronger a year from now because of the steps which had been taken. [45]

Provided with assurances and having lessened the potential of disadvantages for AT&T, Leroy Wilson on July 1, 1949, formally accepted the contract for operation of the Sandia Laboratory. In his letter of understanding and confirmation, he explained for the record the Corporation's rationale for accepting the task and his position on the antitrust suit. "The Bell System," he said, "has always stood ready to do its part in the national defense by undertaking work for which it is particularly fitted." He agreed that AT&T and its methods of operation gave them the special qualifications required. "For these reasons," he stated, "we are willing to undertake the project."[46]

It has been suggested that AT&T's major motive in assuming responsibility for the management of Sandia was to obtain a favorable position in the antitrust litigation.[47] As numerous references to the issue indicate, Wilson did seek early assurance that the government would not use the existence of an AEC/AT&T contract to establish its case in court. Wilson obviously viewed his efforts on AT&T's behalf as a leadership obligation. Yet, there were other determining factors—for example, the System's long history of service in the public interest. Management of the nation's nuclear ordnance lab certainly fell into that category, as Truman had clearly stated. Furthermore, much as AT&T has downplayed its contribution to U.S. defense efforts, a study of its activities during wartime, reveals the System's major involvement.[48] And, finally, there was also common logic in the official AT&T position on the antitrust suit as it related to the government's request to manage Sandia.

The suit did have its ironic aspects. In his letter of confirmation, Wilson reiterated concerns voiced earlier to Lilienthal and McCormack: "The antitrust suit brought by the Department of Justice last January," he pointed out, "seeks to terminate the very same Western Electric—Bell Laboratories—Bell System relationship which gives our organization the unique qualifications to which you refer." A contract with Western Electric to operate Sandia would not change Western's relationship with Bell Laboratories. "[Western] and the Bell Laboratories," he added, "would indeed work as one, as they do now in Bell System affairs, and the effectiveness of their work would depend, as we have explained to you that it always has, upon their close connection as units of the Bell System."[49] In essence, the antitrust suit could destroy the very combination that made the Bell System the best qualified for the management job. There was, therefore, a definite validity in the AT&T argument.

If indeed AT&T officials truly expected the Justice Department to lay aside its suit against them as a tradeoff for accepting management of Sandia Corporation, they were to be woefully disappointed. Even AT&T's request for a postponement of prosecution of the suit would be denied, although a de facto postponement until 1951 did occur. In the interim, the Department of Defense became AT&T's ally in the antitrust battles through the years.

In 1953, Secretary of Defense Charles Wilson reflected this sympathetic stance in a letter to Attorney General Herbert Brownell: "The Department of Defense," he wrote, "wishes to express its serious concern regarding further prosecution of the antitrust case now pending against Western Electric Company and AT&T Co."[50] The DOD's favorable attitude toward AT&T had no doubt been influenced to a large extent by successes of the Sandia Corporation under Bell System management in responding to national nuclear ordnance needs.[51]

Before any of the issues raised by the Department of Justice came to trial, however, AT&T signed a Consent Decree in January 1956. Central to the Decree was a requirement that AT&T divest itself of any business not subject to common carrier regulation. Once again, however, the Bell System would retain Western Electric, despite the Justice Department's desire to have the manufacturing company broken up into three independent and competing firms.[52] Divestiture would be postponed.

In 1949 the Bell System reacted according to its fundamental legacy of meeting the national need. By accepting management of Sandia Corporation, AT&T, because of the nature of its contract, had little to gain except governmental good will. Sandia, on the other hand, after the inevitable period of adjustment, would thrive under industrial/AT&T leadership and expertise. And in the broader view, the entire nation would profit from a drastically improved defense posture. Although critics would view the AEC–AT&T association as the epitome of the controversial "military–industrial complex," others would consider the Bell System affiliation as "a leading national defense asset."[53] But how did this marriage of the military, engineering, science, and industry come about?

Participants in negotiations leading to the changeover in management of Sandia Laboratory from the University of California to AT&T: left to right: Mervin J. Kelly, Bell Telephone Laboratories; George A. Landry, Sandia Corporation; Bennett Boskey, AEC's General Counsel office in Washington; General James McCormack, Division of Military Application (DMA); Colonel R. T. Coiner, DMA; Donald A. Quarles, Bell Telephone Laboratories; Stanley Bracken, Western Electric Company; Paul J. Larsen, Sandia Laboratory (while under University of California); Fred Lack, Western Electric Company; Richard Smith, AEC Procurement, New York Operations Office; George P. Kraker, AEC, Albuquerque.

Chapter XIII.

ON THE BASIS OF "GOOD INDUSTRIAL PRACTICE"

> ... the Government desires to utilize in the operation of the Sandia Laboratory of the Commission and related work, the management, engineering, scientific research and development, and manufacturing skills of Western and Bell Telephone Laboratories, Incorporated . . .
>
> —Contract AT-(29-1)-789 between USAEC, and Western Electric Company, Inc., and Sandia Corporation

On May 12, 1949, Sandia Laboratory department managers, division leaders, and staff assistants filed into the base theater at Kirtland. The atmosphere was charged with anticipation and anxiety. Speculation was rampant, although nearly everyone knew the purpose of the meeting. Rumors had been rife for some time about the prospects of a changeover to industrial management. An article in the local paper to the effect that Sandia was indeed to undergo a change of ownership fueled the fire. Carroll Tyler, thinking that a public announcement would help quell the rumors, had sanctioned the article.[1] Unfortunately, the reverse was true. Personnel were concerned about the ramifications of a change in management—both for themselves and for the future of the Laboratory.

NUCLEAR ORDNANCE ENGINEER FOR THE NATION

Larsen Responds to the Kelly Report

Sandia Director Paul J. Larsen took the podium. Larsen, who had expended considerable energy and organizational talent during his relatively brief tenure at Sandia, had his own agenda. Aware of the tension in the audience, he would use the session as a forum—if not for rebuttal of certain aspects of the Kelly report—at least for clarification of the record. As the Laboratory's leader, Larsen felt no small measure of responsibility for the Lab's success, or the absence thereof. Furthermore, he realized the potential for serious morale problems.

"The purpose of this meeting," Larsen began, "is to acquaint everyone with the plans on the future management of Sandia Laboratory."[2] After giving a brief résumé of events leading to the University of California's request for separation, Larsen continued by mentioning his proposal that a New Mexico corporation be formed to operate Sandia, and how the AEC had decided instead to have Mervin Kelly of Bell Labs conduct an independent evaluation of both the Los Alamos and Sandia Laboratories. Larsen noted that he, Dr. Bradbury, Captain Tyler, and Admiral George P. Kraker of the Albuquerque Operations Office had recently returned from Washington where they attended a special session of the Commission at which Kelly presented his verbal report. Consequently, Larsen added, the purpose of the present gathering was to pass on highlights of that report.[3]

From this point, Larsen's version of the Kelly evaluation makes a subtle, but interesting contrast to the official transcript and illuminates especially the political undercurrents of the Sandia–Military–AEC interface. Realizing the potential for morale problems, Larsen chose to emphasize the positive aspects of the report and downplay the criticisms. He started out by complimenting the quality of personnel. "Recruitment, at both Sandia and Los Alamos," he said, "was as good as any industrial organization could possibly perform." Continuing his positive approach, Larsen reported that Kelly "expressed admiration" for the job being done at Sandia, saying that "it had been one of the toughest jobs he has seen and that great progress has been made."[4]

Larsen proceeded to Kelly's evaluation of specific departments, making generally complimentary observations on Field Test, Military Liaison, and the Manual Preparation Division. He approached negative aspects of the report more gingerly. Although Sandia had "an excellent Engineering Department," Larsen said, it was "extremely limited in performing production or product engineering." Kelly, therefore, had recommended that a separate organization be formed to take over production engineering, which would allow the engineering group to concentrate on development engineering phases. "That is precisely what we ourselves had planned," Larsen observed.[5] Without mentioning names, he moved on to Kelly's concerns about the need for professionalism in the Road Department, acceleration of Surveillance operations, and support for the Applied Physics Department in order to enhance research activities at the Lab.

In regard to the functioning of the various boards at Sandia, Larsen noted that Kelly had recommended that the scope of the Research and Development Board be expanded to include broad research and development problems rather than limiting itself to analyzing "screws and nut–and–bolt items." Indulging in superb understatement, Larsen added: "I might mention that this is the only criticism made by Mr. Kelly of our entire organization . . . and in reality that [the Board] is not a Laboratory, but a joint Laboratory–military organization."[6]

Kelly's observations and recommendations on Sandia interaction with the military touched a sore spot with Larsen, whose comments indicated that he considered Kelly's perspective on that particular issue to be limited. Using the subject of crash programs as an entrée, he proceeded to set the record straight. Although the issue of crash programs is nowhere recorded in the official transcript of the Kelly report, Larsen indicated that the subject *had* been discussed. Ostensibly, Kelly charged Sandia with the establishment of an "orderly, planned program" for weapons development. Larsen admitted that the Lab was indeed conscious of the crippling effect of crash programs, especially in planning and long-range research. "Unfortunately," he continued:

> during the past year and a half there has been no help for it. The international situation was very critical, and for that reason we had to work on a crash basis to meet the military requirements. The pressure from the Washington level was terrific. Frankly, we've done a grand job. We've done more than any other organization in the country could have done under the conditions we found here in the fall of 1947.[7]

There was truth in Larsen's claims. Sandia had maintained schedules and, in fact, put into Road a new weapon nine months ahead of schedule. Larsen acknowledged the benefits of long-range research and planning, but added pointedly:

> The military must . . . not continually initiate urgent requests with dates that are impossible to meet, nor create crash programs to design, procure, and deliver material to them. They must realize that a great deal of time is involved between the inception of a new idea and the development of the final product.[8]

In a related vein, Larsen summarized the AEC's rationale for the selection of AT&T, noting that Kelly was strongly opposed to Sandia being directly operated by the government because he felt "it would affect contractual obligations, bring about eventual compulsory civil service for employees and pose all the other red-tape problems by which government laboratories are bound."[9]

The Sandia director concluded his presentation by reading a teletype from McCormack, assuring them that "there need not be any personnel uneasiness at Sandia," that "the Commission is keenly aware of progress to date by the present operation . . . and the vital necessity for continuity of competent personnel."[10]

The question-and-answer period that followed hinted at some of the problems the changeover would create. Regarding benefits and accrued leave, Larsen reported that Kelly had recommended to the AEC that any contractor taking over Sandia should continue the same employee benefits as those extended by the University of California. Larsen was in strong agreement. "The five weeks vacation . . . must remain," he said. "No personnel will tolerate reduction in benefits."[11] The director spoke prophetic words.

At the conclusion of the question–and–answer period, Larsen asked Admiral Kraker to comment on the Laboratory "from [the] AEC angle." Of paramount concern to Kraker was the continued production "of the end product going into the pile" and the possibility that the changeover might affect "the development progress of the new weapons." Nevertheless, the meeting ended on a positive note. "The Commission generally and everybody back in Washington," Kraker said, "is fully cognizant of the outstanding job that has been performed here. . . . "[12]

Bridging the Gap

The nation learned of plans for the transfer on July 12, 1949. A New York *Times* article, based upon an AEC announcement from Washington, succinctly explained a part of the rationale for transfer; namely, the growth of the Laboratory that, according to the announcement, "has been the result of the Commission's effort to integrate research, development, and production activities in accordance with the best academic and industrial practice and with the most competent available supervision in each technical area."[13] What the press release did not mention, however, was that the change in management also signified the Commission's recognition of Sandia as the hub of the developing defense complex, and the overriding desire of the AEC to ensure the continued mobilization of science and engineering in the name of national readiness.

As a preliminary step, the AEC dispatched a special team of officials from Western, Bell, and the AEC to Sandia to initiate the transfer.[14] In a letter to Senator McMahon, Lilienthal explained that the group would be studying requirements of the project and "laying the groundwork for early consummation of a contract for operation of the Laboratory." This task force would be involved in the first of a series of negotiations in which government and corporation, in the words of author John Brook, "played a game of cat and mouse."[15]

The representatives flew in to Albuquerque on a Special Mission aircraft. A photograph published in the Albuquerque papers on July 14, 1949, showed those present for the AEC to be General McCormack; Colonel R. T. Coiner, Deputy Director, DMA; Bennett Boskey of the AEC's legal staff; Richard Smith, AEC Procurement, New York Operations Office; and Kraker of the Sandia Operations Office. Representing Western Electric were Stanley Bracken, President; Frederick R. Lack, Vice-President Radio Division; and George A. Landry, Manager of the Operations Division. Kelly, Executive Vice-President, headed the Bell Lab group. He was accompanied by Donald A. Quarles, Vice-President in charge of staff and all government-sponsored research in the Bell Laboratories. P. J. Larsen, Assistant Director of the Sandia Branch Lab, hosted the gathering.

A few days later, Norris Bradbury gave UC Regent Robert Underhill a brief summary of the visit. "One thing seems clear from the discussions here," he wrote, "The boys have had their marching orders from President Leroy Wilson of the American Telephone and Telegraph Company to take on this project and make a success of it. Probably within the next two or three weeks, a contract will be drawn up between Western Electric and the AEC."[16]

Shortly after the conclusion of business at Sandia, Bennett Boskey began work on the first draft of a contract with Western Electric. On August 4, 1949, Fred Lack met with General McCormack in Washington. McCormack recalled that "Boskey walked in with a fat contract draft." Lack was not pleased; he thought the contract should be brief, perhaps not over a page. He said he would draft it himself "after [the] lawyers got through arguing."[17]

The issue of fee surfaced again. McCormack felt that a fee *should* be levied so the "AEC could criticize (the) operation and make AT&T pay for blunders." But Lack held firm to the AT&T position of "no–profit, no–fee." Lack also wanted "good industrial practices" inserted in the terminology rather than a specific cost section. He posited that AT&T, in exchange for no fee, should have the freedom to operate within these broad limits. Kimball Prince, later to become Sandia's general counsel, explained the terminology saying that it would "assure the flexibility deemed necessary to accomplish the tasks in a new frontier of science, as well as provide a criterion against which Sandia's operations could be measured."[18]

A precedent for the no–fee, no–profit contract had been established by the AEC's contract for the operation of the Brookhaven National Laboratory with Associated Universities, Inc. Under the no–profit arrangement, the contractor was to be reimbursed for all costs and expenses, direct or indirect, resulting from work under the contract. Paragraph 3, Article IV of the contract proved to be significant in the negotiation of other contracts, including the one for Sandia. Thinking ahead, the Commission, in January 1948, had proposed to use a provision similar to that specific paragraph in the drafting of other no–profit contracts for research and development work where reimbursement on an actual cost basis appeared to be advisable.[19]

Less similar in provision was the cost–plus–fixed–fee arrangement between General Electric and E. I. DuPont de Nemours to carry out a contract for certain production, research, construction, and maintenance services connected with the Commission's installations at Hanford, Washington, and Schenectady, New York.[20] The Western contract draft, as initially submitted by the AEC on August 2, was more similar to the one negotiated by the AEC with Bendix for operation of Project Royal. Like the Bendix contract, it contained a detailed "Cost of Work" clause and a fairly broad indemnity except for "willful misconduct or bad faith." Lack, however, advised the AEC that the Bendix–style contract was unacceptable—largely because of the detail.

The rather general requirement that the Corporation should be run on the basis of "good industrial practice," was duly passed on to Western and became the central theme of the contract, as originally signed. Later, this phrase was changed to adopt "Western practices" as the criteria.[21]

Norris Bradbury had his own suggestions regarding the transfer.[22] His comments, along with others, were discussed in a series of meetings that took negotiators from high-rise offices in New York to the mountains of Los Alamos, New Mexico, and on to Bendix, Kansas City. In addition to contract particulars, the division of responsibilities and procedures for review and handling of production schedules and procurement policies had to be established. At a conference held in Captain Tyler's office on August 29 in Los Alamos, Tyler of the AEC, Underhill for the University of California, and Lack for Western Electric drew up and signed the formal "Takeover Agreement." At this meeting, plans were made to effect the smooth transition to industrial management. The representatives wanted to ensure that the current operations of the Sandia project were not unduly disturbed.[23]

On Wednesday, September 14, Western Electric submitted to the Commission an alternate draft of the proposed contract. The general formula was considered acceptable; however, during subsequent meetings, the draft underwent considerable revisions until a final discussion was held in Washington on September 29. On this same date, Western Electric executives H. C. Beal, Vice-President Manufacturing; Fred R. Lack, Vice-President Radio Division; and Walter L. Brown, Vice-President and General Counsel, signed the Certificate of Incorporation creating Sandia. The new organization's broadly stated purpose was:

> to engage in any kind of research and development, and any kind of manufacturing, production and procurement to the extent that lawfully may be done. . . . [24]

The contract was partially executed by the AEC in Washington on the afternoon of October 4, 1949. The Certificate of Incorporation was filed in the State of Delaware the following day. And on

October 6, the contract was transmitted to New York, where it would be signed by Sandia and Western Electric at the first meeting of the Board of Directors.

Just before incorporation, it was learned that a New Mexico concern owned by John H. Hall and Robert S. Poage held the name "Sandia." When apprised of the situation by Western's attorney, Hall and Poage graciously agreed to change the name of their company so that it would be available for Sandia's use. Legal expenses and filing fees were paid for by the Corporation.[25]

The Corporation's first Board of Directors, composed of Beal, Lack, Brown, and Landry, all of Western Electric, held their initial meeting at AT&T headquarters, 195 Broadway, New York City. With Beal acting as chairman, the Board elected George A. Landry President and P. D. Wesson Secretary. The next order of business included approval of the form of stock certificates and the corporate seal.

Landry then presented Contract AT-(29-1)-789 between the AEC, acting on behalf of the United States of America, and Western Electric Company, Inc., relating to the operation of Sandia Laboratory. In the name of the Corporation, Landry formally endorsed the contract, thereby making Sandia Corporation a party to the agreement and "assuming all of the obligations therein imposed upon it." It was further resolved that the president was authorized to enter into a contract in the name and on behalf of Sandia Corporation with Western Electric in the form set forth in Appendix A. Accordingly, Landry signed the document as signatory for Sandia, and Fred R. Lack for Western Electric. The contract called for the operation of Sandia Laboratory until December 31, 1953.

To capitalize the new corporation, Western Electric paid $1,000, the minimum required by the laws of Delaware and New Mexico, for 100 shares (the entire issue) of no-par-value stock of Sandia Corporation. Sandia then invested the money in Series F United States Savings Bonds.

On November 1, 1949, Sandia Corporation assumed active direction of the Laboratory.[26]

During the first meeting of Sandia's Board of Directors, Henry C. Beal (right) served as Chairman. George A. Landry (left) was elected President of Sandia Corporation. Mervin J. Kelly is in the center.

ON THE BASIS OF "GOOD INDUSTRIAL PRACTICE"

Prime Contract Relations

As formalized, the prime contract between the AEC and Western Electric was an unusual document. Not only did it employ the concept of no-profit, no-fee and the liberal parameters of "good industrial practice," but it was a tripartite agreement involving the AEC, Western Electric, Sandia Corporation, and indirectly, Bell Telephone Laboratories. The signatories of the prime contract were the AEC and Western Electric. However, Sandia and the AEC fully expected to benefit from the expertise of both BTL and Western. As stipulated in the contract:

> The government desires to utilize in the operation of the Sandia Laboratory of the Commission and related work, the management, engineering, scientific research and development, and manufacturing skills of Western and Bell Telephone Laboratories, Incorporated.[27]

In this manner the contract provided for the loan of management and technical personnel from the Bell System to Sandia. If, for example, Sandia needed a technical, financial, purchasing, or accounting expert, or a general manager for overall corporate administration, the Bell System would provide such abilities from any part of its nationwide operations until such time as those services were no longer needed. Not long after Sandia came under AT&T management, it was recognized that full-time legal counsel would be necessary. Subsequently, in March 1950, Phillip D. Wesson was transferred to Sandia from Western Electric to serve as the Corporation's first attorney.[28]

Through the years, the roster of Bell System people at Sandia has varied. On November 1, 1949, when Sandia had 1,742 employees, the number of Bell people transferred or loaned to Sandia stood at fourteen. By April 1953, the number was approximately seventy. Bell Lab transfers in 1949 were well represented in the upper echelons of management since Western has under its contract a broad trusteeship over Sandia

Corporation. Seven of Sandia's presidents have come from BTL, with only two, George Landry and Siegmund P. "Monk" Schwartz, coming from Western Electric.[29]

In addition to specifying provisions for the sharing of professional and scientific expertise, the Western–AEC contract also provided for broad indemnification, one of the most carefully analyzed topics of discussion during contract negotiations. Admittedly, the work specified under contract involved the potential for "unusual, unpredictable and abnormal risks"—a fact recognized by all parties involved. Section 6 of the Atomic Energy Act of 1946 authorized the Commission to produce atomic weapons. Moreover, Congress had passed the Act knowing that atomic weapons had been and were to be produced by contractors. As specifically set forth in the AEC–Western contract, Article VI—Assumption of Risk by Government:

> The parties recognize the work under this contract involves unusual, unpredictable and abnormal risks. In view of these circumstances and the no-profit feature of the contract, neither Western, Sandia Corporation, Bell Laboratories, nor any company of the Bell System, shall be liable for, and the Government shall indemnify and hold them harmless against, any delay, failure, loss, expense (including expense of litigation), or damage (including personal injuries and deaths of persons and damage to property) of any kind and for any cause whatsoever, arising out of or connected with the work, and whether or not any employee or employees of any such company is or are responsible therefor.[30]

As Secretary of the Sandia Board of Directors P. D. Wesson explained to Walter L. Brown: "The indemnity agreement is not technically an independent commitment, but . . . is . . . nothing more than a reasonable and necessary incidental feature of the basic relationship . . . wherein the contractor in effect is an agency or instrumentality of the Government."[31]

The Sandia contract contained other interesting aspects related either directly or indirectly to the Laboratory's mission and position as an instrument of the government. The responsibility for security, for example, was to rest with the Commission, due to the sensitive nature of atomic weapons. Sandia and

Western, in accordance with the Commission's security regulations, agreed to safeguard restricted data and other types of classified matter, and to protect against sabotage, espionage, and theft of documents and materials of intrinsic value.[32]

Still another provision required that "all drawings, designs, specifications, data, books of account, correspondence, and records and memoranda of every kind and description prepared by Sandia Corporation in connection with the performance of work should be preserved except as otherwise directed by the Commission." Furthermore, Sandia agreed to deliver these records to the Commission at its request.[33]

Contractual Modifications

Again, because of the nature of Sandia's business, and because the government considers that inventions made by employees of government contractors are the result of federally funded research, the patent provision of the 1949 contract favored the government. Accordingly, the Commission retained the sole power to determine whether or not and where a patent could be filed, as well as disposition of title rights to such inventions. Western also granted to the government an irrevocable, royalty-free, nonexclusive license for the entire term of its patents for use in Sandia Corporation products that utilized fissionable material or atomic energy in or with a military weapon. Similarly, the government granted AT&T a nonexclusive license to government-owned patents from Sandia.[34]

These patent provisions as agreed upon in the 1949 contract meant that if a Sandia employee, in the course of work, made or conceived an invention or discovery, the AEC would be furnished with the complete information thereon. The Commission would then determine whether or not a patent application should be filed. To effectuate such a contract provision, employees of Sandia would be asked to sign patent agreements. If an individual wanted to obtain a patent in his or her own name, that person would be required to obtain a waiver from the government.

Because such patent provisions have favored the government and because of the basic no-profit, no-fee concept, AT&T traditionally maintained an arms-length relationship toward Sandia during these early years. The 1949 contract, as noted, provided that AT&T would reap no benefit from sales or services, except for communications equipment.[35]

One of the most difficult aspects of Sandia's legal situation for outsiders to comprehend was the fact that Sandia does not own land or property. With headquarters in Albuquerque situated on a military reservation, even ownership of buildings and furnishings remained with the government. Sandia Corporation, in fact, had no assets other than the $1,000 paid-in capital necessary for incorporation in the state of Delaware.

The financial provisos of the Sandia contract contributed to its uniqueness. The contract provided "full cost reimbursement" for expenditures, with a single, narrow exception. This single exception pertained to controlling corporate officers for whose "willful misconduct or bad faith" reimbursement could be withheld. In comparison, other government contractors had a long list of exceptions. All financial obligations incurred by the Corporation were considered government obligations.[36]

From the beginning, it was recognized that the effective operation of Sandia Corporation was all important to the defense of the United States. Its administrative and legal position was unique; there was little precedent for operating guidelines and procedures. The 1949 contract, therefore, had to form the basis for a mature body of knowledge in administration and applicable legal principles that was not yet in the textbooks. The parties involved in the original contract negotiations forged legal tools out of raw facts to accomplish the necessary tasks.

During this formative period, while corporate administrators, the AEC, and their respective counsels ironed out the legal aspects of Sandia's transition to corporate status, events were taking place on the world scene that added a certain urgency to the proceedings.

ON THE BASIS OF "GOOD INDUSTRIAL PRACTICE"

Administrative staff of Sandia Corporation, 1950.

George A. Landry, first President of Sandia Corporation.

Chapter XIV.

MEETING THE CHALLENGE:
The Production Era, 1950-1952

"That the President direct the Secretary of State and the Secretary of Defense to undertake a re-examination of our objectives in peace and war and of the effect of these objectives on our strategic plans, in light of the probable fission bomb capability and possible thermonuclear capability of Soviet Union."

Presidential Directive
January 31, 1950

"The Joint Committee Chairman introduced in the Senate, and I introduced in the House, a concurrent resolution resolving that 'The United States must go all out in atomic development and production. In our resolution, the Chairman and I pointed out that the cost of military firepower based upon atomic bombs is hundreds of times cheaper, dollar for dollar, than conventional explosives.' "

Carl T. Durham, Acting Chairman
JCAE, June 24, 1952

The year 1949 marked a turning point, not only for the Sandia Laboratory, but for America and the world at large. In February and March of that year State Department and Defense officials drafted "National Security Council 68." Referred to as NSC-68, the document established for the next three decades the foundation of American foreign and defense policy. Basically, NSC-68 embodied the ideology that it was the moral and political obligation of the United States as a world leader to preserve freedom around the globe. The rationale decreed that to do so, the United States must adequately arm itself with both conventional and nuclear firepower. America, in effect, would provide military protection to vast areas of territory outside the continental domain.[1]

Statistics recorded in NSC-68 supported the U.S. position: America devoted only six to seven percent of its gross national product to arms; while Russia committed nearly twice as much, or 13.8 percent. The approval of NSC-68 signified a historic reversal of American foreign policy. The results would soon be reflected in military support for forty-seven countries and increased troop commitments abroad.[2]

In the interim, a series of significant events hit with catalytic force. China was overrun by the Communists, and in the United States the public was shocked to learn that Alger Hiss, well-respected State Department official, had given secrets to the Russians. More significantly, on September 23, 1949, a radioactive cloud carrying fission products drifted eastward across American territory.[3]

Some three weeks later, Presidential Press Secretary Charles Ross summoned reporters to his office. Ross's demeanor left little doubt as to the serious nature of the business at hand. "Close the doors," he ordered. "Nobody is leaving here until everybody has this statement." From behind his impressive walnut desk, he handed out copies of an equally impressive presidential announcement: "We have evidence that within recent weeks, an atomic explosion occurred in the U.S.S.R."[4]

United States scientists investigating the data were more specific. The detonation, they felt, took place between August 26 and August 29, 1949. Confirmation that the Russians had exploded their first atomic device made an immediate impact, but the ramifications were long-term. "Joe One," as the device was called, meant that the U.S. atomic monopoly was at an end. This historic event, combined with others, convinced government and military leaders that a showdown with Russia was inevitable.[5]

The result of the Soviet detonation was to be a bevy of new alliances in support of nuclear power, the expansion of the Strategic Air Command to deliver the firepower, and the rapid buildup of the stockpile. At Sandia Corporation it meant a strong emphasis on production during the first two years of the fifties and the continuation of a series of crash programs. For the laboratories at large and the American scientific community, it would serve as rationalization to advance the frontiers of knowledge in the field of nuclear technology.

George A. Landry: First President

The first President of the new Sandia Corporation, George Albert Landry, was a stern-looking, bespectacled man with thinning white hair. At fifty-nine years of age, Landry brought to the young and growing organization many years of experience in manufacturing operations. During World War II, he had served briefly with the War Production Board and in 1945 became Operating Manager of Western Electric's nationwide installation forces.[6]

As President of Sandia Corporation, Landry was to tighten up the operation of the Laboratory, spearhead a strong production drive, install Bell System personnel benefits, and continue the construction efforts started by Larsen. Described by one Vice-President as a "somewhat polished but nonetheless bull-of-the-woods type manager," Landry quickly imposed a Western Electric infrastructure upon the organization.[7]

In keeping with the mission assigned to it by the AEC, one of the principal tasks of Sandia Corporation in the operation of the Laboratory was to provide a smooth transition between the development and the manufacture of atomic weapons. This meant a continuation of applied research, engineering development, and design for production of the ordnance phases of atomic energy, as well as the assembly of weapon products for delivery to the military and to the national stockpile. It also meant assumption of administrative responsibilities and business functions, including the transfer of purchasing and accounting activities from the University of California to Sandia. The transition would not be as "smooth" as the AEC had hoped for.

In taking over the reins of management, Western Electric sent to Sandia a team of hand-picked supervisors from Western Electric and Bell Telephone. At the outset, Landry said, "It was apparent that the University's concept of supervisory responsibility did not agree with ours [Western Electric]."[8] The President made particular reference to the fact that at Sandia functional responsibilities were not centralized and organizational activities pursuing more or less parallel courses were not coordinated to ensure maximum efficiency.

Through reorganization, Landry created a number of broad functional units headed by members of his immediate staff. This, in effect, established a new and overriding level of top supervision.

Not long after formal transfer, Landry called his first staff meeting. Those attending included Fred Schmidt, Vice-President and Operating Manager who, like Landry, came to Sandia with many years of management experience with Western Electric. Before leaving Western, Schmidt had held the position of Labor Relations Manager at the Kearny Works. The new Corporation's Director of Development, Robert E. Poole, hailed from Bell Telephone Laboratories, where he had served as Director of Military Electronics Development. Other members of the first executive "Small Staff" were Frederick B. Smith, Treasurer, Personnel and Public Relations; and Secretary and Comptroller James A. Dempsey.[9]

There was a certain irony in these appointments. Although Smith's background at Western Electric had been in manufacturing, at Sandia he was assigned to personnel; while Schmidt, who had a background in personnel, was assigned to operations.

From the beginning, Sandia Laboratory would provide a fertile training ground for BTL and Western employees, many of whom were on a rotational track.[10]

The influx of Bell management personnel and Landry's operating style quickly made an imprint on the small, loosely structured Laboratory in the Southwestern outback. A survey of personnel conducted in August 1949, two months before changeover, shows an extremely young population, with the average age of male employees being thirty-three years and female employees being twenty-nine years.[11] Not only was the Laboratory characterized by youth, it was also a uniquely informal but pragmatically oriented organization. As one employee explained:

> There were some people who came in with very stilted opinions of themselves and were dismayed to find out that there really wasn't any hierarchy of educational level. The assignment, and who could get the job done, . . . was the real criterion of a man's worth. Everybody was pretty flippant with their conversations; we even greeted the janitors as "Hi, Doc!"[12]

Despite their youth, many members of the Laboratory's homegrown management team, as well as staff members, were Manhattan Project veterans accustomed to working side by side with the world's best-known scientists. It is not surprising, therefore, that normal concerns relating to the transfer of benefits, salary, and seniority, would be compounded by the imposition of a highly structured, "go-by-the-rule-book" style of management. Sandians regarded the new management's emphasis on stratification and protocol with a jaundiced view. The sight of Landry arriving at work in a chauffeur-driven car from his home three or four blocks across the parade ground raised eyebrows. Ray Powell, who in 1954 became Superintendent of Personnel and Public Relations, explained the situation: "Landry was good for the Laboratory because practices were quite loose and sloppy at that time, but he tightened things up to the extreme, and there was vigorous reaction."[13]

Farewell ceremonies for Paul Larsen were held in Norris Bradbury's office at Los Alamos. On hand for the occasion (from left to right) are Sandia Base Commander General Robert M. Montague, Rear Admiral William S. Parsons, Paul Larsen, and George Kraker, Manager of the AEC's Santa Fe Operations Office.

On the other hand, the new management personnel suffered from corporate culture shock in reverse. The structured environment of the eastern manufacturing world had not prepared them for the lack of formality, the first-name basis, of a laboratory whose ethos from its inception had been based upon cooperative interaction and "getting the job done." For example, new executives found to their dismay that there were no keys to the executive washroom. In fact, not only were there no keys, there was no "executive washroom," only "public restrooms" for staffers and executives alike. When Landry designated one of the integrated restrooms on the second floor of Building 800 as the "corporate washroom," it became a source of underground humor, and the arrangement lasted only temporarily.[14]

It was inevitable that the transition to corporate status would have its problem areas, both large and small. To Landry's credit, he realized that he could benefit from his predecessor's experience and guidance; consequently he requested the extension of Larsen's contract. In a letter to R. E. Gibson, Director of the Applied Physics Laboratory at Johns Hopkins, Landry explained his position:

> It is going to be of immense help to us if we can retain the services of Mr. Larsen for another year. . . . It appears that Mr. Larsen has done a grand job in the face of enormous problems. He has a very thorough understanding of the product of the laboratory, of the functions of its various units and of the tie-in to the AEC, with the Los Alamos Laboratory and the Military Services with whom it continually cooperates. Consequently, we expect that his value to us during the first few months of the Laboratory will be very great, and his very excellent attitude of cooperation with us leads us to value the help that he can give us all the more.[15]

Gibson granted Landry's request, but Larsen would serve as a consultant only until March when he was appointed Director of Civilian Mobilization and returned to Washington. If Larsen had been able to remain at Sandia as originally planned, the transition might have been a smoother one. Despite the numerous problems to be solved, the reorganization and work of the Lab proceeded at a rapid pace.

During the first part of the new year, Landry and staff focussed on reorganizational functions and the assumption of responsibility for many services previously held by the AEC and the military. These included operation of the motor pool, housing, and other community facilities, the base at Salton Sea, California, the guard force, and some purchasing operations.[16]

The new Plant Services Department, set up in 1950, helped ease the growing pains of the new organization.[17] Among the new buildings constructed, Building 800 was completed in 1949 as the first of the 1947 long-range program. Originally slated to be used as the Administration Building, it was turned over to Purchasing because its location made it more suitable to conduct business with uncleared suppliers. A warehousing building, designated 894 and constructed out of red brick in a style similar to that of Building 800, housed Shipping, Receiving, and Inspection activities and became for a time the home of General Stores. The last of the group to be finished, 802, although initially designed to centralize technical personnel, was used more and more as the Administration Building.[18]

Construction at the Laboratory received another boost when in July 1950, Congress asked for some $260 million in supplementary appropriations to support the thermonuclear crash program. By mid-October Congress had modified the construction rider to advance by some months the start of the quarter-billion-dollar expansion program.[19] Both Sandia and Los Alamos laboratories reflected the injection of funds. As one observer noted, the drab buildings at Los Alamos began to take on "signs of prosperity"; while at Sandia, the appearance—if not of prosperity—was at least one of more permanence and increased activity. Approximately $32.5 million went into construction at the Albuquerque Laboratory alone.

The AEC Semi-Annual Report for 1950 indicated the majority of the original construction campaign at Sandia had been completed by June 1950, but statistics on increases in square footage reveal that heavy construction efforts for this first era carried over into 1951.[20]

MEETING THE CHALLENGE: THE PRODUCTION ERA, 1950-1952

The Administration Building 800 nears completion in early 1949.

Organization of Sandia Corporation
November 1, 1949

- **DIRECTOR OF DEVELOPMENT** (SLDD) — R. E. Poole**
  - **Research Associate Director** (SLRAD) — R. P. Peterson
    - Applied Physics Lab (SLA)
  - **Technical Associate Director** (SLTAD) — R. W. Henderson
    - Engineering Department (SLE) — R. A. Bice, Mgr.
    - Field Test Department (SLT) — G. A. Fowler, Mgr.
    - Military Liaison Department (SLM) — A. B. Machen, Mgr.
    - Surveillance Department (SLS) — L. J. Paddison, Mgr.
  - **Technical Staff Engineer** (SLTE) — F. Cowan**
  - **Research Consultant** (SLRC) — W. A. MacNair**
  - **Superintendent of Manufacturing Engineering** (SLSM) — W. H. Pagenkopf*
    - Production Engineering Department (SLPE) — L. J. Biskner, Mgr.
    - Development Fabrication Department (SLF) — L. F. Yost, Mgr.
    - Project Royal Liaison (SLRL) — G. C. Buonagurio
- **CONSULTANT TO THE PRESIDENT** (SLCP) — P. J. Larsen

*Western Electric Personnel
**Bell Telephone Laboratories Personnel

MEETING THE CHALLENGE: THE PRODUCTION ERA, 1950-1952

PRESIDENT (SLPRE)
George A. Landry*

VICE PRESIDENT AND OPERATING MANAGER (SLVP)
F. Schmidt*

TREASURER, PERSONNEL, AND PUBLIC RELATIONS (SLTR)
F. B. Smith*

SECRETARY AND COMPTROLLER (SLSE)
J. A. Dempsey*

Road Department (SLR) — F. H. Longyear, Mgr.

Building Maintenance Department (SLZ) — L. J. Heilman, Mgr.

Coordinator Master Scheduling (SLMS) — J. W. Hook

Purchasing Agent (SLPA) — H. G. Ross*

Purchasing (SLPB) — W. F. Dietrich*

Procurement and Supply Department (SLP) — D. B. Miller, Mgr.

Expediting (SLPC) — W. J. Madura

Contract Liaison Department (SLCL) — H. E. Sunde, Mgr.

Assistant Treasurer (SLTA) — C. Olajos

Business Manager Sandia Lab (SLBA) — R. E. Roy

Administration Department (SLX) — C. W. Campbell, Mgr.

Document Department (SLD) — S. Harris, Mgr.

Staff Development Department (SLSD) — S. Harris, Mgr.

Financial Consultant (SLFC) — C. E. Russell*

Fiscal Administration Department (SLY) — E. W. Baldwin, Mgr.

Cost and Accounting (SLCA)

Inventory (SLI) — M. J. Hughes*

Assistant Secretary and Business Methods (SLBM) — H. W. Sharp

Accounting Consultant (SLAC) — R. M. Nicol*

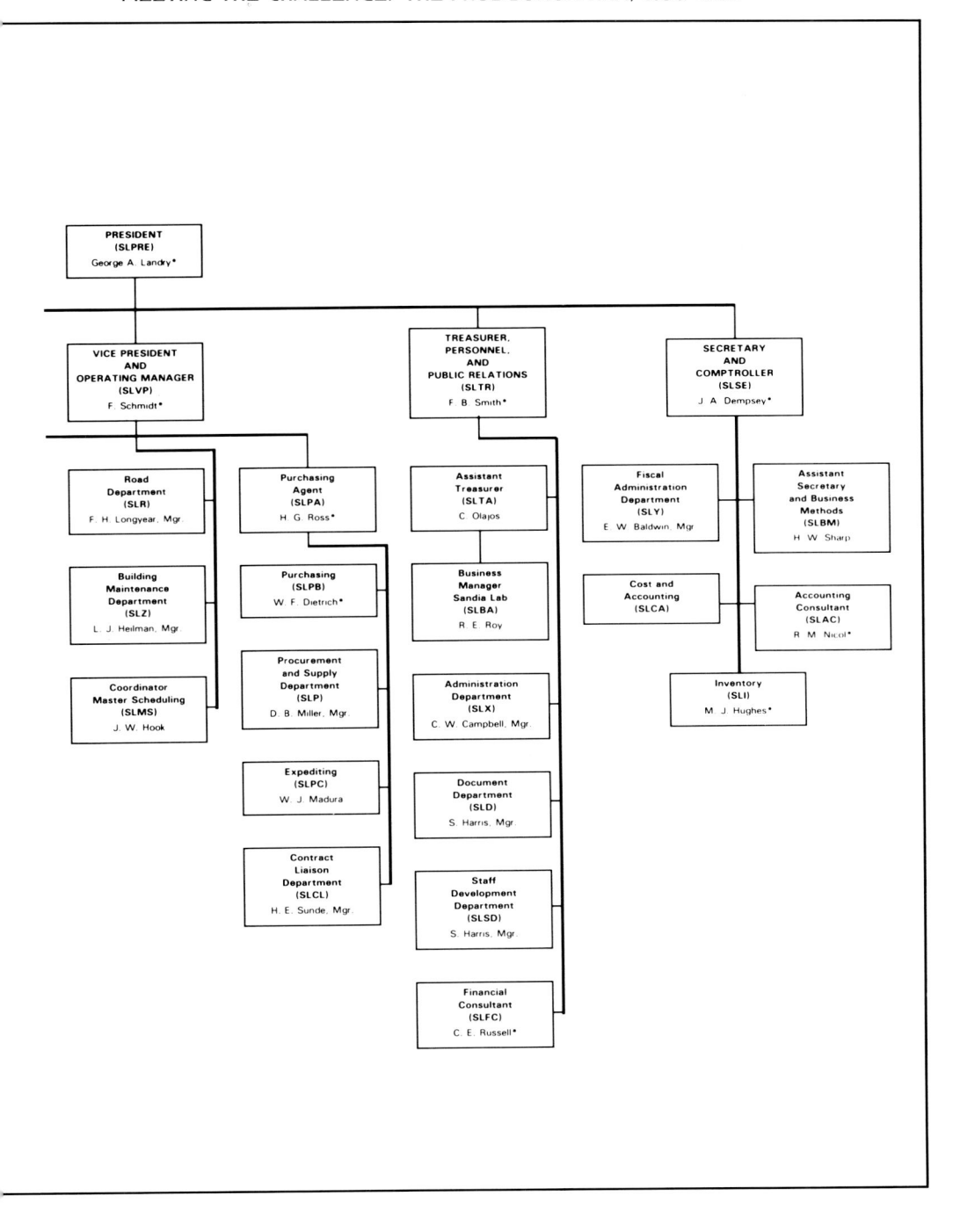

375

On November 14-15, 1950, the management of Sandia Corporation visited BTL. Among those pictured (left to right) are James B. Fisk (BTL), George Hansche, Everett Cox, H. W. Bode (BTL), R. P. Petersen, W. A. MacNair, Mervin J. Kelly (BTL), R. E. Poole, Donald A. Quarles, R. Brown, and Ken Erickson.

MEETING THE CHALLENGE: THE PRODUCTION ERA, 1950-1952

When George Kraker was promoted to Deputy Manager of the AEC's Santa Fe Operations Office, Daniel F. Worth, Jr. (left), who had been Kraker's assistant, succeeded him as Sandia Field Manager. Worth, who was known for his sense of humor and a deep and gravelly voice, often referred to Sandia as "The Cooperation." He listed knitting among his hobbies.

Expansion of Production Facilities

In addition to improvement of facilities at the nuclear laboratories, international tensions led to increased emphasis on the construction of production facilities around the country. Carl T. Durham, Acting Chairman of the Joint Committee on Atomic Energy, in a speech to the House of Representatives, explained the background to the atomic plant expansion program and urged the speedy passage of additional funding. He noted that on June 7, 1951, the Joint Committee requested and received a report providing cost estimates on the proposed expansion. Subsequently, a concurrent resolution introduced in the House and the Senate resolved that "the United States must go all-out in atomic development and production." Durham reiterated the opinion held by the JCAE that "the cost of military firepower based upon atomic bombs is hundreds of times cheaper, dollar for dollar, than conventional explosives."[21]

Detailed testimony from the Secretary of Defense, each of the Service Secretaries, and the Joint Chiefs of Staff verified the need for more atomic weapons.[22] Thus, the combined strength of the military, AEC, and Congress mustered the support necessary to acquire appropriations for the establishment of five more production facilities during the fifties era.

Even before the latest supplementary funding, on June 12, 1950, the AEC asked that E. I. du Pont de Nemours, manufacturer of black powder since 1802 and a national leader in explosives technology, undertake the construction and operation of a plant to produce plutonium and tritium. Having designed, built, and operated the world's first plutonium production reactor at Hanford, Washington, during World War II, DuPont was a logical choice. By 1953, DuPont's Savannah River Plant was in operation. Situated on the pine-covered coastal plains of the historic Savannah River, near Aiken, South Carolina, and Augusta, Georgia, the plant produced plutonium and tritium.

In 1951, the AEC started construction on the facility known as "Apple," at Rocky Flats, Colorado. The Rocky Flats plant, located twenty-one miles northwest of Denver, became operational in early 1952. Operated by Dow Chemical, the plant fabricated plutonium and manufactured various stainless steel, beryllium, and uranium alloy components.

In 1952 American Car and Foundry established a plant in the southwest valley area of Albuquerque, New Mexico. This small facility, bordered by farm and ranch lands and a meat packing company, manufactured case parts for the thermonuclear bomb and weapon handling equipment.

The General Electric Company, under a subcontract with Sandia Corporation, completed construction of the first building for what would be the Pinellas Plant near Clearwater, Florida, in 1956. Purchased by the AEC, the plant had a single purpose at the beginning of operations in 1958—to manufacture neutron generators for the initiating of nuclear weapons.

Components from all facilities originally came together for final assembly at the Burlington Plant in Iowa and later at Pantex in the Texas Panhandle (hence the name). Pantex, located amid the windswept isolation of the high Texas plains, traditionally presented an appearance of anonymity incongruous to the importance of its function.

The facility itself was situated in a pastoral setting about twenty-five miles northeast of Amarillo. Cows browsed around the perimeter of the wire fence that encircled an assortment of warehouses, administrative buildings, and a maze of underground bays topped by sealed mounds of sand, gravel, and earth-covered bunkers that erupt from the ground like molehills. For safety reasons, the underground bays, called "gravel gerties," were designed to collapse inward, entrapping radioactive contamination in the event of an accident.

Under a memorandum of agreement with the AEC, the Army Ordnance Corps took responsibility for administrative control of Pantex until 1951 when Ordnance contracted with Proctor and Gamble Defense Corporation to operate the plant. In 1956 Proctor and Gamble withdrew from Pantex, and Mason and Hanger-Silas Mason Co., Inc., the Kentucky-based firm renowned for engineering the Grand Coulee Dam and the Lincoln

THE INTEGRATED

CONTRACTOR	PLANT NAME & LOCATION	ESTABLISHED
UNION CARBIDE (Replaced by Martin Marietta Energy Systems in 1983)	Y-12 Plant Oak Ridge, TN	1947
MASON & HANGER (Silas Mason Co., Inc.)	Burlington Plant Burlington, IA (Consolidated with Pantex July 1975)	1947
MONSANTO RESEARCH CORPORATION	Mound Facility Miamisburg, OH	1948
BENDIX, KC (Acquired by Allied in 1983)	KANSAS CITY, MO	1949
DOW CHEMICAL (Replaced by Rockwell International in 1975; curently Atomics International)	Rocky Flats Golden, CO	1951
MASON & HANGER	Pantex Plant Amarillo, TX	1951
AMERICAN CAR & FOUNDRY (ACF Industries)	Albuquerque, NM	1952
DuPONT COMPANY	Savannah River Plant Aiken, SC	1953
GENERAL ELECTRIC COMPANY (GEND)	Pinellas Plant Largo, FL	1958

CONTRACTOR COMPLEX

<u>PRINCIPAL MISSIONS</u>

Production of test and stockpile assemblies; fabrication of heavy case parts; fabrication and research in uranium; machining.

Fabrication of chemical explosives and assembly of nonnuclear components, final assembly.

Fabrication of small H.E. units, explosive components such as detonators, timers, transducers, firesets, and pellets; pyrotechnic devices; responsible for stockpile sampling on boosting systems.

Fabrication and assembly of complex electromechanical and precision mechanical devices, rubber, plastics, and nonnuclear components; heavy machining; electronic systems.

Fabrication of beryllium, plutonium, and uranium alloy; plutonium recovery and research; fabrication of pressure vessels—e.g., boosting systems.

Production of large H.E. units; assembly of final product; heavy machining; preparation for shipment to the military; disassembly and retirement of weapons.

Machining of case parts for the thermonuclear bomb.

Production of tritium and plutonium.

Production of neutron generators, thermal batteries, RTGs, lightning arrester connectors, capacitors, neutron detectors.

Tunnel, became the operating contractor. From Pantex, completed weapons were transferred to the military and eventually returned to the plant for disassembly and retirement. Thus, the Pantex facility was to serve as the hub of a defense network spanning the nation.[23]

Through this system of integrated contractors, Sandia's products—largely designs and prototypes—were to be converted into military hardware. Between 1947 and 1958, the integrated contractor complex would grow to incorporate nine plant facilities.

Management of the complex, in accordance with provisions of the original Atomic Energy Act of 1946, remained under civilian control.[24] The Santa Fe Operations Office, the civilian field office in charge of administering the nuclear weapons program, became the Albuquerque Operations Office (ALO) of the AEC in 1956.

Although headquarters remained in the Washington, D.C. area, the decentralized concept of production administration made possible the effective management of the geographically dispersed facilities.

The integrated nature of the weapons complex fostered a synergism unique among complex and diversified industrial and government endeavors. Diverse capabilities brought to bear on common interests, and dedication to singular objectives were quickly recognized as advantages of the integrated structure.[25]

At the inception, however, organizational roles and responsibilities at Sandia and elsewhere had to be defined. Although charged with a mission that specifically included research and development, the organization for the first two years of its existence bore the imprint and strong production orientation of its Western Electric operator.

Engineering for Manufacture and Production

The new organizational structure established by President Landry in late 1949 clearly reveals the strong emphasis that Bell

MEETING THE CHALLENGE: THE PRODUCTION ERA, 1950-1952

System senior management would place on weapon production responsibilities. For the first time, Sandia Corporation had a Vice-President of Operations, Fred Schmidt, who reported directly to the President. During the 1949-1951 period, the Western Electric management team systematically implemented modifications of production routines to meet stockpile quotas.

Development work, placed on a project basis, fed design information on individual weapons to the production organization. The scope of Production Engineering was expanded and a standard nomenclature system was devised to identify and differentiate equipment fabricated to constantly changing military requirements. Management made considerable progress in introducing these changes and others during the first fourteen months of the Corporation's activity.[26]

To take over the expanded Production Engineering Department, L. J. Biskner, an experienced production specialist, was recruited from Emerson Electric Company's St. Louis plant. Tool and gauge designers along with manufacturing development engineers, also joined the organization. Other supervisory and staff-level personnel were brought in from Western Electric to assist in setting up the Engineering for Manufacture program and to train staff already on roll. To add effectiveness to the program, these individuals worked on improving procedures for change order services, production and inspection, technical record keeping, and data review.[27]

Production Analysis under A. K. Leupold and Technical Service under C. E. Foster also became a part of the Production Engineering Department. A 1951 Sandia brochure, in homespun style, compared Production Analysis to "a nagging wife," whose job was "to remove flaws and make improvements in design of weapon components . . . until they are as perfect as that querulous lady would like her husband to be."[28]

To keep the military abreast of these new weapon developments and to provide training guidelines, the Technical Service Division prepared various illustrated publications. Artists, draftsmen, and writers worked together to produce comprehensive catalogs of each weapon type, described down to the smallest detail necessary for assembly operation.

That same 1951 brochure on Production Engineering described the overall functions of the department accordingly: "While it [Production Engineering] does not create the weapon, it sort of mothers it, erases the flaws, dresses it up and packs it off to the Military in its best bib and tucker, knowing it is all primed to perform the lethal task to which it may be assigned."[29]

The expanded functions of the Production Engineering organization provided for closer liaison with the contractors and resulted in a higher degree of quality control. Sandia could devote more time to actual design work and the building of prototypes. By 1950, the original engineering, assembly, and production-related functions had evolved into a combined Manufacturing Engineering organization (2100) with directorate level status.

Walter H. Pagenkopf, a highly experienced manufacturing expert from Western Electric, served as the group's "Superintendent." By mid-July 1950, Frank H. Longyear had been promoted from Road Department Manager to Superintendent of Production, a move that elevated the Road Department to directorate level as well.[30]

While the title "Director" was reserved for Research and Development activities, the title "Superintendent" was used for non-R&D activities at the level immediately above Department Manager. A carryover from the Western Electric manufacturing plants, the use of the Superintendent designation was indicative of the corporate schizophrenia of the times.

As the workload grew, the AEC Santa Fe Operations Office found it increasingly difficult to cope with the many activities of weapons program planning, particularly in the areas of hardware design and development and coordination with production plants. Therefore, by early 1949, the AEC Santa Fe Operations Office had delegated responsibility for weapons program planning and scheduling to Sandia. This activity became more formalized when John T. Kane from Western Electric became supervisor of the Production Planning Division later this same year.

MEETING THE CHALLENGE: THE PRODUCTION ERA, 1950-1952

By late 1949, the Fat Man design had evolved into the Mk 4 and was stockpiled. Production Planning was assigned the task of establishing development and production milestones for this and other weapon programs. All preproduction and production weapons, test assemblies, test and handling equipment, spares, and interproject shipments between AEC agencies had to be scheduled. Since legally Sandia could not exercise direction over other AEC agencies, the published plans and schedules were approved and issued as AEC/ALO documents.[31]

A third department in the Manufacturing Engineering organization, Development Fabrication, was headed by Luther F. Yost. Manned by skilled craftsmen, this department stemmed from Manhattan Project days when the Army operated a small fabrication shop at Sandia Base. At the time, the shop consisted of one trailer, equipped with a lathe, a small mill, and a drill press. Eventually, shop work was performed in barracks or semipermanent buildings spread over a wide area, a condition that interfered with efficiency. When the Lab became a corporation in 1949, this was one of two major difficulties to be solved. A second involved defective work by suppliers of the Laboratory.[32]

To address the latter problem, it was decided that receiving inspection and inspection at vendor's plants were required to eliminate defective and unusable products. As a result, a department for receiving inspection under R. J. Bradford was established. Elsewhere, resident inspector stations were set up in various parts of the country, and teams of inspectors visited the plants to monitor quality.

G. C. "Corry" McDonald, who would become well known for his contributions in the manufacturing development area, credits Luther Yost with consolidating the Laboratory's shop and fabrication activity—an effort that also involved establishment of a number of machine shops in the Albuquerque community to handle unclassified work. "When it became obvious that we needed a good machine shop, we started to pull together a unified activity from all the different projects," McDonald recalled. "You can imagine the objections that arose when we took people and equipment away from the individual project groups," he added.[33]

To house production and manufacturing activities and staff, a highly specialized facility—the first major building (892) to be built in Tech Area I—was constructed in 1950. The "high bay" parts of the building accommodated the overhead cranes needed to assemble the heavy sections of the early bombs then in production. (Assembly of the H.E. spheres was done at two new buildings in the remote Technical Area II.)

By 1951, the Lab had seven machine shops in operation, as well as miscellaneous shops for the production of items ranging from plastics to field test assembly units. The Field Test Assembly shop—organized primarily for the purpose of readying full-scale and model units for drop and other tests—handled one of the most complex jobs. Extreme care had to be exercised to make sure that all parts of the unit functioned properly during a test. A bent arming wire, a small flake of paint in a pressure pickup, or a ground on a piece of electronic equipment could nullify a test.[34] For production in quantity, breadboard models turned out by the Electronics Division, as well as designs of development and prototype models, were farmed out to other manufacturers, soon to be referred to as the "integrated contractor complex."

Plant Engineering, organized in December 1949, made up another department in the Manufacturing Engineering organization. Luther Heilman, the Plant Maintenance and Service Department Manager, in 1951 became the Laboratory's youngest Superintendent in charge of Plant Services. Although only thirty-four years old, Heilman, a former Army officer, had worked at Sandia Base since June of 1946, when he was assigned as Post Engineer. During that early period, he arranged for the design and construction of many of the first permanent buildings in the Tech area. With the transfer of Sandia Base Engineering from Army jurisdiction to the University of California, Heilman joined the Laboratory staff where he set up the Plant Maintenance organization.

After Heilman's promotion to Superintendent of Plant Services, Robert E. Hopper, a Mississippian who had previously worked at AEC installations at Hanford and Oak Ridge, took over

MEETING THE CHALLENGE: THE PRODUCTION ERA, 1950-1952

as the new manager of Plant Engineering. In 1956, Hopper would succeed Heilman as superintendent. Under the leadership of Heilman and Hopper, the department worked to implement the expansion program necessary to accommodate physically the growing responsibilities of the Laboratory.

Another young manager was David S. Tarbox, who as an Army 1st Lieutenant received the Silver Star for gallantry in action, as well as the Purple Heart. Tarbox first came to Sandia Base as the Army's Assistant Post Engineer working for Heilman, then in 1948 joined the Sandia Laboratory staff in the Field Test Department. He became manager of Plant Maintenance and Services when his former supervisor, Heilman, was promoted to Superintendent.

The youth of the administrators in Plant Engineering was typical, in many respects, of the management that would characterize the Laboratory for years to come. Profiles reveal that leadership in the organization was predominately male and youthful (under the age of forty-five). Many were former servicemen. During the early years, female employees were oftentimes wives of employees or former WACs.[35]

Two new department managers, Robert Hopper (left, Plant Engineering) and David Tarbox (right, Plant Engineering and Service), pause to talk shop in the foyer of the new Road Building.

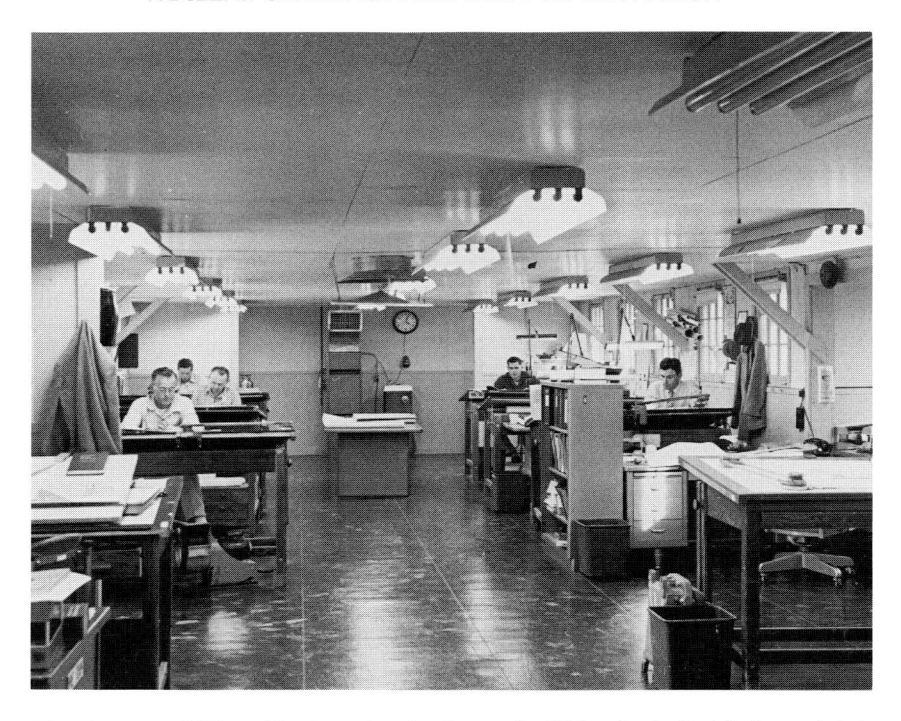

Employees of Plant Engineering in the early fifties included (left to right) Ray Richardson, Omar Heins, Hal Baxter, Bill Brown, and Don Knott.

TWA Engineering

Recruiting efforts of the time channelled many of the newcomers to the organization into production-related activities. Some of these personnel were involved in production liaison with the developing contractor complex. To handle the task of liaison with the Bendix facility, the Laboratory hired G. C. Buonagurio from the Pullman Company. Sandia, beginning to come into its own as a design lab, utilized its own extensive shops or looked to the outside world for production centers. Sandians would generate the design, including specifications, find a firm to manufacture the components, fund the activity, and follow the device through assembly to production. As one director observed: "The Sandia ethos in the late forties and fifties has been described uncharitably, but not inaccurately, as 'TWA Engineering,' which meant you got on an airplane to talk to a commercial

MEETING THE CHALLENGE: THE PRODUCTION ERA, 1950-1952

outfit that would do the job we needed to have done. We went to private industry." However, the connotation of "TWA Engineering" need not be construed as a necessarily negative one. Given the resources available and the youth of the organization, this mode of operation was a means of fulfilling a mission. Robert Henderson, Technical Associate Director in charge of Engineering in 1949, explains further: "If someone had a good product, we got them to improve it, control it, and placed our SA (Sandia Apparatus) number on it. (Batteries were originally in that category.) We required them to do certain things in the production process they didn't originally do."[36]

Such interaction meant that management had to become more conscious of public relations. Division leaders, for example, were told to advise their people that as representatives of the Laboratory, they were expected to maintain their professionalism both in their appearance and their presentations. "Everyone representing the Lab . . . ," wrote Dick Bice, "should be dressed as a business man, . . . have his business well organized and be ready to present it in a business-like manner."[37]

As a means of improving product quality and facilitating interaction with the contractors, formalized procedures in both drafting and standards also had to be developed. Management learned that the business of trying to *inspect* quality into a product, according to one executive, "was a lost cause." The solution was to find the causes of variabilities and correct those.[38]

Strange drawings coming in from the various shops were an obvious problem. Corry McDonald, Department Manager for Drafting and Standards in 1950, recalled that "these drawings, in some cases, depending on where they were originated, looked like they were done by aircraft draftsmen or engineers."[39] Action had to be taken to consolidate and establish a unified drafting system. To free McDonald to devote more time to the consolidation effort, the Laboratory's first administrative assistant, Henry W. "Hank" Willis, Jr., was hired. It was Willis' job to obtain through procurement channels the necessary materials for the drafting and design activities.[40]

Design responsibilities at this time were not clearly delineated. In many cases, the draftsmen had more experience in electronic or mechanical design than did some of the engineers. On the other hand, the engineers at times did their own drafting. When they finished the initial design phase, they would frequently take their drawings directly to the shop and have the prototype built.

It became apparent that there was need for a chief draftsman to carry out a more formalized system. McDonald took advantage of the transition to Bell System management to visit the Whippany and Murray Hill facilities to find such an individual. Tom Robertson was selected as the candidate of choice.[41]

Materials and Standards

Management also recognized that Sandia's drafting effort would be enhanced by developing a standard method of making the drawings so they could be understood unequivocally by the shops and the integrated contractors. Director of Development Robert E. Poole addressed this issue by recruiting Robert Townsend from Murray Hill. Townsend, who had worked on many national and international committees on standardization, was well qualified to provide direction in the area of materials, metallurgy, and standards. Under Townsend's direction, the Standards organization functioned as a "Transition Engineering" group between design and production. A system of standardized callouts was designated to benefit weapons projects across the board and as a means of interpreting product definition to the contractors.[42]

From a modest chemical lab run by Ralph Fisher, the materials and standards function developed into an important and contributing facet of the Laboratory's quality assurance program. As Lou Paddison, Director of Quality Assurance (QA) by 1954, observed: "The establishment of a working relationship with Bell Labs was a positive factor."[43] Through Townsend, Paddison arranged for Harold Dodge, and later George Edwards, both of Bell Labs, to serve as consultants to evaluate the QA program and make recommendations.

Edwards submitted a memorandum of understanding detailing operational procedures for the Laboratory's QA efforts, but the document's major significance rested upon his designation to Sandia of the responsibility normally exercised by the customer for the acceptance function of weapons production. "In effect," Paddison explained, "we were in the position of accepting weapons as an agency for the government."[44] This formula, although somewhat unique, followed the Bell System pattern whereby Bell Laboratories held responsibility for making a decision on the acceptability of Western Electric products for AT&T.

Sandia's interest in preventing the deterioration of weapons in stockpile also led to close collaboration with Bell Labs experts in the area of materials sciences. In 1950 when Townsend was director of Materials Engineering at BTL, he and Bernard S. Biggs were asked to go to Albuquerque as consultants. Biggs, a former football coach and science teacher, had earned his Ph.D. at the University of Texas, then went to work for Carnegie Tech. In 1936, he was hired by Bell Laboratories where he became part of a team that chose the materials for the first transatlantic submarine telephone cable. As a part of this group, Biggs helped pioneer the use of black polyethylene as a replacement for lead in the sheathings or outside protective coverings for telephone cables. At the time Sandia came under management of AT&T, Biggs was directing a group studying the deterioration of materials. When Townsend was badly injured in a fall from a horse, Biggs succeeded him at Sandia.

Under Biggs' direction, the Standards activity was separated organizationally and more emphasis placed on materials and metallurgy. The size and weight of the early Marks decreed that weapons had to be designed to the limits; therefore, firsthand information on materials, stress capabilities, and physical characteristics was an important consideration. Documents composed by the Technical Writing organization detailed the specifications—how materials, weapons configurations, and components should react under test conditions.[45]

The job of translating Sandia designs to other contractors fell to Sandians such as George Hildebrandt. In 1949, Hildebrandt, a member of the Navy Reserve, was studying for a degree in Mechanical Engineering at the University of New Mexico. Like many student engineers, Hildebrandt wanted to work at Sandia because this would give him a chance to stay in the area. At the base, he was interviewed by a man from Handling Equipment. The interviewer looked at the burly young man's credentials, heaved a deep sigh, and said: "George, I'm sorry, but I really don't think there will ever be a place for you in a design organization."[46]

Hildebrandt continued at the University of New Mexico, got his Master of Science degree, and moved on. Two years later, while passing through Albuquerque to visit in-laws in Amarillo, Texas, he saw that Sandia was interviewing for new employees at the Copper Kettle Restaurant, situated in a small row of offices in back of the Triangle Bar near the University. The openings, he learned, were in what was then called Production Engineering. Hildebrandt recalled the experience:

> The man who interviewed me was John Kane. . . . He wasn't too interested in me either until I told him how I had torn down my Model A Ford and rebuilt it. He made me an offer and I was hired in Production Engineering at Sandia as a nongraded engineer, which was a step lower than an MTS (Member of Technical Staff). My first assignment was to provide manufacturing layouts (MLs) for the Assembly of War Reserve and Trainer versions of the Mk 5 bomb. This was my first assignment, and I was the only engineer on the project.
>
> It was not until years later that Harvey Mehlhouse, who later became a President of Western Electric, came in as Director of Production Engineering at Sandia and decided that what we were doing was the equivalent to what was being done in the design organization. Consequently, he was responsible for upgrading us and facilitating transfers to the design group. . . . He did us a big service.[47]

Hildebrandt's experience reflects the identity crisis that the Laboratory was going through during the first decade of its existence. Was it to be a manufacturing organization, a research institution, or a field test contingent? A count of persons on roll as of July 1950 reveals that production-related activites in the 2000 organization had a staff of ninety-eight in two directorates, whereas the single Engineering Development Directorate for weapons had a staff of sixty-three.[48] As indicated by Hildebrandt's case, Sandia's recruiting campaign was directed toward meeting the urgent goal of building a stockpile of nuclear weapons.

The establishment of an effective working relationship with the integrated contractors would help resolve the identity issue and pave the way for the eventual transfer of production responsibilities—an action that would ultimately change the complexion of the Laboratory. Within ten years, the organization would be transformed from a factory-style ordnance facility to a research and development lab working closely with industry.

The budget reflects this transition. In fiscal year 1951, the total operating budget was $62 million, split about fifty-fifty between weapons production and research and development. This relatively high production emphasis continued through fiscal year 56. Beginning in FY 57, and thereafter, research and development and test activities took an ever-increasing portion of Sandia's budgeted resources. Production budgets generally showed corresponding decreases.[49] Similarly, between 1947 and 1958, the integrated contractor complex grew to include nine plant facilities.

The well-planned growth and successful operation of the integrated contractor complex, combined with strong leadership and an improved organizational structure at Sandia, were to have a significant impact on local economies, the future of the Laboratory, and the state of the nation's nuclear arsenal.

BOMB PROGRAMS IN DEVELOPMENT, PRODUCTION, AND STOCKPILE

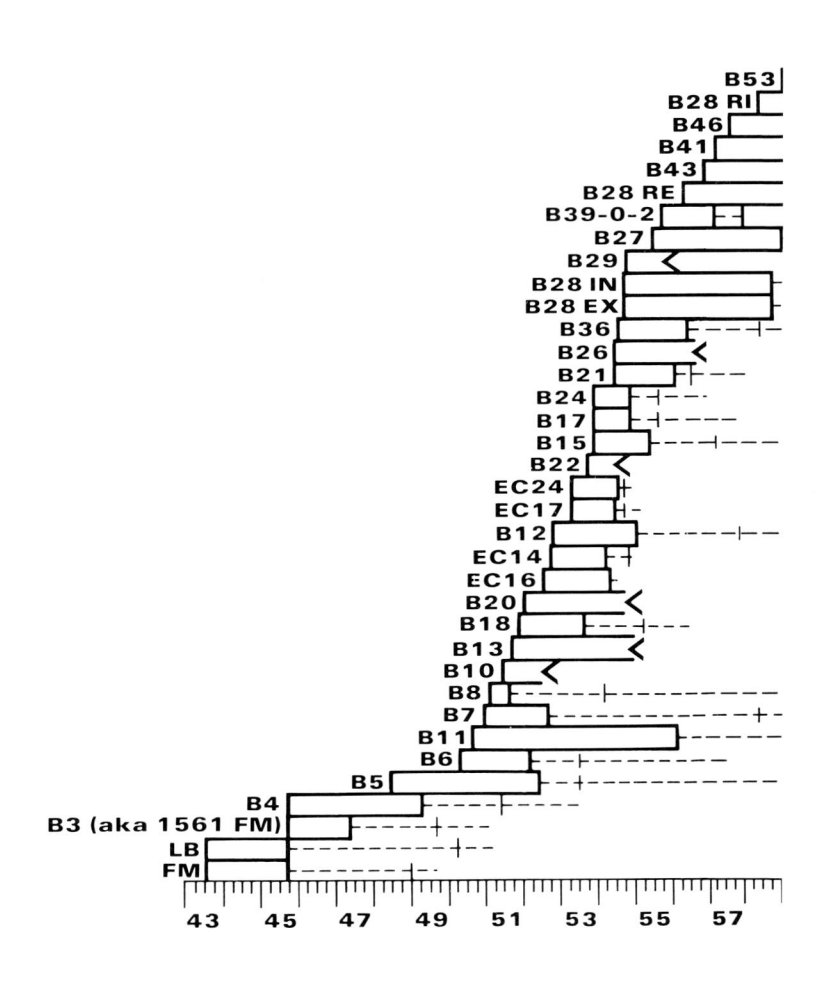

< = PROJECT CANCELLATION DATE
------ = PRODUCTION & STOCKPILE
—+—+—+ = END OF PRODUCTION

Chapter XV.

A SHIFT IN EMPHASIS:
Toward Research and Development

"I wish to be clear . . . in light of the time that has now elapsed since the Sandia Corporation moved in, and in light of current international events, that first priority must be placed on expedition in development and engineering."

James McCormack, Jr.
Director of the Division of Military Application
July 31, 1950

For Sandia's President, George A. Landry, the construction campaign and establishment of the new corporate structure were formidable tasks in themselves, but the outbreak of hostilities in Korea also increased the scope and intensity of the Laboratory's technical activities. The war in Korea not only meant that a number of employees were "called to the colors," as the Sandia *Bulletin* termed it, but also that AEC and Washington officials expected certain goals to be met rapidly and without fail.[1] The coordination of procedures established between the development and production organizations would soon be tested.

The military, operating under a tremendous sense of urgency, passed on the crisis atmosphere to the Laboratory. General McCormack, head of the Division of Military Application, minced few words in voicing his expectations of the Laboratory organized in Albuquerque just nine months earlier:

> I wish to be clear ... in light of the time that has now elapsed since the Sandia Corporation moved in, and in light of current international events, that first priority must be placed on expedition in development and engineering. I do not know that there is any excessive devotion to routine procedures, minor economies, or exaggerated perfection in final development work, but if there is any tendency in this direction, it should be checked.[2]

McCormack went on to stress the need for flexibility in Sandia's procedures to meet military requirements. Santa Fe Operations, McCormack felt, should encourage Sandia along these lines. "We will not be displeased here in the Division of Military Application," he added, "if occasionally news comes from Sandia that a mistake has been made because someone pushed too hard or ran too fast."[3]

Despite the urgency of McCormack's words, Santa Fe Operations and Laboratory management did not interpret the message as a sanction for expediency. Neither did the General's message make much of an impact on Sandia employees, such as G. Howard Mauldin, who recalled that during the early period people did not quote reliability goals: "The weapons simply should not fail."[4] There would continue to be strong emphasis placed on quality of the stockpile and reliability in the field. From the beginning, Sandia would play a leading role in these areas. Moreover, shortly after the changeover to Western management, the surveillance function at the storage sites was formally assigned to the Sandia Laboratory on January 1, 1950.

Sandia personnel were charged with surveillance of material during storage in stockpile and modernization of stockpile items to keep abreast of improving technology. The Laboratory continued to furnish technical instructions and consultation to the Armed Services in operational procedures relating to weapons and storage sites, modification of aircraft, and other delivery vehicles.[5]

Early on, Los Alamos and Sandia Laboratories agreed upon a basic division of technical responsibilities. Los Alamos retained primary responsibility for designing the explosive and nuclear

A SHIFT IN EMPHASIS: TOWARD RESEARCH AND DEVELOPMENT

components; Sandia would incorporate the components into weapons designed for specific military uses, arrange for production, and stockpile the complete weapons in accordance with schedules furnished by the AEC. The "complete weapon" meant the primary weapon itself, the packaging required for transport and storage, and the test and handling equipment required in the field during operational use by the Armed Forces. During the early fifties, concern over packaging criteria and design considerations for seals and shock mitigation systems led Auxiliaries Engineering Department Manager Lessel E. Lamkin to investigate such factors. Publication of the results illustrated that Sandia's requirements were more stringent than those specified by military requirements. R. W. Henderson, in his direct and inimitable style, commended Lamkin's efforts and advised him to "keep slugging for a rational approach to design." "It's about time design should be dictated by service rather than preconceived dogma," Henderson added.[6]

Postwar developments also made readily apparent to the AEC and Sandia's new management the need for simpler, more reliable bombs, produced in much larger quantities. Furthermore, the military services had expanded their focus in the application of atomic explosions to military situations. Initially, military atomic planning activities centered around the Strategic Air Command and its retaliatory capability. Early atomic weapons were thought of as "strategic" because they were designed to effect the destruction of a country's ability to produce the weapons of war.[7]

In this regard, Sandia accomplished one of its first milestones as a corporation by completing a weapons system analysis. Known as the Bode–MacNair Report, the study dealt with strategic uses of atomic weapons with special reference to the fuzing problem.

The emphasis then turned to tactical weapons designed to meet the requirements of tactical warfare, that is, for direct deployment against the Armed Forces of a country. Because the requirements for the two types of weapons are quite different, Sandians had to find solutions for many new engineering and

RESEARCH 1951

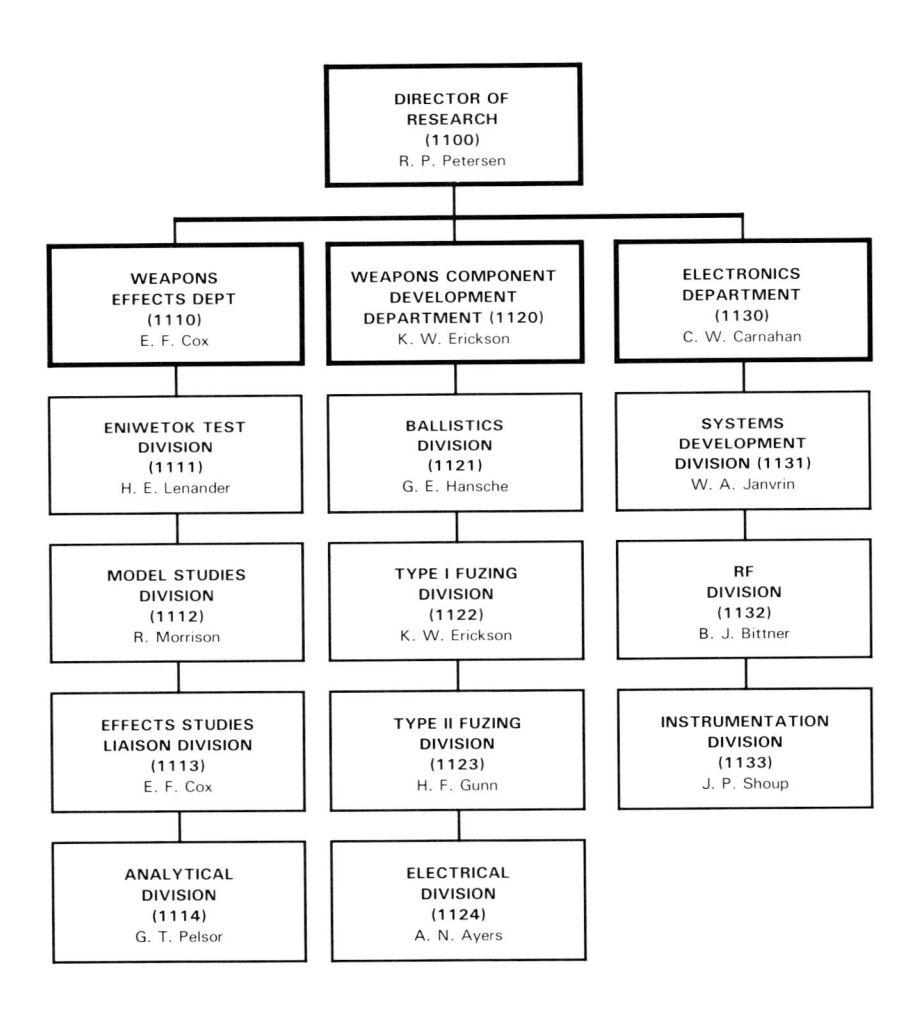

- **DIRECTOR OF RESEARCH (1100)** — R. P. Petersen
  - **WEAPONS EFFECTS DEPT (1110)** — E. F. Cox
    - ENIWETOK TEST DIVISION (1111) — H. E. Lenander
    - MODEL STUDIES DIVISION (1112) — R. Morrison
    - EFFECTS STUDIES LIAISON DIVISION (1113) — E. F. Cox
    - ANALYTICAL DIVISION (1114) — G. T. Pelsor
  - **WEAPONS COMPONENT DEVELOPMENT DEPARTMENT (1120)** — K. W. Erickson
    - BALLISTICS DIVISION (1121) — G. E. Hansche
    - TYPE I FUZING DIVISION (1122) — K. W. Erickson
    - TYPE II FUZING DIVISION (1123) — H. F. Gunn
    - ELECTRICAL DIVISION (1124) — A. N. Ayers
  - **ELECTRONICS DEPARTMENT (1130)** — C. W. Carnahan
    - SYSTEMS DEVELOPMENT DIVISION (1131) — W. A. Janvrin
    - RF DIVISION (1132) — B. J. Bittner
    - INSTRUMENTATION DIVISION (1133) — J. P. Shoup

Source: *Sandia Corporation Telephone Directory*, July 1950

design problems.[8] The responsibilities of the Laboratory's new Research Directorate would expand accordingly.

Formation of the Research Directorate

Formally established in late 1949 as a product of the Laboratory's reorganization under AT&T, Sandia's Research directorate was first headed by Robert P. Petersen. Petersen, along with Everett F. Cox, had been a colleague of Paul Larsen at the Applied Physics Laboratory. With Larsen's encouragement, both men joined Sandia's fledgling research organization not long before the changeover to corporate status.

As initially constituted, the organization included a Weapons Component Development Department under Ken W. Erickson, an Electronics Department under C. W. Carnahan, and a Weapons Effects Department headed by Cox.[9] The growth of research functions within Sandia's engineering assignments was directly related to the atomic weapons task. Sandia's research efforts would be driven by the continuous introduction of ever more sophisticated techniques in ordnance engineering. Response to immediate need established priorities.

The strategic weapons study, known as the Bode–MacNair Report, was a prime example. The events leading to the investigation went back to January 10, 1950, when the Field Command Armed Forces Special Weapons Project requested that Sandia put forth an all-out effort on fuzing the Mk 4 with a pressure-sensitive, nonradiating component called a barometric switch. The Laboratory had been expending considerable funds and manpower on radar fuzes, but AFSWP questioned Sandia's prudence in view of the National Military Establishment's belief that barofuzing was entirely feasible. Interest in the matter was prompted by AFSWP's desire to prevent enemy jamming of radars, resulting in a "dud," and also by the desire to reduce bomb weight by replacing the radar and its power supply with a direct-acting baroswitch network.

The following day, January 11, during a meeting of the Sandia Research and Development Board, Walter A. MacNair, who had been Director of Military Systems Engineering at Bell Telephone Laboratories (BTL), and Board Chairman Robert P. Petersen pointed out that the barofuzing question was much more complicated than indicated by the military and that further detailed study was needed. The outgrowth of this concern was a task force effort undertaken jointly by BTL and Sandia. Participants from BTL included Hendrik W. Bode, then Director of Mathematics Research at BTL and a national pioneer in the discipline to become known as "systems engineering." He was assisted by future BTL Vice-President Donald P. Ling and Robert C. Prim, who would later become a Vice-President at Sandia. The team also included B. McMillan and G. T. Pelsor of Sandia.

Among the questions investigated were: At what height should an atomic bomb burst to maximize the damage resulting from blast? What physical factors determine this choice, and what is the status of present knowledge of these factors? How critical is the choice of burst height; i.e., how large is the sacrifice in military damage for bursts above and below the optimum?

The team, aware that the answers to these questions were basic to an understanding of the fuzing problem and to a realistic statement of the functional requirements of the fuzing system, enlisted the aid of Sandia weapon designers. As part of their investigation, Sandia's designers used the height-of-burst (HOB) or fuzing altitude accuracy versus damage area data from the standard reference, Los Alamos document LA-743. This document, based upon data from empirical measurements of blast waves, indicated existence of a pronounced optimum HOB that dominated thinking about fuzing methods. However, the unanticipated results of the Bode-MacNair study, presented to the Board on August 9, 1950, revealed that HOB was not such an important parameter and that improvements in bombing accuracy would be more effective in causing target damage than increases in bomb yield.

Another report, *Technical and Cost Aspects of Radar Fuze Countermeasures*, prepared concurrently under the direction of R. C. Newhouse, dealt with the jamming problems for radar fuzes.[10] In this way, a weapon's need, fuzing, determined the future direction of Sandia's research activity, including a focus on the improvement of components, on weapon blast effects, and target damage criteria. Overall, these studies had a further significance in that they set the stage for the Laboratory's change in emphasis from production toward research and development.

Origins of the Systems Analysis Function

By 1950 reliability requirements, that is, the probability of success for postwar nuclear bombs, had been reduced significantly from the extemely small failure allowance of 1 dud in 100,000 drops. Credit for this more realistic ratio goes to Los Alamos Director Norris Bradbury. The numerical requirement was significant because it directly affected the maximum use of scarce nuclear materials.

In early 1951, Major General Kenneth Nichols, AFSWP, and Bradbury suggested that the Weapons Reliability Committee, established by the joint Sandia/military coordination body for nuclear weapons development, address the issue. Walter A. MacNair, who joined Sandia in November 1949 as Technical Associate Director, chaired the committee. Hendrik Bode was once again called upon as a BTL consultant; Robert P. Petersen and William E. Boyes represented Sandia. The Committee was to conduct a study of reliability requirements and establish the level of reliability for the atomic bombs based on an evaluation of the entire weapon system.

On December 4, 1952, the Weapon's Reliability Committee issued its report. The Committee's findings, well-recognized for their long-term validity and pertinence, declared that nuclear weapons have a character that is indeed "special" in relation to conventional weapons; namely, they possess a "fundamental and unavoidable complexity" and their actual reliability "is not subject to verification" in peacetime. The report's recommendation would remain essentially unchanged.[11]

Sandia, recognizing the imperative need for determining reliability, added a new function to the Research Directorate in 1951. The Systems Evaluation Department, as it was known, had the initial task of developing statistical models. Within a few years, however, the department developed a true systems analysis capability as reflected in reports that significantly influenced the design of nuclear weapons. By 1952, R. A. Liebler headed a Systems Evaluation Department that consisted of two divisions—Physics under supervisor Gene T. Pelsor and Statistical under P. Maxwell.[12] In the interim, the Laboratory's method of operation continued to be largely reactionary, its course determined by the crash programs and emergency capabilities levied upon it.

Aerodynamics, The Mk 4 Case Controversy, and Timer Problems

The focus of the earliest Lab effort was the Mk 4. The poor ballistic qualities of the 1561-type (i.e., Mk 3 Nagasaki) atomic bomb had been recognized since its initial use; therefore, aerodynamic improvements and overall reduction in weight and size were major objectives. The basic body configuration had been selected through a series of low-speed, wind tunnel tests, after which a suitable fin design was chosen. But project personnel continued to have problems in getting the fins to maintain stability.

By 1952, Ken Erickson had transferred over to manage the newly established Weapons Analysis Department and George E. Hansche took his place as manager of the Weapons Component Department. Reporting to Hansche were R. F. Brodsky, who supervised the Analytical Aerodynamics Division and A. Y. Pope, who headed Experimental Aerodynamics. When Randy Maydew came to Sandia that year after having worked for the National Advisory Committee for Aeronautics (NACA), Ames Laboratory in Mountainview, California, he joined the Analytical Aerodynamics Division. Maydew and other Sandians such as Warren Curry, Ed Clark, Warren Myer, Bill Stephenson, and Paul Rowe began investigating the aerodynamic fin stability problems of the Mk 4 and Mk 6 bombs.

In search of a solution, Maydew sought the advice of a consultant in the East who advised them to use double-wedge fins, meaning a wedge forward and a wedge backward. Fortunately—as it turned out—there was a communication problem. They interpreted what the consultant said as meaning "a wedge forward, then a wider wedge behind it." This design interpretation worked beautifully—a masterpiece of misunderstanding that solved a serious problem. Adaptation of a double-wedge fin improved the bomb's aerodynamics, but weight had been a matter of long-standing concern. From a military standpoint, development of a lightweight external case was considered to be of "vital importance."[13]

Members of the Los Alamos Lab and aerodynamicists from several major aircraft manufacturers first met to discuss weight-saving measures on September 2-3, 1948. At this time, J. K. Northrop of Northrop Aircraft Company suggested that 1,000 to 4,000 pounds could be saved by replacing the steel case with a magnesium or aluminum alloy. In the original design, the military required that a considerable degree of flak protection should be incorporated into the bomb itself, hence the use, initially, of steel armor plate and later mild steel boiler plate as the structural material for the case.[14]

Representatives of the Air Force were enthusiastic about Northrop's suggestion, but Sandia personnel approached the design change cautiously. They first contracted with Northrop for a prototype bomb case and proceeded to make a detailed study of the case design. Such caution led the Air Force

APPROXIMATE WEIGHT COMPARISONS
of the Mk 4 Mod 0 (Steel)
and the Mk 4 Mod 1 (Aluminum)
Welded and Riveted Cases

	Northrop "41" Design (Al, Riveted) (lbs)	ACF "41" Design (Al, Welded) (lbs)	"40" Design (Steel) (lbs)
Forward Case	120	133	520
Split Band	70	70	707
Rear Case	292	414	1617
Rear Cover Plate	11	11	43
Total Case Weight	493	628	2887
"40" Fins and Bolts	88	88	88
Total Case Weight with Fins	581	716	2975

A SHIFT IN EMPHASIS: TOWARD RESEARCH AND DEVELOPMENT

Northrop Mk 4 riveted case, rear quarter view.

Northrop Mk 4 riveted case, front quarter view.

ACF Mk 4 welded aluminum case, rear quarter view.

ACF Mk 4 welded aluminum case, front quarter view.

to accuse Sandia of being slow to respond and less than cooperative.[15]

Wrangling over design problems related to the external case continued for about two years. Air Force personnel, in particular, wanted the lightweight riveted design of aluminum, while Sandia engineers favored the slightly heavier welded case produced by American Car and Foundry. In mid-February 1949, the Division of Military Application directed Sandia to subcontract the design of an aluminum case. Subsequently, a parallel production program was launched to gain production time. Northrop Aircraft was invited to design the riveted case, and American Car and Foundry was asked to design the welded case. Ballistic testing proved both cases to be satisfactory, and since cost and weight comparisons were roughly equivalent, the decision was made to procure both types. Production of the riveted design was turned over to "Royal" (Bendix) at Kansas City. ACF's New York plant fabricated the welded case.[16] Before the deadline could be met, however, drop tests showed the Mk 4 to be dynamically unstable within a certain range.[17] Still another technical concern related to the need for a modified arming timer, the MC-4.

The timer was already in production when Howard Mauldin returned to Sandia in November 1949, after three years with the Physics Department at the University of Michigan. Mauldin was working with Leon Smith in Electronic Systems when he learned that there were serious problems with the Mk 4 timer. Stockpile sampling revealed that the connector bushings on the oil-filled capacitors were leaking. With a deadline to meet, immediate remedial action was required.[18]

Mauldin, assigned as a troubleshooter, searched through manufacturers' catalogs to locate a firm that made the particular kind of bushing that was needed. He found an item listed that apparently would fit and immediately contacted the company to see if it could furnish Sandia with 100 or so on an emergency basis. Mauldin then made arrangements for the bushings to be shipped to a capacitor manufacturer in Chicago and indicated that he would arrive at a certain time to discuss production for Sandia.

A SHIFT IN EMPHASIS: TOWARD RESEARCH AND DEVELOPMENT

The trip was to be a memorable one and indicative of the fast footwork necessary to meet the demands of the times. Mauldin arrived in Chicago, spent the night in a hotel, and the next day called a cab. When Mauldin gave the driver the address of the capacitor manufacturer, the cabbie shook his head: "Gee, I don't believe I know where this is," he muttered.[19] Finally, the driver called the dispatcher for directions and took off on a circuitous route that eventually brought them to the railroad tracks. The driver then turned and drove parallel to the tracks toward a group of deserted-looking warehouses. Just when Mauldin thought he had been sent on a "wild goose chase," the driver stopped the cab in front of a door with the name of a company written across it in chalk. Mauldin climbed the stairs to the second floor, which consisted of one large, open room with a desk over in one corner where the President sat and a desk in another corner occupied by the Chief Engineer. These two individuals made up the total management of the concern.

Despite the "low overhead" environment, the men knew their business and together quickly worked out the processes necessary for production of a true hermetic seal that would solve the problem of the leaking capacitors. The ceramic stud and rubber washers, which degraded with time and allowed the leakage to occur, were replaced by a hermetic seal electrode soldered into the hole, bonding the glass to metal.

Mauldin thought the production problem had been solved, but three weeks after he returned to Sandia, appalling news arrived. The capacitor factory had burned down! There was conjecture that the cause was arson. On the other hand, Mauldin recalled that the floor of the plant was soaked with oil that could have ignited easily with a spark. The destruction of the capacitor factory was a blow to Sandia as well as to the plant's owners. Because production of the capacitors was critical to the Laboratory's progress, the AEC made the decision to set up the company in another facility.[20]

Still another problem involved the vacuum tube used in the Mk 4 timer. The Sylvania Company had been producing the tubes for Sandia according to joint Army and Navy specifications when they discovered that their product was unable to meet specified requirements. In response to a frantic call for assistance, Mauldin again packed his bag and boarded a train for the factory in Emporium, Pennsylvania.

After meeting with Sylvania personnel, he learned that they were subjecting the tubes to 500 hours of torture testing at extremely high temperatures. He pointed out that the Sandia application required that the tube be subjected to maximum current for only thirty seconds; therefore, the Sylvania test was totally unreasonable. The tubes were indeed acceptable product.[21]

4N Program

Part of the pressure surrounding perfection of the Mk 4 timer and its components resulted from a decision by the military and the AEC to levy an emergency requirement for a specific number of bombs to be produced ahead of schedule and used for training and test purposes. In late July 1950, Landry went to Washington to discuss the issue with the AEC. At that time the AEC indicated that the April 1951 stockpiling date would have to be moved up. Sandia was asked to review the Laboratory's capabilities with this objective in mind.[22]

The emergency prototypes were to be used to solve the instability problems associated with the Mk 4 and to test the new fuzing devices. Designated the 4N program, this accelerated program within a crash program had as its objective the production of a certain number of engineering models of complete bombs together with a limited amount of handling and test equipment. The deadline for completion of the special test models was set as November 1950, hence the symbol—4 for the Mk number and N for November. Since it was necessary to handle the work orders for those specific bombs separate from the regular Engineering Department work orders, a special stamp denoting a large 4N was used.[23]

A SHIFT IN EMPHASIS: TOWARD RESEARCH AND DEVELOPMENT

Sandia weapons engineers were involved in the development of two kinds of auxiliary support equipment: mechanical devices to facilitate assembly and loading of bombs (called H-Equipment for Handling), and mechanical or electrical devices to facilitate the testing of bombs and components (referred to as T-equipment for Test.) The first H-1, developed by the Technical Facilities Division, was used to remove the bomb's noseplate and manipulate the high-explosive charges. A trapdoor effect was created, allowing insertion of the nuclear capsule.

Prototypes of the H-1 hardware were then provided to the local U.S. Air Force Weapons Center (AFWC) for testing under simulated use conditions. The Aircraft Liaison section acted as coordinators. Problems arose when AFWC issued an unsatisfactory report, noting that the formation of frost at high altitude on the surface of the H.E. charges prevented the vacuum cups from holding the charges securely. The displeasure of the military was a factor in a subsequent decision made by designers at Los Alamos Scientific Laboratory to develop a "cored" H.E. charge that would significantly ease nuclear arming operations in the cramped, unpressurized bomb bay.[24] Future directors C. H. "Hilt" DeSelm, Lessel E. Lamkin, Eton H. Draper, and William A. Gardner were among those engineers who faced technological challenges in these areas.

Another product of AT&T recruiting efforts, Robert P. Stromberg, joined Sandia in time to participate in the 4N program. His first assignment, indicative of the temper of the times, was to sort out the good electrical contact elements from a batch of MC-5 Barometric Pressure Sensing Switches in order to assemble the requisite number of Mk 4 bombs on demand. Sandia was extremely short of personnel for this crash program; therefore, the Field Command Armed Forces Special Weapons Project resident on Sandia Base pulled the strings necessary for emergency assignments of military personnel to assist. Stromberg recalled that testing exercises associated with the program sometimes got to be exciting affairs as when the tail of their low-flying B-29 tore out a good section of the barbed wire fence along Gibson Avenue before landing safely.[25]

By 1950 the Mk 4 had evolved into the Mk 6. Design changes included substitution of the improved fuzes with resistance to jamming as well as remote setting of fuzes for desired height-of-burst. This particular improvement allowed the choice of alternate targets and was, therefore, of major importance to the Armed Services.

Nevertheless, the press of world events resulted in stockpiling of the Mk 4 without incorporation of all desirable changes. Consequently, many of the design problems of the Mk 4 carried over into the Mk 6 development program. Adoption of the aluminum case to replace the heavier steel casing reduced overall weight, allowing the carrying aircraft less takeoff distance, but the ballistic performance of the lightweight case suffered in comparison. Sandia aerodynamicists developed spoiler bands as a solution. A major emphasis was also placed on making the weapon easier for the military to handle through improved methods of assembly at the storage sites and by in-flight insertion (IFI) gear.[26]

Design changes such as the IFI had ramifications for safety as well as for improvement of performance. Nuclear testing conducted in 1948 had first introduced the concept of a removable capsule design whereby the fissile material could be inserted or removed manually from an otherwise fully assembled weapon. Subsequently, early meetings of SWDB, chaired by Dick Bice, devoted considerable attention to development of a system for in-flight nuclear extraction. This design, used for both the Mk 4 and Mk 6, had the advantage of being absolutely nuclear-safe since the capsule could be inserted manually in the bomb bay while on the way to the target, or removed before landing if the mission was cancelled or aborted. It also improved reliability since the weapon could be examined on a continuous basis and shortened the time required to respond to a strike order.

With the Mk 5, however, the technology soon advanced to include remotely controlled mechanically inserted capsules. The advent of unmanned cruise missiles such as the MATADOR and REGULUS meant that, prior to motorized IFI, capsules would have been placed in the weapons prior to launch, a frightening idea. The Mk 7, which was small enough to be carried externally on fighter aircraft, required similar considerations for bomb applications. The solution was an in-flight insertion device to hold the capsule in a position external to the H.E. sphere where it could be inserted electromechanically shortly before release. The device, therefore, seemed to preserve both the safety advantages gained by the Mk 4 and 6 designs, and it also increased operational capability.[27]

The required number of prototype Mk 6 bombs was delivered to storage during January and February 1951. The majority of the Mk 6's, which incorporated the first satisfactory impact fuze, were slated for use by AFSWP for training and testing purposes. Much of the work on contact fuzing took place in Ken Erickson's Weapons Component Department. At the time, Leo Gutierrez was Section Supervisor for the Pressure Analysis Section of the Aerodynamics Division. When Richard S. Claassen came into the organization in March 1951 as a bright young Ph.D. from the University of Minnesota, he and Dave Webb were asked to work on the contact fuzing project under Gutierrez. Webb and Claassen originated the use of barium titanate, a ferroelectric ceramic for contact fuzing elements. Subsequently, the firing set design was changed to accept these kinds of signals, an innovation that added substantial flexibility to weapons systems.[28]

In addition to the Mk 4 and Mk 6, other bombs were placed in design and development as crash programs. Among these were the Mks 5, 7, and 8.

Other Marks Meet Emergency Capability Requirements

On July 11, 1950, the Division of Military Application directed that a plan be formulated "using all facilities at your disposal to deliver to War Reserve at the earliest possible date service models of the TX-5."[29] Sandia had organized a new division specifically for development of the TX-5 early in 1949. Initially, however, the DMA had not required substantial stockpiling of the weapon, but the outbreak of the war in Korea generated a reversal in plans. By 1950, the pressure was on. The TX-5 was regarded as the answer to the military services' expressed desire for a smaller, lighter-weight bomb that could be delivered by fighter bomber aircraft.[30]

Eton H. Draper in 1949 was promoted to supervise the Mk 5 development project. Draper, then thirty-one years old, previously had worked as a project engineer for an aircraft company during the war and brought to Sandia Laboratory management a charismatic leadership style that would make a positive impact. He was to rise rapidly through the ranks from Department level (1950) to the Directorate level (1957) and to Vice-President (1961) before his death in 1966, due in part to complications from a near-fatal automobile accident while on a Sandia business trip.[31]

In addition to the development of a motorized insertion mechanism for the TX-5, engineers went to work on ballistics problems by conducting wind tunnel tests on twenty-eight different designs of nose and tail configurations. Results showed the bomb shape to be stable except upon leaving the B-47 bomb bay, where air flow around the bomb caused violent oscillations. The time-dictated solution was to eliminate the B-47 as a carrier.

The arming and firing subsystem also had to be significantly reduced in size and weight to fit inside the smaller bomb casing. To date, the firing set had been redundant; that is, there were two identical channels, each containing the same elements, and either channel capable of firing the bomb. This redundancy caused high penalties in size and weight; yet eliminating redundancy meant having to have higher reliability in the single channel that remained. Because the electrical capacitor bank accounted for much of the firing set's weight, capacitor development was of primary interest. Sandia engineers, therefore, selected and procured to specifications a commercially available high-voltage capacitor, termed the SA-112. The SA designation identified it as a Sandia Apparatus, whereas components designed by the Laboratory were denoted "MC" for Major Components.[32]

A SHIFT IN EMPHASIS: TOWARD RESEARCH AND DEVELOPMENT

Awkward nose-jack loading system (above), an early approach to the loading problem, was soon abandoned in favor of an in-ground pit equipped with a hydraulic hoist (below).

Little Boy.

Art Machen with the Fat Man.

A SHIFT IN EMPHASIS: TOWARD RESEARCH AND DEVELOPMENT

MK IV MOD 0 FM

MK III MOD 0 FM

There were significant changes in the shape from the Mk 3 Fat Man to the Mk 4 Fat Man.

Corry McDonald with the Mk 5. The Mk 5 was the first nuclear bomb compatible with shipboard naval aircraft and was designed for electrically actuated insertion of the nuclear core in-flight, thereby greatly enhancing safety in the operational environment.

NUCLEAR ORDNANCE ENGINEER FOR THE NATION

Al Fite with the Mk 7/CORPORAL.

Dick Bice with the Mk 6 bomb.

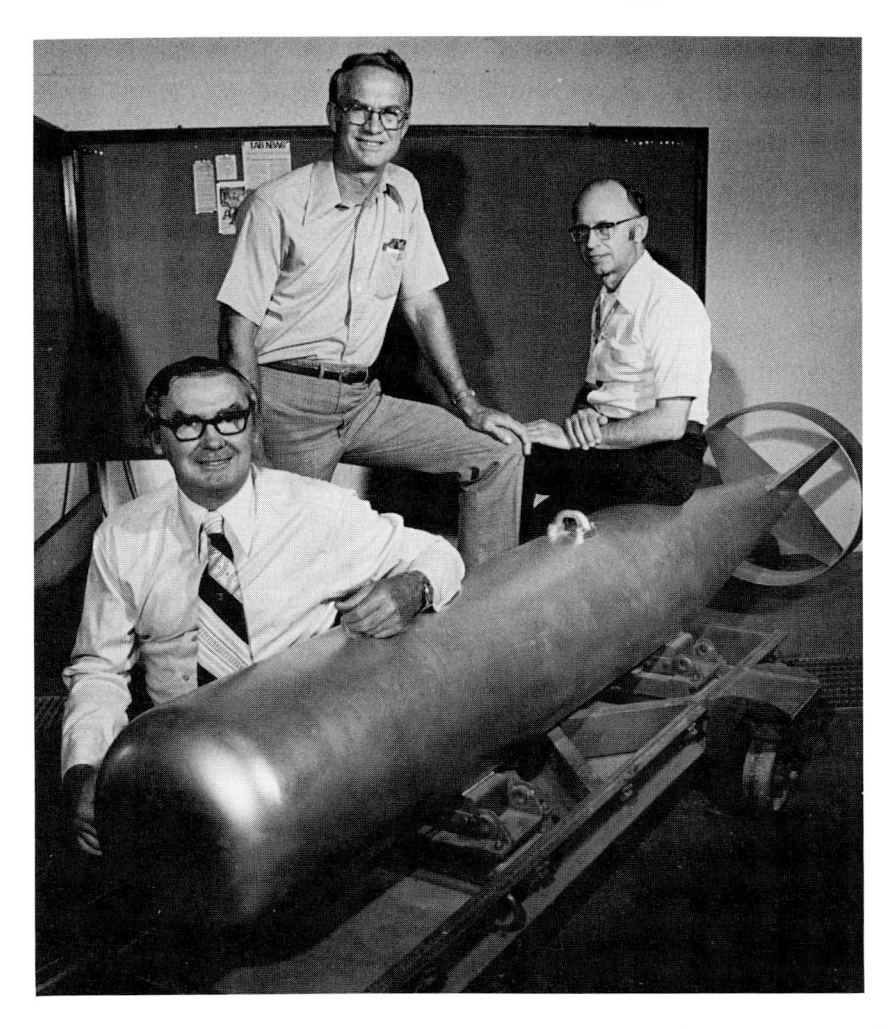

Charlie Runyan, Dick Kidd, and Bill Lawrence with the Mk 8 bomb. The Mk 8 was developed as a penetration weapon for use against concrete aircraft runways, submarine pans, and heavily fortified underground structures.

Ralph Wilson with the Mk 12. The Mk 12's fins folded flat for aircraft and ground clearance.

A SHIFT IN EMPHASIS: TOWARD RESEARCH AND DEVELOPMENT

As a precursor to Operation Greenhouse, testing of the Mk 5 took place at Frenchman Flat north of Las Vegas, Nevada, in early 1951. Rear Admiral William S. Parsons, anticipating the need for a continental proving ground to test certain developmental theories, recommended the site. The Mk 5, described as a free-fall, air-burst, implosion-type radar and baro-fuzed strategic bomb, was released for production June 1, 1951, and stockpiled in March 1952. Later modifications included the baro-armed and baro-fuzed system with contact backup.[33]

Other devices reflecting the trend toward smaller, more flexible weapons included the TX-7, which was then in the design stages, and the TX-8. On May 26, 1950, the Los Alamos Scientific Laboratory presented to the TX-5 Steering Committee results of its work on a small-diameter bomb for tactical use by the military services. The Committee reacted favorably and assigned the code name "Thor" to the nuclear subassembly device. The bomb was predicted to weigh about 2,000 pounds, with a diameter of about three feet or less. This design stood in sharp contrast to the 3,300-pound, forty-five-inch diameter of the TX-5 then in full development. The Korean War also resulted in acceleration of this program with the objective of a production date of January 1, 1952. This placed the TX-5 and the Thor on essentially concurrent time scales. Plans were to use the TX-5 components on the Thor.

The Thor's streamlined shape made possible for the first time external carriage on fighter-type aircraft. External carriage naturally demanded an automatic in-flight insertion system (AIFI) for the nuclear capsule. Similarly, the aerodynamics aspects represented an area new to Sandia. The Laboratory, therefore, chose to award a development contract to Douglas Aircraft Company of California. By September 8, 1950, the Thor had become the TX-7 bomb, and the Air Force's F-84 jet aircraft was selected as the first delivery aircraft.[34]

Early wind tunnel tests showed the external shape to be acceptable, but the first full-scale drops at Salton Sea Test Base resulted in a bomb and tail separation. Misalignment of the fin was a critical factor. To prevent the bomb from falling through

Mach I, Sandia engineers developed dive brakes. Eton H. Draper's Engineering Department, with an Electrical Division under E. J. Bruda and a Mechanical Division supervised by C. F. Robinson, had responsibility for the design.

Because of the international situation, the Lab went on a six-day-week work schedule. The short time between design initiation and production demanded a pragmatic and dedicated approach. The Mk 7, Mod 0 was design-released September 1, 1951, with production achieved by July 1952. Once again, an emergency capability effort designated the "7N" program was levied, and the assigned quota was produced successfully between January and April 1952. Modifications of the Mk 7 included the capability to operate radar and battery heaters from external power, a neoprene-coated radome to prevent icing, a fin assembly that could be rotated to fit different aircraft, nickel cadmium batteries, new fuze designs, and improved ballistics, among other improvements.[35]

In 1950, a "38N" program was also instituted with the objective of producing an interim stockpile of TX-8 bombs for operational use. The TX-8, the first atomic weapon designed to achieve penetration before explosion, was a gun-type weapon for both tactical and strategic use. Its predecessor was the "Elsie" (the phonetic nickname for "LC," a follow-on to "LB" or the Little Boy gun-type weapon).

For a while, the gun method of assembling fissile material had been largely abandoned because of its inefficient use of nuclear materials in comparison to the implosion method of the Fat Man. However, following Operation Crossroads *Baker* full-scale test detonation at the Bikini Atoll, the U.S. Navy became interested in a bomb that could be dropped from an aircraft and detonated underwater. The Sandia Research and Development Board afforded the Elsie high priority, next to production of the Mk 4.

The shape of the TX-8, particularly its blunt nose, limited carriage to the bomb bay of the aircraft. The method of inserting the capsule required manual operations. All of these factors precluded the TX-8 from being carried externally on the Navy's

A SHIFT IN EMPHASIS: TOWARD RESEARCH AND DEVELOPMENT

F4U, AD, and A2D aircraft and at supersonic velocities. Consequently, on October 3, 1950, the AEC authorized development of an externally carried TX-8 for subsonic carriage and development of a new bomb, the TX-11, for supersonic external carriage.

The bombs required for emergency stockpiling by March 15, 1951, were hand-built under Sandia's engineering direction and stockpiled on schedule. This 38N Emergency Capability Project, Sandia's first, was completed two weeks before the 4N project was design-released in November 1950. The original TX-8 entered stockpile in January 1952.[36]

The TX-9, the first atomic artillery shell; the TX-10, the first gun-type weapon designed for external carriage; the TX-11 gun-type penetration bomb; and the TX-12, a smaller diameter implosion weapon, were all in the design stages during this period. The TX-9 evolved from the Army's interest in developing an artillery-delivered nuclear weapon. Los Alamos followed up on this by proposing to change the Army's 240-mm conventional artillery piece to provide a 280-mm barrel (allowing an eleven-inch-diameter shell). In May 1950, the Joint Chiefs of Staff established a requirement for development of the weapon system and assigned responsibility to the Ordnance Corps of the Army. Picatinny Arsenal at Dover, New Jersey, was placed in charge of the design, and the Army nomenclature T124 was assigned.

On June 19, 1950, Sandia was requested to provide design consultation and to assist in solving problems of storage, surveillance, and handling. A committee called the "Button Coordination Committee," so-named because of the notion that the shell could be delivered accurately "right on the button," held its first meeting July 21, 1950. Three months later, the program was realigned by the AEC's Santa Fe Operations Office, with Los Alamos and Sandia tasked to assign normal project development roles. The TX-G or Gun Committee, including members from Los Alamos and Sandia, were assigned the job of project coordination. Its counterpart was the TX-N Committee for implosion-type weapons. The TX-9 nomenclature was assigned on December 18, 1950.

Sandia assumed responsibility for nonnuclear components, development of an evaluation program, certification of all nonnuclear parts, establishment of a training program, and design and procurement of nonnuclear test and assembly equipment. The Army would develop the fuze and the shell structure. By mid-1951, the TX-9 gun-type artillery shell underwent nonnuclear firing tests of prototype shells, and the design was frozen on November 1, 1951. The work remaining on this weapon included tests of a new type of initiator recently developed by Los Alamos, tests of fuzes, and full-scale nuclear testing several years later.

Quality Assurance engineer Douglas A. Ballard, assigned to track development of the TX-9, viewed one of these tests from the vantage point of an Army helicopter. His objective was to assist in pinpointing the exact place of impact to facilitate recovery of fragments for examination by QA personnel. Staging for the test involved simulation of a battlefield array of tanks attacking through an artificial forest set in concrete with tanks placed strategically, both under the trees and out in the open. This particular nuclear shot, fired at 8:30 in the morning on a bright sunny day, created a display of all the colors in the rainbow in the burst and in the cloud. "We were very interested because this was one of the first gun-type shots since Hiroshima," Ballard observed, "and an opportunity to observe the technology at close range." Initial deliveries of the TX-9 artillery shell to stockpile occurred in April 1952.[37]

The first program not to reach production status was the TX-10. This was a gun-type, lightweight bomb for airburst delivery, carried internally or externally by supersonic "tactical" aircraft. Sandia had studied the apparent duplication in weaponry capabilities that would result from developing the TX-12 (a scaled-down successor to the TX-7 and TX-10), and the Laboratory's report influenced the Sandia Weapons Development Board in late December 1950 to recommend cancellation of the TX-10 in favor of the TX-12. In March 1951, however, the Military Liaison Committee (MLC) rejected the suggestion, and the development program continued. Sandia contracted with the Navy's Bureau of Ordnance for development and prooftesting of

A SHIFT IN EMPHASIS: TOWARD RESEARCH AND DEVELOPMENT

the gun assembly. By midyear, it was clear that the bomb would exceed the diameter and weight specified by the Military Characteristics without revision of the gun design by Los Alamos. The TX-10 program, therefore, had to be returned to a study phase, to include Sandia's miniaturization of the fuze. Cancellation of the program by the MLC on May 7, 1952, in favor of the TX-12 marked the first program at Sandia to fail to reach production status.[38]

On November 29, 1950, the TX-11 gun-type penetration bomb for external carriage on high-velocity aircraft was approved. The Navy's Bureau of Ordnance handled the development phase with Sandia assisting by fitting the design to the various delivery aircraft. By midyear, the program was directed toward a weapon that would survive impacts with reinforced concrete (as in submarine pens) at a delivery speed of 1,500 feet per second.

The TX-12 implosion-type bomb was conceived by Los Alamos Scientific Laboratory to extend the weight and size reduction that had been achieved by the Thor (TX-7) concept of 1947. The relation of the TX-12 to the Thor was implied by its code name "Brok," the mythical dwarf who forged the hammer of Thor. By mid-1951, the TX-N Steering Committee officially recognized the weapon program as the TX-12.[39]

Sandia's challenge was to establish a ballistic shape by aerodynamics and aircraft compatibility studies for the dozen-or-so prospective military aircraft that could carry the bomb externally at transonic speeds. Since the TX-12 and the TX-7 bomb development programs involved the same organization in several areas, the TX-12 program was delayed in the fall of 1951 when the third crash program, the "7N" program, was begun. The TX-12 was regarded as the "ultimate" bomb design.[40]

Project Greenfruit

In this era of crash programs and emergency capabilities, manufacturing difficulties in radar devices, which resulted in a low acceptance rate, had to be dealt with. A correction task force called Project Greenfruit ("GF" or Grouped Forces, as some top management types preferred to call it) was organized to address the problem. The code name for the task force gave a graphic representation of the situation. Essentially, components, like fruit, were being plucked for use before they were ready.

Results of the study showed a need for closer cooperation between designer and manufacturer, as well as better manufacturing process control. Sandia Apparatus specifications were developed as a part of the effort to improve overall quality. Close-tolerance items were selected from commercial stock or were specially manufactured, as in the case of the capacitors obtained for the Laboratory by Howard Mauldin. A notable exception to the "design by contract" modus operandi were the radar fuzes first developed by Sandia's Applied Physics Department, and later by the Component Development Directorate, established in 1952.[41]

In addition to radar technology, Sandia engineers and scientists contributed significantly to advances being made in the area of power supplies. In April 1950, weapons project engineers made a survey of state of the art in this field and adopted a new type of nickel-cadmium storage battery, which became known as the MC-193. The MC-193 reduced preparation time, including charge, from the twenty or so hours required by its predecessor, the ER-12-10 lead-acid battery, to one hour. Furthermore, charged storage life was increased from 1 day to 240 days, depending upon the ambient temperature. Beginning in May 1952, Sonotone Corporation produced the MC-193 for later versions of the Mk 6 and 7 bombs.[42]

In retrospect, problems with radars and the resulting Greenfruit Task Force contributed to a full-fledged investigation of components in the stockpile, and an increase in testing. Emphasis on improved battery technology and other components with longer shelf life provided the technological base for what would be known as the "wooden bomb concept." This concept, as the name implied, referred to the goal wherein nuclear weapons would be designed to be maintenance-free. "Like an inert log in a forest," they would not require regular testing.[43]

A SHIFT IN EMPHASIS: TOWARD RESEARCH AND DEVELOPMENT

To make these gains, the Laboratory during 1951 increased research and development personnel from 733 to 1,163, while the total number of employees increased from 2,686 to 4,044. Thus the first two years of the Laboratory's existence under AT&T were characterized by important advances in research and development. The year 1951, therefore, marked a change of emphasis. Management began discussions that subsequently led to a plan, to be worked out over a period of time, whereby research and development would receive greater importance, and manufacture of materiel for the AEC's overall production program would be limited in amount.[44]

In the interim, while Sandia scientists and engineers worked long hours to meet technological demands, the administrative side of the house faced challenges in which the major component was the human factor.

February 21, 1950: The Sandia Board of Directors was increased by two members. Left to right, Walter L. Brown, Frederick R. Lack, Donald A. Quarles, Henry C. Beal, George A. Landry, and Stanley Bracken.

Chapter XVI.

UNIONIZATION:
Aftermath of Incorporation

"As to the dignity of all employees, I have often observed . . . that class distinctions tend to grow up in organizations—and especially that having a supervisory or professional position tends to make us look down on other people. This, where it occurs, is a very grave mistake. Never underestimate the inherent worth of the mechanic in the shop, the stenographer who does your typing, the guards around our premises, or even the janitor that cleans at night."

Vice-President Timothy E. Shea,
Sandia Corporation Supervisory Dinner Meeting,
February 17, 1953

One month after taking over as President of Sandia Corporation, George Landry was back in New York. His limousine crossed Fulton Street near St. Paul's Chapel and pulled up to the AT&T headquarter's building at 195 Broadway. He got out in front of an imposing neoclassical style structure, passed through the columned entry, and on up to the twenty-sixth floor where he joined other members of Sandia Corporation's Board of Directors. The presence of Directors Beal, Lack, and Brown in addition to Landry constituted a quorum.[1] In an effort to comply with the wishes of the AEC, and unaware of the ramifications of their actions, the Directors were about to make a decision that would ensure a rocky transition for the new corporation.

A meeting among Bell Lab and Sandia officials included (from left to right) Donald A. Quarles, Sandia President; Mervin J. Kelly, President of Bell Telephone Laboratories; James E. Dingman, Vice-President and General Manager of Bell Telephone Laboratories; and Timothy E. Shea, Vice-President and General Manager of Sandia.

Administrative business was primary on the agenda. Board members agreed to execute a supplemental agreement to the contract beween the AEC and Western Electric, increasing from $10 million to $20 million the total amount that Sandia was authorized to expend or commit under the contract. The meeting proceeded. On motion duly seconded, the Board adopted a resolution to change the twenty-four-day vacation granted Sandia employees under the University of California to coincide with Western Electric's standard plan. This would mean a sharp reduction in the number of vacation days, particularly for shorter-service employees. The decision resulted from a critical review of Sandia's vacation policy, which showed that in contrast to most industrial concerns, the Laboratory's plan was extremely liberal. Employees accrued two days of vacation for each month of employment, up to a cumulative limit of thirty-six days. The AEC, therefore, wanted Sandia's plan brought more in line with traditional practice.[2]

Formation of Metal Trades Council and Sandia Employees' Association

Shortly after his return to Albuquerque, Landry issued a memorandum to all employees announcing the revised vacation plan. He called attention to the fact that the plan compared favorably with vacation payments made by other progressive companies and stressed that the change was necessary in view of "conditions" [emergency preparedness] existent at the Lab. Nevertheless, the company position failed to persuade a substantial number of Sandia employees. Although Western made the change, with a corresponding six percent adjustment in base pay to compensate for lost vacation, employees figured out that the six percent would eventually disappear as other adjustments occurred over time. According to Superintendent Luther Heilman, who was involved in the negotiations, "This was a strong point."[3] The vacation issue, in combination with other factors, triggered the unionization of Sandia's graded employees. Three unions would be established within a matter of months.

Although there had been no unions at Sandia before incorporation, as Ernest Peterson, head of the Labor Relations Department, explained: "This was not so much a matter of lack of desire for union organization but was due more to the newness of the atomic energy program. An additional reason was the fact that Sandia Laboratory . . . was operated by the University of California."[4] The University of California was, in effect, a political subdivision of the State of California. Therefore, a union at the Sandia Laboratory in New Mexico simply could not be recognized legally except with concurrence of the University of California, and the chance of that happening was slender.

Basic to the developing controversy, however, was the fact that from the beginning, employees had enjoyed two days of vacation for each month of employment. Recruiters used this extremely liberal plan to attract prospective employees to join the Corporation. Once on board, employees did not look favorably on the loss of such an attractive fringe benefit.[5]

Representatives of various labor groups, largely members of different American Federation of Labor (AF of L) craft unions, saw in the situation an excellent opportunity to organize Sandia employees. Among the first were the International Chemical Workers who appeared on the scene to try to organize production and maintenance personnel into a union affiliated with the AF of L. Almost simultaneously, a group of employees from both shop and office decided to form a union to keep out unions. The Sandia Employees' Association, as the group came to be known, was unaffiliated with any national union and wanted to represent all Laboratory employees, with the exception of Security guards, management personnel, and professionals.

On January 4, 1950, the El Paso Office of the National Labor Relations Board (NLRB) notified the Corporation that the Sandia Employees' Association had filed a petition for election. At the same time, the NLRB advised that an unfair labor practice charge had been filed against the Corporation by the International Chemical Workers Union of the AF of L. The charge cited five unfair practices, namely, that an employee had been discharged because of his union activities; that this same employee was discriminated against in the assignment of overtime work because of his union activities; that the Corporation had aided and abetted the organization of an independent employees' association; that the Corporation had granted an increase in wages in the face of organizational activities; and that the Corporation was querying prospective employees concerning their membership in unions.[6]

The immediate effect of the unfair labor practices charge was to stop action on the Association's petition for election. The NLRB ruled that it could not authorize an election for a union charged with being company-dominated.[7]

The burst of union activity immediately raised the question as to whether unions were legally permissible at Sandia under the Labor Management Relations Act of 1947. The Taft–Hartley Act, as it was commonly referred to, allowed labor organizations to petition the NLRB to hold an election to determine whether or not a corporation would be unionized. Even more specific to Sandia's case, however, was a ruling made as a result of a strike at the Oak Ridge Laboratory in Tennessee. The strike had prompted President Truman in September 1948 to investigate labor relations in the atomic energy field. The investigating committee, known as the Davis Commission, recommended that all eligible employees engaged in atomic-energy work be permitted to organize and bargain collectively with management.

At first glance, one would think that this designation automatically included Sandia. But the Sandia Laboratory was unique because of its location on a military base and its exclusive concern with military weapons. A ruling was required. Ultimately, management referred the matter to the AEC Industrial Relations Office, which ruled that Sandia was indeed subject to the Taft–Hartley Act and granted permission for employees to organize.[8]

When it became apparent that no one craft union could muster enough votes to carry an election for all Sandia production workers, the various craft unions pooled their resources and formed the Atomic Project and Production Workers, Metal Trades Council of the American Federation of Labor. This first group to take advantage of the ruling, the Metal Trades Council, rented space in a building on Wyoming Avenue opposite the main entrance to the base. Union organizers went to work distributing literature at the technical area gates and urging attendance at their meetings.[9]

Meanwhile, the Sandia Employees' Association learned that the NLRB considered it inappropriate for a bargaining unit to contain both production and office workers. The Association, therefore, had the choice of splitting into two bargaining units or presenting their case in formal hearing before the Board. After a meeting in the Albuquerque High School auditorium, the Sandia Employees' Association filed a petition requesting that office and clerical workers be segregated into a separate bargaining unit; the Association would confine its organizational efforts to production personnel.[10]

After a series of conferences, the Metal Trades Council and the Sandia Employees' Association agreed to hold a joint election with both groups on the ballot. The Metal Trades Council won the election, held May 11, 1950, with seventy-five percent of the vote and was certified on May 19.[11]

Federal Mediators and the Davis Panel Intercede

Collective bargaining began soon thereafter, with the Council proposing that the Corporation negotiate with the fifteen or so different unions. Realizing the tremendous problems this would entail, the Corporation strongly resisted such a split. The National Labor Relations Board decided in favor of the Corporation, and by July 1950, an agreement had been reached—the Corporation would deal with the Council as a body and not with the separate entities. Throughout July and into August, bargaining continued as the Corporation, and the Council attempted to come to terms on wages, benefits, and working conditions.

As of mid-August, a total of fourteen articles had been tentatively approved, most of these having to do with relationships between the Council and the Corporation. Very little had been accomplished concerning agreements on wages, hours, and working conditions; therefore, at the request of the Council, a representative of the Federal Mediation and Conciliation Service entered negotiations on July 20 and attended most of the bargaining sessions until August 16. On that date the agent withdrew, saying that nothing further could be accomplished by his office in bringing the parties closer together. Subsequently, on August 18, the Metal Trades Council requested the intervention of the Atomic Energy Labor Relations Panel.[12]

The function of the Federal Mediation and Conciliation Service is to serve as a catalyst or arbitrator in disputes where interstate commerce is involved. The overall objective of the service is to prevent strikes. However, the AEC Panel, known as the "Davis Panel," operates in a different manner. This Panel enters union disputes only when the Atomic Energy program is involved and only after the Conciliation Panel has certified that they cannot resolve the dispute through normal conciliation activities, that is, at the point where the union issues a strike vote. The criticality of the situation is also a factor.[13]

At Sandia, therefore, when intercession became necessary, a series of hearings were held that extended into the fall. The Panel eventually recommended that the Corporation restore the twenty-four-day-per-year vacation plan of the University of California. Although the Metal Trades Council accepted all recommendations, the Corporation balked at inclusion of a vacation clause. Nevertheless, on January 15, 1951, with the encouragement of the AEC, the Corporation signed a labor contract that included the twenty-four-day vacation provision, with the stipulation that the issue could be reopened at a later date. The Panel had also recommended continuation of the sick leave plan in existence under the University of California.[14]

As agreed upon, a second election had to be held among office and clerical employees. This group, organized by the Office Employees International Union of the AF of L in 1945, was

NUCLEAR ORDNANCE ENGINEER FOR THE NATION

The agreement between Sandia Corporation and the Office Employees International Union, AF of L, was signed in April 1951 by officials representing the two groups. Shown here at the brief ceremonies which accompanied the signing of the agreement are representatives of Sandia Corporation and officials and union officers. Seated left to right: E. J. Domeier, Labor Relations Department; Robert Hawk, Union President, and Kenneth Shinn, Secretary-Treasurer, Local 251, OEIU. Standing, Jack Hart and Ernest Peterson, Labor Relations Department; F. B. Smith, Personnel Director and Treasurer, Sandia Corporation; Frank Morton, international representative of OEIU, and Brad Shaw, Ralph Ridenour, James Stoll, and John Stark, union officers.

434

relatively new. Its origins at Sandia can be traced to two Sandia employees, a man and his wife, who felt strongly that the Laboratory needed a clerical union. In 1950, their names were given to union organizer Frank Morton. The success of the Metal Trades Council spurred sympathizers among office and clerical workers to take similar action. Names of those interested were distributed to organizers, and sentiment favoring a white collar union developed rapidly. With assistance from the Metal Trades Council—in the form of advice, office space, and supplies—the office workers launched their own campaign; and on September 8, 1950, the Office Employees' International Union (OEIU), Local 251, was certified.[15]

Bargaining followed, and when discussions reached a stalemate, the Federal Mediator and the Panel were called in to arbitrate. The Office Union hoped to gain liberalization of some of the clauses previously accepted by the Metal Trades Council. On April 13, 1951, the points of contention were finally resolved, largely in favor of the union. The April 27, 1951, issue of the Sandia *Bulletin* announced signing of the agreement between Sandia Corporation and the Office Employees' International Union of the AF of L. Taking part in the ceremonies were E. J. Domeier, Labor Relations Department; Robert Hawk, Union President; Kenneth Shinn, Secretary Treasurer, Local 251, OEIU; Jack Hart and Ernest Peterson, Labor Relations Department; F. B. Smith, Personnel Director and Treasurer, Sandia Corporation; Frank Morton, international representative of the OEIU; and Brad Shaw, Ralph Ridenour, James Stoll, and John Stark, union officers.[16]

In the early months of 1953, the Corporation again opened bargaining on the contract. With the assistance of the Federal Mediation Conciliation Service, a two-year agreement was negotiated on April 21, 1953. However, a strike had been planned previously; therefore, when word of the agreement failed to reach all union members, a partial walkout occurred. Among revisions to the contract were elimination of the provision allowing "buy-back" of unused vacation and a general increase of seven cents per hour in pay.

Strikes occurred on two other occasions. In 1955, when the Corporation and the unions reached a stalemate, the Panel sent both parties a notification of intervention, which the unions rejected. A nine-day strike followed, extending from April 26 to May 5. A third strike occurred, again during extended periods of contract negotiations, on July 1 to July 18, 1955. The unions were invited to Washington to meet with the AEC and the Davis Panel. After lengthy bargaining, an agreement was reached on August 14, 1955.[17]

Impact of Unionization on Nonunion Employees

In retrospect, the unionization efforts of the Metal Trades Council and the Office Employees' union had positive benefits for nonunion employees at Sandia as well. Management recognized that concessions granted the unions would have to be extended across the board. Unionization forced management to address issues such as payment for overtime and the advisability of raising entry-level pay scales for applicants for work, especially highly trained people. Furthermore, after the Atomic Energy Panel had twice recommended restoration of the twenty-four-day vacation plan, the Board of Directors at their annual meeting in New York on April 24, 1951, resolved that the President be authorized to apply to the Federal Wage and Stabilization authorities for approval to extend the plan to "nonunion employees on the same terms extended to union members."[18]

On June 1, 1951, a Lab-wide bulletin announced reinstatement of the twenty-four-day vacation plan. The bulletin also noted that procedures were being adopted whereby payment could be made in lieu of vacation at the employee's option. Management explained a part of the rationale for the option accordingly:

> The President of the United States has declared that a National Emergency exists. Not until this emergency is over

can much hope be held out that all of us can take vacations up to the full extent of our vacation accruals, and for this reason provisions for making cash payments in lieu of unused vacation seems advisable.[19]

By 1951, the push for unionization at the Laboratory extended to the Security force. For the guard force, however, the issues at stake were somewhat different.

Sandia Security Guards Unionize

In May 1950, Sandia Corporation agreed to take over the security responsibility from the AEC. Previously, the AEC Security Office, using their own and military personnel, manned the gates and towers. Until this time there was no guard force *per se* at Sandia, although there were Security and Fire Inspectors who patrolled the buildings in pairs. Sandia's new Plant Security Division took over these functions in a step-by-step process.[20] Despite the change in management, the Laboratory guard force continued to show the influence of the military for some time.

Hal F. Gunn, an employee in the Purchasing Department, was chosen to head the new Security Liaison Division and begin the hiring process. Gunn was assisted by Charles A. "Buck" Weaver of AEC Security, who was one of the first recruited to join the force. Weaver, a former professional wrestler from Arkansas, stood six feet seven inches tall and weighed close to 300 pounds. Remembered as "one hell of a pistol shot," he was also regarded as "tolerant but a disciplinarian."[21]

Weaver and Gunn interviewed approximately 1,000 men, and of that number, hired 150. Although they looked for men with a military or police background, the crew as selected turned out to be a heterogeneous lot. Verne Honeyfield was working as a bus driver when he was hired; Ted Varoz had been a cattle inspector with the U.S. Department of Agriculture, and Ed Sims was working as a carpenter for Lemke Construction building the Coronado Club. Robert L. Stewart came to Albuquerque from the logging camps in Washington state; Albert Joe Angel was a local boy from Albuquerque. William Bramlett had been a school teacher.[22]

In addition to hiring, Gunn had the responsibility for outfitting the new hires with uniforms and equipment, "everything from whistles and chains to uniforms that were tailor made," he recalled. A local firm, Simons, provided the original uniforms, which were an attractive beige color with Eisenhower campaign-style jackets, shirts, slacks, and Stetson hats, commonly referred to as "Smokey Bears."[23]

To get the uniform accepted by management, Gunn came up with the idea of having a sample uniform measured and made to fit one of the better-looking young guards. Verne Honeyfield, dark-complexioned, handsome, and well-built, fit the bill. Gunn took Honeyfield, decked out in the new uniform to see Vice-President Fred Schmidt. "Mr. Schmidt, a tough old boy," according to Gunn, liked what he saw. "That's it," Schmidt said. "That's what we want," and the uniform was adopted.[24]

Once the guards were hired, they had to wait for Q clearances to be conducted by the FBI. In the meantime, the men began orientation and worked at the gates assisting the MPs with property control. As Robert L. Stewart, who would later hold the post of union president, recalled: "PR didn't mean much to these military types."[25] Gunn wrote the training manual for the new guard force, and assisted by Nick Tarnawsky, a former member of the New York Police Department, taught the orientation classes. James S. Hinson, who would be promoted to Supervisor in 1954, was one of the original group of fourteen hired to participate in a crash training program that included the use of firearms.

The fifties were tense times; it was the era of McCarthyism, Klaus Fuchs, and the Rosenbergs. The tension was reflected in the no-nonsense attitude toward security. Military guards still manned Base areas, and their influence carried over to the supervisors' treatment of force members. "We organized the group on a military pattern with lieutenants supervising a group of men," Gunn said. "There was some military discipline."[26]

One of these lieutenants was N. "Pop" Jones, formerly a Provost Marshal with the U.S. Army. "The first day Sandia assumed control of the guard force, Jones had us fall-in and step forward military style," Robert Byrd recalled. "But that was the first and last time he had us do that because we all bowed our backs. We were no longer in the military, but they tried to carry over procedures."[27]

By June 1950, Corporation guards had assumed responsibility for manning the security posts at West Lab, and by October, they were performing functions for Areas I and II. By November, Sandia's guard force was monitoring all areas. The approach to security during the fifties has been described as "mechanical." In making rounds, for example, guards used Detex clocks to indicate a post had been checked. According to procedure, the guard on patrol would turn the key inserted into the clock, causing an imprint. This device not only showed that a secured area had been checked, it also afforded management a record proving that employees were on the job. Patrols were frequent since there was no electronic surveillance until the mid-fifties when the Dudley electronic system was installed.[28]

A part of the Security Inspector's function was to check the numerous safes on the Sandia premises. Every night they patrolled the buildings "shaking the safes." Those unfortunate individuals who failed to lock up properly would find an ominous anouncement in their safes in the morning: "The Hairy Hand has been here. Please report to the Security Office." The offenders, faced with the prospect of a lecture on the importance of respecting security measures, failed to appreciate the black humor of the Security Inspector's Hairy Hand notices. Gunn recalled that "all types were guilty [of security infractions] from clerks to top engineers. The engineers were the worst," he added.[29]

Sandia's Security force attempted to improve on the manners of their military predecessors with a friendly, courteous attitude; however, there were those who felt that, in some cases, the courtesies they extended to other employees were not extended to them by guard force management. Strict regimentation was the rule rather than the exception. Smoking, eating, or gum chewing on the job were not allowed. And although the guards were encouraged to be courteous, a too casual or too friendly attitude would result in demerits. Members had to be clean-shaven, have their hair neatly cut and their shoes shined. The dress code was strict. James Hinson recalled that management encouraged a clean, well-groomed appearance by paying for laundry and dry cleaning and the time it took to change into uniform. Guards working the gates as a team had to be dressed identically. According to Honeyfield, "Everyone wore the same style hat, same coat. Didn't make any difference how cold it was. If somebody in an office decided the sun was out and it had warmed up, they'd call up and say, 'Okay, take all the parkas off.' "[30]

Supervisors enforced the dress codes strictly. Ed Sims remembered a humorous incident involving the prescribed uniform. "Back then you had to wear your hat even in the escort shack," Ed said. "We were sitting around waiting to escort somebody, and Stuart Breeding had his hat off. The Sergeant, Digger Dance, came in and told him to put his hat back on. Breeding did, but took it off as soon as Digger left. Digger came back and again told him to put the hat back on. As soon as Digger left, off came the hat. The third time, Digger just took the hat and jammed it on Breeding's head down over his ears. Meanwhile, Charlie Graves was logging in trucks. He thought Digger had left and said, 'Well, you old S. O. B., seein' as you're gone, I think I'll take my hat off.' And he did. But then he looked up and saw Digger standing there. Without batting an eye, he said, 'And seein' as you're back, I'll put it right back on,' and kept right on logging."[31]

Having a good sense of humor helped, but it didn't improve working conditions. The guard shacks were cold and drafty; drinking fountains and lavatories nonexistent. Everyone had to drink from a communal bag or gallon jug and clock in and out of the various posts. The water bags, made of cactus fiber, leaked through the exterior and were cooled by the wind. Starting pay was $1.46 per hour. More important than the working conditions, however, were what some members of the force regarded as unfair treatment and a lack of recourse for grievances. Old time union members recall certain practices in particular that caused them to support unionization, namely, raises not always awarded on the basis of merit, the manner in which assignments for interior and exterior patrol were awarded, time off without pay as punishment, instances of abusive behavior, and the latitude to fire people almost "at will."[32]

By 1951, the perception of unfair treatment and primitive working conditions encouraged certain members of the guard force actively to consider unionizing. Marvin Brown was sympathetic to the union cause and assisted workers in the Metal Trades and the OEIU in their unionization efforts by providing addresses to the local organizer. In November 1951, Brown volunteered for duty at the Nevada Test Site near Las Vegas. There he worked with the Hanford guards, who were in Local 21, Region 1, of the International Guards Union of America (IGUA). From the Hanford guards, Brown obtained the name of Roderick MacDonald, national president of the IGUA.

Brown wrote to MacDonald and asked him to make a trip to Albuquerque to talk to the guards. Meanwhile, another member of the force, Corbett P. Krause, had contacted the Plant Guard Workers of America, which was affiliated with the CIO (Congress of Industrial Organizations). As 1952 approached, union talk was strong around the Lab. When McDonald arrived, he met with the guards, told them of other AEC plants where IGUA had locals and how they could go about signing up for an election. Local organizers were impressed with MacDonald and the points he made.[33]

In the fall of 1952, the Sandia Security guards in Albuquerque formally petitioned for a union election. The Salton Sea Test Base in California also had eighteen guards, but organizers in Albuquerque planned on the Salton Sea group's forming their own local. A problem arose when the NLRB ruled that they had to vote with the Albuquerque force. Corbett Krause promptly flew to Salton Sea and came back with fifteen signatures. In the election, held December 4, 1952, all but two of the Salton Sea guards voted for the union. The union won by a substantial majority, was certified on December 12, 1952, and in January 1953, started negotiations on a contract for Local 27, IGUA.[34]

According to Superintendent Luther Heilman, three major concerns seemed to surface during negotiations: seniority, time off without pay as punishment, and the scheduling of vacations. The practice in force made it extremely difficult to change previously scheduled vacations even to meet emergency occurrences. Heilman remembered a situation where a guard requested time off to visit with an Army buddy coming through town. The only alternative was to get someone to trade vacations, but people were not eager, or oftentimes able, to accommodate. Heilman later conceived the idea of an "Extra Board" to be made up of individuals on call, who could be tapped to fill in when needed.[35]

A contract agreement was signed on April 22, 1953. Charter members were William Bramlett, Marvin Brown, Webb Shaffer, Corbett B. Krause, Malcolm Ward, Tony V. Gallegos, and Marcos J. Phelps.[36] Some union members such as Thomas Chiado, Corbett B. Krause, and John Wahlen–Maier, among others, were promoted eventually to supervisory positions. By 1952, E. P. Hutson succeeded Hal Gunn as Department Manager.[37]

The unionization process took less than a month—from November 1952, when representatives of the guard force petitioned for union representation, to the election held on December 4, 1952, which resulted in certification eight days later. Subsequently, members of the guard force were officially designated Security Inspectors and upgraded monetarily. Their gains through organization included a higher base pay, enforced seniority, guaranteed sick leave, proper shift premiums, workable grievance procedures, and protection against unfair discharge. A major difference in the contract negotiated with the guards, as Heilman observed, was inclusion of a no–strike clause. "It is understood," the contract read, "that in the event of any strike, or threatened strike, . . . the plant protection employees will continue to report for duty, remain at their posts and discharge their duties in the regular manner . . . as are necessary."[38]

Because of its mission, Sandia's Security force has functioned when needed as a "peace-keeper" during tense negotiation periods with other unions. William Bramlett, who would be elected the union's first president and national president in 1958, explained that this is understandable considering that the Security force is, in actuality, the "eyes and ears of the organization."[39]

In retrospect, the Board's decision to reduce the twenty-four-day vacation plan triggered the unionization of Sandia employees; but other factors, such as a general climate of tension and unrest, made the Laboratory receptive to organizing efforts. As soon as the University of California announced its desire to be relieved of its contract with Sandia in 1948, rumors began to circulate. Employees, concerned about their future, were suspicious of changes in the offing. In a speech to management at the Skytop Conference in 1950, President Landry acknowledged:

> We were unaware of the tension and suspicion among employees until we set about making certain changes in operating policies and procedures. We took over the personnel as a group, but the "inherited" had no firsthand knowledge of Western Electric, and quite naturally felt no real sense of loyalty to the new management. The morale situation became capital for labor organizations seeking to organize.[40]

With this statement, Landry pinpointed the basic cause underlying the more overt factors such as fringe benefits, or the lack thereof, and primitive working conditions. The campaign at Sandia, in retrospect, appears to have been a textbook case wherein employees unionized mainly in protest to management practices. In addition to the problems associated with unionization, the Western Electric management team faced other administrative challenges.

Above, Military Police performed guard functions at Sandia Base until May of 1950 when Sandia agreed to assume these responsibilities. Below, two Sandians walk toward the guard station located at the truck delivery entrance to the Z-Division technical area.

UNIONIZATION: AFTERMATH OF INCORPORATION

In 1948 Art Jimenez was a member of the 8450th MP group, the security force for Areas I and II on Sandia Base. Jimenez became a Sandia employee following a tour of duty in Korea.

Wooden guard towers stood like sentinels in strategic positions throughout the technical area. The tower in this picture overlooks Building 828. At right is an Area III guard shack, circa 1953.

MANPOWER REQUIREMENTS FOR MANNING GUARD POSTS

(December 19, 1952)

SYMBOL: SAD-2

The following is a list of the Guard Posts for the 7:00 AM to 3:00 PM shift and the number of personnel required to man them.

* * * * * * * * * * * * * * * * * * *

7:00 AM to 3:00 PM, Shift 1, Area I

One lieutenant — Coordinates guard activities — Areas 1 and 2
Two sergeants — One shift sergeant and one escort sergeant
Two communicators — Located at Building 801 — man radios and telephones

Post G-1 — Two men — West "H" Street gate, 24 hours
Post G-2 — Two men — Gate between 800 and 802 — 7 AM to 6 PM
Post G-3 — Two men — West "G" Street gate, 7 AM to 6 PM
Post G-4 — One man — North 6th Street gate (Pedestrian), 7 AM to 6 PM
Post G-5 — One man — North 8th Street gate (Pedestrian), Rush hour gate
— 7 AM to 8:15 AM, 12 noon to 1:15 PM, and 4:30 PM to 5:30 PM —
This post is usually manned by the tower relief or an extra man when one is available.
Post G-6 — Two men — East "H" Street gate — 7 AM to 6 PM
Post G-7 — Two men — South 9th Street gate — 7 AM to 6 PM
Post MP-1 (Motor Patrol) — One man, 24 hours — Perimeter fence vehicle patrol
Post FP-1 (Foot Patrol) — One man, 24 hours — Perimeter fence foot patrol
Post FP-2 (Foot Patrol) — One man, 24 hours — Perimeter fence foot patrol
Post A-13 — One man, 24 hours — Building 892 guard
Post A-23 — One man, 7:30 AM to 5:30 PM — Building 804 lobby
Post A-28 — One man, 24 hours — Administration Building lobby

* * * * * * * * * * * * * * * * * * *

Area II

One sergeant
Post B-1 — One man, 24 hours — Gate guard
Post B-4 — One man, 24 hours — East tower
Post B-5 — One man, 24 hours — West tower

* * * * * * * * * * * * * * * * * * *

The above listed are actual requirements. In addition to these there is one tower relief, who mans post G-5 during rush hours, and also ten escorts.

This makes a total of thirty-seven men.

Officials of the International Guards Union of America included Albuquerqueans, left to right, Marvin H. Brown, Vice-President and Regional Director of Region 6; William Bramlett, International President; Robert L. Byrd, Assistant Regional Director of Region 6; Leonard Decker, president of Local 99.

UNIONIZATION: AFTERMATH OF INCORPORATION

Early members of the Sandia guard force size up vulnerable points on a police silhouette target.

Members of Sandia's first guard force, established in 1950, included (front row, from left) Les Bauman, Milt Lesicka, Ted Varoz, Roy Brett, Noel Kent,

Robert Byrd, and Fred Lopez. In the back row are Tony Uszuko, Mac McMurfrey, Frank Martin, Clyde Sealey, Jim Hinson, Ed Sims, Verne Honeyfield, and Bo Ellis.

1952 exhibit used in the recruitment of personnel.

Chapter XVII.

MEETING THE ADMINISTRATIVE CHALLENGE:

Expansion of Services

There are two significant frontiers—the frontier of technical understanding and the frontier of human understanding.

J. W. McRae, President, Sandia Corporation
Commencement Address, University of New Mexico,
June 6, 1956

We serve the AEC and the DOD. If Sandia delivers good and timely service this public will regard us well. If not, no amount of speech making or advertising will improve the situation.

D. A. Quarles, President, Sandia Corporation
Speech, Annual Dinner Meeting, Sandia Supervisors,
February 17, 1953

Soon after incorporation, Sandia launched a vigorous hiring campaign to meet the demands of crash programs and an increasing number of new engineering projects. In November 1949, employees numbered 1,742. New hires joined the Corporation rapidly, and by 1955 Sandia employed 5,752 people.[1] The increase in personnel created an urgent need for a more formalized approach to business operations, an expansion of personnel and purchasing activities, and the creation of a public relations division.

OPERATIONS
1950 – 1952

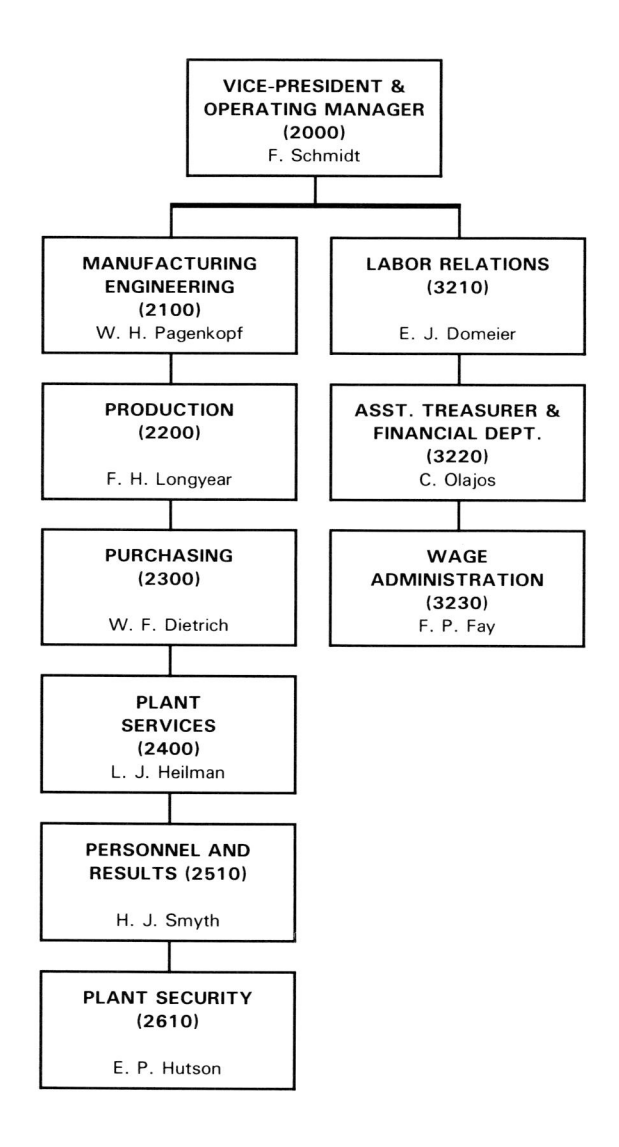

- **VICE-PRESIDENT & OPERATING MANAGER (2000)** — F. Schmidt
  - **MANUFACTURING ENGINEERING (2100)** — W. H. Pagenkopf
    - **PRODUCTION (2200)** — F. H. Longyear
    - **PURCHASING (2300)** — W. F. Dietrich
    - **PLANT SERVICES (2400)** — L. J. Heilman
    - **PERSONNEL AND RESULTS (2510)** — H. J. Smyth
    - **PLANT SECURITY (2610)** — E. P. Hutson
  - **LABOR RELATIONS (3210)** — E. J. Domeier
    - **ASST. TREASURER & FINANCIAL DEPT. (3220)** — C. Olajos
    - **WAGE ADMINISTRATION (3230)** — F. P. Fay

PERSONNEL AND PUBLIC RELATIONS 1956

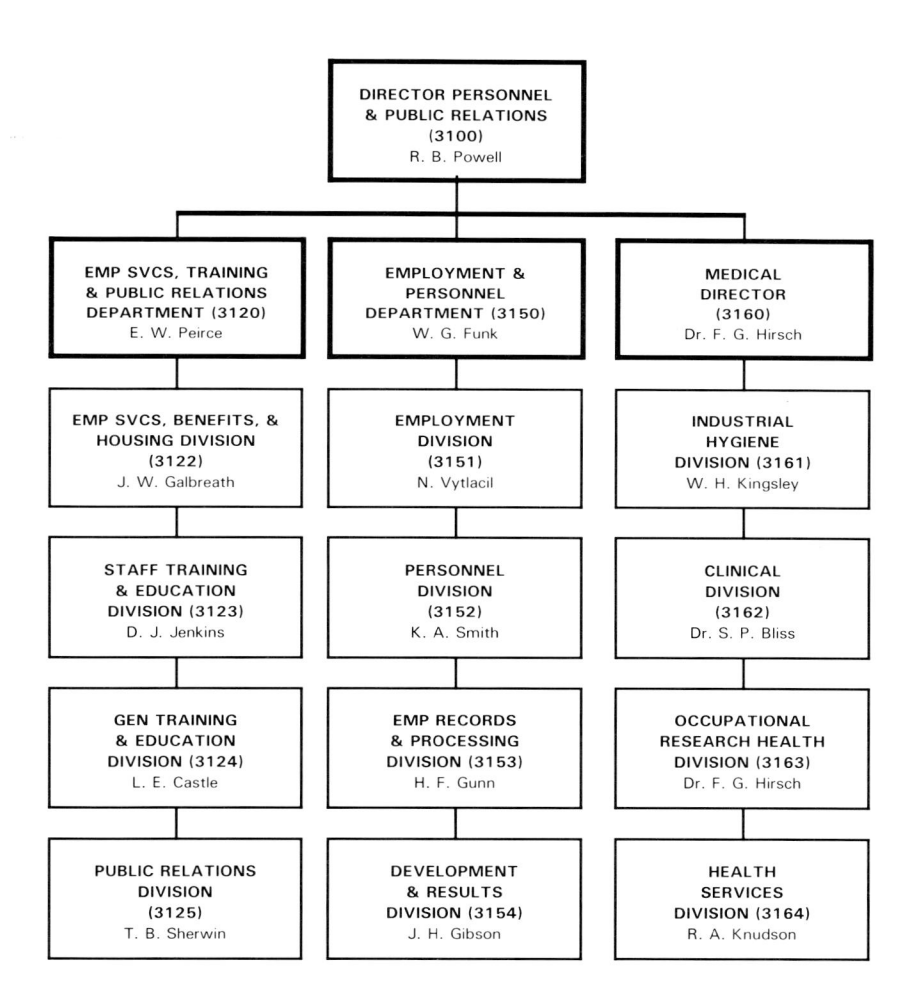

- **DIRECTOR PERSONNEL & PUBLIC RELATIONS (3100)** — R. B. Powell
  - **EMP SVCS, TRAINING & PUBLIC RELATIONS DEPARTMENT (3120)** — E. W. Peirce
    - EMP SVCS, BENEFITS, & HOUSING DIVISION (3122) — J. W. Galbreath
    - STAFF TRAINING & EDUCATION DIVISION (3123) — D. J. Jenkins
    - GEN TRAINING & EDUCATION DIVISION (3124) — L. E. Castle
    - PUBLIC RELATIONS DIVISION (3125) — T. B. Sherwin
  - **EMPLOYMENT & PERSONNEL DEPARTMENT (3150)** — W. G. Funk
    - EMPLOYMENT DIVISION (3151) — N. Vytlacil
    - PERSONNEL DIVISION (3152) — K. A. Smith
    - EMP RECORDS & PROCESSING DIVISION (3153) — H. F. Gunn
    - DEVELOPMENT & RESULTS DIVISION (3154) — J. H. Gibson
  - **MEDICAL DIRECTOR (3160)** — Dr. F. G. Hirsch
    - INDUSTRIAL HYGIENE DIVISION (3161) — W. H. Kingsley
    - CLINICAL DIVISION (3162) — Dr. S. P. Bliss
    - OCCUPATIONAL RESEARCH HEALTH DIVISION (3163) — Dr. F. G. Hirsch
    - HEALTH SERVICES DIVISION (3164) — R. A. Knudson

Source: *Sandia Corporation Telephone Directory*, March 1956

AVERAGE AGE & SERVICE

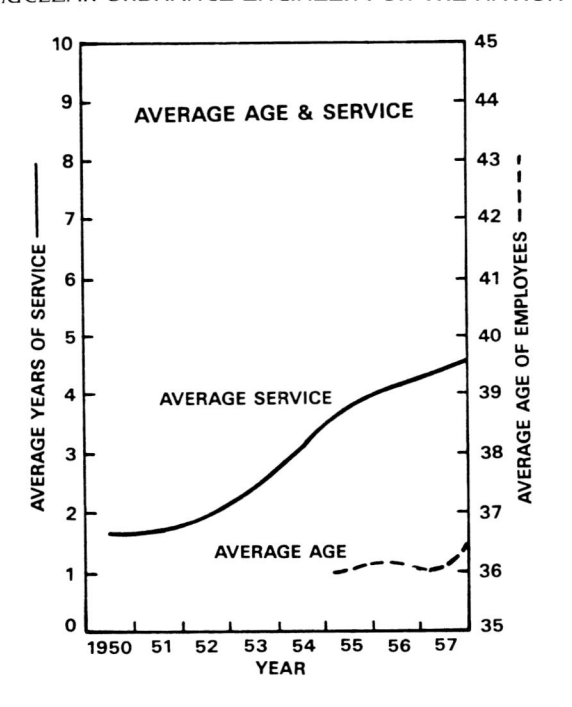

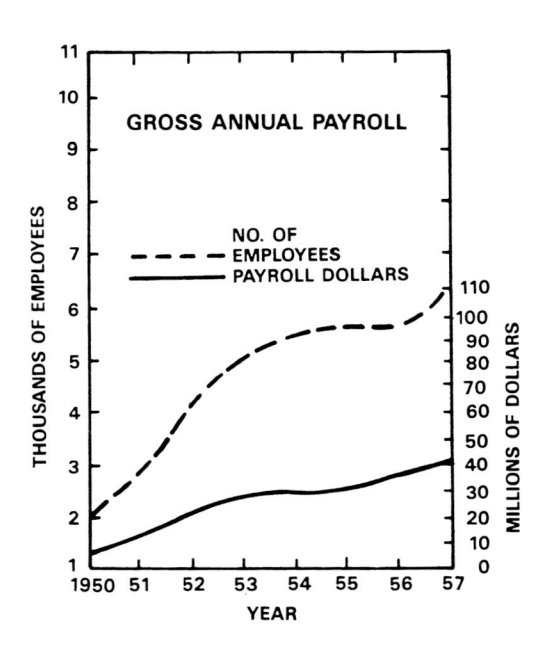

GROSS ANNUAL PAYROLL

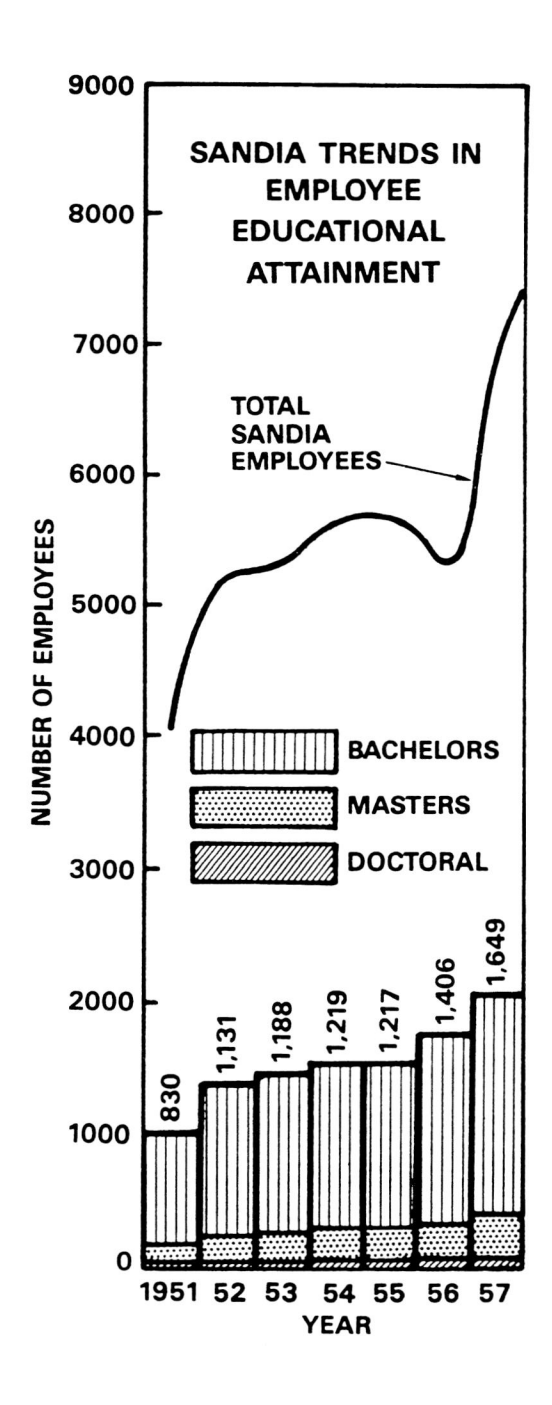

SANDIA TRENDS IN EMPLOYEE EDUCATIONAL ATTAINMENT

TOTAL SANDIA EMPLOYEES

BACHELORS
MASTERS
DOCTORAL

Year	1951	52	53	54	55	56	57
	830	1,131	1,188	1,219	1,217	1,406	1,649

In addition to top-level management, Western Electric placed Bell System men at various supervisory levels throughout the organization, principally in charge of functions not previously performed by Sandia. These included purchasing, expediting, payroll, accounting, auditing, business methods, finance, and labor relations. Certain functions had to be implemented on a priority basis. Western Electric sent in three teams of experts to assist—one to conduct a product and plant inventory, one to review job evaluation and wage administration problems, and one to help establish accounting, cost, and payroll procedures. The inventory job alone, with a team of men from Hawthorne, took five months to complete.[2]

Financial, Personnel, and Business Operations

Among the first activities necessary to the ongoing of the organization were accounting operations, transferred from the University of California to Sandia. Books of accounting, methods for control of expenditures, and auditing procedures had to be developed. A Superintendent of Accounting, E. J. Cooney, was appointed in 1951.

Under terms of the basic contract between Sandia and the AEC, budget estimates for fiscal year operations were submitted to the Finance Division of the AEC and, when approved, formed the basis for inclusion in the overall AEC budget. After Congress passed the appropriation bill, Sandia was advised of the specific amount of money allocated to its operations. Authorization providing a specified sum of money for a given period was then transmitted to Sandia, which allowed the Laboratory to commit for payroll, materials, and other procurement. From time to time, the Corporation made requests to the AEC Fiscal Office for deposit of certain amounts on specified dates to fund payments coming due during these periods. Monies in Sandia's special bank accounts remained the property of the U.S. Government at all times, but the Laboratory was authorized to draw on those monies in the same manner as it would on its own corporate bank account in ordinary industry.[3]

In 1951, escalation of military activity in Korea required several upward revisions of the budget. During the 1950-1955 period, military involvement of the United States and Sandia's contributions to the defense effort were reflected in a Laboratory budget for Research and Development and Production that rose from a total of $41,861,000 in 1950 to a total of $102,608,000 in 1955.[4]

In general, Sandia's administrative functions were reorganized to conform with those of Western Electric. Certain activities were handled by both a centralized administrative organization and by production and research and development groups. Early on, management questioned this arrangement. Within the Research and Development organizations, for example, an attempt was made on May 1, 1950 (as previously mentioned), to consolidate the nontechnical personnel into a unit called Development Staff that reported to Director Robert E. Poole. As explained in the *Annual Report* for that year: "The broad objective in separating these nontechnical and service employees from the various individual groups in the Research and Development departments was to permit technical supervisors to concentrate . . . their entire attention and effort on technical problems."[5] The change also made it possible to carry out the evaluation of job performance and to select individuals for promotion on a more equitable basis.

By February 1951, the Development Staff Department had been transferred to the organization of the Secretary and Comptroller, J. A. Dempsey (4000). F. Cowan, Superintendent, Staff and Business Methods (4200), headed the organization, which included secretarial services, personnel administrative assistance, patent liaison, the technical library, and budget and administrative services. In 1955, Charles W. Campbell became superintendent, and the directorate–level organization was designated Development Staff Services.[6]

Although certain functions reported to the Production organization or to Research and Development groups during this period, the various administrative activities were coordinated with the Personnel organization under Vice-President 3000, Fred B. Smith, and Superintendent and Assistant Treasurer, H. W. Sharp (3000). In 1954, Ray Powell succeeded Sharp as Superintendent of Personnel and Public Relations. This basic arrangement whereby Personnel functions were split would exist until the late fifties when Sandia phased out its manufacturing operations.[7]

Expansion of the Laboratory's responsibilities and assignments made it obvious that additional personnel of all kinds were needed. In 1950, the Laboratory started a college recruitment program, supplemented by national newspaper and magazine advertising. At the end of 1951, twenty-five percent of all employees on roll held college degrees. For three years, 1950-1953, Sandia maintained a branch office in the Albuquerque heights area for the recruitment of nonprofessionals.[8]

The Personnel Department had to develop training and orientation programs for both new hires and supervisors. Starting in 1951, supervisory meetings were held to discuss issues of mutual concern, and a plan for rotation of supervisors was initiated to provide broader Laboratory experience. An out-of-hours educational development program gave employees the opportunity to keep abreast of new technological developments. In-hours studies on job-related subjects were also offered through the joint efforts of the line organization and the Technical and Trades Training Division. With the encouragement of the AEC, corporate training and education would be substantially increased as the years went by. In 1956, a separate division was created to handle management, staff training, and education.[9]

Not long after formation of the Corporation, a review of personnel policies was conducted. As part of this process, management studied a number of retirement plans and selected a group annuity policy offered by the Prudential Insurance Company of America to replace the one in effect under the University of California. The plan provided for payment of a retirement income at age sixty-five at the rate of one-third of the employee's contributions. The paid sick leave plan held under the University was transferred intact. Under this plan, employees could accumulate twelve hours per month up to a maximum of 720 hours, although this number was later increased to 1,000.[10]

For the first few months of the Corporation's existence, the Laboratory's new payroll organization had to prepare the payroll manually. Even after tabulating equipment arrived, the dusty conditions of the poorly constructed building caused malfunctioning. According to historian Frederick Alexander, "It was not until Building 830 had been sealed that supervision could release a payroll with full assurance that it was correct."[11]

Sandia also needed more formal procedures to cover the conduct of business and a group to produce and publish these policies. Again, personnel from Western Electric assisted the Corporation in setting up the Business Methods Organization and in creating procedures called "Sandia Corporate Instructions." These "SCIs," as they were called, provided supervisors with a uniform approach to policy and standardized procedures for dealing with personnel and the operation of the Laboratory.[12]

Wage and Salary Administration

When the Corporation was formed, R. H. Landes was brought in from the parent company to organize and operate a wage and salary program based upon Western Electric policies, an analysis of systems used by the various contractors, and a wage survey of the local area. The unionization efforts at the Laboratory encouraged management to make a concerted effort to finalize a job classification structure and implement a wage and salary program as soon as possible.[13]

Under the University of California system, Sandia's wage administration program had been based upon the SCP plan, where S stood for skills, C for crafts, and P for production. A quartile system based upon the numbers of years since receiving the Bachelor's degree designated the position for engineers hiring into the organization, whereas graded employees hired on as research assistants and worked for promotion to staff. Staff member designations usually went to employees with a Bachelor of Science degree and four years of applicable experience. In contrast to graded employees, staff members were assigned to professional or managerial positions.[14]

Landes went to work developing job evaluation plans and a salary scale. As a basis for planning, he made a tour of various AEC contractors to determine how they were handling wage and salary administration for technicians and staff members. Where graded people were concerned, Landes applied Western Electric philosophy and procedures. Russ Waley, also from Western Electric, assisted by Ted Anderson of Sandia, conducted a local survey of hourly paid occupations, which provided management with information for use in discussions with the unions.

Early in 1951, Frederick P. Fay replaced Landes, who returned to Western Electric. About this time, Lloyd Fuller joined the organization and continued the wage and salary study. D. T. Anderson, D. Worthen, and J. C. Hart also worked on job evaluation and wage and salary administration. In March of 1952, partially to streamline bookkeeping functions, the Corporation converted to a wage control based on the age of the employee rather than the number of years since receiving the degree. Technicians from the research and development organizations were converted from hourly to salary paid employees. Largely owing to the support of Robert E. Poole and President Landry, and over the objections of the wage administration people, these technicians were not placed under any job evaluation plan.[15]

Passage of the National Wage Stabilization Act in the spring of 1951 soon had an effect on the organization. Under this act, wages were frozen, raises received close review, and the administration had to conduct a historical study to prove that the Laboratory's wage pattern for union and nonunion employees had not been materially altered. There was concern that increases granted would cause a distortion between salaries for nonunion and union employees. As a result of the study, four grade categories were established affecting union hourly, union salaried, nonunion hourly, and nonunion salaried groups. The Corporation, taking care to keep union representatives carefully informed, assigned an automatic wage progression to graded job categories.[16]

Job descriptions were also reviewed and evaluated, and on February 15, 1951, a Staff Member Committee, chaired by Glenn Fowler, was set up to consider employees eligible for promotion to staff member category. Significantly, the Staff Member Committee represented all applicable organizations on a Lab-wide basis. To be considered for promotion, a staff member had to possess a college degree and have a B average or better. Those individuals with practical experience, but less academic education, were designated as "Engineer." The Staff Member Committee and a committee of upper management personnel, after much discussion, made the decision to segregate staff members into technical, administrative, and drafting categories. By 1955, however, the Committee had decided to limit staff member categories to Staff Member Technical and Staff Member Administrative.[17]

Upgrading took place rapidly among Research and Development Technicians, sometimes without regard to job content; consequently in January 1952, the establishment of a Technical Review Committee provided a mechanism for control. The Committee assigned the title "Research and Development Assistant" to these technicians, but in 1954 the designation was shortened to "Staff Assistant." In this year a Salary Committee was also created to provide overall guidance on salary structure, level, and classification of nongraded jobs.[18]

Initially, the wage and salary people in the 2000 organization were the only so-called professional employees on the administrative side of the house. Accounting personnel and other nonengineering organizations offered only graded positions. Gradually these organizations converted to professional status, with Personnel in the lead.

During the early fifties, there was support for the establishment of a single company salary program, but this was changed in 1958 so that there were two curves, one for administrative staff and one for technical staff. As a result of market-driven competition, research and development work became a higher-paid occupation at Sandia. In 1956, the functions of job evaluation, wage and salary administration, and organization engineering were consolidated into an Industrial Engineering Department under the supervision of Lloyd Fuller.[19]

Legal, Medical, and Safety

As the Corporation grew, so did the need for full-time legal counsel. Western Electric provided the services of attorney Phillip D. Wesson to assist in solving the many problems encountered in setting up the operations of the Corporation. President George Landry, much impressed with Wesson, expressed his appreciation to W. L. Brown of Western Electric:

> My main purpose in writing you . . . is to let you know how deeply grateful we are to Phil Wesson for the very valuable work he has done for us. He injected himself completely into this new interprise [sic] of ours and his contribution to the successful start of the operation has been invaluable. I am equally grateful to you for anticipating our needs and for sending Phil here to be at our right hand during these critical first days. I little realized beforehand what a multitude of things there are to do.[20]

At the top of the list of action items was the establishment of credit for the new corporation, which created a minor crisis of a legal and financial nature. The AEC decreed that Sandia could not prepare a balance sheet for external use. Dunn and Bradstreet, in turn, informed the Corporation that without a balance sheet, a credit rating would not be issued. Problems arose when Sandia suppliers turned to Dunn and Bradstreet to check the credit rating of this largely unknown Laboratory and found none available. After much correspondence on the subject, and a special meeting of the Dunn and Bradstreet Board, the firm agreed to give Sandia a top rating.[21]

Wesson assisted in the initial establishment of the Corporation until the Office of General Attorney was set up in March 1950. At that time Wesson returned to Western Electric, and Frank L. Dewey became the Corporation's first full-time counselor.[22]

One of the first problems to be resolved by the new legal department was to remove the AEC areas of Sandia Base from federal jurisdiction to provide residents of these areas with access to state courts and voting rights. Termed the "Federal Island issue," the problem became apparent in 1952 when residents went to register in Bernalillo County for the national election. Voting authorities informed them that the land they lived on had been ceded by the State of New Mexico to the U.S. government; therefore, they were not residents of the state of New Mexico and were not entitled to vote.

This decree raised another issue. How can New Mexico collect state income taxes from residents of an area that is not considered to be a part of New Mexico? The answer to that question was found in a law known as the Buck Act, passed July 30, 1947, which provided that "no person shall be relieved from liability for any income tax levied by any State, or by any duly constituted taxing authority therein."[23]

A similar situation involving exclusive jurisdiction of land in Los Alamos had been settled a few years earlier in the case of *Arledge vs. Mabry*. Using this case as a precedent, Sandia's attorney presented the situation to the Santa Fe Operations Office of the AEC. Subsequently, on January 24, 1951, Manager Carroll L. Tyler recommended to headquarters that retrocession be made to areas over which exclusive jurisdiction had been obtained by the Federal government. In support of this recommendation, Tyler submitted correspondence from the Field Manager, AEC, Sandia Base, President George Landry, and a detailed report by Sandia's corporate attorney, Frank L. Dewey. The AEC endorsed the proposal and formally recommended to Congress that the area under exclusive jurisdiction be retroceded to the state of New Mexico. Subsequently, an amendment to the Atomic Energy Act was passed to effect such a result.[24]

Negotiations leading to renewal of the contract occupied the attention of the legal department during four months of 1953. The AEC had prepared a new and more detailed contract similar to that held with Bendix, but the Corporation rejected it. The problem was finally resolved in favor of an extension of the previous contract.[25]

In August 1952, attorney Frank Dewey returned to Western Electric, and Phillip Wesson rejoined the Corporation as General Counsel. Wesson would remain at Sandia until November 1, 1955, when he was succeeded by Kimball Prince from Western Electric headquarters in New York.[26]

In addition to legal services, Sandia and Western Electric management recognized early on that the Laboratory would require the services of a medical staff. For some time, even before November 1, 1949, the University of California had searched for a qualified industrial physician to be assigned to the Laboratory. Recruiting efforts were made both locally in Albuquerque and throughout the country. Discussions were eventually held with the Lovelace Foundation for Medical Education and Research in Albuquerque, which agreed to provide the services of a part-time plant physician, Dr. F. G. Hirsch.[27]

It was apparent to Hirsch soon after his arrival that the explosion in employee population required a full-time physician. "Not only this," he recalled, "but it was by then apparent that the inescapable by-product of the type of work which the corporation was engaged in required not only a full-time physician, but 'the whole ball of wax' insofar as occupational medicine was concerned." For a while, Hirsch was placed in the anomalous position of being a department manager, yet not officially an employee of the Corporation. "You can imagine the dilemma this presented to inspectors and auditing groups," Hirsch added with some humor. "As a result, one of Landry's last tasks as President was to negotiate with the Lovelace Foundation for an amicable cancellation of their contract to provide . . . my services."[28]

Subsequently, in 1951 Hirsch became Sandia's first full-time medical director. Another part-time physician, Dr. R. C. Powell, was also added to the staff; and Mildred Whitten from the Salton Sea facility became Sandia's visiting nurse. The Health and Hazards Division under W. H. Kingsley moved from Labor Relations, Health and Safety to the Medical Department. An early goal was to have a medical department approved by the Industrial Medical Association.[29]

One of the more unique aspects of the program was the inauguration of an Employee Review Committee. The Committee, which included the President, attempted to ensure adequate and consistent treatment in cases involving determination of liability and payments under Workmen's Compensation laws. Dr. Hirsch observed that "the Committee served as a device whereby matters were . . . settled without the necessity of a grieved employee having to employ counsel."[30]

After going through a period "when we jerry-rigged," according to Hirsch, the Laboratory obtained the necessary clinical and diagnostic equipment to conduct laboratory examinations, chest X rays, and full-scale physicals. In addition, periodic examinations were given to employees exposed to unusual hazards. Medical also began a program providing physicals for top-level supervision. To structure Sandia's medical program, management used the policies of Western Electric as a guidepost because of the company's successful history in the area of industrial medicine.[31]

The Laboratory first started its Industrial Hygiene efforts in 1949 by establishing a film-badge program and a laboratory for performing bio-assays on personnel exposed to toxic substances. In connection with the early tests in Nevada, Sandia initiated an extensive radiation safety program to monitor fallout. The Lab also designed and produced radiation counting instruments and participated in the establishment of health physics programs at other AEC contractor facilities. In 1951, Sandia implemented a disaster response system for emergencies involving nuclear material. Pioneering efforts in this area were adopted as a model by the AEC for application to other facilities.[32]

Yet Sandia's record in the area of industrial safety during its first two years was far from exemplary. The Lab's record for 1950, for example, was rated as "poor compared to good industrial performance." As a result, the Laboratory began a comprehensive campaign to train employees in proper safety precautions and practices. Among other actions taken, management made arrangements to secure the services of a safety specialist from Western Electric. "W. H. Kingsley of the Health Hazards Division," Hirsch said, exerted considerable "interest, vigilance, and energy toward improving Sandia's record."[33]

In 1952, the Laboratory established a formal Safety Directorate headed by A. "Burt" Metzger. Efforts to improve Sandia's safety record were soon rewarded. In 1953, the Lab received an award from the National Safety Council; and in 1960, after 417 days without a disabling injury, the Laboratory received the AEC safety trophy for 14,939,169 accident-free hours. Thus by the end of the fifties decade, the Corporation safety record rose from last place among the twenty-two AEC contractors to the top of the list.[34]

Purchasing "Keeps the Wheels of Progress Turning"

The *Sandia Bulletin* reported that Sandia Purchasing personnel in 1951 "kept the wheels of progress turning" by placing more than 32,000 orders totaling many millions of dollars with 3,570 suppliers throughout the country. Western Electric, recognizing that a Purchasing organization was integral to the function of the Laboratory, had sent in their first cadre of purchasing professionals shortly after incorporation in November and December 1949. Hardy Ross, the Lab's first Purchasing Agent, stayed six months, then was succeeded by William F. Dietrich. Dietrich, who was interviewed for the *Bulletin* article, observed that the search for products to supply Sandia's needs had taken buyers to forty-one different states. Many of the purchases, however, were made locally. Orders placed with New Mexico businesses in 1951 totaled $3,476,821, which represented an increase of approximately eighty-five percent over 1950, when $1,877,088 of orders were placed. The year 1951, therefore, marked Sandia's first multimillion-dollar purchasing year.[35]

SANDIA PURCHASING ACTIVITY IN NEW MEXICO

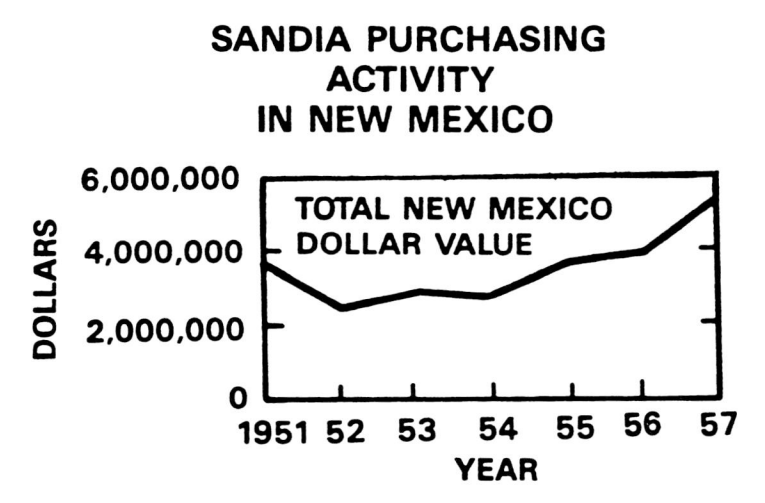

[Graph showing Total New Mexico Dollar Value from 1951 to 1957, with dollar values ranging from 0 to 6,000,000]

When Dietrich arrived in 1950, he was assisted by Department Managers H. V. Ahl, who headed a general Purchasing Department, and Sanford K. "Tommy" Thompson, in charge of Expediting and Traffic. The organization grew rapidly, and by 1952 departments had been established for Electrical Purchasing under D. H. Sampson, Mechanical Purchasing under J. E. McGovern, and Reclamation under J. W. Gray. The Contract and Traffic Department was separated from Expediting and placed under Ken Spoon. A buying staff of some fifteen expediters and clerical personnel provided support. Sandia engineers worked closely with buyers in the organization and vice versa. Engineers assisted by locating supplies throughout the country, making plant surveys, and reporting on the quality of work produced. Buyers, in turn, worked directly with the engineers on the various weapon projects.[36]

Like other organizations at Sandia, Purchasing felt the imprint of Western Electric management style and practices. The early Purchasing directors and managers, according to Roy Crumley who joined the Expediting group in 1952, "were obviously purchasing professionals."[37] Contracts were short, sometimes no longer than two to three pages, although the contract might be for a million dollars worth of product. The front page of the contract contained a description of the item, an auditing clause, and the delivery schedule. Terms and conditions were listed on the back. There were few procedures to follow, and buyers had considerable freedom in the design of contracts.[38] Nevertheless, Western Electric managers had definite means of quality control.

One of the Western Electric managers, for example, was renowned for initiating a practice referred to as "the Witch Hunt." At the Witch Hunt, engineers with critical components in the system gave progress reports. The meetings were also attended by buyers from Purchasing who were expected to provide updates on the delivery status of major weapon components. If an engineer indicated that he was experiencing problems with delivery, it was this manager's practice to get to the source of the problem in public forum. Waylon Ferguson, who transferred from the Accounting organization to the Reclamation Division of Purchasing in 1950, recalled these periodic status report meetings and noted that the items in trouble were listed on pink slips that reached upper management. "You got a great deal of attention when your components appeared upon the pink sheet," Ferguson added with some understatement.[39]

During 1952–1953, department managers handled large item contracts personally, reflecting a management style that did not allow delegation to buyer level. In 1952, for example, Department Manager Ken Spoon personally handled the Motorola contracts for certain electronic components used in radar units. The focus on placing weapons in stockpile made the fifties a hectic time for the Purchasing Organization.

Expediting, or ensuring the timely and accurate filling and delivery of orders, was a significant part of the Purchasing function. Sanford K. "Tommy" Thompson was imported from Western Electric to be the Department Manager for the expediting group. Frank Duggin, who became an expediter during the first year of the organization's existence, remembers Thompson as "the controlling force of Purchasing during the 53–55 time period," and as "a strict disciplinarian who wanted reports of deliveries down to the most minute detail."[40] For example, Leroy Crumley made the mistake of writing on his weekly activity card the notation "in transit tentative." Thompson promptly jumped all over him: "What in the hell does in transit tentative mean?" he demanded.[41] Thompson, again according to Western Electric philosophy, also demanded quality service from contractors. If they could not produce, Sandia did not continue to do business with them.

Expediters such as J. W. Hughes and Crumley came into the purchasing process as soon as a requisition for the purchase of material was made out. Expediters, in effect, monitored the progress of an order from the time a purchase requisition left the shops until the material was received. Conferences with technicians were required to determine schedules and to locate materials. Buyers spent many hours in the marketplace during the production era.

From 1950 to 1955, although procurement policies at Sandia largely bore the imprint of Western Electric, the AEC gradually imposed its own requirements. In 1952, for example, the AEC placed into effect a procurement monitoring policy, which required a fourteen-point justification for each purchase order being submitted to the AEC for approval.[42] In general, however, government regulations were not always strictly enforced, and problems were dealt with in a direct manner. Ed Herrity, who became a Department Manager in Purchasing, recalled that it was Dietrich's style to take the initiative in dealing with bureaucratic bottlenecks. "Dietrich," he said, "would tell them to sign the G . . . d thing or send it back and *you* take responsibility for the schedule."[43] A characteristic of the Purchasing organization, shared by other groups at this time, was the overall emphasis on getting the job done—and in as expeditious a manner as possible. The lack of social legislation and extensive government regulations made it easier than in later years.

In 1955, the structure and mode of operation of the Purchasing organization was changed when management decided to eliminate its expediting staff and increase the buying staff. According to the new rationale, the buying staff should also learn to expedite. Four expediters were brought in from Western to train buyers and effect the change in operations.

At this time, management also initiated a "task force approach" to purchasing. Personnel from Purchasing, Engineering, Quality Assurance, and other pertinent areas became involved with a particular weapons program. These individuals then worked together in the Lab and in the field for the duration of the project—usually three months to a year. Duggin recalls that this approach to purchasing and problem solving "gave everyone a sense of belonging, and helped cement relationships between the technical and administrative staff. . . . We were close enough to it [the work] to see what it was all about."[44]

The changing emphasis of the Laboratory from production toward research and development had an impact on the Purchasing function. Whereas in the fifties, Sandia's Purchasing organization bought practically all the MC components and furnished them to the complex—Bendix, Kansas City; Rocky Flats, etc.—the gradual relinquishment of this activity, in the opinion of some, had its negative aspects. The trend toward research and development, for example, obviated much of the need for interaction between buyers and engineers. The resulting separation destroyed the closeness that technical and administrative staff previously enjoyed.[45]

There was also a new "breed" arriving on the Corporate scene. According to Frank Duggin:

> Where we had oldtime manufacturing engineers that were concerned about getting the product on time, now we had a new breed . . . that was concerned about building a better mousetrap; the timeframe required to build it was . . . secondary. Our goal at the time we were in production was that 95–98% of all orders placed would be delivered on time and acceptable. After the change . . . our on-time delivery was like 70%. . . . Multiple in-field task forces had to be developed.[46]

The closer working relationships of the fifties era, according to Duggin, were reflected not only in after-hours socializing, but in a dedication to getting the job done regardless of the number of hours it took.[47]

MEETING THE ADMINISTRATIVE CHALLENGE: EXPANSION OF SERVICES

In 1955, Bill Dietrich was pulled back to Western Electric, and Ken Spoon, who had transferred in as a Department Manager in 1954, was promoted to Director as his replacement. Spoon, remembered as "a Purchasing pro," served as director from 1953 to 1967, during which time the organization enjoyed an extended period of stability and high productivity.[48]

Public Relations

From the inception, management's stance on public relations would be influenced by the Laboratory's situation as a government contractor involved in highly classified work. This important aspect of the mission, therefore, helped to establish the underlying precept of the Corporation's public relations policy.

In general terms, management viewed public relations "as an estimate of what the public thinks about Sandia."[49] The distinguishing factor was that Sandia's "public" encompassed not only Albuquerque, New Mexico, but those the Lab served—the Atomic Energy Commission and the Department of Defense. Although Sandia worked for the AEC, its assignment was to serve the military. As one executive of the fifties era explained:

> This reduces the public relations job to very simple terms. If we deliver good and timely service, our public will regard us well. If we don't, no amount of speech-making and advertising can improve the situation.[50]

This philosophy, destined to be enduring—if somewhat limiting—became the basis for the Laboratory's approach to public relations.

For the first year and a half of its existence, Sandia Corporation had no Public Relations function *per se*, despite the efforts of a far-sighted young man who joined the Lab in May 1949. Ted Sherwin hired on as a manual writer in Sylvan Harris' Documents (SLD) Department, then situated in the Technical Information Building 818. Sherwin recognized that the Corporation needed some kind of an information program and mentioned the matter to his immediate supervisor, Purdy Meigs.

At the time, however, separation from the University of California was in the offing. Both Meigs and Harris suggested that it would be best to wait until negotiations with AT&T were completed. Sherwin agreed, and in the interim put his ideas on paper in the form of a memorandum formally proposing establishment of a Public Relations program.[51]

The five-page memo, as Meigs indicated, was "well-presented" and gave numerous reasons justifying the establishment of a public relations function. The local news media, for example, periodically carried stories on the Lab. Furthermore, Congressional inquiry concerning the AEC and a flood of activity dealing with all declassified phases of the atomic weapons program heightened the interest of the local populace. Without the coordination of an adequate PR program, Sherwin warned, media coverage could place the security program in jeopardy. A formal public relations program, he posited, would improve relations with the press by making available to them factual news stories and by assisting reporters in carrying out story assignments.

Sherwin also envisioned the public relations office as a contact point for civic groups interested in having officials of the Laboratory appear at public functions. This service would relieve upper management from handling routine press inquiries. Public relations personnel could prepare news releases concerning changes in management or administration and coordinate public functions such as a Laboratory open-house or tour of facilities.

Sherwin suggested that the public relations office could be established organizationally as a division of an existing department such as Documents. He felt that a qualified individual with a newspaper or public relations background should head the group.[52]

Amid the activity surrounding the transfer to AT&T management and the pressure of the urgent operational demands, no action was taken on the proposal. Sherwin persisted, however, and at the opportune time, discussed the matter with Fred Smith. Smith, who had the somewhat misleading title of Superintendent, Personnel and Public Relations (since such a function did not exist), thought it was a good idea, but advised that the proposal be "kept on a back burner."[53] Indeed, that is where it stayed until 1951.

At this time Harold Sharp succeeded Smith as Superintendent of the 3100 organization, and Smith was elevated to Personnel Director and Treasurer. Sharp, recognizing the organization's need for an infomation program, initiated a search for candidates qualified to direct such an effort. "My 4001 file fell out," Sherwin recalled, "and he called me in for an interview."[54] Sherwin produced his proposal, and the upshot was that Sharp offered Sherwin the job.

The July 1951 phone directory for the Laboratory shows for the first time a Public Relations Division (3125) under supervision of Ted Sherwin. For a while, as Sherwin phrases it, "Public Relations was a one-man show."[55] Before long, however, he had hired Bob Gillespie from the University of New Mexico to serve as editor of the Lab newspaper. Originally called the Sandia *Bulletin*, the newspaper was renamed the Sandia *Lab News*. On February 16, 1951, the first letterpress copy of the paper was produced.[56]

During the first decade of the Corporation's existence, management carefully censored all news copy. Landry started such a practice early on while the newspaper was still a small mimeographed bulletin. The talented, if somewhat irreverent, cartoons of Felix Padilla caught President Landry's attention and motivated him to establish an Editorial Review Board. The President himself was a member of the Board as well as the Superintendent of Personnel and Public Relations Harold Sharp. Sharp also served as Secretary. As such, one of his duties was to read aloud every word of the *Lab News* in the presence of Landry and the rest of the Committee before release of copy for publication. Landry would sit in his chair, hands folded across his chest, eyes closed, as he listened carefully to the impact of the words. As Sharp intoned the news releases, Landry would suddenly raise his hand: "Stop, repeat that!" Sharp would repeat the last sentence. "Strike that!" Landry would say, and the reading would continue. In this manner, Sherwin recalled, news releases were filtered through the management concept of what should be said and what should not be said.[57]

When Landry became too busy to take such an active part in the editorial review process, he appointed Sharp as chairman. Sharp attempted to get representation from different organizations around the company to serve on the committee. Eventually Dick Claassen became chairman. Claassen, however, became disenchanted with his role because management never accepted the few recommendations suggested. Finally, Sherwin recalled, Claassen made a speech at a Large Staff meeting where he pointed out that "the Editorial Review Committee had become nothing more than a rubber stamp for top management . . . " and indicated that he wanted no further part of it.[58] Siegmund Schwartz, who was president at the time, agreed, and the Editorial Review Committee was dissolved. Review procedures continued, for both content and classification issues, but never again in formal committee format.

For some time, the Public Relations Division served more as an adjunct to the Technical Information program. "It was clear to me," Sherwin explained, "that there was a much more obvious mandate for a technical information program than there was for a public relations program."[59] The AEC established clear procedures and requirements for the release of unclassified technical information, but showed considerable reticence when it came to the release of unclassified information of the news variety. Nevertheless, as Personnel Director Fred Smith warned, "the essential secrecy of our work must not be used as a cloak to conceal information which is of legitimate public interest and concern."[60] Smith felt that one of the best inherent aspects of the Lab's PR program was its recruiting activity. "In that work," he said, "we are selling not only our Company and the Atomic Energy Program, but the Albuquerque community as well, the Land of Enchantment." "Money also talks" Smith added, pointing out that the Laboratory injected some three to five million dollars into the local economy in 1952.[61]

In an attempt to come up with a charter for expansion of the division's functions into the realm of public information, Sherwin located a pertinent section in a guide for AEC contractors. The paragraphs, headed "Public Information," read as follows:

> It is the policy of the AEC to encourage public interest in all aspects of the atomic energy program. To this end, contractors are expected to supply, on their own initiative, information to the public when such release does not endanger security and when it adds to the public knowledge of the results of expenditures of public funds by the AEC.
>
> The withholding on the basis of a security representation of information not involving national security is not only detrimental to security but is a serious denial of the obligation of a government agency to the people of the United States.[62]

These paragraphs, exemplifying as they do the AEC's conception of how a public information program should work, eventually became Sandia's Public Relations charter.

A series of meetings led to formalization of the Laboratory's public relations function. In 1954, President James McRae, Vice-Presidents Max H. Howarth and Tim Shea, Superintendent Ray Powell, and Ted Sherwin met to discuss the program. Shea expressed the view that the heart of the Public Relations program ought to be the activities of employees in community affairs, an approach that Powell and Sherwin had already agreed upon. Almost a year later in September 1955, Sherwin and E. W. Peirce, Manager Employee Services, Training, and Public Relations, met in Ray Powell's office to discuss further the proposed Public Relations program. The group reviewed in detail the proposal Sherwin had first submitted in 1949 and agreed upon certain modifications. They decided that the proposal should be redrafted in the form of an operational guide to be put into effect after review and approval by Small Staff, namely the Lab's Vice-Presidents and President.[63]

In 1956, Max Howarth, who was then Executive Vice-President, approved release of the Public Relations guide as a chapter in the Supervisor's Manual. The chapter set forth for the first time the framework for review and release of public information independent of the Technical Information program.[64]

By this time, an Employee Benefits and Civic Relations section, headed by R. A. Knudsen, had already been established, signalling the increasing involvement of the Laboratory in community affairs. The decision was made to add another staff member and move the division to Building 800 outside the Tech Area. From the standpoint of Public Relations, this location was felt to be ideal because the public recognized the building as corporate headquarters and also because there was adequate space in the lobby for visitors.[65]

Sherwin described the informality of the Public Relations mode of operation accordingly: During the fifties, he said: "When an individual needed some help with a letter or a talk they were going to give, they could come in informally, sit down with me or anyone, and we would straighten out spelling and add a sentence or two in about fifteen minutes, then the person would go on his way."[66] This one-to-one relationship with people, Sherwin maintains, resulted in high morale among staff writers and the feeling that they were appreciated. Inevitably, however, management imposed procedures that in his opinion, "diminished output, diminished interest, and despirited people."[67]

Public Relations was carried as an overhead expense. Hank Willis, who worked in the Technical Information organization for a short period of time in 1955 and in later years served as its Director, added that charging overhead case by case "dramatized how much it cost to get a writer to spend three days on a report, which often encouraged line organizations to have a technical staff person do the same work less efficiently and probably more slowly."[68]

Amid the changes taking place, the Public Relations Division began to expand its program to the benefit of both the Laboratory and the community. Because security restrictions limited the use that the Lab could make of newspapers and other media, management from the beginning supported the involvement of Sandia employees in business, civic, and service organizations.

MEETING THE ADMINISTRATIVE CHALLENGE: EXPANSION OF SERVICES

Left to right, Gene Peirce, Department Manager; Ken Smith, Supervisor of Employee Services; and Ted Sherwin, Supervisor of Public Relations stand outside Sandia's Employee Services and Public Relations office in this 1954 photo.

This one-to-one approach to public relations helped to foster good community relations. In 1950 the President of the Corporation, George Landry, served as Director of the Albuquerque Community Chest and also as a member of the Advisory Board of the Lower Rio Grande Flood Control Association. Vice President of Research Walter A. MacNair served as a member of the Albuquerque Chamber of Commerce; and staff members served in various capacities for other civic, charitable, and welfare organizations.[69]

In October 1950, the Public Relations Division participated in the Albuquerque Community Chest fund drive and helped implement a plan for the sale of bonds to employees through payroll deductions. An Employee Contributions Plan, founded by Sherwin some years later, eventually replaced the Community Chest activity.[70]

With the Corporation looking forward to its fifth anniversary in 1955, Public Relations launched a contest to select an emblem for a service pin to be awarded to employees. A secret committee was appointed to screen the entries; however, employees voted on those entries making the final round of the contest. By this time, more than 1,000 employees were eligible for five-year awards, and approximately a dozen employees, counting Z-Division service, were eligible for ten-year awards; therefore, the contest generated widespread interest. On June 24, 1955, H. C. "Clyde" Walker won the contest and a $75.00 U.S. Savings Bond when employees selected his artistic rendition of a thunderbird as the winning entry. Walker's design featured a colorful and symbolic thunderbird in brilliant turquoise enamel on a copper background. William E. Caldes, a ten-year veteran with the longest service record in the Corporation, was the recipient of the first service recognition pin.[71]

One of the division's first attempts at improving public relations outside the Laboratory involved Indians on a nearby reservation where Sandia was conducting field testing. To help the Indians understand what was taking place, the division made a film featuring test operations. Sherwin hired a Native American to translate the English script so that the sound track could be

MEETING THE ADMINISTRATIVE CHALLENGE: EXPANSION OF SERVICES

Sandia engineer H. Clyde Walker proudly accepts a seventy-five dollar savings bond and congratulations from President James W. McRae after his thunderbird drawing was selected as the winning entry in the company-sponsored contest to design a service recognition pin. Walker, who nearly missed the entry deadline, had been looking for a Southwestern image that would reflect Sandia's mission. He chose the thunderbird after learning that it symbolized the spirit of protection for the Indian Nation.

dubbed in the native language. At the suggestion of the head of the field test operation, Glenn Fowler, the new Public Relations Division premiered their film at the Indian Tribal Ceremonial at Window Rock, Arizona. With some pride and anticipation, the operator started the film and stood back in the wings to catch the crowd's reaction. When the Indians started shaking their heads and looking at one another as if to figure out what was going on, someone walked out into the audience to see what was wrong. "We discovered that the film and sound track had been put on in reverse," Sherwin said. "It looked and sounded like gobbledy-gook." Despite the snafu, the film proved to be a successful public relations tool and assisted in getting concessions from the Indians for the use of their land for testing purposes.[72]

During the years 1951-1954, and also in connection with the Laboratory's field testing activities, Public Relations initiated the Full-Scale Test Observer Program. The program, designed to improve morale by showing employees the end result of all their classified labors, involved making arrangements for approximately 1000 employees to travel to the Nevada Test Site to witness testing operations. From 1951 to 1956, ten full-scale test operations would be conducted in the Pacific and at the continental test site in Nevada.[73] The strong emphasis of the AEC and the military on testing during this period would occupy the efforts of not only the Public Relations Division but the entire laboratory.

Under Sherwin's direction, the Public Relations function during the fifties grew to include a highly successful Employee Contribution Plan, support for increasing civic involvement, establishment of a more professional Laboratory newspaper, Family Day activities, and the production of a film, the "Sandia Story," featuring the Dean Thornbrough family, as well as information publications such as the "Employee Bulletin" and "Management News Briefs."[74] Established primarily as a means of improving employee communications during a time when morale was low, the function by 1955 had expanded into the realm of a full-fledged Public Relations program. In retrospect, Sandia's PR program continues to bear the strong imprint of Ted

Sherwin. Sherwin not only introduced the public information concept to management, but developed the program's basic philosophy, diligently worked for its acceptance, and expanded the function into a viable program whose scope has not changed significantly over the years.

MISSILE-ATOMIC WARHEAD PROGRAM
May 1954

WARHEADS	ARMY	STAGE	NAVY	STAGE	AIR FORCE	S
			AA ROCKET		AA ROCKET	
XW-12			TALOS W (JOHNS HOPKINS UNIVERSITY)	2	F-99 (BOMARC) (BOEING)	
XW-7	NIKE B (BELL TELEPHONE LAB.)	2	BETTY (NAVAL ORDNANCE LAB.)	3		
	CORPORAL (CALIF. INSTITUTE TECH.)	3*				
	HONEST JOHN (DOUGLAS)	3*	BOAR (NAVAL ORDNANCE TEST STA.)	2		
	SERGEANT (CALIF. INSTITUTE TECH.)	0				
XW-5			REGULUS (CHANCE VOUGHT)	3*	B-61A (MATADOR) (GLENN L. MARTIN)	
			REGULUS 2 (CHANCE VOUGHT)	1	B-63 (RASCAL) (BELL)	
			TRITON (JOHNS HOPKINS UNIV.)	1	ATLAS (3/8 SCALE) (CONVAIR)	
XW-13	REDSTONE (REDSTONE ARSENAL)	2			B-62 (SNARK) (NORTHRUP)	
					B-64 (NAVAHO) (NORTH AMERICAN)	
XW-8			REGULUS (CHANCE VOUGHT)	2		

WARHEAD INSTALLATION STAGE
0 APPROVAL EXPECTED
1 UNDER STUDY
2 FULL DEVELOPMENT
3 PRODUCTION (AUTHORIZED)
* IN PRODUCTION

Chapter XVIII.

A MISSILE REVOLUTION IN THE MAKING

One new weapon was started into design in 1948, two more in 1949, and three in 1950; the floodgates opened in 1951, when many designs for guided missiles were introduced.

Frederick C. Alexander, Jr.
<u>History of Sandia Corporation</u>

On October 27, 1950, Clarence N. Hickman, the Corporation's recently hired consultant on rockets and missiles, boarded a single-engine Carco aircraft in Albuquerque for a flight to Fort Bliss near El Paso, Texas. At Fort Bliss, Hickman witnessed the demonstration of a rocket fire control system and proceeded the following day to White Sands Missile Range near Alamogordo, New Mexico.

The trip, as Hickman described it was one "never to be forgotten."[1] Blustery winds, up to 65 miles an hour with gusts up to 100 miles, buffeted the little plane so that it tumbled rather than flew down the White Sands runway. During the postwar years, Hickman had taken many such trips to the White Sands proving ground where he watched the firing of various rockets and missiles under the direction of a seemingly misplaced contingent of German scientists and engineers.[2]

The implausible hegira of German missile scientists from the small peninsula of Peenemünde on the Baltic Sea to the scorching desert of the White Sands Missile Range began during the waning days of World War II. At that time, German Chancellor Adolf Hitler had launched a barrage of rockets against London and Allied seaports. Known as V-1 "buzz bombs" and V-2 "vengeance weapons," these rockets represented the technological expertise of a select cadre of German scientists and engineers under the leadership of Wernher von Braun and Walter Dornberger.[3]

The twenty-seven-foot long V-1 was a large winged air-breathing cruise missile that weighed some 5,000 pounds. Unreliable, inaccurate, and sounding much like a "motorcycle without a muffler," only fifty percent of the 8,000 bombs launched reached their intended targets. In contrast, the V-2 ballistic missile, steered by an inertial guidance system, was even more of a Goliath at forty-six feet in length, and it was more accurate.[4]

Here in the United States, Robert H. Goddard, America's own rocket pioneer, periodically ventured forth from his prairie sanctuary near Roswell, New Mexico, to perfect some of the 200 patents he held in the field of rockets, guided missiles, and space exploration. Before his retreat to the Southwest, Goddard taught physics at Worcester Polytechnic Institute, Clark University, and later at Princeton University. At Auburn, Massachusetts, on a cold overcast day in March 1926, he launched his first liquid-propelled rocket, and in 1942 delivered an operational rocket to the U.S. Army. In the interim, however, Goddard moved from New England to New Mexico where the climate and isolation were more conducive to research on the rockets that occupied his life and career.[5]

Not surprising in retrospect, the end of the war found the Americans lagging far behind both the Russians and the Germans in the rocket field. While the Germans were willing to gamble on new advances in rocketry, the Allies had relied on air power and the certainty that they could deliver even larger explosives with greater probability of a hit. However, the advent of nuclear weapons motivated a reassessment of rocketry and its potential. Weapons designers began to anticipate the benefits to be gained by combining the improved accuracy of guided

A MISSILE REVOLUTION IN THE MAKING

Kenner Hertford (right), Manager of the Albuquerque Operations Office, visits with rocket expert Wernher von Braun during a visit to Sandia. Von Braun came to this country as part of Operation Paperclip, bringing German rocket experts to the United States.

missiles with the larger destructive radius made possible by a nuclear armed device. There were those who, optimistically and not realizing the technical complexities involved, envisioned a marriage of missiles and nuclear weapons leading to an era of push-button warfare.[6]

From German Rockets to Redstone Arsenal

Faced with the prospect of annihilation by the Nazis to prevent their expertise from falling into the hands of the Allies, the Peenemünde team of scientists and engineers cast their lot with the Americans. "Operation Paperclip," carefully orchestrated by U.S. Army Intelligence personnel, brought the German contingent to America where they continued to expand the frontiers of rocket technology under the aegis of the United States Government. With V-2s and technical papers intact, they came first to Boston, Massachusetts, where they provided instruction in the use of the V-2 rockets, then on to Fort Bliss at El Paso, Texas. Fort Bliss, located in convenient proximity to White Sands Missile Range, became headquarters for the German scientists, who in 1946 began a guided missile development program.[7]

If the Intelligence branch showed foresight, U.S. policy makers did not. In June 1947, Congress established funding limitations that stifled the development of intercontinental ballistic missiles in the postwar period. A Pentagon study titled "Operational Requirements for Guided Missiles," placed severe restraints on the fledgling U.S. missile program. This document specifically cancelled government support for the Consolidated Vultee Aircraft Corporation program to design a large liquid-fueled intercontinental ballistic missile; thereby postponing U.S. entry into the large long-range rocket program and preventing the Americans from competing more successfully with the Russians in the race for space.[8]

In the interim, the U.S. Air Force established the RAND (research and development) project in 1946 under direction of the Douglas Aircraft Company. The RAND team envisioned a satellite vehicle using liquid fuel and multistage rockets. Dr. Theodore von Karman, one of the founders of the RAND project, proposed that the most feasible method of increasing the range of the V-2 was to add wings to the missile. This led to consideration of low-flying "pilotless bombers" and the development of air-breathing missiles rather than pure rockets.[9]

By 1949, other events were taking place that influenced the direction of missile development activities in the U.S. When it became obvious that the missile program was outgrowing the Fort Bliss location, headquarters for the Peenemünde team was moved to the Redstone Arsenal in Huntsville, Alabama. Situated on government property and near the power resources of the TVA, the site was also close to the long-range proving grounds of Cape Canaveral. In 1949, 130 members of the von Braun contingent transferred to Huntsville, where they began work on development of a longer-range ballistic missile, with a nuclear warhead to be known as the Redstone.[10]

Meanwhile, the number and complexity of missiles grew. Advanced rocketry combined with nuclear weapons began to alter the concept of warfare as previously known—a realization reflected in interservice rivalry among the branches of the U.S. military. The Navy would strive to extend its strike force through use of submarine-fired missiles that spanned continents. The Air Force, although still championing the airplane, would eventually entertain methods for long-range bombardment. The Army would vie for control of long-range missile development based upon the concept that missiles were "extensions of artillery." The result would be a period of duplication, backbiting, and interservice feuding.

Early on, the steady proliferation of missile designs was a cause for concern. To institute controls and eliminate duplication, the Research and Development Board in January 1949 established the Committee on Guided Missiles. Chaired by Cal Tech aerodynamicist Clark Millikan, the Board found it difficult to bring order to what scientist and author Herbert York, former Director of Defense Research and Engineering of the Department of Defense, has termed "an increasingly chaotic situation."[11] Undersecretary of the Air Force John McCone and Army General Kenneth Nichols, among others, recommended establishment of a project focussed on missiles as the Manhattan Project had been focussed on development of the atomic bomb.

With this objective in mind, President Truman on October 26, 1950, announced the selection of Kaufman "K. T." Keller of the Chrysler Corporation as Director of Guided Missiles. Keller would be responsible for organizing the program and implementing controls. Referred to as "The Guided Missile Czar," Keller reported to the Secretary of Defense but had direct access to the President.[12] His reporting relationship reflected the importance accorded the project and the decision to cut through the bureaucratic red tape.

Kenneth Nichols, former deputy to General Groves on the original Manhattan Project, became Keller's second in command. Although Keller decided that establishment of a project the scope of the atomic bomb effort would not be feasible, he maintained close rein on the guided missile program. According to York, "A recommendation by Keller had the force of a decision by the Secretary of Defense."[13] It was Keller who, after a visit to the Redstone Arsenal in Huntsville, recognized the potential of von Braun and the Peenemünde team, and subsequently gave the go-ahead for the scientists to proceed from feasibility studies to the building of actual rockets—a decision that would have important ramifications for the future of the United States in space.[14]

During the fifties, several air-breathing cruise missiles would undergo development, but only two—the MATADOR and the REGULUS—would be added to the U.S. arsenal. Both the SNARK and the NAVAHO would become obsolete before deployment. The sudden onset of the Korean War and the knowledge that Russia was developing a significant missile program raised the level of alarm and led to a number of crash programs to build long-range strategic missiles as well—missiles such as the ATLAS, TITAN, and MINUTEMAN. The objective was massive retaliation in the event of attack; the result—a missile revolution in the making.[15]

A MISSILE REVOLUTION IN THE MAKING

Sandia Establishes Directorate for Warhead Development

Interest in this new branch of weapons development first touched the nation's nuclear weapons laboratories in 1946. In December of that year, General Leslie Groves informed the Los Alamos Scientific Laboratory of an Air Materiel Command project to put an atomic warhead into an air-to-ground guided missile. "We have been requested to furnish such assistance as we are able," Groves told Bradbury, adding that the project was to be known as "MASTIFF." Z Division was assigned to make the preliminary analysis of the experimental missile, although development of the MASTIFF was to be deferred because of the work load.[16]

Budget cuts during the next few years reduced military participation in missile development; however, in 1949, the Division of Military Application notified the Military Liaison Committee to reopen consideration of the possibility of developing atomic warheads for guided missiles. This led to the establishment on June 21, 1949, of an Ad Hoc Committee to review the concept of "marrying" nuclear warheads to guided missiles.[17]

The committee, which included Dr. Norris Bradbury of the Los Alamos Lab, Lt. General J. E. Hull, Director of the Weapons System Evaluations Group, and Dr. F. D. Hovde, President of Purdue University, submitted its report to the Secretary of Defense and the Joint Chiefs of Staff in January 1950. Committee members concluded that four missiles could be adapted to carry warheads—the Air Force's SNARK, the Navy's REGULUS, Army's HERMES, and unspecified warheads for the U.S.A.F. RASCAL. The Research and Development Board, the postwar replacement for the Office of Scientific Research and Development, later added the Army's CORPORAL to the list.[18]

The Ad Hoc Committee agreed that the Sandia Laboratory should have principal design responsibilities for marrying the nuclear warhead to the missile. The Laboratory's main tasks were to develop new arming and fuzing subsystems and to modify components to withstand the accelerations, temperatures, and vibrations attendant to missile launch and flight.

However, ultimate responsibility for fuzing of warheads was destined to become a topic of much discussion and controversy during the 1950-1952 period.

The Joint Chiefs of Staff and the Secretary of Defense accepted the report of the Ad Hoc Committee, and in March 1950 a symposium was held at Sandia to provide guided missile contractors and warhead designers the opportunity to exchange information. Those attending determined that the Mk 4, TX-5, and airburst and penetrating versions of the TX-8 were of immediate interest for use as warheads. It was decided that the Special Weapons Development Board would serve as the forum for interaction between the designer and user of atomic weapons and act as coordinator for guided missile and atomic warhead development.[19]

The assignment of numerous projects related to guided missile development required expansion of the Corporation's research and development activities to accommodate the rapidly increasing work load. The effort of the Laboratory shifted from concentration on one or two weapon projects to effectively pursuing concurrently two or three times that many.

In October 1950, the Laboratory took certain actions that further committed Sandia to support of the nation's guided missile program. First, management set up a new organization, initially a department of one, under Louis A. Hopkins, Jr., specifically for the purpose of integrating warheads with missiles. Hopkins, who held a master's degree in electrical engineering from the University of Michigan, spent five years in the military during World War II. His tour of duty included service as a ground radar specialist and as a major in the Air Corps. After obtaining his degree in 1946, he joined Sandia as a section supervisor for the development of electronic components.[20]

Second, at the request of Robert Poole, the Corporation hired well-known rocket expert Clarence N. Hickman to serve as a consultant. Hickman, in 1940, had been one of the principal directors of rocket research and development for the National Defense Research Committee. As a graduate student at Clark University, studying under Robert H. Goddard, Hickman and

H. S. Parker participated in the development of short-range rocket projectiles that could be fired by infantrymen at tanks. These devices were progenitors to the World War II bazooka, which Colonel Leslie Skinner of Army Ordnance, Lieutenant Edward G. Uhl, and Hickman later developed for the U.S. Army. As a Bell Laboratories consultant, Hickman arrived at Sandia Corporation October 2, 1950. He brought with him an expertise invaluable to an organization attempting to set up a guided missile warhead program. Between visits to missile programs around the country, Hickman acquainted himself with Sandia Corporation methods of operation and discussed organizational plans for the department with Hopkins.[21]

Still another significant event occurred in October when the Sandia Weapons Development Board appointed a Guided Missiles Committee. The Committee, which met for the first time October 16, 1950, discussed the general environmental conditions that guided missiles were expected to encounter and concluded that atomic warheads could indeed withstand such conditions. The Committee proceeded to set up Working Groups to address technical solutions to problems that might arise. Hopkins served as chairman. In December, Hickman, Hopkins, and Glenn Fowler met with AFSWP liaison officers Colonel James Bain, Army; Lieutenant Commander Charles Rush, U.S. Navy; and Lieutenant Colonel John Harris, Air Force, to discuss full-scale tests of the HERMES and CORPORAL missiles at White Sands, the REGULUS at Point Mugu in California, and the MATADOR and SNARK at Patrick Air Force Base near Cocoa Beach, Florida. Of these, the CORPORAL, REGULUS, and MATADOR, along with the HONEST JOHN became major programs for the laboratory during this early period.[22]

Throughout the following year, both Hickman and Hopkins worked diligently to structure the department to meet the work load. On December 7, 1950, Hopkins submitted his first proposal suggesting establishment of three divisions within the Guided Missiles Department, but management rejected the idea on the basis that it was contrary to company policy to have divisions before obtaining adequate support personnel. As of

December, only two men reported to Hopkins: G. F. Heckman and J. W. Jones. Meanwhile, Hickman continued on a consultant basis until March 1951 when he joined the Corporation as a staff member reporting to Walter A. MacNair, who held the directorate-level position of Bell Labs Research Consultant.

Undaunted, Hickman and Hopkins continued their efforts to build an enlarged organization for support of the missile program. In April and in October, Hopkins resubmitted his recommendations, again unsuccessfully. By this time the department had twenty-five employees.[23] It had been a year since the first organizational proposal had been made, and the department was continuing to grow. The work load was also growing; in fact, the warhead program at this time reached the magnitude of the expanded bomb development program. By the end of 1951, the Laboratory had thirteen missile warhead combinations in active development.

On November 5, 1951, Hickman visited Poole personally to make another bid for expansion. Subsequently, a formal presentation on the Laboratory's Atomic Warheads-Guided Missile Program was made to Sandia management, to General Fields, and to Mervin G. Kelly, who had remained actively involved in Sandia affairs. Shortly thereafter, the decision was made to split the 1000 organization into two engineering directorates—one for warheads and one for bombs, both reporting to Vice-President of Development Robert E. Poole. Robert Henderson, Director of Engineering I, remained in charge of bomb development, and Louis Hopkins, at age thirty-four, was promoted to head the new directorate for warhead development.[24]

Thus, after four attempts and a year of persistent effort, Hickman and Hopkins were successful in expanding the organizational structure of Sandia's guided missiles program. Under the new format Hopkins had three department heads reporting to him. By July 1952, each department had been set up to include both an electrical and mechanical division, as well as project leaders for each missile system. The Engineering Department headed by Hilt DeSelm, for example, handled warhead applications for the Mk 5 and Mk 7, such as the MATADOR and

A MISSILE REVOLUTION IN THE MAKING

**ENGINEERING I
1952**

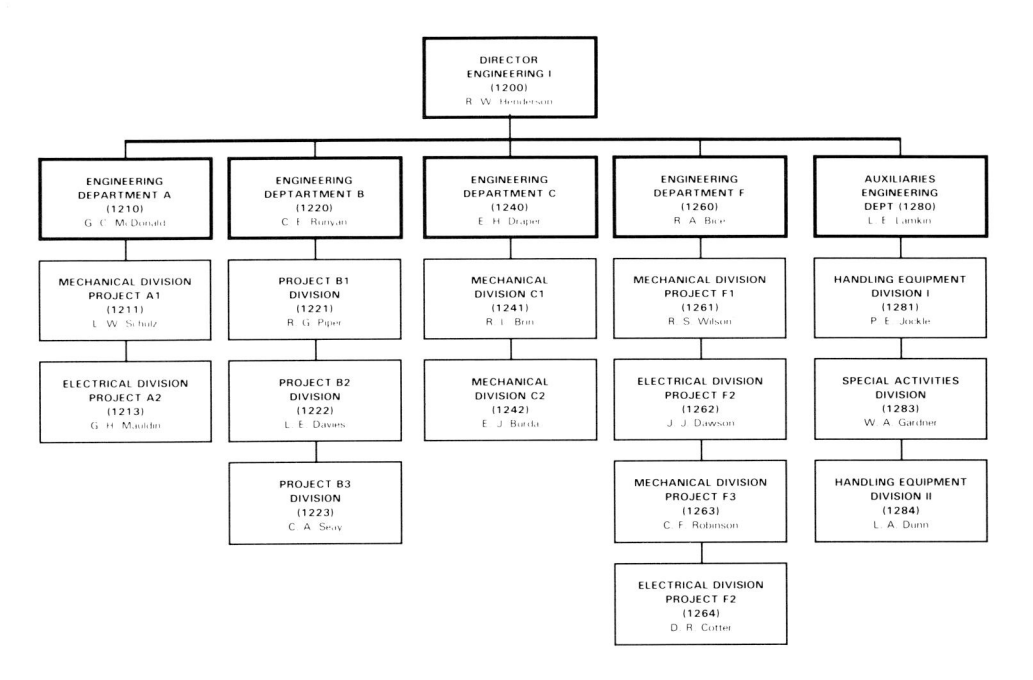

- DIRECTOR ENGINEERING I (1200) — R. W. Henderson
  - ENGINEERING DEPARTMENT A (1210) — G. C. McDonald
    - MECHANICAL DIVISION PROJECT A1 (1211) — L. W. Schulz
    - ELECTRICAL DIVISION PROJECT A2 (1213) — G. H. Mauldin
  - ENGINEERING DEPARTMENT B (1220) — C. F. Runyan
    - PROJECT B1 DIVISION (1221) — R. G. Piper
    - PROJECT B2 DIVISION (1222) — L. E. Davies
    - PROJECT B3 DIVISION (1223) — C. A. Seay
  - ENGINEERING DEPARTMENT C (1240) — E. H. Draper
    - MECHANICAL DIVISION C1 (1241) — R. L. Brin
    - MECHANICAL DIVISION C2 (1242) — E. J. Burda
  - ENGINEERING DEPARTMENT F (1260) — R. A. Bice
    - MECHANICAL DIVISION PROJECT F1 (1261) — R. S. Wilson
    - ELECTRICAL DIVISION PROJECT F2 (1262) — J. J. Dawson
    - MECHANICAL DIVISION PROJECT F3 (1263) — C. F. Robinson
    - ELECTRICAL DIVISION PROJECT F2 (1264) — D. R. Cotter
  - AUXILIARIES ENGINEERING DEPT (1280) — L. E. Lamkin
    - HANDLING EQUIPMENT DIVISION I (1281) — P. E. Jackle
    - SPECIAL ACTIVITIES DIVISION (1283) — W. A. Gardner
    - HANDLING EQUIPMENT DIVISION II (1284) — L. A. Dunn

Source: *Sandia Corporation Telephone Directory*, June 1952

ENGINEERING II
1952

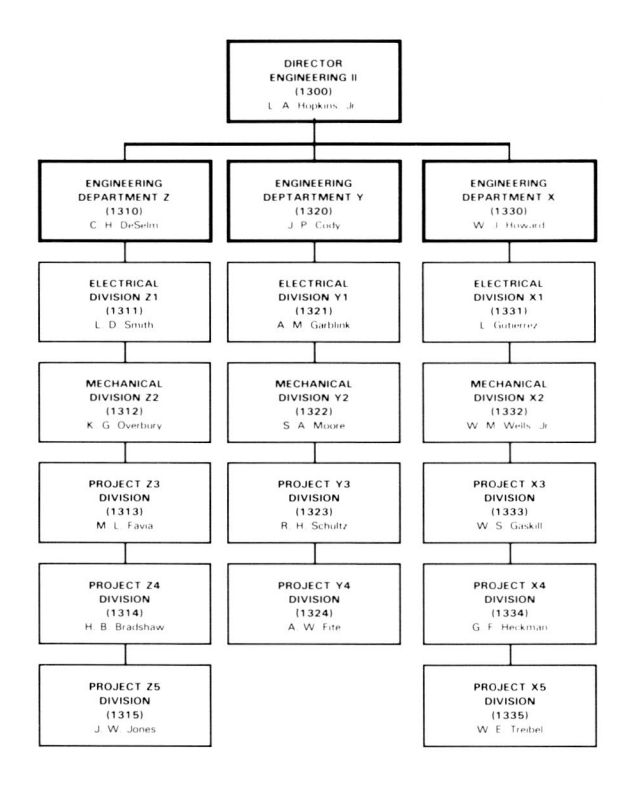

```
                    DIRECTOR
                  ENGINEERING II
                      (1300)
                   L. A. Hopkins, Jr.
                        |
     _____|_____
     |                  |                  |
 ENGINEERING        ENGINEERING        ENGINEERING
 DEPARTMENT Z     DEPTARTMENT Y      DEPARTMENT X
    (1310)            (1320)             (1330)
  C. H. DeSelm       J. P. Cody        W. J. Howard
     |                  |                  |
  ELECTRICAL        ELECTRICAL         ELECTRICAL
  DIVISION Z1      DIVISION Y1        DIVISION X1
    (1311)            (1321)             (1331)
   L. D. Smith     A. M. Garblink      L. Gutierrez
     |                  |                  |
  MECHANICAL        MECHANICAL         MECHANICAL
  DIVISION Z2      DIVISION Y2        DIVISION X2
    (1312)            (1322)             (1332)
 K. G. Overbury     S. A. Moore      W. M. Wells, Jr.
     |                  |                  |
  PROJECT Z3        PROJECT Y3         PROJECT X3
   DIVISION          DIVISION           DIVISION
    (1313)            (1323)             (1333)
   M. L. Favia      R. H. Schultz      W. S. Gaskill
     |                  |                  |
  PROJECT Z4        PROJECT Y4         PROJECT X4
   DIVISION          DIVISION           DIVISION
    (1314)            (1324)             (1334)
 H. B. Bradshaw      A. W. Fite        G. F. Heckman
     |                                     |
  PROJECT Z5                           PROJECT X5
   DIVISION                             DIVISION
    (1315)                               (1335)
   J. W. Jones                         W. E. Treibel
```

Source: *Sandia Corporation Telephone Directory*, June 1952

REGULUS. Leon Smith was the electrical division supervisor, and K. G. Overbury had the mechanical division. Department Manager J. P. Cody also handled applications for the Mk 7, namely the CORPORAL, HONEST JOHN, and RASCAL. A. M. Garblik headed the Electrical Division for Cody's department, and Sam Moore headed the Mechanical Division. W. J. Howard was Department Manager for both large and small systems, the TX-13 and TX-12, used for the REDSTONE, SNARK, and NAVAHO. Howard's Electrical Division supervisor was Leo Gutierrez. W. M. Wells, Jr., had the Mechanical Division. The project leaders within each department also held division supervisor status.[25]

Warheads for Guided Missiles

The growing pains that characterized the first year of the Guided Missile Department's existence did not appreciably impede technical progress. When K. T. Keller, Director of the Guided Missile Office, visited Sandia Laboratory early in the year to hear a report on the status of the Lab's Warhead-Guided Missile Program, he was favorably impressed. As "Missile Czar," he had toured the country visiting the various sites associated with guided missile development. Of these, he found the work going on at Sandia Laboratory commendable. "I am not going to worry about a place as obviously competent as this one," he reported.[26]

Work on warhead designs intensified after the AEC made its budgetary decision in May 1951. Among the initial efforts made to provide structure and organization to the missile program was the selection of a nomenclature system. The prefix "XW" would be used to identify warheads under design, with the "X" standing for "experimental" and the "W" for "warhead." The warhead identification, or Mark number, followed, coupled with the missile designator, for example, the XW-5/REGULUS.[27]

The Committee ruled out the possibility of using the Mk 4 as a warhead when it became evident that considerable redesign would be required to strengthen it adequately, and also because it would be years before missiles capable of carrying such a heavy load could be developed. The decision was made, therefore, to rule out Mk 4 and Mk 6 weapons for adaptation to current missile applications.

Priorities were established and an effort made to reduce in number the list of missiles, and in some instances, add others. The CORPORAL and the SNARK, for example, were temporarily deleted and the HERMES, RIGEL, NAVAHO, and TRITON added. The air-burst Mk 8 was also eliminated, but the Mk 8 warhead was retained despite the knowledge that it would require a major development program to incorporate adequate resistance to shock and impact. Earlier, the MLC had added the Mk 7 warhead to the development list.[28]

From the inception of the missile development program, nuclear safety and reliability were important considerations. At its November 22 meeting, the Special Weapons Development Board formally accepted the charter for the Guided Missiles Committee and noted the Air Force requirement for automatic nuclear safing. In regard to such military requirements, Sandia representatives pointedly expressed the hope that the military "would state the performance or the result desired, . . . [but] will not state the detailed technical method of accomplishing the desired result."[29]

The issue of safing was a primary topic of discussion at the second meeting of the Guided Missiles Committee on November 26, as well, especially the point at which in-flight insertion, or nuclear arming, should take place. Since some missiles would be launched from ground bases, committee members felt that nuclear insertion should take place only after the missile crossed into enemy territory. The objective was to prevent the possibility of an accidental crash of an atomic warhead in friendly territory, and also to prevent the enemy from gaining control of active material.

William E. Boyes, who headed a reliability group during the early fifties, recalls that the emphasis on safety and the demand for reduced weight and space also meant increased emphasis on the reliability of components. As Boyes explained: "In large bombs there was more room for components. But we had to minimize redundancy to conserve space. The remaining components had to work."[30]

Still another problem was the proliferation of missile and warhead designs. As the Division of Military Application recognized, the large number of designs made standardization almost impossible to achieve. Some progress toward control was made when studies revealed that designs for the Mk 5 and Mk 7 had progressed to the point where their use with missiles would eliminate the need for alternatives. The smaller diameter and lighter weight of the Mk 7 bomb over the Mk 5 made it a logical choice for missile application.

As head of the warhead development program at Sandia, Hopkins found that the numerous technical problems associated with standardization and the development of fuzing systems specifically suitable to guided missiles were compounded by political ramifications. No definitive ground rules existed, and competition for technical control between Sandia and the military escalated. The question arose: Who was going to be responsible for which facet of the task?

In certain areas, the Guided Missile Committee proved to be a forum for agreement. The Committee, for example, decided upon the formation of Ad Hoc Working Groups as the organizational pattern for all missile programs and, in view of time constraints, agreed to modify existing systems rather than design new ones. For the U.S. Army's CORPORAL, the first warhead application program at Sandia, the Committee decided that the Mk 7 could be adapted.[31]

CORPORAL

The program for the U.S. Army's CORPORAL began on June 12, 1951, with the appointment of an Ad Hoc Working Group. Lou Hopkins introduced Al W. Fite, Project Engineer for the CORPORAL warhead, at the meeting held on the following day. By November 1951, Military Characteristics for the XW-7 warhead had been released by the Division of Military Application.

CORPORAL/Mk 7 Warhead.

A MISSILE REVOLUTION IN THE MAKING

Under Fite's leadership, group members made maximum use of components from the TX-5 and TX-7 programs and armed electrical subsystems with barometric switches that closed on missile descent. The Working Group recommended that Sandia develop the interim fuzing system; however, both Sandia and the Army's Ordnance Corps were to continue parallel design of an optimum fuze for a ballistic missile. It was agreed that responsibility for development of the interim arming system would also be divided between the Ordnance Corps and Sandia.[32]

The CORPORAL, a surface-to-surface liquid-fueled transonic missile with a range between twenty-five and seventy-five miles, is described by Sandians who worked with it in less than favorable terms. Basically German V-2 technology, although much slimmer, the missile was fueled with aniline and red fuming nitric acid, which did not represent the ideal for ease of handling or tactical application. Being liquid-fueled, the CORPORAL lifted very slowly, so slowly, according to one observer, "that you feared it would topple before it ever got up to a reasonable speed. During this time, it was wriggling and correcting so that your fears were almost justified." The sound of the missile, described as "a simultaneous scream and a roar," had to be heard to be believed.

Mishaps occurred. Andrew Lieber, who was a participant in the program recalled humorously that a weapon once veered off toward Las Cruces—"but the range safety officer blew the fuel lines and brought the missile down short." Another time, ignition failed, and according to Lieber, "we scoured the White Sands desert with a reddish cloud of fuming nitric acid."[33] Because of the need for fuel handling and other equipment, those who worked with the cumbersome weapon claimed that at normal road march intervals, a CORPORAL battery required over 100 miles of highway space to move. Hopkins agreed that logistics were "terrible," adding that it took forty trucks to support the missile battery. "It was a major breakthrough," he said, "when the change was made from liquid-fueled to a solid-booster rocket that required little maintenance."[34]

One test flight of the CORPORAL was especially memorable. In order to have a backup for telemetry signals, the crew would fire a smoke puff out the side upon receipt of the signal to fire. This ensured that the photographer would still be able to capture the event in case telemetry was lost. However, the test crew observed that each time the smoke puff spotting charge was fired, the telemetering went off the air. One of the team members noted that the bottom of the smoke puff appeared to be blowing out right at the bottom of the junction box.

Project members took the problem to the resident expert, C. N. Hickman. Waving his hand (with fingers missing) in the air, Hickman discounted the theory: "Oh no," he said, "that little bit of explosive couldn't do that; you could hold that amount in your hand!" But the telemetry group was not convinced.

Just to make sure, they went out to the canyon, set up the charge on a steel plate, and fired it vertically, which promptly blew a hole in the plate. From the height of the rise, it was computed that there was sufficient energy to cause the smoke puff housing to shear the heads off the mounting screws. The crew then went back to Hickman to report.

"How much did you say the weight of that charge was?" Hickman asked in astonishment.

". . . grams," was the reply.

"Oh, I thought you said *grains!*"[35]

Nevertheless, the CORPORAL had its moments of glory. In 1956, in the midst of the Redwing test series in the Pacific, interest developed in the effects of high-altitude nuclear detonation. The CORPORAL, it turned out, was the only vehicle capable of lifting the payload to the required altitude of 100,000 or more feet. Sandia, the Jet Propulsion Laboratory (JPL), and the Army worked rapidly to provide the warhead, fuze, and missile—all required for the test—within a thirty-day period. "JPL," Lieber recalled, "turned to and revised the guidance system—cutting new cams." Sandians devised the command arming and fuzing system, and since the missile did not normally operate at such altitudes, they also devised seals to assure high-voltage functioning in the near vacuum of 100,000 feet. Shortly before the end of the thirty-day period, the test was cancelled; but Sandia and JPL, after completing their work ahead of schedule, proceeded to conduct a successful test firing at White Sands to prove the design and hardware as a hedge against future needs.[36]

REGULUS

In 1951, Sandia began work on the second missile in its warhead application program—the XW-5/REGULUS, which was developed for the U.S. Navy by the Chance-Vought Aircraft Company. The REGULUS, a heavy bombardment, surface-to-surface subsonic missile, was capable of delivering a warhead to a range of 500 nautical miles. Constructed in the shape of a streamlined jet fighter plane, the missile could be launched from submarine, ship, or land base, using a short-rail launcher and two jettisonable solid-fuel Jet Assisted Take-off (JATO) rocket boosters. Test versions of REGULUS included a landing gear. Thus, a single missile could be used over and over for developmental testing.

From April to early May 1951, Military Liaison officers, several Washington officials, and representatives from the Division of Military Application, together with Lou Hopkins and Clarence Hickman from Sandia, made an extensive tour of contractor plants and military installations where work on the REGULUS was in progress. The tour included visits to Salton Sea, the North American plant at San Diego, Edwards Air Force Base, and Mare Island. To study the effects of pressure changes on warhead components, group members took a submarine ride, and at Mare Island Clarence Hickman provided on-the-spot assistance with problems related to the submarine hangar. When the hangar door was opened, the submarine would list due to the change in the center of gravity. Hickman solved the problem by suggesting that the door be hinged at the top, which not only eliminated the lateral shift in the center of gravity but also greatly reduced the angle through which the door had to be opened. The proposal was later incorporated into submarine design.[37]

The Ad Hoc Working Group for the REGULUS first met in September 1951. Project leader was W. J. Denison. Again, because of the urgency of the program, the decision was made to maximize use of components already in development for the arming and fuzing subsystem. These included the radar fuze from the TX-5 and an arming switch composed of a combination of a motor-driven timer, which began operation at launch, and a barometric pressure switch that permitted arming of the X-unit and reception of this signal in the missile. Sandia was responsible for supplying the arming signal to the warhead and also for development of the fuzing system. Del Olson, who joined the Corporation in 1953 to work as an electrical system designer on the Mk 5 warhead, recalled that the design work on the REGULUS was done at Sandia, with testing conducted at Point Mugu on the coast near Oxnard, California.

The test procedure involved a submarine launch in a trajectory that would carry it to the target area before a nose dive. At a specified altitude, the baroswitch would give the signal to fire and, theoretically, the warhead would detonate. Toward the end of the test program, however, problems developed when the baroswitch failed to close before the missile hit the water. Two months of intense detective work revealed that the problem stemmed from a minor improvement in the jet engine made earlier without the Lab's knowledge. The engine modification boosted the speed just enough to make the rubber hoses collapse and prevent air from passing through to indicate to the baroswitches that the pressure was going up. The problem was easily solved by installing stronger hoses.[38]

Serious consideration was also given around this time to adapting the TX-8 and the TX-11 penetrating warheads to the REGULUS missile. The concept for development of a penetrating missile can be traced to the immediate postwar period of the Crossroads test in the Pacific when Roger Warner and others of Z Division tried to persuade the Navy that there was value in developing a missile (starting with the Little Boy) that could penetrate certain distances through concrete, earth, or water before detonating. Project Elsie was subsequently initiated by the Bureau of Ordnance to proceed with development of a missile designed to survive impact on hard surfaces.

Since Sandia had been assigned the XW-5/REGULUS arming and fuzing job, the Ad Hoc Working Group agreed that the Lab should also handle the XW-8 REGULUS. The Guided Missile Committee recommended that the Navy be responsible for the transmission of radio command signals to the missile and for reception of these signals in the missile. Sandia was to take responsibility for applying these signals to the warhead installation to effect arming and safing. Thus Sandia was responsible for the entire warhead installation, which included the monitoring structure, the terraballistic tail, and arming and fuzing components. Component evaluation and warhead systems tests were conducted in September 1953, and warhead design was released in August 1954; however, the possibility of developing a guidance system to deliver the XW-8 REGULUS effectively with pinpoint accuracy appeared remote. Therefore, on May 20, the Navy suspended activity in the program, together with work on the TX-11.[39]

In August 1952, the Military Liaison Committee through a letter to the Division of Military Application, initiated a related program called RAM (REGULUS Assault Missile). This program provided the capability for launching the REGULUS missile from a surface ship, guiding the missile to target, and arming and detonating the warhead by command from carrier-based fighter aircraft. After a total of seven flight tests, the warhead installation was design-released in mid-May 1953. Ultimately, however, it was decided that the Mk 5 warhead would be replaced by the Mk 37. Sandia activity on the Mk 5/REGULUS Warhead Installation was suspended March 1, 1956.[40]

The MATADOR and Other Air-Breathing Missiles

When the Air Force began to explore the possibility of strategic warfare, consideration was given to other air-breathing missiles, including the MATADOR, the NAVAHO, and the SNARK. Of the three, the MATADOR, which became operational in 1955, enjoyed the longest life span. The Mk 5 warhead for the MATADOR represented still another program accelerated because of the Korean War. Characterized as a surface-to-surface turbo-jet-powered transonic missile with a range of 600 nautical miles, the MATADOR was designed and built by the Glenn L. Martin Company for the Air Force. Constructed in the shape of a streamlined fighter plane with swept-back wings, the missile was launched by a single, solid propellant rocket, which accelerated it until the turbojet engine could attain enough thrust to sustain flight.

The Guided Missiles Committee named the Ad Hoc Working Group for the XW-5 MATADOR in October 1951. M. L. Favia was appointed as Sandia's project leader for the program. In September of the following year, the Division of Military Application requested a modification of the fuzing system of several missile warheads, including the Mk 5 MATADOR, to incorporate a contact fuze and provide a means of continuing the safed condition of the warhead during ground handling. The MATADOR, however, was plagued by a series of flight test failures and missile production difficulties, which postponed the design release date. After missile modifications, warhead installation flights were resumed in September 1953.[41]

In 1954, Operation "Black Swan" established a MATADOR Interim Capability Program under which bomb-to-warhead conversion components were supplied on an expedited basis, in association with a number of Mk 5 bombs for use in the advent of national emergency. Final evaluation on the Atomic Warhead Installation was held in July 1955. The warhead, although not a universal design in entirety, incorporated many components from other weapons.[42]

The NAVAHO, developed for the Air Force by North American Aviation, was tested successfully in 1956. Ninety-five feet long, it weighed 300,000 pounds. Boosted by three large liquid-propellant rocket engines, the NAVAHO could be accelerated to three times the speed of sound. Although this missile was cancelled before reaching operational status, its development contributed a significant technological base from which the U.S.

would launch further exploration into the realm of rocket and space technology. Sandian W. C. Gaskill served as project leader for both the NAVAHO and the SNARK.

The SNARK, a subsonic winged cruise missile boosted by JATO rockets, became functional, after some delay, in 1958. After watching several of the missiles swivel and dump into the ocean at Cape Canaveral, observers began referring to the test site as "the *SNARK*-infested waters." Field tester Jim Scott noted that "in the early days some of the missiles really didn't work very well; however, ninety percent of the test objectives were achieved. Of course, it all depended on how you wrote your test objectives," he added. The SNARK was declared obsolete almost as soon as it was introduced.[43]

MATADOR XW-5 Prototype.

HONEST JOHN

The HONEST JOHN represented another major warhead program for Sandia. The Laboratory first became involved in development of the HONEST JOHN rocket in January of 1951 when the Department of Military Application authorized informal discussion between Sandia and the Douglas Aircraft Company. The HONEST JOHN, a free-flight, solid-fuel rocket with a maximum range of about fifteen miles, was described as "wingless." It had a pointed nose and a four-finned tail with several spin rockets located on the periphery of the missile for flight stabilization.

Within eight months, the AEC notified Sandia that development work on the HONEST JOHN warhead had to be accelerated to the status of a crash program. Under Project Leader Ray Schultz, Sandia was to pursue the possibility of adapting the Mk 7 as the warhead. The Laboratory undertook a feasibility and test program in cooperation with Los Alamos to determine if the missile could survive certain environmental stresses. Because of the emergency nature of the program, the HONEST JOHN, therefore, would have an extended two-year development phase involving multiple tests. In particular, the proposal for the HONEST JOHN missile to carry an implosion-type warhead raised serious doubts as to whether it would stand the high accelerations during launching.[44]

In the spring, Hickman visited the Allegheny Ballistic Laboratory in Cumberland, Maryland, where the HONEST JOHN rocket was being developed. While there, he proposed that an overtest vehicle be made by attaching booster rockets to the HONEST JOHN. By this means, increases in acceleration could be made according to the number of rocket motors used. The Ballistic Laboratory had a rocket named Deacon that could be used for this purpose. It was determined that as many as ten of these could be mounted on the HONEST JOHN. Hickman proposed the name "FATHER JOHN" for the combination of an HONEST JOHN with Deacon rockets, referring to the attached boosters as "children."[45]

A MISSILE REVOLUTION IN THE MAKING

Army personnel remove thermal blankets from an HONEST JOHN in preparation for firing.

HONEST JOHN/Mk 7 warhead on weapon carrier in the field.

Page 1 of 2

HONEST JOHN LAUNCHING SCHEDULE
Rounds 40 and 41

Mountain Standard Time	Operations	Group
0500 Hrs	1. Round 40 ready to leave N-77.	DAC
0515 Hrs	1. Round 40 loaded on West Launcher; secured against stops; sway braces in place. Remove hoist beam. Work stands in place. Elevate West Launcher. 2. Round 41 ready to leave N-77.	DAC
0530 Hrs	Telemetry checks start on Round 40.	SC
0545 Hrs	Round 41 loaded on East Launcher; secured against stops; sway braces in place; hoist beam and nose cone removed. Work stands in place.	DAC
0600 Hrs	Telemetry checks start on Round 41.	NBS
0600 Hrs to 0700 Hrs	Breakfast.	
0700 Hrs	Nose cone replaced, Round 41.	NBS
0715 Hrs	Start final checks, Round 41.	DAC
0815 Hrs	Telemetry on; field strength check.	SC
0830 Hrs	Start final check, Round 40.	DAC
0845 Hrs	Check motor squib lines, both rockets.	DAC
X–60 Mins (0900 Hrs)	1. Install motor igniter and flight head cap. 2. Close and secure access doors on both rockets.	DAC
X–30 Mins (0930 Hrs)	Close roadblocks, Highway #70.	Range Safety
X–25 Mins (0935 Hrs)	Connect motor squib lines.	DAC
X–20 Mins (0940 Hrs)	Clear area and have guard patrol area.	PO
X–15 Mins (0945 Hrs)	1. Red smoke flare. 2. All personnel in area enter blockhouse or firing control room, except as authorized by Project Officer. 3. Guard lock gate leading into launching area.	PO

Source: Andrew Lieber, Sandia National Laboratories

HONEST JOHN LAUNCHING SCHEDULE
Rounds 40 and 41

Mountain Standard Time	Operations	Group
X−10 Mins (0950 Hrs)	Telemetry on.	SC
X−5 Mins (0955 Hrs)	Ready signal all stations.	PO
X−2 Mins (0958 Hrs)	1. 2 Min X−time announced all stations. 2. Red flare from blockhouse.	PO
X−45 Secs	Ground power plug release switch thrown.	SC
X−20 Secs	Countdown (to continue to X−plus 45 secs). Area in front of firing panel clock will be kept clear.	PO
X−15 Secs	Bowen-Knapp cameras on.	FDL
X−2 Secs	Fastax cameras on.	DAC
0 Secs (1000 Hrs)	Fire Round 40.	DAC
X+5 Secs	Fastax cameras off.	DAC
X−2 Mins	1. 2 minutes X−time announced all stations. 2. Red flare from blockhouse.	PO
X−20 Secs	Count down (to continue to X−plus 45 secs). Area in front of firing panel clock will be kept clear.	PO
X−15 Secs	Bowen-Knapp cameras on.	FDL
X−2 Secs	Fastax cameras on.	DAC
0 Secs (1003½ Hrs)	Fire Round 41.	DAC
X+5 Secs	Fastax cameras off.	DAC

HONEST JOHN east and west launchers.

A MISSILE REVOLUTION IN THE MAKING

FATHER JOHN/Mk 7 Warhead.

Essentially, the FATHER JOHN was simply a test version of the HONEST JOHN. Andrew Lieber recalled that one of his first design jobs was the ignition circuit for the boosters. Maximum acceleration could be achieved only if the boosters *and* the main rocket burned out simultaneously. The nominal burn time for the HONEST JOHN was about four seconds and, for the boosters, about one and one-half seconds; however, these varied with temperature. Using existing components, Lieber devised a scheme that was self-compensating. The idea earned him the dubious privilege of performing the last-minute checks and helping the safety engineer connect the igniters. On one memorable occasion, after connecting the first igniter, the engineer yelled: "The circuit's closed!" The warning came while Lieber was standing about a foot from the nozzles of all those rockets. After a millisecond lifetime, Lieber realized what had happened and with great relief explained to the engineer: "Of course. Your meter is reading the igniter we just connected!"[46]

Many of the firings of the HONEST JOHN took place at White Sands Proving Ground. One day in late 1952, the newly hired Sandia engineer assigned to the project learned that he had to be at Holloman Air Force Base near Alamogordo in the morning, then at White Sands in the afternoon for tests on the HONEST JOHN. After speeding down Highway 70, the young engineer, who still had Indiana plates on his car, ran into a roadblock set up by the military police (MPs). The Sandian found himself about twenty cars back in the roadblock behind a group of tourists and local ranchers. Recognizing that he was in a Catch-22 situation, he had to go to some extent to convince the MPs that the firing would never occur until he reached the launch site.[47]

Environmental Test Facilities Expanded to Meet Missile Needs

The HONEST JOHN warhead was also tested on Sandia premises as well as at White Sands. Project Leader Ray Schultz and group members staged a series of acceleration tests at the

Coyote Test area and on two towers erected in Tech Area I. According to the test procedure, a warhead would be pendulum-suspended from an overhead frame. A bazooka rocket attached to the unit would be fired, producing a momentary but high-g impulse. The objective was to determine if the Mk 7 warhead would hold together when hit with accelerations produced by the rocket.

William A. Little, who was assistant Project Leader on the HONEST JOHN pendulum tests, recalled that group members built a tower near Coyote Canyon, hung the pendulum, and secured the bazookas. Then Security came out and told them that they would have to erect a wooden fence to prevent people from observing the test from the hills. The first test proved the pendulum to be too heavy; therefore, modifications had to be incorporated to make it much lighter and longer. In the process, they forgot all about the fence. However, when two of the Laboratory's directors came out to view the test, they were soon reminded. The rocket thrust zapped the warhead upward according to plan, and then down—right on top of the security fence. "I never did get over trying to explain that one," Little noted. But the test was successful. In fact, the test force, estimated to be about 2 g's, turned out to be closer to 60 g's due to the fact that the pendulum structure was momentarily distorted when the Bazooka fired, then made an elastic return, whiplashing the test unit. The warhead withstood the severe overtest without damage.[48]

The need for tests such as these led to an expansion of Sandia's environmental test facilities. Rocket-propelled sled tests were being conducted at Holloman Air Force Base, New Mexico, but the high costs involved prompted Hickman to propose a rocket-powered centrifuge for Sandia Corporation. The proposal was accepted, and in 1952 a centrifuge was quickly constructed in what was to be known as Tech Area III.

Construction of the centrifuge, built from available materials, took 120 days from conception to first test. The warhead was mounted on the end of a fifty-foot boom, which was accelerated by rockets. After a number of rockets were tried, with limited success, the Thiokol T40 was finally selected as the propulsion device.

Data were transmitted through three forty-pin umbilical kickout connectors that came out the top of the centrifuge arm at the point of rotation. Cables from the connectors were supported overhead by a steel cable suspended between two telephone poles. The centrifuge was rotated backward ten to fifteen revolutions, putting a pretwist in the instrumentation cables. When the rockets fired, the cables would untwist and then turn in the other direction. After ten to fifteen revolutions, the test would be completed and the instrumentation cables would be blown free of the centrifuge. The rocket-powered centrifuge was used to test various rockets and missile components, including those for the HONEST JOHN, the CORPORAL, and the ALIAS/BETTY depth bomb. A sliding arrangement was devised that allowed the warhead to move outward on the arm, striking a stop that generated the effect of a water entry spike. In 1954, this original centrifuge was replaced by one run hydraulically.

Centrifuge under construction in Area III in the Fifties.

Area III Test Facilities.

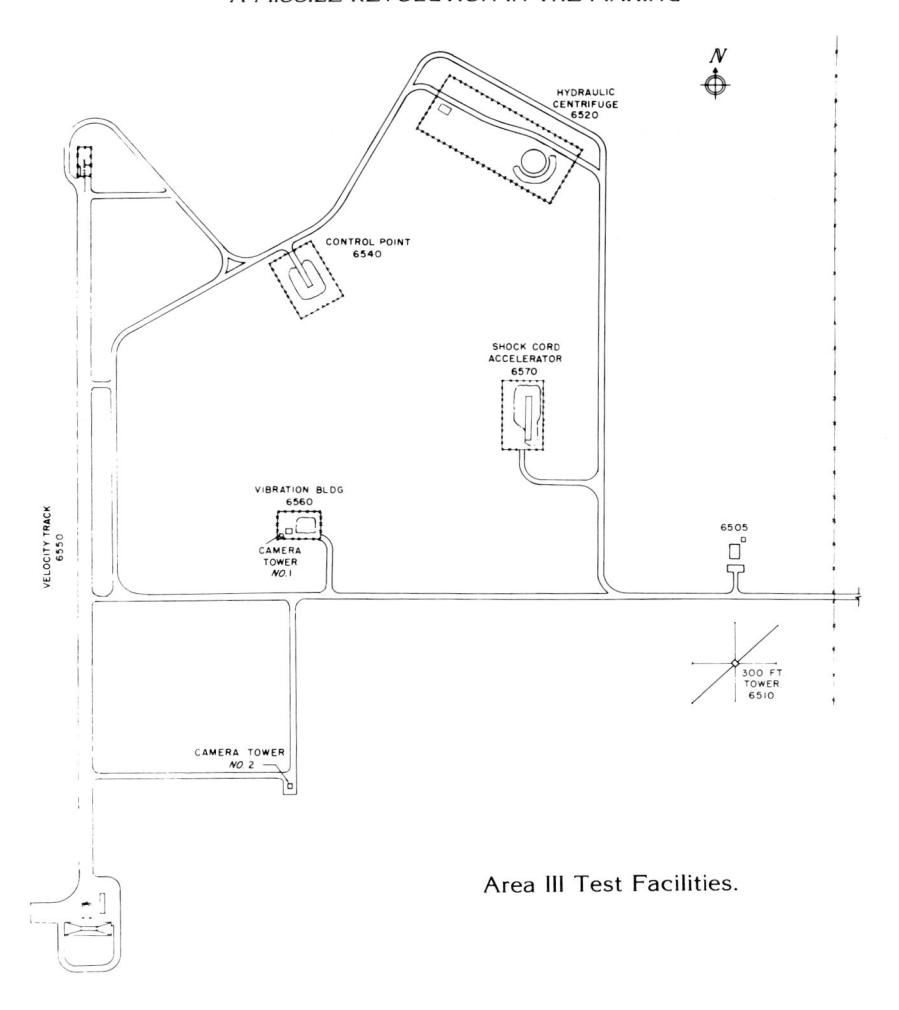

The need for other types of environmental tests encouraged the Laboratory to provide a specialized Area III for hazardous testing operations some seven miles south of the main Tech Area I. In this area was a somewhat rudimentary vibration facility, essentially a quonset hut covered with dirt. Paul Adams, who spent the majority of his career in environmental testing, recalled that the conditions of this facility lost the group a number of good secretaries. "It was an interesting place," Adams said. "The girls would come out and open the door, look in . . .

turn around and walk off. They weren't about to work in a building without any windows and dirt all over."[49] In addition to the pendulum, centrifuge, and vibration facility, the Laboratory needed a sled track to test impact crystals. The track, equipped with high-velocity sled, tested missile nose cones and the effects of impact upon different kinds of targets.

Meanwhile, there was still a question as to the HONEST JOHN warhead's ability to operate when subjected to the spin that stabilized the rocket in flight. It was felt necessary to overtest the warhead by firing it in Big Stoop rockets, which were essentially HONEST JOHNs fifty feet long, fitted with rocket motors. However, efforts to obtain these Big Stoop vehicles were not successful, and eventually Sandia fabricated its own overtest rockets using Hickman's Father John-with-Children concept. By this means, a twenty-five percent overtest was produced, which was absorbed by the warhead without damage. Successful firings of the HONEST JOHN were first held at the White Sands Missile Range during the summer of 1951. Often a "matched pair" was fired only a few minutes apart to obtain nearly identical meteorological conditions. As with artillery, it was necessary to develop so-called firing tables.[50]

In September, the Division of Military Application issued interim authorization for development of a Mk 7 warhead for free rocket application, followed by full development authorization after the spin tests were completed. Rocket marriage programs were to be conducted under the same auspices as those for guided missiles, with the Sandia Weapons Development Board functioning as coordinating agency. The Joint Chiefs of Staff authorized a production program for the XW-7 HONEST JOHN on February 1, 1952; and in late spring, the Army urged the Military Liaison Committee to expedite the availability date for the rocket because of the international situation.[51]

The Guided Missile Committee recommended that responsibility for development of fuzing and arming for the XW-7/HONEST JOHN be assigned to Sandia Corporation. The accelerated time scales made it mandatory that only readily available components be used. Early production was successfully reached on schedule in September 1953, and stockpile entry took place in November. Kenneth E. Fields, Brigadier General and Director of Military Application, wrote to Carroll Tyler, Manager, Santa Fe Operations in commendatory terms:

> It is with appreciation that I have noted the admirable performance of Sandia Corporation in meeting accelerated development project dates for the HONEST JOHN and CORPORAL. Once again Sandia Corporation has come through in splendid fashion and I would be grateful to you if you relate to them in your office my sincere thanks for a job well done.[52]

In June 1954, an improvement program for the HONEST JOHN was initiated, although a four-month delay was authorized not long after to allow design changes to be used with an adaption kit called the Demi-John. The Demi-John, essentially an HONEST JOHN with "elephant ear" dive brakes, was used to facilitate a steep impact angle at short ranges. However, developmental flights revealed that the brakes caused incredible vibration; subsequently, the concept was cancelled.

All responsibility for the HONEST JOHN was transferred to the Army in January 1955. The Army then contracted for Sandia to develop the fuze as well as the warhead for the improved version.[53]

XW-7/ALIAS/BETTY

The Mk 7 warhead was also applied to the ALIAS/BETTY depth bomb. The underwater test in Operation Crossroads demonstrated in 1946 the lethality of this type of nuclear detonation, but it was not until the advent of the small Mk 7 warhead that serious attention was paid to the possibility of designing a high-yield depth bomb. The Joint Chiefs of Staff wanted to adapt such warheads for subsurface use, with particular reference to tactical use in antisubmarine operations. The immediate objective was design of a parachute-retarded, deep submergence weapon using the Mk 7 warhead, but the ultimate objective was development of an impact-resistant, implosion-type depth bomb that could be dropped without parachute and used to mine shallow coastal waters and harbors.[54]

NUCLEAR ORDNANCE ENGINEER FOR THE NATION

START SWITCH (AT RELEASE)
1. Starts safe sep. time
2. Starts inverter

IFM

IFI

FAIRING RELEASED

PARACHUTE DEPLOYS

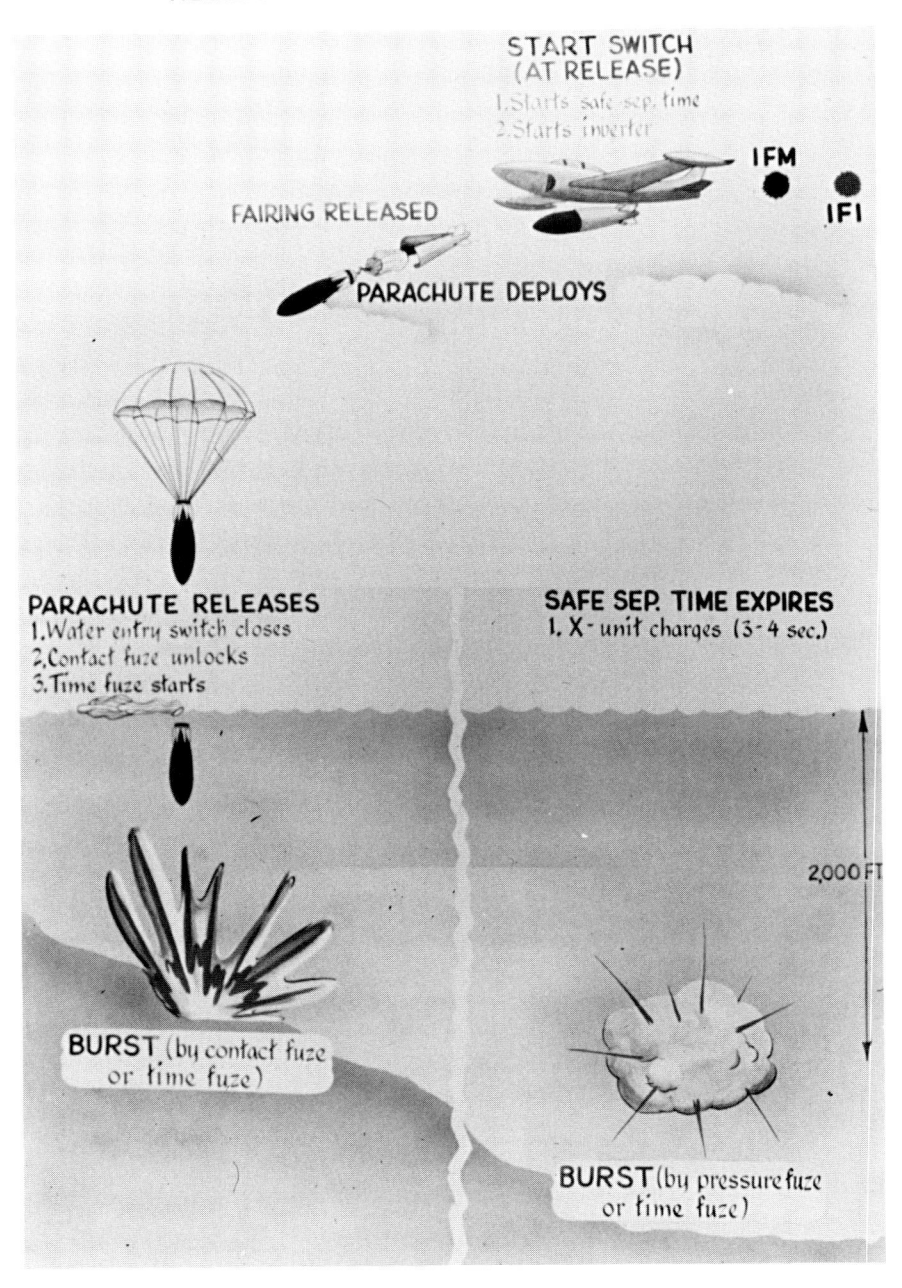

PARACHUTE RELEASES
1. Water entry switch closes
2. Contact fuze unlocks
3. Time fuze starts

SAFE SEP. TIME EXPIRES
1. X-unit charges (3-4 sec.)

2,000 FT

BURST (by contact fuze or time fuze)

BURST (by pressure fuze or time fuze)

ALIAS/BETTY Trajectory.

A MISSILE REVOLUTION IN THE MAKING

A secret letter from Admiral Kraker to heads of the Laboratories in April 1952 initiated Sandia's participation in the project. The Ad Hoc Committee for the BETTY, with representatives from the Naval Ordnance Laboratory, Los Alamos, and Sandia, held its first meeting June 11, 1952. Sandia's major task in the program was to determine whether the CORPORAL warhead could meet all the ALIAS/BETTY environmental requirements.[55]

Consequently, Sandia's pendulum device again came into use. When the Navy Ordnance Lab requested that the warhead be subjected to acceleration that would simulate water entry, Hickman outlined a plan for making the tests on the pendulum. The proposal was approved and a contract given to Douglas Aircraft Company to construct the rocket carriage and that part of the pendulum needed for mounting the warhead. The Laboratory released its portion of design responsibility on the warhead and bomb-to-warhead conversion components in July 1954. First production occurred in June 1955.[56]

BOAR

Another Mk 7 warhead application program was the BOAR, an air-to-surface unguided missile that had transonic speed and a range between five and seven miles. Douglas Aircraft, the rocket's contractor, requested Sandia's assistance on June 29, 1951, but little action was taken until July 1952 when the BOAR Ad Hoc Committee first met. The Committee recommended that the Naval Ordnance Test Station assume primary development responsibility. NOTS would design the fuze using a Sandia timer and baroswitch, but Sandia would be responsible for the power supply. Douglas Aircraft, under subcontract to NOTS, provided mechanical components. After design release in November 1954, drop tests of the BOAR were conducted at the Naval Ordnance Test Station. At Sandia, the BOAR successfully passed a bevy of environmental tests, including resistance of the weapon to sand, dust, salt spray, rain, fungus, dynamic ejection, vibration, and flyaround. In July 1955, the Air Force discontinued use of the BOAR design.[57]

RASCAL

The Laboratory also became involved in warhead development for the U.S. Air Force's RASCAL, an air-to-surface rocket-powered, supersonic missile with a ninety-mile range. The RASCAL, whose name was derived from the title <u>Ra</u>dar <u>Sc</u>anning <u>L</u>ink, was developed by Bell Aircraft Company. After launch from the carrying bomber at an altitude between 30,000 to 40,000 feet, the RASCAL was intended to climb to 60,000 feet and cruise at Mach 2.5, with midcourse guidance provided by the launch aircraft. An inertial system and guidance radar directed terminal guidance during dive to target.

The Ad Hoc Working Group for the RASCAL met on November 7, 1952. J. W. Jones served as project leader for the Laboratory's participation in the program. At this first meeting, Sandia was designated to design and provide the initial fuze. However, disagreement developed between Sandia representatives who favored a radar system, and military participants, who preferred a nonradiating fuze. Although Sandia agreed to develop both types of systems, the military ultimately proposed that the entire arming, fuzing, and firing task be assigned to the Air Force.[58]

The Air Force received responsibility as requested, but fuze design moved so slowly that it became apparent that the design would not be ready by operational date for the missile. Meanwhile, Sandia had developed a baro-contact fuze appropriate for the RASCAL. Ironically, the Air Force, although they felt the fuze was more complex and expensive than necessary, authorized Bell Aircraft to procure the Sandia-designed fuze and assemble it for use in the RASCAL.

In addition to problems of a political nature, the RASCAL missile experienced flight schedule postponements and missile malfunctions. Finally, in August 1955, the Air Force Special Weapons Center notified Sandia that plans were being made to provide a thermonuclear warhead capability for the missile; the RASCAL Operational Squadron had been cancelled. On November 15, 1955, the Mk 5 warhead program for the RASCAL was officially terminated.[59]

NIKE HERCULES

Sandia also contributed to adaptation of a warhead for the NIKE, a ground-to-air-guided supersonic rocket with a range of fifty nautical miles. The NIKE B, later called the NIKE HERCULES, was a missile designed for defense against formations of aircraft, tanks, and enemy manpower. In May 1953, the NIKE B Project was established; and in November, Picatinny Arsenal wrote to the Ordnance Liaison officer at Sandia Base requesting that he negotiate with Sandia Corporation for assistance in adaption kit responsibilities.

Representatives of Douglas Aircraft, Western Electric as a NIKE system contractor, and Army Ordnance contractors met at Sandia to prepare a feasibility study recommending that the NIKE missile be provided with an atomic warhead. Preliminary studies considered the Mk 5, Mk 7, and Mk 12 warheads, but in April 1953, the Division of Military Application reported that the Joint Chiefs of Staff had approved selection of the Mk 7.

The AEC was to adapt the warhead to the NIKE system; the Army would provide fuzing, launcher, and fire-control equipment. The universal Mk 7 warhead was to be used, together with components produced for the CORPORAL and HONEST JOHN; however, neither Sandia nor Los Alamos representatives felt that the Mk 7 was entirely suitable for application to the NIKE HERCULES. Both Labs pointed out that the requirement for longer periods of nuclear readiness and stockpile storage made the use of the newly "optimized warhead," designated the Mk 31, a better choice. The Division of Military Application accepted the recommendations, but decided to continue the Mk 7 warhead as well to assure availability compatible with the missile production timescale. Components for the NIKE were released in June 1957.[60]

Reorganization

During the early years of the missile revolution, the multitude of warhead programs undergoing simultaneous development at Sandia resulted in a competition for support from the electrical and mechanical divisions. Smaller programs had to vie with major programs such as the HONEST JOHN. To help alleviate this problem, Leon Smith in March 1956 proposed that an electrical department be set up to coordinate electrical activities across all systems.

Organizationally, the timing for Smith's proposal was fortuitous. The Laboratory had just undergone a major reorganization, including promotion of two of its directors to vice-president positions and consolidation of the warhead and bomb engineering groups under Bob Henderson, who was named Director of Systems Development. Lou Hopkins was made Director of Components.[61]

Smith's proposal to eliminate the bifurcated system for electrical design by combining that function under one electrical department serving both bombs and warheads was significant in establishing a more effective mode of operation. Not only was Smith's proposal positively received; he was assigned to carry it out. As the group's new Department Manager, Smith had reporting to him supervisors Don R. Cotter, Milton E. Bailey, George "Howie" Mauldin, and Rudy O. Frantik.[62]

In the interim, a series of national and international events foretold of the changes to come in the U.S. missile development program. The years 1952–1953, for example, marked the beginning of the thermonuclear era and the end of the war in Korea. Dwight D. Eisenhower, the general with the engaging grin, was elected President of the United States. And an alarming number of intelligence reports revealed that the U.S.S.R. had in operation an important program for the development of intercontinental ballistic missiles and rockets—information that motivated the U.S. to launch a number of crash programs to produce long-range nuclear-tipped strategic missiles.[63]

Meanwhile, as a result of the Korean threat, von Braun's Peenemünde group in Huntsville, Alabama, had been given the go-ahead to proceed with development of a long-range ballistic missile for use in tactical warfare. Slated to be used by combat troops in the field, the large liquid-propelled rocket was called the REDSTONE after the arsenal where it was built. At Sandia, support for the REDSTONE was provided in Jack Howard's engineering department under Project Leader G. F. Heckman. The REDSTONE, although ineffective as a weapon, nevertheless provided a launching pad for related missile technology including the ATLAS.

The mid-fifties ushered in the next stage of the U.S. missile revolution, directed largely by the powerful von Neumann Committee. Under the leadership of the brilliant John von Neumann, this group, which also included such well-known figures as George Kistiakowsky, Jerome Wiesner, Herb York, Darol Froman, Charles A. Lindbergh, Simon Ramo, and Dean Wooldridge, determined that the United States should place "the highest national priority" on a program to build intercontinental and intermediate-range ballistic missiles (ICBMs and IRBMs).

To implement this decision, the Air Force set up the Western Development Division of the Air Research and Development Command under Brigadier General Bernard A. Schriever. Schriever, a pioneer in systems development, coordinated the personnel and facilities to accomplish the program. Ramo and Wooldridge, in the meantime, had set up their own systems engineering corporation. The Ramo-Wooldridge group, which later merged with a hardware concern called Thompson Products, was asked by the von Neumann Committee to direct their engineering and scientific expertise toward development of the ATLAS ICBM.[64] Sandia would become involved through intra-lab committees, such as that composed of Mike May, Harold Brown, Darol Froman, Jane Hall, Bob Henderson, and Leon Smith. This group was set up to work with the Ramo-Wooldridge group to review development going on at General Electric and AVCO. Through this association, Sandia would be asked to submit a fuzing proposal for the ATLAS.[65]

As it became apparent that thermonuclear devices could be packaged in smaller form, their application to warheads for both tactical and strategic use were investigated. Soon the TITAN, THOR, MINUTEMAN, POLARIS, and JUPITER would join the ATLAS as part of the Lab's programmatic focus.[66] While the technical side of the Laboratory attempted to keep in step with the missile revolution, Sandia management became directly involved in working out the parameters for interaction between the laboratory and government agencies.

Donald A. Quarles, second President of Sandia Corporation.

Chapter XIX.

PARAMETERS FOR INTERACTION:
DOD-AEC-Sandia

On a wintry day in January 1952, supervisors from Sandia Laboratories traveled to Santa Fe for a management retreat. They rendezvoused at the Rotary Club headquarters, where President George Landry presented the opening talk. It soon became apparent that Landry would use the opportunity to explain the influx of outside management personnel into the Sandia system. He reassured his supervisors that although it was "policy" to place Bell System employees in certain positions, the objective was gradually to develop management personnel of Sandia Corporation from Sandia sources. However, as he explained:

> The national emergency that we are in cannot wait for the development of people if there is any possibility at all of making up present deficiencies in experience and maturity by drawing from other organizations. And so for a time, because of the expansion in force, we will increase our numbers of Bell System people in key positions.[1]

By 1952, this was indeed the trend.

Mervin J. Kelly, Executive Vice-President of Bell Telephone Laboratories (BTL), had just completed another survey of the organization and performance of Sandia Corporation. As a result, he recommended to the AEC on January 11, 1952, that "Bell Laboratories . . . accept the whole research and development responsibility." He advised that Walter A. MacNair, who had been serving as a Laboratory consultant for BTL, "go to Sandia and assume responsibility for the research half of Mr. Poole's job."

"If sore spots developed," he added, "the Bell System was prepared to move still more people out to Sandia."[2]

Kelly's plans were soon implemented, and by the time the Weapons Development group had been divided into responsibilities of the bombs and warheads, a parallel organization called Systems Research was set up under MacNair's direction and elevated to a vice-presidency. Formerly Director of Military Systems Engineering at Bell, MacNair had a professional background that involved experience in various fields of physical research, particularly those relating to problems in acoustics, switching, and military equipment. During the war years, he devoted much of his efforts to military and civilian government agencies and contributed particularly to the study of fire control problems and guided missiles.

As Vice-President of Systems Research at Sandia, MacNair had reporting to him Directors Glenn Fowler, Field Test; Frederick J. Given, Apparatus Engineering; L. G. Abraham, Electronics; and Stuart C. Hight, Research. As promised, the Bell System responded to expanding R&D demands by transferring in additional personnel to managerial positions. Hight, Givens, and Abraham, for example, were all BTL employees; and in May 1952, a Vice-President and General Manager post was established to supervise all nonresearch and development functions. The first to occupy this position was Timothy E. Shea.[3]

Such organizational changes were indicative of the changing complexion of the Laboratory. As the production role decreased, so did the Western Electric influence. Conversely, the increase in research, development, and testing activities meant that the role of BTL at Sandia would expand. "The bifurcation of the Labs," according to Henderson, "had created real tensions and was recognized as more and more of a problem." Mervin Kelly's initial recommendation that Sandia should be established "as a production-type organization tuned to systems work with a great ability in scientific fundamentals, had been translated too literally," Henderson added.[4]

By 1952, dominance of the Western Electric production-type ethos was considered to be detrimental to the future of the Labs. Furthermore, the expanded mission of the Labs required

PARAMETERS FOR INTERACTION: DOD-AEC-SANDIA

**SANDIA CORPORATION
1952**

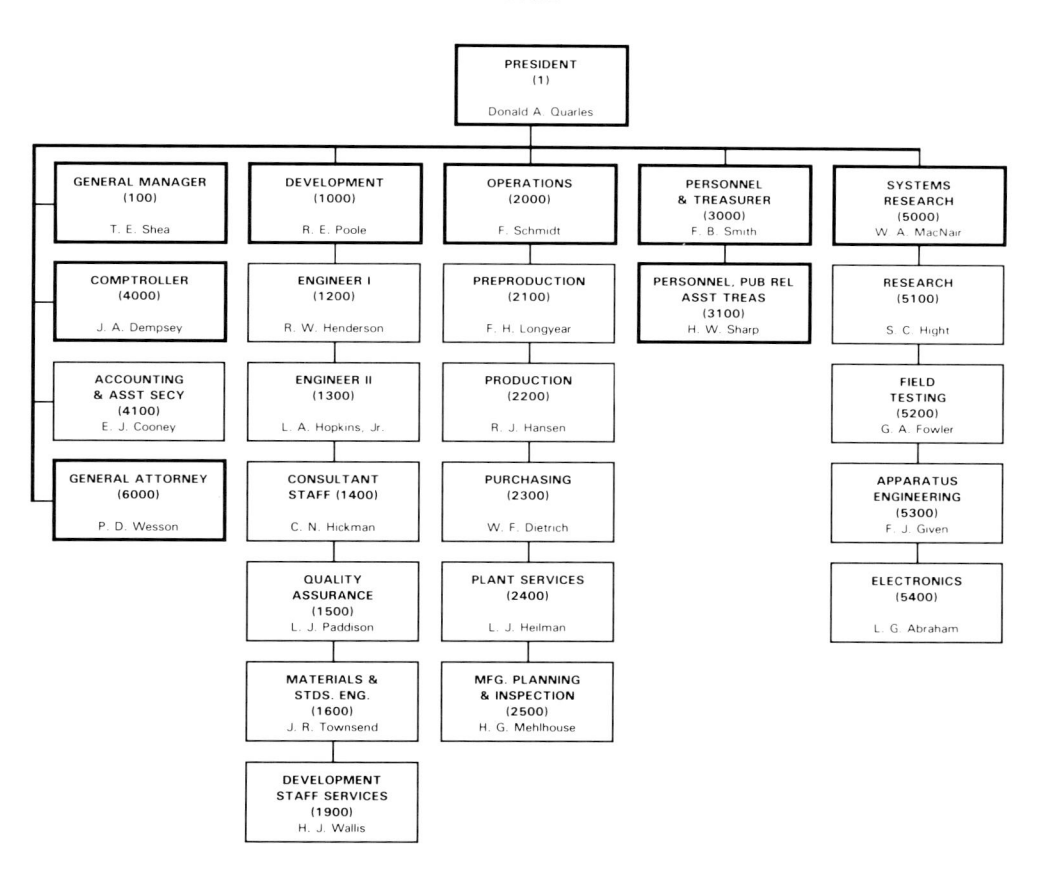

- **PRESIDENT (1)** — Donald A. Quarles
  - **GENERAL MANAGER (100)** — T. E. Shea
    - **COMPTROLLER (4000)** — J. A. Dempsey
    - **ACCOUNTING & ASST SECY (4100)** — E. J. Cooney
    - **GENERAL ATTORNEY (6000)** — P. D. Wesson
  - **DEVELOPMENT (1000)** — R. E. Poole
    - **ENGINEER I (1200)** — R. W. Henderson
    - **ENGINEER II (1300)** — L. A. Hopkins, Jr.
    - **CONSULTANT STAFF (1400)** — C. N. Hickman
    - **QUALITY ASSURANCE (1500)** — L. J. Paddison
    - **MATERIALS & STDS. ENG. (1600)** — J. R. Townsend
    - **DEVELOPMENT STAFF SERVICES (1900)** — H. J. Wallis
  - **OPERATIONS (2000)** — F. Schmidt
    - **PREPRODUCTION (2100)** — F. H. Longyear
    - **PRODUCTION (2200)** — R. J. Hansen
    - **PURCHASING (2300)** — W. F. Dietrich
    - **PLANT SERVICES (2400)** — L. J. Heilman
    - **MFG. PLANNING & INSPECTION (2500)** — H. G. Mehlhouse
  - **PERSONNEL & TREASURER (3000)** — F. B. Smith
    - **PERSONNEL, PUB REL ASST TREAS (3100)** — H. W. Sharp
  - **SYSTEMS RESEARCH (5000)** — W. A. MacNair
    - **RESEARCH (5100)** — S. C. Hight
    - **FIELD TESTING (5200)** — G. A. Fowler
    - **APPARATUS ENGINEERING (5300)** — F. J. Given
    - **ELECTRONICS (5400)** — L. G. Abraham

Source: *Sandia Corporation Telephone Directory*, June 1952

research-oriented leadership. On March 1, 1952, Landry, who epitomized the Western Electric manufacturing leadership, returned east to take charge of Western's Purchasing and Traffic activities. AT&T sent as his successor Bell Labs Vice-President Donald A. Quarles.[5] Whereas Landry's tenure had been characterized by a restructuring of the Laboratory, implementing operating procedures patterned after the parent company, Western Electric, and by making initial attempts to iron out interaction problems with the military, his successor Donald Quarles would devote an increasing portion of his time to negotiating Sandia's position at the national level.

Donald A. Quarles, Second President, Sandia Corporation

Small of stature, quiet, and unassuming, Donald A. Quarles was described as a remarkable man. Recognized as an outstanding authority in the electronics field, Quarles had joined the Bell System in 1919 where he advanced to such positions as Director of Outside Plant Development, Director of Transmission Development, and Director of Apparatus Development.[6] As Western Electric's Director of Transmission Development from 1940 to 1941, he was in charge of carrier telephone systems, broadband telephone and television systems, and improved voice frequency transmission systems. He became a Vice-President of Bell Laboratories in 1947. When he came to Sandia five years later, he was serving as Chairman of the Committee on Electronics for the Research and Development Board of the Department of Defense and had just been elected President of the American Institute of Electrical and Electronic Engineers (AIEEE).[7]

The engaging, person-to-person style of Don Quarles quickly made a positive impression on the Laboratory. It was not unusual to see the new president and his wife Nona joining in activities at the Coronado Club and interacting with employees on a social as well as professional level. Before long, according to one executive, "one could sense a change of attitude at the Labs. Like a fresh breeze, the feeling of camaraderie spread through the Laboratory."

A typical work day for the Sandia President began at 6:00 in the morning and ended after 6:30 in the evening. His executive secretary, Rosalie Franey Crawford (who would serve in that capacity for seven other Sandia presidents), recalled that he would dictate four hours "every day of the world," with little changing or editing of text. According to Crawford: "After dictating some twenty-four pages of notes, Quarles would say, 'Back there, about the third paragraph where I said so and so,' and he would repeat word for word what he had said, adding: 'I would like to insert such and such.' "[8] Quarles' astute intellect and writing ability stood him in good stead as he tried to resolve the standardization and fuzing issues.

In the technical realm, advanced designs for capacitors, spark gaps, electronic tubes, and other devices were beginning to flow from the design boards of Sandia engineers. In time these items became smaller, more efficient, more reliable, and more tolerant of severe environmental conditions. The results were weapon systems that required less and less maintenance. In the interim, however, there were numerous problems to be resolved relating to the Laboratory's technical and administrative interface with the DOD and the AEC.

Among the issues demanding the new President's attention were two that would greatly influence Sandia's method of operation and the Lab's division of weapons responsibilities with the military services. From the beginning, the matter of how the DOD and the AEC would partition responsibilities for the development, production, and standardization of nuclear weapons was contentious. It became obvious that an agreement was essential to orderly progress of the national security program. Controversy in this area was exacerbated by the major DOD development program for missiles and rockets to deliver nuclear warheads to the target, especially in the area of responsibility for the fuzing of the warhead. When Quarles joined Sandia, negotiations on these matters were in full swing.

Sandia President Donald A. Quarles visits with Executive Secretary Rosalie Crawford (seated at left). Seated at right is Mavis Randle, Secretary to R. E. Poole. Executive Secretary Virginia Potter, Robert Henderson's Secretary Beulah Amole, and Secretary Helen Russo stand behind Quarles (left to right).

PARAMETERS FOR INTERACTION: DOD-AEC-SANDIA

The Development and Production Controversy

During 1952, Quarles found himself spending more and more time negotiating at the Washington level, attempting to establish the parameters of Sandia's technical involvement in relation to the DOD. The need to arbitrate such issues was due partly to the fact that both the AEC and the DOD had been in existence a relatively brief time. During these early years, working arrangements between the two had evolved somewhat informally. Sandia derived its authorization to conduct research and development from powers granted the Atomic Energy Commission by Congress through the Atomic Energy Act of 1946. The Act authorized the AEC a broad charter "to conduct experiments and do research and development work in the military application of atomic energy and engage in the production of atomic bomb parts or other military weapons utilizing fissionable material."[9]

The President, according to the Act, had the authority to direct the Commission to turn over weapons or fissionable materials to the military whenever he deemed it necessary in the interest of national defense. As a result of the Korean crisis, the President used this authority to release nine complete atomic bombs to the Air Force in 1952 and would do so again in 1956, when he completed transfer of custody to the military.[10] In the ensuing years, however, interface problems such as standardization of weapons and missiles had to be negotiated.

A part of this particular problem derived from the fact that the military employed no clearly defined system to acquaint the AEC with requirements of the Department of Defense. The lack of an orderly system for channeling of pertinent military information frequently resulted in misunderstandings.

Recurrent difficulties revolved around procedures, or the lack thereof, used to develop the technical details of nuclear weapons. While the Military Liaison Committee (MLC) recognized the statutory responsibilities of the AEC for development, members of the Committee regarded the relationship between the Commission and the Department of Defense as that of "contractor and buyer, respectively." The MLC, therefore, strongly encouraged the establishment of technical specifications for weapons with the Special Weapons Development Board serving as the medium for discussion. They envisioned that specifications would then be forwarded to the Chief, Armed Forces Special Weapons Project, who would obtain concurrence of the using services. The original proposal, titled "Procedure for Providing Military Guidance to the AEC in the Atomic Weapons Field" initiated a two-year negotiating process directly involving Sandia management and eventually leading to agreement between the AEC and DOD for the development, production, and standardization of atomic weapons.[11]

On March 25, 1952, the secretaries of the three Services and the Atomic Energy Commission met in Washington, D.C., to discuss the third revision of the proposed draft of the document. AEC Chairman Gordon Dean reiterated the AEC's responsibilities in the weapons field, namely, that the AEC was charged with keeping the U.S. ahead of other nations in the development of military uses of atomic energy and, along with the DOD, for assuming responsibility for complete weapons and weapon systems that would meet the needs of the services. Development and production, he stressed, should occur as rapidly as possible. To achieve those ends, Dean recommended that the AEC and DOD "cooperate fully," with the objective of utilizing their combined facilities and talent most effectively. In this vein, Dean noted that the AEC intended to continue "vigorous efforts at LASL, Sandia Corporation, and elsewhere to insure the greatest progress in nuclear weapons development."[12]

Subsequently, the MLC appointed General Herbert B. "Doc" Loper to represent the DOD, and the AEC selected Brigadier General Kenneth E. Fields as its representative in the negotiations. Assisted by the staffs of the Armed Forces Special Weapons Project and DMA, Loper and Fields worked out a proposal that was then sent to Sandia Corporation for comment. Included in the proposal were provisions for "Standardization," defined as "a formalized acceptance by DOD of a new type of nuclear weapon produced by the AEC as being suitable for military use without limitations—an action traditionally accomplished by DOD for conventional weapons."[13]

Had the draft been accepted as written, the AEC's role in the nuclear weapons program would have been drastically diminished to include only the "nuclear system" of the warhead or bomb. To prevent this occurrence, Quarles offered a complete rewrite of that particular section, substituting the words "stockpiling of atomic weapons" for "nuclear systems" and indicating that the AEC and DOD would share the responsibility jointly. The Sandia president also clarified the DOD's role in determining whether or not the weapon met requirements by specifying that the DOD would determine "military characteristics, suitability, and acceptability (standardization)." His careful dissection of the document, known unofficially as "The 1953 AEC-DOD Agreement," reveals his awareness of its long-term ramifications.[14]

On January 16, 1953, Gordon Dean wrote to Robert LeBaron, Chairman of the MLC, informing him that the AEC had reviewed the proposed agreement, along with the changes submitted by the Military Liaison Committee, and had accepted all of them. The AEC added a few changes of its own, incorporating especially Quarles' stipulation of joint responsibility. Ironically, in the process of working out the standardization and fuzing issues, the AEC self-limited its broad charter as originally granted under the Atomic Energy Act.[15]

The Fuzing Controversy

Historically, origins of the controversy on warhead fuzing went back to the late forties. In January 1950, the Ad Hoc Committee of Bradbury, Hovde, and Hull had assigned Sandia Corporation the principal responsibility for marrying nuclear warheads to guided missiles, but the issue was far from settled. The ensuing controversy would span a three-year period and involve a national-level decision-making process.

At conferences held in the Washington, D.C., area the following May, representatives of the U.S. Army Ordnance Corps, the AEC's Division of Military Application, and Santa Fe Operations met to discuss the guided missile programs. The fuzing issue quickly surfaced as a topic of concern. The AEC reiterated its position in the context of the U.S. Army's HERMES missile program, which employed the XW-8 warhead derived from the Little Boy of World War II. AEC representatives posited that the fuze should be designed by the agency building the missile since fuze characteristics would be "delineated by missile performance and environmental conditions imposed by the missile." A strong part of the military's argument stemmed from the fact that the missile's navigational system inherently contained within it the basic information needed for fuzing at the end of its flight. The Army, however, envisioned the AEC as the provider of funds, with nominal control of development, which would be carried out by the Army Ordnance Laboratory or its contractor.[16]

The next month, at its thirty-ninth meeting, the Special Weapons Development Board appointed the Guided Missiles Subcommittee and acknowledged that the question of fuzing responsibility had to be settled. The Board agreed that the missile was simply a carrier, that it replaced the aircraft carrying bombs, and that the fuze was part of warhead development. However, it was recognized that there was no easily defined line of demarcation between warhead and missile as there was between bomb and bomber.

At this session, the AEC representative maintained that each missile project should be examined on its own merits to determine where fuze development should be carried out, subject to the provision that the organization that budgets and controls expenditure of funds should also have responsibility for fuze development.[17] Clearly, the AEC did not favor the U.S. Army's proposal to have the AEC fund and the Army develop the fuze.

The management of Sandia Corporation held a similar view. When asked to comment on the subject for the benefit of the newly assigned director of the AEC's Division of Military Application, Brigadier General James McCormack, Landry pointed out that the Lab considered itself "the best qualified to undertake the fuzing responsibility as well as the overall guided missile-atomic warhead systems analysis."[18]

From the beginning of its involvement in the missile program, Sandia strongly advocated a systems analysis of the complete weapon as a prelude to any full-scale warhead development program. In this regard, President Landry had called attention to a major systems study conducted jointly by Sandia and the Bell Telephone Laboratories on the effects of atomic explosions. The study, he said, revealed that fuzing was a function of both the guidance system (over the target) and the fuzing system (height-of-burst above the target). For this reason, Landry observed, collaboration of "highest order" was required among the missile, warhead, and fuze designers. "The AEC," Landry advised, "should delegate fuzing responsibility to its contractor, Sandia Corporation."[19] Sandia, he stressed, was particularly skilled and experienced in solving fuzing and firing problems and also possessed expertise in the areas of quality assurance, surveillance, and preparation of manufacturing and production information on atomic weapons for delivery to stockpile.[20]

Despite the advice of Sandia management, DMA Director McCormack rejected any notion of expanding responsibilities in order to analyze the adequacy of missile warhead systems. McCormack maintained the position that overall guided missile-atomic warhead studies were the responsibility of the Department of Defense and that such studies should parallel warhead development. The DMA director agreed, however, that the AEC should budget for all atomic warhead fuzing development, and that fuzing responsibility should be assigned on a case-by-case basis.

The SWDB, with representation from the military, the AEC, and Sandia, was designated as part of standard operating procedure to make these evaluations until experience dictated the need for change. Sandia management agreed with McCormack's proposal, providing that the definition of fuzing include "signals for automatic inflight [nuclear] insertion, primary fuze arming, and so forth" and that the AEC contractors would be continuously informed of the status and results of the military's warhead systems evaluations.

When the Special Weapons Development Board learned of its assignment, the Board charged its Guided Missile Committee with recommending a procedure for establishing fuzing requirements for guided missiles. Following discussions between Sandia representatives and Guided Missile Military Liaison officers, the Committee agreed that the Services would submit to Sandia Corporation through normal channels pertinent information on missile characteristics and expected types of targets.[21]

Therefore, from the latter part of 1950, Sandia operated in the atomic warhead fuzing field in accordance with the proposed policy with modifications as stipulated. After considering the fuzing and arming problems in detail, the Ad Hoc Working Groups set up by the Special Weapons Development Board assigned responsibility for the first atomic warhead fuzing system—the XW-7 CORPORAL—to Sandia. Modified atomic bomb fuzes also had to be employed on the REGULUS and MATADOR, with interim fuzing responsibility again going to Sandia. Until 1952, fuzing decisions made by the Special Weapons Development Board were based primarily on the fact that modified atomic bomb fuzes were the only fuzes satisfying the technical requirements that could meet missile time scales.[22]

In the interim, the military continued in its bid to control fuzing. The DMA in the spring of 1951 requested that the Special Weapons Development Board assign responsibility for fuze development for Army guided missiles carrying warheads to the Army's Ordnance Corps, which was already developing fuzes for missiles carrying conventional warheads.

The DOD Assumes Responsibility for Fuzing

After Quarles took office in March 1952, one of his first actions was to draft a letter recapping Sandia's posture on the fuzing issue. He pointed out that fuzing reliability and fuzing accuracy could be "best controlled by the AEC." Sandia, he said, planned to remain in the forefront of air burst and contact burst fuzing research and development; therefore, the AEC should continue to hold "primary interest" in warhead fuzing to effect efficient use of fissionable materials. Fuzing information, he maintained "is not inherent in the current guided missile guidance system."[23] In essence, fuzing was of primary interest to the

AEC and to Sandia because of its direct relationship to weapon effectiveness.

Quarles also took the opportunity to chide the DOD for not having followed through on its obligation to inform the Laboratory on the status or results of systems studies, despite Sandia's cooperation and offers of assistance. "Unless an active overall guided missile atomic warhead systems analysis program is initiated and vigorously pursued in the immediate future," he warned, "the missile warhead program could suffer the identical pains now forced on the airplane bomb program."[24]

At this point, however, Quarles showed a willingness to give ground on the fuzing issue. "The overall guided missile atomic warhead program was . . . [currently] much better understood than at the time the joint AEC–DOD fuzing responsibility was originally proposed," he pointed out; consequently, he believed that it would be appropriate "to establish that the AEC should be responsible for all airburst and contact burst atomic warhead fuzing with the DOD responsible for missile guidance system fuzing."[25] Essentially, Quarles was suggesting a division of responsibility whereby the AEC would continue to budget for all atomic warhead fuzing development programs and the AEC and DOD would work "in closest harmony" to scrutinize the problems involved.[26]

By this time, however, the general situation at the working level had changed, reflecting a new policy where the U.S. Air Force, and perhaps the entire DOD's missile contractors, would become solely responsible for fuzing new programs. During the process of the Special Weapons Development Board's appraisal of the military characteristics for the XW-5 nuclear warhead being considered for the U.S. Air Force MATADOR and the U.S. Navy's REGULUS guided missiles, it became clear that items such as "warhead," and "warhead installation," needed to be defined. Terms were agreed upon, but the DOD/MLC added still another definition—"adaption kit," which referred to all items except the warhead, namely "the arming and fuzing systems, power supply and all hardware required by a particular installation." This definition, in the opinion of some, suggested a move by the DOD to facilitate assignment of such responsibilities to its own agency. Not surprisingly, in August 1953 the DOD/MLC recommended that those elements of missiles not included in the atomic warhead . . . be developed and procured by the Department of Defense." The reference was to adaption kits.[27]

The DOD position gained ground when the AEC/DMA rejected Quarles' definition of "nuclear system" and "atomic weapon," in effect supporting movement of the U.S. Air Force and the Army to wrest fuzing responsibility from the AEC. On January 22, 1953, the AEC Commissioners ended the controversy by approving a paper prepared by the AEC's DMA that divided task assignments for rockets and missiles between the AEC and the DOD. The document, referred to as "AEC Missile and Rocket Responsibilities," awarded fuzing to the DOD and arming and firing to the AEC. The terminology "rocket" was in deference to the U.S. Army's HONEST JOHN, which was not guided, but a ballistic trajectory rocket.[28]

As a result of this action, the Guided Missiles Committee of the Special Weapons Development Board was phased out and the tasks of its Ad Hoc Working Groups were taken over by new joint weapons development committees. These committees were the first of the types known as Project Officers Groups, which coordinated activities of the AEC (DOE) with its laboratories and military service agencies and contractors.[29]

Thus, despite the persistent efforts of Presidents Landry and Quarles from 1950 to 1952 to maintain Sandia's role in fuzing, the decision was made ultimately to award fuzing for guided missiles and rockets to the DOD exclusively, rather than on the merits of a case-by-case review. Contrary to the recommendation of the AEC's designated technical agent, Sandia Corporation, the AEC/DMA and the Commissioners approved the DOD proposal, essentially as written. In retrospect, the determining factors for the decision appear to have been political and administrative, rather than technical, since no demonstrated technical design capability then existed within DOD laboratories or the aircraft companies serving as U.S. Air Force contractors.

From Quarles' policy statement of March 1952, which marked a sudden change in posture, it is apparent that fuzing responsibilities were not wrested from Sandia control without the cognizance of management. It may be that Quarles realized, pragmatically, that the Laboratory's technical staff was inadequate to handle the proliferation in nuclear warhead application programs being contemplated by the military services. It has also been suggested that another factor was the influence of the military industrial complex, especially U.S. Air Force contractors, who were realizing the potential magnitude of opportunities for deeper involvement in the U.S. national security program.[30]

As agreed, Sandia completed development of arming and fuzing subsystems for the programs then underway—namely, the XW-7 CORPORAL and the XW-7 HONEST JOHN, the XW-5 REGULUS I and the XW-5 MATADOR, which entered stockpile in September 1953 and April 1954. The arming and fuzing system for the XW-7 ALIAS BETTY depth bomb, which entered stockpile in June 1955, was the first for which the DOD assumed responsibility that was not cancelled before production.[31]

Politics surrounding the fuzing issue did not abate appreciably. Shortly after the U.S. strategic ballistic missile development program was accelerated in late 1954 under the Air Force's Western Development Division, the Air Force requested that Sandia submit a proposal for a fuzing system for the ATLAS intercontinental ballistic missile. The formal request read:

> It is understood that as an operating contractor for the Atomic Energy Commission, your organization has had considerable experience in the design of fuzing systems for atomic bombs and warheads and has developed a broad understanding of nuclear weapons effects which appear essential to the solution of the fuzing problems involved in the ICBM. . . . Your organization may be in an excellent position to investigate certain phenomena associated with the fuzing of the ICBM warhead contemplated by this project.[32]

A bright young engineer and former Texas Aggie by the name of Robert L. Peurifoy, Jr., was asked to generate the arming and fuzing proposal. However, the formal proposal to explore fuzing characteristics, submitted by Sandia in December 1955, was rejected by the Air Force in favor of proposals submitted by AVCO and General Electric—proposals that, ironically, were almost identical to Sandia's. This decision established a pattern of awarding arming and fuzing subsystems for Air Force ballistic missiles to either AVCO or General Electric. Similarly, the U.S. Army's Picatinny Arsenal was awarded arming and fuzing responsibility for Army programs after the W-31/HONEST JOHN (1955). Nevertheless, Sandia interaction with the U.S. Navy eventually resulted in an approach whereby the Lab would be awarded reimbursed funding for development of arming, fuzing, and firing subsystems for fleet ballistic missile programs.[33]

Implementation of the 1953 Agreements

Development Phases

The 1953 agreements did much to establish the respective roles and parameters for interaction among the AEC, the DOD, and the nuclear weapon laboratories, although the debate over military-civilian control—which lay at the heart of both issues—would continue to surface. For the AEC and the DOD, the agreements specified the functions to be performed by each. For Sandia, as one of the AEC agencies, the agreements formalized the Laboratory's cradle-to-grave responsibility for nuclear weapons development.

The original six (later seven) phases outlined in the document trace the life cycle of a weapon from conception or birth through development and production to stockpile and retirement. In practice, new weapon programs would not develop precisely in accordance with these phases, or in the chronological order spelled out, but the agreement effectively formalized a workable process to be generally followed in the development, production, and standardization of weapons. In some cases, phases would of necessity be merged, omitted, or deferred with the understanding of both the AEC and DOD.[34]

AEC-DOD WEAPON PHASES

Phase 1: WEAPON CONCEPTION	This phase consists of continuing studies by AEC laboratories, DOD agencies, and others. A continuous exchange of information, both formal and informal, is conducted among individuals and groups. This results in the focusing of sufficient interest in an idea for a new weapon or component to warrant a program study.
Phase 2: PROGRAM STUDY (DETERMINATION OF FEASIBILITY AND RESPONSIBILITY)	This phase includes the determination of feasibility and desirability of undertaking the development of a new weapon or component, the establishment of military characteristics for the article, and the determination of respective responsibilities between the AEC and the DOD for the various tasks involved in its development and procurement.
Phase 3: DEVELOPMENT ENGINEERING	This phase includes events beginning with the launching of AEC's development program, through the determination of development specifications, and culminating in the design release by the development agencies.
Phase 4: PRODUCTION ENGINEERING	This phase covers activities that adapt the developmental design into a manufacturing system which can produce weapons and components on a production basis.
Phase 5: FIRST PRODUCTION	This phase comprises the delivery of the first weapons from production facilities. The production rate is limited, but increases as the various production facilities come into operation. These first weapons are evaluated by AEC and DOD agencies. During this phase, AEC makes a preliminary evaluation of the weapon pending its final evaluation and subsequent approval as to suitability for standardization. This phase terminates in the DOD's formal standardization action.
Phase 6: QUANTITY PRODUCTION OF MARK WEAPONS FOR STOCKPILE	During this phase the AEC undertakes the necessary quantity production of Mark weapons for stockpile. This includes the phased production of components, spare parts, and ancillary gear. Previously produced weapons are redesignated as Mark weapons if they meet the criteria for a standardized weapon. If not, an appropriate modification program may be undertaken.
Phase 7: RETIREMENT*	In this final phase, a program for the physical elimination from stockpile of an atomic weapon or major assembly is initiated.

*Not defined in the 1953 AEC-DOD Agreement.

COMPARISON OF VIEWS ON RESPONSIBILITIES IN THE NUCLEAR WEAPONS PROGRAM

Proposed Draft, Views of the Military Services, October 1952	Proposed Rewrite, by Sandia Corporation, November 1, 1952

PART II

GENERAL OUTLINE OF FUNCTIONS AND RESPONSIBILITIES

1. The functions, responsibilities, and procedures established by this agreement are based on the premise that, unless otherwise provided by law or by agreement between the Atomic Energy Commission and the Department of Defense, the development and production of nuclear systems are functions of the AEC,

are based on the following premises:

a. that, unless otherwise provided by by law, or by agreement between the Atomic Energy Commission and the Department of Defense, the development, production, *and stockpiling of atomic weapons will be the joint responsibility of the AEC and the DOD;*

b. that the development, production, *etc.*, of nuclear systems are functions of the AEC;

c. *that the division of responsibility for the development, production, etc., of atomic weapons, exclusive of nuclear systems, will be by joint agreement on each weapon or by classes of weapons between the AEC and DOD; and*

that the determination of military suitability and acceptability (standardization) is a function of the DOD

d. that the determination of *military characteristics*, suitability, and acceptability (standardization) is a function of the DOD.

Source: W. L. Stevens

COMPARISON OF VIEWS ON RESPONSIBILITIES
IN THE NUCLEAR WEAPONS PROGRAM

Proposed Draft, Views of the Military Services, October 1952	Proposed Rewrite, by Sandia Corporation, November 1, 1952
This agreement does not define an atomic weapon, nor is such a definition necessary to its purpose at this time. Should an agreed definition of an atomic weapon be reached at any future time, this agreement will be subject to re-examination and amendment as necessary.	
Part IV	
DEFINITIONS	
1. *Nuclear System* – The nuclear system is comprised of the fission and/or fusion material, together with those components required to convert the system from the safe condition to an explosion. This definition specifically excludes the fuzing system of the weapon.	*Omit the last sentence.*
	2. *Atomic Weapon – An atomic weapon is the combination of a nuclear system with such fuzing components, housings, and the like as may be required for effective delivery. The atomic weapon does not include those features of the delivery system such as airplanes and guided missiles that are common to the delivery of other weapons.*

NOTE: Italics have been used to emphasize significant differences.

CLASSIFICATIONS FOR ENGINEERING CHANGE ORDERS

CLASS A	A failure or hazard rectification change which must be accomplished before items can be shipped or retained for readiness in the stockpile. Immediate incorporation was required for new production.
CLASS B1	A capability change which is not critical enough for category A, but one that is sufficiently important to warrant a separately scheduled retrofit when new material is available, after which the change is required for retrofitting and for new production.
CLASS B2	A lesser capability change that is only of sufficient importance to include its accomplishment in a subsequently scheduled retrofit change.
CLASS B3	An improvement change which is of sufficient importance for early incorporation in new production as soon as changed material can become available but is not of sufficient importance to be included in a retrofit. Changes from "A" through "C1" are considered to be important enough to be made without regard to cost of reworking or scrapping material, piece parts, or assemblies.
CLASS C2	A minor improvement change which is the only true in-process change since they are intended only to be incorporated as previous issue materials, piece parts, or assemblies are used up in new production. It is never included as a retrofit item, nor is any other "C" change.
CLASS C4	A clarification of specifications or to correct drawing errors of a minor nature.

PARAMETERS FOR INTERACTION: DOD-AEC-SANDIA

At Sandia, Director of Development Bob Henderson in an intra-Sandia committee presentation explained how the provisions of the agreement were being implemented at the time. Before describing the process to his audience, however, he summarized the overall impact of the 1953 Agreement at the agency level by observing that his engineers had adopted "a new policy by which they accept continuing responsibility from feasibility to stockpile." Henderson indicated that a typical weapon program could be initiated by a letter from the Division of Military Application, AEC, by a request from the military, or, he added: "We may wish to investigate an idea which is proposed internally within the Corporation." "If the idea concerns an implosion bomb," Henderson continued, "the proposal is presented to the TX-N Committee for consideration." (He noted that on all proposals Sandia maintained a close relationship with Los Alamos.)[35]

During the years 1952-1954, the TX-N Committee membership included—in addition to Henderson—the Director of warhead development, Lou Hopkins, and bomb and warhead design department managers, as well as representatives from Los Alamos. Chaired by Sandia's Vice-President for Systems Research, Walter MacNair, the Committee reported directly to the President of Sandia Corporation and the Director of Los Alamos. As an inter-Laboratories group, the Committee had no military representation. The operating procedure for the Committee was to forward recommendations to the Sandia President and the Los Alamos Director, who jointly decided upon a course of action. If the decision was made to proceed, a recommendation was then forwarded concurrently to the AEC through the Santa Fe Operations Office and through military channels. As Henderson explained:

> These channels meet at the MLC-DMA level, and, if agreement is reached, the Corporation receives notice of military requirements which come from the Military Liaison Committee.[36]

Henderson stressed that the Corporation accepted military requirements only from the MLC, which he said, "has proved to be a very effective filter that prevents duplicate requirements from different agencies of the armed forces."[37] The requirements, issued to the Corporation through the AEC and the Santa Fe Operations Office, provide for a Stage I study.

The Stage I referred to by Henderson, also known as Phase I, initiates the "Concept Formulation" part of the process during which time the nuclear labs compete for programs. Henderson proceeded with his explanation. "At Sandia," he said:

> the job is given to a development department and a case is written. A fairly recent procedure in the case system allows us to write a preliminary case covering Stage I only. After the feasibility study has been made, a report is written by Sandia and Los Alamos which is presented to the Special Weapons Development Board who, in turn, forwards its recommendations to Washington.
>
> AFSWP also forwards the recommendation through military channels. Normally these recommendations are paperwork only, and we proceed with early development if the Special Weapons Development Board thinks it is a good idea. With approval for Stage II ["Program Study for Feasibility and Responsibility"], we proceed with development and at this time we have in hand the "Desired Military Characteristics" of the weapon. These come to us via the MLC–DMA channel.[38]

The desired Military Characteristics referred to by Henderson specified performance, physical characteristics, and a stockpile-to-target sequence that enumerated potential weapon environments. The Development Phase began when DMA issued a directive to the Santa Fe Operations Office of the AEC. The Development Authorization named the nuclear laboratory, established time scales, and provided other pertinent information. The first step in Contract Definition was the release by DOD of a Request for Proposal (RFP), which defined the AEC–DOD interface and established tasks to be addressed in the contractor's proposals.[39]

Continuing to explain the procedure as it existed in the mid–fifties, Henderson said:

> When we feel that development has proceeded to the point that we have valid information, we write a "Proposed Ordnance Characteristics" report which is the first report that describes a layout of the system. At that time we have established a weapon system, location of components, and a ballistic case. We include photographs of such mockups as are available.[40]

In response to a question regarding the time at which the component and test equipment groups were brought into the process, Henderson indicated that this was done approximately four months after the initiation of Stage II. He mentioned also that within the development organization an Electrical Systems Coordination Group had been established. This group included all electrical division supervisors in the warhead and bomb development organizatons, as well as representatives from the Test Equipment Organization. The ESCG, as it was known, held frequent meetings where members were kept informed of all programs. By this arrangement the development groups learned of developments in other areas, and as Henderson termed it, "re-invention was minimized." The ESCG functioned in an advisory capacity only; responsibility on specific programs rested with the line organization.[41]

Henderson continued:

> In 1200 [Bombs] a Department Manager is the Project Engineer, responsible for all aspects of a program from Stage I forward. The same pattern is true in 1300 [Warheads] except that a Department Manager may be assigned several programs and has a staff of Project Engineers to assist him. The TX-N Committee assigns responsibility on a component basis to Sandia and Los Alamos and this assignment covers all development and logistic phases. At that time a development schedule is prepared. All through the development stage, correspondence flows back and forth, but we accept MLC as the only official source—all others we take under advisement. Sometimes we disagree with MLC and request changes in the desired characteristics and state reasons why. We insist on the word "desired."[42]

During Phase III, referred to as "Development Engineering," Sandia engineers laid out the design on the drawing board as completely as possible in an attempt to minimize cut-and-try development. Parameters derived from the Military Characteristics, from the feasibility study, and those requirements established as standard by Government Standards and Specifications were used and expanded. Environmental and other type testing was performed at this stage to verify functions, materials, and processes. The Lab then developed prototypes for evaluation and determined the developmental design release date.[43]

The weapon system development groups first submitted reports to joint interlab committees, such as the TX Committee. The report was then sent to the Special Weapons Development Board, which functioned as a joint AFSWP–Laboratory technical review board, for approval. With the sanction of the Special Weapons Development Board, the report was then transmitted to DOD through DMA. This report provided something tangible for the DOD to review to determine if the new weapon met all the requirements established by the Military Characteristics. When the DOD furnished the AEC its quantitative requirements, the AEC issued the authorization for procurement and began preliminary planning and scheduling for production rates and deliveries to the DOD. Design release signalled completion of the Development Phase.[44]

Production and Standardization Phases

Without waiting for formal comments of DOD on the developmental design, Phase IV, "Production Engineering" began. This phase covered those activities which adapt developmental design into a manufacturing system that can produce weapons and components on a production basis. At this time, the Lab prepared product specifications for production release and furnished these specifications to the DOD for review. The DOD maintained liaison with the various AEC agencies on product design changes, and provided appropriate guidance.

Until its demise, the Manufacturing Development Engineering organization at Sandia reviewed all drawings and specifications for new weapons. Contracts were then let to development manufacturers having the capability of final production. As Corry McDonald, a member of the Manufacturing Development organization, observed: "Past experience proved that the lowest bid is not always the least expensive in the long run as a new entry [contractor] in the field may not be fully aware of all ramifications of production."[45] Although Sandia employed the three-bid system, single-source procurement was allowed in justifiable situations. Occasionally, local shops were used for the fabrication of preproduction models.

The Laboratory performed its own evaluations of product, and on that basis released weapons to the DOD for testing, training, and other purposes. Phase IV ended with delivery of the First Production Unit (FPU). As Henderson explained, the Final Evaluation Report had to be submitted and approved within ninety days after FPU. Rebuild or retrofit operations required by New Material System and Stockpile Sampling Tests were conducted at Modification Centers where Sandians provided technical assistance.[46]

Efforts to establish programs for efficient interaction did not stop at the agency level. Since the underlying objective of all such action was to improve the safety and reliability of the stockpile, formalized procedures to ensure uniformity at the operational level were also implemented. One such procedure involved development of a classification system for indicating changes to product being manufactured or already in stockpile. The classifications, which established a uniform pattern of action, ranged from Class A—denoting a hazard requiring immediate incorporation—to C4—indicating the need for clarification or corrections of a minor nature.[47]

Carried out as part of the Lab's Quality Assurance program, the procedures employed in the classification system were patterned after those of Bell Telephone Laboratories and Western Electric, recognized as a national leader in the quality assurance field. The major difference was that Sandia's check inspections assessed demerits to product rather than to individuals. A committee, including representatives from the production centers, was set up to determine the seriousness of defects and allot the demerits. A Class A, or "critical" classification, for example, earned 100 demerits; while the least serious, Class D, rated "incidental," had a demerit value of 1.[48]

During Phase V, "First Production," also known as "Manufacturing," the AEC initiated the manufacture of weapons according to military specifications. After the Lab conducted product evaluation, it released the weapon to the DOD for testing, training, and other purposes. The AEC approved the weapon model as suitable for standardization and for Initial Release to stockpile. Certain restrictions could be placed on use of Initial Production weapons that would not be acceptable to the using agency on a long-term basis. A General Major Assembly Release for the new weapon was normally issued six months after Initial Release.

The Major Assembly Release, prepared by Sandia and approved by the Albuquerque Operations Office, indicated that Mk material was satisfactory for release to the DOD on a designated effective date for specified uses with stated limitations as necessary. An Interim Release was issued to advise of changes in the limitations on use specified in the Initial Release; whereas, an Emergency Capability Release, employed on so many of the early Mks, advised that specific preproduction or engineering-type material met operational concepts. Theoretically, it was possible to issue a General Release first; however, either form of release signified the culmination of Phase V in the production process.[49]

During Phase VI, known as "Quantity Production" of Mark Weapons for Stockpile, the Special Weapons Development Board received an Engineering Evaluation Report indicating that the stockpiled weapons were ready for general use. At the same time as evaluation was being made by the Quality Assurance organizations, the DOD started Operational Suitability Tests. Doug Ballard, who became supervisor of the Tool-Made Sample and Major Assembly Release Division in 1954, recalled that "the authority given Sandia became a powerful tool for controlling quality output of the production agencies." The group's responsibilities included evaluation of product at the various supplier's plants. Samples were selected, then tested at Sandia's environmental test laboratories. Field usage could indicate that further development was desirable, leading to Mod designations. Ongoing evaluation of product in stockpile ensured that reliability was maintained. Phase VI, therefore, included retrofitting. After development of the original Agreement of 1953, a formal Phase VII, "Retirement," was added to this final stage in the life of the weapon.[50]

Sandia Strives to Improve Organizational Effectiveness

Implementation of the 1953 Agreement on development, production, and standardization helped to delineate procedures and responsibilities of the DOD, the AEC, and its laboratories from the national to the local level. Sandia Corporation, however, had been striving to improve internal organizational effectiveness since November 1952. At this time, President Don Quarles assigned Vice-Presidents MacNair, Poole, Schmidt, and Shea to consider problems related to development and production. The objective of the study was to determine ways to improve the flow of design information from the development groups through Manufacturing Engineering to the model shops making the production units. The proper relationship of Transition Engineering to Manufacturing Engineering appeared to be at the center of the issue, although there were other factors involved.[51]

According to Vice-President of Development Bob Poole, a part of the problem with Transition Engineering was that it was being affected "by a universe that was expanding because of the addition of new and relatively inexperienced development people."[52] Tim Shea observed that this was only part of the problem. The fact that "the Lab was under great pressure from the Military to complete developments quickly," according to Shea, resulted in insufficient emphasis on readiness for manufacture. "Unless designs permit manufacture in accordance with desired schedules and desired quality," Shea added, "military capability is not adequately achieved, and we fail to a degree in the primary purpose of development."[53]

After weighing the recommendations of the Vice-Presidents, Quarles concluded that "Transition Engineering should function as a service to the design engineer, leaving to the latter the full responsibility in a product design sense."[54] He urged a strengthening of the Product Engineering organization and a reorientation of Transition Engineering toward more of a Specifications Engineering function closely integrated with drafting standards and materials standardization. As a means of developing the "team play" desired, Quarles proposed that the Lab adopt a conscious policy of assigning both production engineers and transition engineers into the product development group "as a means of picking up the baton and entering the relay race as early as possible." Specifically in regard to "Transition Engineering," Quarles advised:

> We must realize that the conditions that call for this setup as between the Whippany and North Carolina factories of Western, are in large measure absent in our setup. The geographical separation between development and manufacture does not apply in the sense that things which are manufactured in our shop are in the same location as the development engineers, and the transition engineers are no nearer to other manufacturers than are the development engineers. For this reason, I think we must be careful to assign Transition Engineering those functions that are consistent with and dictated by our conditions. In fact, we should be ready to discard the function unless it is justified by our conditions.[55]

In this manner, Quarles identified an issue that would continue to be debated as part of the introspective corporate analysis he had initiated. The whole issue of relations between Manufacturing Development, Manufacturing Engineering, and Standards Development (Transition Engineering) would become the focus of a year-long committee effort; however, this study would be directed by his successor.

PARAMETERS FOR INTERACTION: DOD–AEC–SANDIA

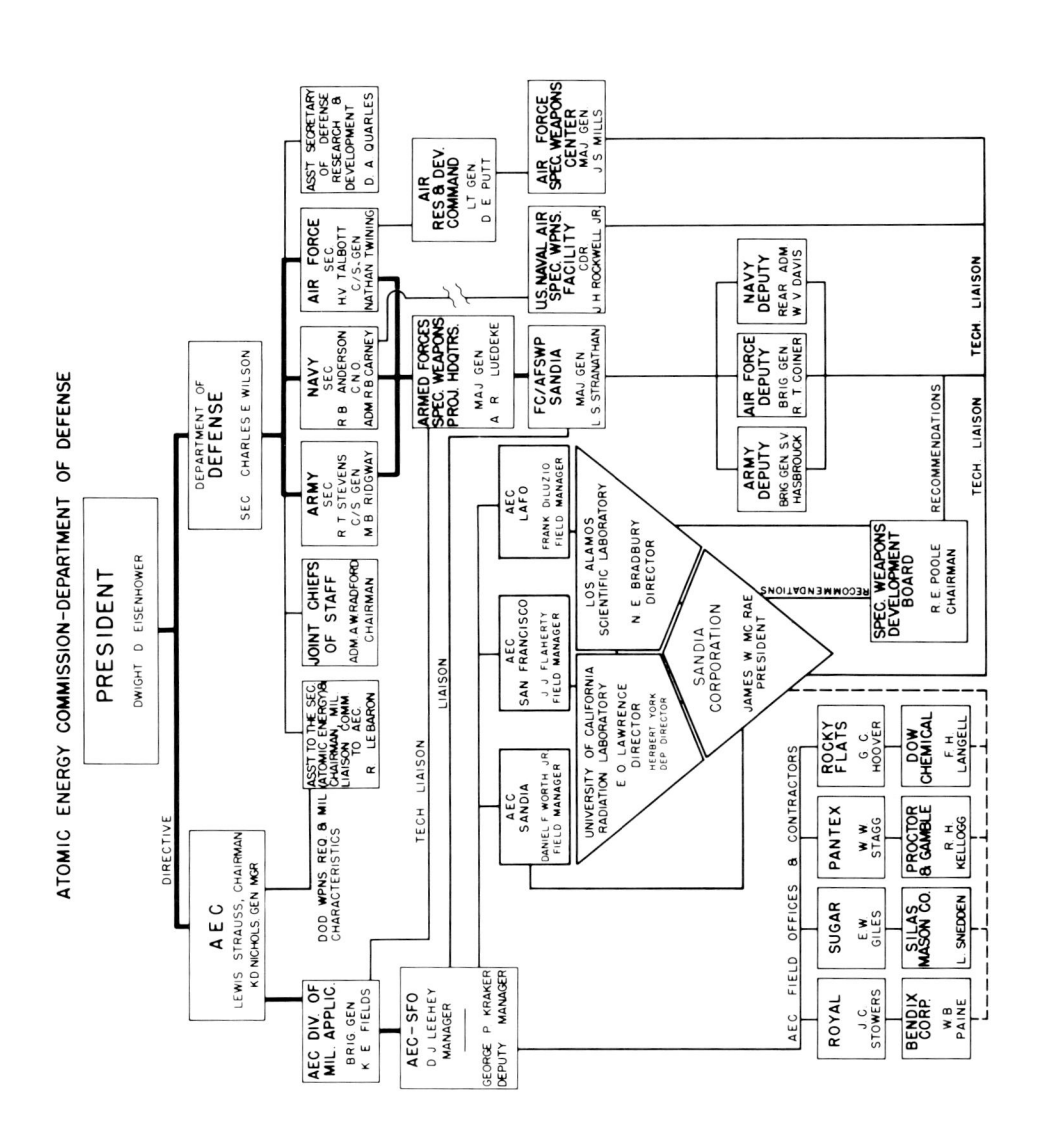

Meanwhile, Quarles' contributions at the national level had not gone unnoticed. He had been a participant in the DOD's Research and Development Board from the beginning, first as a member and later as Chairman of the Board's Committee on Electronics. By 1953, however, certain events had occurred to influence the way R&D was managed. Among these, cessation of hostilities in Korea and the inauguration of Dwight D. Eisenhower as President motivated a reappraisal of the nation's defense policies and planning. Results of the "New Look" at military strategy, as the program was called, were quickly seen in a reduction in military manpower, increased influence and size of the Air Force, and declaration of the doctine of "Massive Retaliation."[56]

As Herbert York and Allen Greb point out in their analysis of this period, these events and reappraisals resulted in the abolition of the military's Research and Development Board and its system of committees. In its place, two separate full-time staffs were appointed, each headed by an Assistant Secretary of Defense. The practice of handling the missile program separately from other facets of military R&D, the authors observe, "was continued but greatly strengthened and expanded."[57]

By 1953, these changes at the national level were formalized in the Defense Reorganization Act, which resulted in the appointment of Donald A. Quarles as Assistant Secretary of Defense for Research and Development and Frank Newberry as head of Applications Engineering. Thus, after serving a very effective, but brief tenure as President of Sandia Corporation from March 1952 through August 1953, Quarles returned to Washington, D.C., to take up his new post.

During his six years in the Pentagon, Quarles would also serve as Secretary of the Air Force from 1955 to 1957, at which time he became Deputy Secretary of Defense—a post he held until his death in 1959. His elevation to Secretary of Defense was to have been made the day he died. York and Greb summarized Quarles' impact by observing that he "exercised very great influence over both the content and the style of defense R&D during a period of exceptionally rapid change."[58]

PARAMETERS FOR INTERACTION: DOD-AEC-SANDIA

In addition to changes at the national and international level, still another major event had occurred to influence the direction of military R&D and the technical focus of the Sandia Laboratory. In 1951, scientific and military personnel carried out the world's first test proving the feasibility of a hydrogen device. Operation Greenhouse, as it was known, was the first of an extensive series of tests conducted at both continental and Pacific test sites before the moratorium of 1958.

Former Sandia President Donald A. Quarles is sworn in as Deputy Secretary of Defense by Frank Sanderson, White House Administrative Officer, as President Dwight D. Eisenhower looks on.

George shot, Operation Greenhouse, May 8, 1951.

Chapter XX.

ERA OF THE SUPERBOMB:
Politics to Prooftesting

> *In view of the Korean situation and its impact on national war readiness, it is evident that increasing emphasis must be placed on the presentation of the tests of atomic weapons now scheduled for the Spring of 1951. It is apparent that the results of these tests will have an important effect on the production of weapons for the War Reserve and on the development program for the thermonuclear weapon. . . . The Commission desires, therefore, to make known to the Department of Defense its strong feeling that there be no delay in the presently scheduled tests.*
>
> Gordon Dean, Chairman, AEC to Robert LeBaron, Chairman, MLC, 13 July 1950

> *The first Soviet A-bomb and the U.S. determination to react to it led by a somewhat complex path to the creation of a second weapons laboratory at Livermore, California and hence eventually to a doubling of the size of the American nuclear weapons development program.*
>
> Herbert York, The Advisors: Oppenheimer, Teller, and the Superbomb

The explosion of "Joe One," named after Stalin, caused an immediate sensation in both military and scientific circles. In October 1949, a memorandum circulated by the Joint Chiefs of Staff admitted that "the Soviets had developed the atomic bomb three years earlier than the date estimated by the American-British Intelligence." In view of the advance of the Soviet timetable, the Joint Chiefs decided that "a review of the military participation in the program should be undertaken."[1] Los

Alamos scientist J. Carson Mark expressed the reaction of many in scientific and political communities: "The Russian shot seemed to demand a dramatic step as a counter-balance," he observed.[2] AEC Commissioner Lewis L. Strauss agreed. Writing to his colleagues, he said:

> It seems to me that the time has now come for a quantum jump in our planning (to borrow a metaphor from our scientist friends) that is to say, that we should now make an intensive effort to get ahead with the Super.[3]

To Hungarian-born scientist Edward Teller, this show of support must have been welcome news. The brilliant scientist with the bushy eyebrows had been trying to "get ahead" with the Super since the early days of the Manhattan Project. Now it appeared he was making headway.

Until the spring of 1951, however, the hydrogen bomb remained a hypothetical weapon. Scientists had considered the possibilities of developing a bomb that would derive its force from "fusion" (as opposed to fission), meaning the coalescence of nuclei of deuterium and tritium (isotopes of the lightest element, hydrogen). However, to generate the extremely high temperature needed, a fission, or atomic, bomb with power derived from the splitting of the heavy elements of uranium and plutonium, had to be perfected first. For years the H-bomb remained a theoretical brainchild in a few scientific minds.[4]

Steppingstones to the Super

In his Los Alamos lecture series on the subject in 1951, Edward Teller noted that the possibility of producing a terrestrial thermonuclear reaction had been discussed since 1928, when Atkinson and Houtermans in *Zeitschrift für Physik* suggested that the source of stellar energy was thermonuclear. But serious consideration of the phenomenon did not occur in the United States until 1933 when George Gamow, a native of

Russia, joined the physics department at George Washington University.

Before his escape to the United States, Gamow had reported to the Soviet Academy of Sciences on work conducted by the British and German physicists suggesting that stellar energy was created by the collision of atomic nuclei. Extremely high temperatures in the interior of the stars, he observed, caused a thermal agitation permitting collision of the nuclei and fusion of small nuclei into larger units—a process the opposite of fission. Excited by the lecture, a high Soviet official offered to allow Gamow to conduct experiments related to thermonuclear energy at the Electric Works of Leningrad, but Gamow refused. After his escape, however, he pursued theoretical aspects of the subject at George Washington University, where he was joined in 1934 by Edward Teller. Graduate student Charles Critchfield, in daily meetings with Gamow and Teller, contributed exciting suggestions. By 1938, Gamow, Critchfield, and Hans Bethe were making what Teller would later term "some marvelously enlightened guesses about thermonuclear reactions." Subsequently, Bethe proposed to Critchfield that they jointly publish a paper on the subject.[5] Nevertheless, the concept remained strictly theoretical since no device existed that contained enough energy to initiate the reaction.

Theory came closer to reality in the late summer of 1941 when Edward Teller met with friend and fellow scientist Enrico Fermi in Chicago. Over lunch Teller listened with fascination as Fermi surmised that a fission bomb might be used as a trigger to ignite a fusion reaction. "Perhaps it was possible," Fermi proposed, "that the detonation of a nuclear fission (atomic) bomb would create heat on the earth comparable to that in the interior of stars." Fermi went on to conjecture that so much heat might cause fusion of the hydrogen atoms, resulting "in a colossal release of energy."[6] Although Teller initially discarded the idea that deuterium could be ignited by an atomic bomb, the concept remained in his subconscious, to be revitalized not long after.

Arthur H. Compton moved closer toward the real development of a terrestrial thermonuclear reaction in May 1942 when he requested that Robert Oppenheimer take over the fission program previously headed by Gregory Breit, a consultant to the Uranium Committee. Oppenheimer was to continue to study and compile data covering the basic nuclear reactions produced by fast neutrons.

To address the subject, he assembled a group of theoreticians at the University of California in Berkeley. Among them was Edward Teller. The Berkeley group—auspicious because of the brilliance of those present and the ideas proposed—reached the pessimistic conclusion that the amount of nuclear material needed for a weapon might be prohibitively large. But there was one startling possibility. The calculations bandied about suggested that a reaction more powerful than nuclear fission might be generated by thermonuclear fusion of deuterium, the heavy-hydrogen isotope.

In July 1942 Oppenheimer visited Compton's summer retreat in Michigan to report on the Berkeley conclave and to express the participants' concern about the possibility of a cataclysmic reaction. Despite conscientious efforts to the contrary, news spread of a theoretical high-yield weapon that might use a more easily acquired material than uranium-235 or plutonium-239. Oppenheimer subsequently arranged for basic nuclear studies of light elements, using cyclotrons at Harvard and the University of Minnesota.[7]

After establishment of the Manhattan Project at Los Alamos, Teller joined the Theoretical Division and continued investigation on the thermonuclear design referred to as the Super. Work on the Super proceeded during the war years, although Oppenheimer placed priority on development of fission weapons. As the atomic bomb became more of a reality, there was further credence for the idea that the temperatures created in the detonation of such a bomb might be used to start a thermonuclear reaction.

Among other problems, scientists discovered that they needed to find a way to ignite the deuterium more rapidly, and in 1944 conceived the idea of adding tritium. But tritium was not readily available. Teller, therefore, urged that the production of plutonium be restricted and all available neutrons be concentrated on tritium production. Oppenheimer took the position that the nation could not then afford to divert plutonium production.

The lack of computing power also impeded progress on the thermonuclear program. The Eniac (Electronic Numerical Integrator and Calculator) computer at Aberdeen, Maryland, provided early calculations; however, the results were inconclusive. Nevertheless, a laboratory review of progress on the thermonuclear bomb held in mid-April 1946 was encouraging. During 1946 and 1947, Teller, Robert Davis Richtmyer, and Lothar W. Nordheim carried out extensive theoretical work, including the "Alarm Clock" concept. Between July and September 1946, Fermi presented a series of six lectures that summarized progress to date.

Invention of the "Booster Technique" by 1948 represented a milestone in thermonuclear research, but the postwar exodus of scientists from the Manhattan Project had a negative effect. By this time, three approaches had been proposed and theoretically explored: the classical Super, also called the Runaway Super; the Alarm Clock, so named because it was hoped that it would awaken people from postwar apathy to the prospects of thermonuclear weapons; and the Booster, which contained a small amount of deuterium-tritium in the center of a fission bomb to boost its efficiency.

In the interim, the Air Force implemented two study projects, Brass Ring and Caucasian, directed toward adapting aircraft to carry the large heavy bombs that were anticipated.[8] Such was the status of work on the hydrogen bomb when the President learned that the Russians had broken the United States atomic monoply.

Politics of the Thermonuclear Program

Detonation of the Soviet device provided the impetus needed to garner support for development of the H-bomb, but expansion of the nuclear arsenal into the fusion field would not take place without conflict and controversy. Teller found ardent supporters for his campaign in fellow scientists Ernest O. Lawrence and Luis Alvarez, and in AEC Commissioner Lewis Strauss. As Strauss wrote to other AEC members:

> I am thinking of a commitment in talent and money comparable, if necessary, to that which produced the first atomic weapon. That is the way to stay ahead.[9]

It was a persuasive argument, but not everyone agreed. The General Advisory Committee, chaired by Oppenheimer, discouraged the effort, citing an array of scientific, political, economic, and moral arguments. With the support of David Lilienthal, Albert Einstein, and Harvard University President James B. Conant, Oppenheimer led the opposition. As a position statement, members of the Committee generated a report in which they based objections to the Super largely on the effects of the weapon on civilians. Despite their opposition to the H–bomb, the GAC supported research leading to tactical atomic weapons designed for battlefield use and the expansion of a stockpile of atomic bombs, which they felt would be equally as effective as a hydrogen weapon against the Russians. They also hoped for negotiations that would lead to control of the nuclear arms race.[10]

The ensuing debate had far-reaching impact on the lives of those who were involved. Oppenheimer's position on the H–bomb decision would be used against him in the security hearings of 1954; Lilienthal, in protest, would resign as chairman of the AEC; and ultimately, the power of the GAC would be diminished.

In 1949, however, the initial vote of the Commission was 3–2 in favor of the GAC decision not to build the bomb, but the Joint Chiefs critiqued the report severely, calling the scientists' moral objections "irrelevant and dangerous." General Omar Bradley supported the Super on grounds that not to do so would give the Russians an advantage in the arms race. Strauss wanted the thermonuclear bomb developed with all haste to keep the U.S. arsenal equipped with "the most potent weapons that . . . technology can devise."[11]

Coincidentally, the arrest of British spy Klaus Fuchs and his American cohort, Harry Gold, made headlines. For those in favor of the Super, the timing was fortuitous. The revelation that Fuchs had provided the Russians with information on the atomic bomb gathered while he worked on the Manhattan Project added to speculation he also might have provided the Russians with information on the Super.

On November 18, 1949, Truman appointed a special Committee of the National Security Council (NSC) to assist him in making the decision to sanction or reject the thermonuclear program. The Committee consisted of three members: Secretary of State Dean Acheson and Secretary of Defense Louis Johnson, both of whom recommended development, and AEC chairman David Lilienthal, who expressed disapproval.

The majority recommendation of the NSC, combined with pressure from a hawk-like public exposed to the mass hysteria of McCarthyism, and the Joint Chief's severe criticism of the GAC report, convinced Truman that he should support development of the H-bomb. On the afternoon of January 31, 1950, only two months after the Bell System took over operation of the Sandia Laboratory, the President announced publicly that he had directed the Commission "to continue its work on all forms of weapons, including the hydrogen or Super bomb."[12]

In the May 1950 *Scientific American*, Dr. Robert F. Bacher commented on the announcement's impact: "The President's decision to go ahead with the development of the hydrogen bomb created a tremendous stir in the nation." It was Bacher's personal opinion that "while it is a terrible weapon, its military importance seems to have been grossly overrated in the mind of the laymen ... pumped full of hysteria by Red scares, aggravated by political mud-slinging."[13] Dr. Harold C. Urey, the discoverer of deuterium, expressed his own views. Commenting on the balance of power issue in the event that both the U.S. and Russia built the bomb, he said: "I am very unhappy to conclude that the hydrogen bomb should be developed and built. I do not think we should intentionally lose the armaments race."[14]

The outbreak of the Korean War in late June 1950 had a paradoxical effect on hydrogen bomb development. On the one

Robert F. Bacher.

Edward Teller.

hand, it vindicated to a degree the President's decision on the H-bomb and calmed the consciences of many of the scientists who had originally opposed it. On the other hand, the crisis atmosphere contributed to the decision to direct the nuclear weapons program toward development of smaller fission weapons for tactical use. As a result, chairman of the AEC Gordon Dean feared a lack of focus. "In view of the Korean situation and its impact on national war readiness," he warned,

> it is evident that increasing emphasis must be placed on the presentation of the tests of atomic weapons now scheduled for the Spring of 1951. It is apparent that the results of these tests will have an important effect on the production of weapons for the War Reserve and on the development program for the thermonuclear weapon. . . . The Commissioners, therefore, wish to make known to the Department of Defense its strong feeling that there be no delay in the presently scheduled test."[15]

Meanwhile, work at Los Alamos continued amid growing conflict between Teller and Bradbury over the amount of effort that should be devoted to the thermonuclear program. J. Carson Mark, leader of the Theoretical Division at Los Alamos, recalled that "Edward was emotionally certain that if we scientists at Los Alamos only worked harder we would find out how it should be done."[16] And at Princeton John von Neumann made progress toward development of a new and advanced electronic computer that would be needed if the hydrogen bomb were to become a full-fledged reality.

But the thermonuclear program ran into one roadblock after another. On the technical side, problems related to ignition of the Super and the high cost of the tritium made the outlook for success gloomy at best. The project reached a new low when mathematicians Stanislaw M. Ulam and Cornelius J. Everett (and later, Enrico Fermi) proved Teller's original concept to be implausible. Ironically, the solution would be found in Teller's rival for attention at Los Alamos—the fission program.

NUCLEAR ORDNANCE ENGINEER FOR THE NATION

John von Neumann.

J. Carson Mark.

During his intermittent sojourns at Los Alamos, Teller provided a contagious enthusiasm and vision, but it was J. Carson Mark of the Theoretical Division who actively directed work on the project. Among the well-respected physicists and mathematicians reporting to Mark were von Neumann's protégé, the quiet, unassuming Stanislaw M. Ulam, and G. Foster Evans. By 1949, with Evans' assistance, George Gamow and Ulam had established important calculations using Monte Carlo probability theories. Teller, in the meantime, promoted plans for the 1951 weapon test series and met with Lawrence and Alvarez in Santa Fe to make arrangements to gather support from the Commission for construction of a production reactor that would use heavy water as a moderator.[17]

In late February 1951, not many months before the Greenhouse test operation, Ulam, while conducting studies related to fission bomb detonation, determined more accurately the amount of tritium required and considered whether soft (long wavelength) X rays could be used to initiate a fusion reaction. This idea, combined with a more effective staging technique to generate high yields, gave Teller and Bradbury enthusiasm and hope.

Teller then developed a parallel or alternative version to Ulam's. As Ulam wrote in his autobiography: "Teller made some changes and additions, and we wrote a joint report quickly. It contained the first engineering sketches of the new possibilities of starting thermonuclear explosions." The jointly produced report would become, in Ulam's words, "the fundamental basis for the design of the first successful thermonuclear reactions and the test in the Pacific called *Mike*."[18]

Teller carried the theory of radiation implosion one step further with still another initiator concept. Expert calculations and projections on the idea, later to be termed the "New Super," were made by mathematician Frederick de Hoffman, who graciously signed only Teller's name to the more detailed report showing how the bomb could be constructed.[19]

The Frenchman Flat area of the Las Vegas Bombing and Gunnery Range was the site selected for the Ranger operation, the second atmospheric test on the North American continent.

Meanwhile, the feasibility of the thermonuclear concepts developed to date had to be proven. Physicist Alvin C. Graves, Director of the Los Alamos Test Division, and Deputy Commander for Scientific Operations, Joint Task Force 3, felt the need to justify in a public statement the frequency of test operations during this era. "We are today," he said, "in the normal and desirable situation in which we are improving and extending the range and usefulness of our product at such a rate that frequent tests are a necessity."[20] Throughout the fifties, full-scale field testing both in the Pacific and on the continent, as Graves had indicated, would play an essential role in the development of nuclear technology.

Ranger

With the Greenhouse test operation planned for the spring of 1951, Los Alamos scientists became concerned over theoretical calculations that showed the possibility of surprisingly large weapon yields. Yet the validity of the calculations was questionable. Therefore, on December 6 and 11, 1950, conferences were held at Los Alamos to discuss ramifications of variations in the compression of critical materials and their effect upon yields. The answers, the scientists concluded, could be determined only by conducting a series of small nuclear experiments to reveal if the yields calculated were attainable in practice. Data were also needed to improve design criteria.

While construction of facilities for the Pacific operation proceeded under the auspices of Holmes and Narver and the 79th Army Engineering Construction Battalion, AEC officials hurriedly considered the selection of a continental test site. A preliminary report by AFSWP, "Project Nutmeg," held that physical drawbacks and domestic political concerns precluded establishment of a test site within the confines of the United States, but it did not rule out the use of a continental site in an emergency. By 1950, the Korean War made such a national emergency appear imminent. The Pacific Proving Grounds were no longer secure from enemy attack, and the logistics of transporting personnel and supplies remained an expensive item. In July, therefore, AEC Chairman Gordon Dean proposed a joint effort by the AEC and DOD to select a continental test site. On December 18, 1950, President Truman approved the selection.[21]

Of the six locations surveyed, the Frenchman Flat area of the Las Vegas Bombing and Gunnery Range was chosen as the most suitable based upon the location's favorable meteorological conditions, distance from populated areas, and proximity to operational facilities. Site construction for Ranger—as the operation was to be known—took place on a crash basis under the code name Project Mercury. Mercury, the initial name for the site itself, would evolve into Nevada Proving Grounds, and later the Nevada Test Site.[22]

On January 11, 1951, the same day that the White House officially approved the operation, the AEC distributed handbills in the Las Vegas/Indian Springs area to alert residents to the coming tests. Conducted under tight security, the Ranger test series represented the integrated efforts of the AEC—in sole command of the operation—the military, and various supporting contractors such as Los Alamos, Sandia Corporation, and EG&G. In this test, as in others held during the fifties, the Los Alamos Scientific Laboratory would play the major scientific role, with Sandia Corporation providing the safing, arming and firing, assembly, and instrumentation. DOD participation in Ranger would be minimal in comparison to later test series.

Ranger, one of a total of fourteen full-scale test operations conducted during the first decade of the Laboratory's existence, took place in January and February 1951. For Glenn A. Fowler, Sandia's Director of Field Test from 1950 to 1955, this meant juggling almost simultaneously the related activities for several operations ranging geographically from the frozen wastes of Amchitka, Alaska, to the barren terrain near Las Vegas, Nevada, and the islands of the Pacific. The Ranger operation itself, which included a test run and five airdrop detonations, originated from Kirtland Air Force Base in Albuquerque, New Mexico.

ERA OF THE SUPERBOMB: POLITICS TO PROOFTESTING

On January 27, 1951, Sandians Don Shuster (third from right) and Carroll McCampbell (fourth from right) were among military and civilian personnel taking part in blast effects testing during Operation Ranger. The photo above records the return of the crew following the *Able* shot.

A copy of the handbill at right was given out prior to the test.

WARNING

January 11, 1951

From this day forward the U. S. Atomic Energy Commission has been authorized to use part of the Las Vegas Bombing and Gunnery Range for test work necessary to the atomic weapons development program.

Test activities will include experimental nuclear detonations for the development of atomic bombs — so-called "A-Bombs" — carried out under controlled conditions.

Tests will be conducted on a routine basis for an indefinite period.

NO PUBLIC ANNOUNCEMENT OF THE TIME OF ANY TEST WILL BE MADE

Unauthorized persons who pass inside the limits of the Las Vegas Bombing and Gunnery Range may be subject to injury from or as a result of the AEC test activities.

Health and safety authorities have determined that no danger from or as a result of AEC test activities may be expected outside the limits of the Las Vegas Bombing and Gunnery Range. All necessary precautions, including radiological surveys and patrolling of the surrounding territory, will be undertaken to insure that safety conditions are maintained.

Full security restrictions of the Atomic Energy Act will apply to the work in this area.

RALPH P. JOHNSON, Project Manager
Las Vegas Project Office
U. S. Atomic Energy Commission

Santa Fe Operations Manager Carroll Tyler and Test Director Alvin Graves of Los Alamos look at a German newspaper during Operation Greenhouse in the Pacific.

ERA OF THE SUPERBOMB: POLITICS TO PROOFTESTING

The January 25, 1951, entry in Fowler's logbook made note of the success of the trial test using high explosives in capital letters: "HE SHOT FOR PROJECT RANGER COMPLETED THIS MORNING." That same day, after a meeting with R. P. Petersen, Bob Henderson, Bill McCord, and Art Machen, Fowler added: "First word about underground shot Site Mercury in April." In other words, while Ranger was taking place as a precursor to Greenhouse, plans were already in the offing for two additional continental tests. By January 28, Fowler had arrived at Mercury site in Nevada to witness a part of the Ranger series before flying on out to the Salton Sea Test Base.[23]

Another Sandia participant in the Ranger operation was Don Shuster, Supervisor of a Special Projects group. Well known throughout the Labs as "an extraordinary idea man," Shuster had developed a method for obtaining unobtrusive and precise measurements in the Greenhouse *George* experiment.[24]

The challenge, according to Shuster, was not only to have the instrumentation very precise, but also to be able to monitor remotely over a long seventeen-mile cable that extended to a station located on one of the southern islands in the Enewetak chain. Shuster worked with Los Alamos personnel to perfect the instrumentation, which required commuting almost daily to the Hill in a single-engine Carco Bonanza aircraft—a trip he termed "an adventure in itself."[25]

Among Shuster's innovative concepts was a new way of measuring transit time—that time between initiation of the explosive in the bomb and the time the nuclear reaction "takes off." Previously, the only method of measurement involved the use of long coaxial cables and detectors, which was extremely costly. Test Director Alvin Graves was especially fascinated by Shuster's idea of measuring transit time by remote control. In essence, he proposed the use of radio telemetry, which allowed the signal to be set off precisely at the time of detonation. To conduct the experiment, the crew installed the radio telemetry in the weapon's case and the monitoring system in the rear end of the drop aircraft where high-speed oscilloscopes recorded the data and the time interval between the two pulses. Test results from Ranger would prove Shuster's radically new approach to be precise and effective.[26]

Logistics for the operation were conducted under the highest security. Shuster recalled that the plans for Ranger were so secret that he merely told his wife that he was going to work, and on January 27 boarded a plane for Indian Springs, Nevada. At the test site, the strategy was for the pilot to make one circle around the target while the crew inserted the nuclear core into the weapon. However, a problem developed when the weaponeer tried to remove the Allen screws around the nose plate. The man in charge of assembly, Tiny Hamilton (so named because of his powerful physique), had rotated the screws to what he considered a snug fit, but to everyone else they were practically impossible to remove. After two extra circles around the test site and a lot of "blue" language, the bolts were removed and the operation proceeded. "We saw a great flash and felt the shock wave hit the airplane," Shuster said, "then headed back for Albuquerque in a sky that was brightly lit by the flash of the detonation."[27]

By the time Shuster returned, still dressed in flight gear and carrying an oxygen mask, it was breakfast time. "When I got back," he recalled, "my wife said pointedly, 'I know where you've been.' She had heard of the detonation on the news and figured it out." "The system worked," he added. "We got transit time, and Al Graves told us that he wanted similar instrumentation for Greenhouse, which meant more flights back and forth to Los Alamos to get the instrumentation integrated into the system."[28]

For the Ranger test series, Sandia Corporation employees also assembled weapon systems components, prepared devices for delivery, and trained personnel in handling of the devices. In addition, Sandians assisted the military by setting up experiments related to gamma radiation exposure. However, a Sandia proposal to stage experiments involving blast effects was rejected because of the short time frame for the test operation.[29]

The radiation exposure experiment dealt with the measurement of gamma radiation at different distances during and immediately following the nuclear detonation. The test crew obtained this measurement by placing forty-one film badges at ninety-meter intervals from Ground Zero. To measure the fraction of initial gamma radiation reaching the film badges, "mousetrap gadgets" with thick lead walls and doors designed to close after initial exposure to shield the film badges from residual radiation were used.[30]

Ranger, the third atmospheric test series, provided design data for the Greenhouse operation and revealed that the basic program planned was indeed feasible. In his overview of the thermonuclear program, Hans Bethe commented on the serendipitous nature of their success. The concept, he said, "came about by a series of accidents, the accidental choice of one particular device for the Eniwetok test . . . and the invention of the radiation implosion just at the right time." Bethe went on to credit Teller's "persistent belief in the practicability of thermonuclear reactions" for the novel concepts shown in 1952.[31]

The positive results of the Ranger test encouraged Teller to push even harder for Los Alamos personnel to give their "undivided attention to the thermonuclear program." He was particularly interested in tests involving boosted weapons, the possibility of a heterocatalytic explosion (that is, implosion of a bomb using the energy liberated from an auxiliary bomb), and tests on mixing during atomic explosions, which were of particular importance in connection with the Alarm Clock.[32]

Increasingly disgruntled with Bradbury's administration of the thermonuclear program, Teller pushed strongly for a separate thermonuclear division and campaigned behind the scenes for establishment of a second laboratory. He was supported in his efforts by political friends such as Commissioners Thomas Murray and Lewis Strauss and Arkansas Senator William Fulbright. Even before the test series began, Gordon Dean recorded in his diary in September 1950 that he had received a call from Fulbright, saying that "top people were getting anxious about the location of the H-bomb plant" and wanted the AEC to make a clarifying statement on the matter; perhaps " 'no decision had been made yet'—or something of the sort."[33]

In February 1951, not long after the Ranger tests had been completed, Strauss visited Gordon Dean in his Washington office. Referring to a written outline of the remarks he wished to make, Strauss voiced his concerns regarding support for the Super program, his belief that Oppenheimer was "sabotaging" the project, and the need for something "radical" to be done to prevent those working at Los Alamos from leaving. Dean pointed out that "the real hard problem was the question of what to do about it." Setting up a brand new weapons laboratory was one of several alternatives. Dean then asked Strauss if he would leave the notes for his reference. Strauss said he preferred to burn them, walked over to the fireplace, and proceeded to do so. While the political situation was unfolding in the nation's capitol, plans proceeded for Greenhouse, scheduled for April through May 1951 at the Enewetak Atoll in the Pacific.[34]

The Greenhouse Operation

Success of the Ranger tests followed by the Ulam-Teller-de Hoffman contributions to nuclear design provided the go-ahead for Operation Greenhouse. The scientific and political communities anxiously awaited the outcome, hoping to have the feasibility of the Super proven once and for all. Among the tests planned, two would involve thermonuclear experiments. The *George* test, in particular, was considered "an important way station on the path to development of thermonuclear devices."[35]

Joint Task Force 3 under the command of Lieutenant General Elwood Quesada had been activated in November 1949 to coordinate planning for the operation, which involved the military services, Los Alamos, Sandia, and other contractors. Alvin Graves, as Quesada's Deputy Commander for Scientific

Matters, directed the Task Force, composed largely of personnel from Los Alamos' J Division. Although Graves' contingent represented the heart of the operation, seventy-eight persons from Sandia Base participated. Weapon assembly and placement in the test configuration was assigned to Bob Henderson in his role as Assistant Scientific Director under Alvin Graves. A Sandia *Bulletin* press release cleared by Los Alamos Classification Officer Ralph Carlisle Smith announced that "Sandia test and Research personnel played a major role in the operation reporting directly to Dr. Alvin C. Graves of the Los Alamos Scientific Laboratory." A significant part of this role would be in the study of blast effects.[36]

Unexpected blast damage to windows in downtown Las Vegas during the Ranger test influenced the AEC to incorporate weapons effects testing at Greenhouse in addition to the testing of thermonuclear concepts. Since the damage in Las Vegas occurred ninety miles away from the test site, scientists determined that the distance was too great for there to be an effect from normal air blast. Analysis seemed to indicate the damage had to be from the air blast refracted and focussed by the upper atmosphere.[37]

The National Academy of Sciences sent out a query asking for information on individuals with expertise in this sort of phenomena and were pleasantly surprised to find that they had one "in their own backyard"—Sandia's Everett Cox. Cox, they learned, had set up a line of pressure-measuring instruments across Europe to study the very sort of phenomena that caused trouble on the Ranger tests. The decision to use the expertise of Cox resulted in the start of Sandia's microbarographic program and expansion of the role of Sandians at Greenhouse to include the fielding and analysis of nuclear blast effects.

In 1951 the Weapons Effects group was a part of the Research directorate under R. P. Petersen. (Stuart Hight would not replace Petersen as Director of Research until 1952.) Not to be confused with Don Shuster's Telemetering Division, which was organizationally a part of Henderson's Engineering directorate, the Weapons Effects effort under Department Manager

Everett Cox, was directed toward study of the response of structures to blast waves.[38]

The Effects Department had two major thrusts: model tests, which took place at Sandia and employed the same general phenomena but used high explosives as the source of the air blast waves, and full-scale tests, such as those planned for Greenhouse. Jack Howard supervised the Model Testing Division at the Lab's Coyote Canyon in Albuquerque, while Byron Murphey, one of Sandia's first Ph.D. physicists, and Harlan Lenander shared responsibility for the Full Scale Effects Measurements Division at the Pacific site. In 1949, Cox had hired Lenander away from the Navy's China Lake installation where he was working on guided missile projects and immediately assigned him the task of recruiting support personnel for the Full Scale Test Division.[39]

The following year in March 1950, a panel consisting of Stanley Burriss of Los Alamos, representatives from the DOD, and Everett Cox and Harlan Lenander met at Sandia Laboratory to discuss instrumentation for the structures program. The panel made fairly definite plans relating to location, type, and make of instruments to be used. On March 16, the Santa Fe Operations Office instructed Holmes and Narver that their participation in this portion of the program consisted of design, "in accordance with information furnished by Sandia Corporation, procurement, and installation of fixtures and conduits necessary to connect the instruments to the recorders." Sandia personnel, in support of the AEC, were to supervise the technical and operational aspects of the installation. It was also agreed that a representative from each service project would be sent to the contractor, Holmes and Narver, who would provide drafting facilities for production of the designs. David Narver, Jr., was to act as project leader.[40]

Luke Vortman, Sandia's project leader for the planning of the structures program, worked out of the Holmes and Narver home office, where he was assisted by a fourteen-member staff. Specifications called for 900 electrical recording gauges and 500 self-recording gauges to be placed on eighteen different types of mounts. Meanwhile, Lenander's whirlwind recruiting tour netted

Bill Blythe prepares rockets for impact tests in Coyote Canyon.

some twenty-five contract personnel, including Jim Scott, who would join Sandia after the test series. Holmes and Narver employees Francis "Tommy" Thompson, George Reese, Larry Witt, and Bob Bunker would also transfer to Sandia. The Sandstone operation had proven the disruptive effects of having employees assigned overseas for extended period of time; therefore, management looked to the hiring of contract personnel as a possible solution.[41]

Just before Thanksgiving 1950, the test crew arrived on the island of Enjebi and began digging trenches, installing cables, and constructing the recording shelters. As the largest of the upper islands, Enjebi afforded room for placement of the structures. The nearby islet of Mijikadrek was also used to test weapons effects on structures placed at longer ranges.

Sandians arrived at Enjebi to find that one of the structures erected was a model of a Russian apartment house. The Sandia crew took over the building and installed a complete calibration system, but not before having a contest to select a name for the place. "The name we finally picked," Lenander recalled, "was Sandia's House of Correction."[42] A lighted sign with the name emblazoned across the front of the building created quite a stir, especially from visiting dignitaries. In the interim, Lenander was called back to Sandia on an emergency basis to handle the purchase of generators and other equipment for an underground shot planned to take place at Amchitka, Alaska. While in Albuquerque, he contracted the mumps, and the day after his arrival back on the island of Engebi, found himself directing his crew from quarantine in the hospital.[43]

Although Sandians would work with Los Alamos to instrument the structures experiments at Greenhouse, Sandia Laboratory's only administratively assigned task was in Weapons Assembly. This task unit, under Bob Henderson, consisted of fifty-five men from Sandia and AFSWP.

On February 23, 1951, the Weapons Assembly Task Unit set sail along with the components, aboard the U.S.S. *Curtiss*. At San Francisco, the ship met surface and air convoys and, after refueling at sea off Pearl Harbor, resumed its voyage across the

Pacific under blackout and close security. During the evening of March 7, reports of an unidentified submarine within a possible striking distance created no little anxiety. On March 8, however, the ship anchored safely in the lagoon off Enewetak.[44]

An approaching typhoon added drama to an already exciting situation. Enjebi, a low-lying island only seven feet above sea level, lay directly in the path of the storm. Since the Task Force on Parry Island and Enewetak was not able to monitor the weather station in Honolulu, the Sandia House of Correction on Enjebi, which was able to receive the Honolulu weather station, became the transmittal point to relay messages from Honolulu to the Task Force Commander on Parry. George Ruiz, a member of Lenander's group, manned the radio around the clock transmitting messages. Fortunately, the storm veered off when it was about fifty miles from the island, but left torrential rains in its wake and forced postponement of the first shot.[45]

The first phase of Operation Greenhouse began three days before zero hour. Under intensified aerial and surface surveillance of the Danger Area, Task Force personnel were withdrawn from the camps near the shot islands and their whereabouts verified. The second phase began at midnight the day before the shot, and continued through shot day. Henderson's crew completed assembly of the device and, with some difficulty because of the rough seas, moved it from the assembly area aboard the U.S.S. *Curtiss* to an LCT for transport to the island of Runit. The typhoon had changed wind-flow patterns and sent swells sweeping into the LCT carrying the arming party. Finally, with the large test device ensconced in the cab atop the 300-foot tower, the team made last-minute instrument adjustments and calibrations for the initial shot in the series, code-named "Dog."

After a two-day postponement because of the weather, the device was successfully detonated April 7, 1951. A select cadre of VIPs, including AEC Chairman Gordon Dean, watched from behind dark glasses as the "blinding light" and "boiling clouds" ascended skyward. The tower vanished in the blast and left in its place a crater that was soon filled with water from the lagoon.[46]

The weapons effects experiments were scheduled for the second test in the series, shot *Easy*. While *Easy* was the most elaborately instrumented of the series, the *George* and *Item* shots that followed were the most significant in proving the new thermonuclear concepts. For the third shot, *George*, the expected yield required "absolutely safe fallout conditions—southerly winds aloft at all levels—the stronger the better." As if to cooperate, typhoon Joan put in a timely appearance and "moved as if it were part of the operation."[47]

As a preliminary to the *George* test, Bill Wells and Joe Penzien had prescribed the pressure measurements, while Harlan Lenander and Byron Murphey and crew prepared the instrumentation to make those measurements. Back at Sandia, Mel Merritt would interpret the data.

High swells again made it difficult for the arming crew, which included Stan Burriss from Los Alamos, Barney O'Keefe from EG&G, and Don Cotter, Bob Henderson, and Don Shuster from Sandia. The waves prevented them from landing at Parry to inspect the control station and required transfer to a landing craft. The group then left for the island of Eleleron, where in the shadow of the *George* device they put in place the vast array of instrumentation. Shuster described the scene accordingly: "A tube for Herb York's x-ray experiment ran down the length of the tower to the bottom and cable trailed everywhere, giving the effect of a monster from a sci-fi movie."[48]

As the crew evacuated the island, the tower stood waiting like a surreal symbol in the semidarkness. May 9, 1951, marked still another technological milestone—the earth's first successful thermonuclear test explosion.[49]

The fourth shot in the series, *Item*, was almost an afterthought. Permission from the AEC to proceed was not granted until it had been decided that an enhanced neutron or boosted experiment was essential to the progress of the thermonuclear program. On this particular shot, members of the arming party, as usual, were the last people on the island.

Item, the first boosted shot and the last of the Greenhouse operation, represented culmination of a year of advances described by Oppenheimer as "technically so sweet that you could not argue about it." In comparison to the 1949 program, which he termed "a tortured thing" that didn't "make a great deal of technical sense," the new advances made the path clear.[50] Success of the operation confirmed the superbomb as a distinct reality and led to increased emphasis on the thermonuclear program.

Shortly after the success of Greenhouse, in June 1951 the AEC organized a round-table conference at the Institute of Advanced Study at Princeton University to discuss the Super. At this meeting, Teller generated a certain scientific electricity by introducing in public forum the newest concepts for development of the thermonuclear weapon. Gordon Dean later testified that the participants, including Drs. Bethe, Teller, Fermi, and Oppenheimer, were enthusiastic "without exception ... now that they had something feasible."[51]

Subsequently, leadership in Washington soon approved plans for two separate operations—Ivy and Castle—to further verify the Teller–Ulam configurations. As spelled out by the AEC's Division of Military Application, the interim objective was demonstration of a deliverable thermonuclear weapon "in order to provide an emergency military capability in this field." But the long-term objective was the development of thermonuclear weapons of reduced weight and size and improved efficiency and utility. Sandia's role at Ivy and Castle continued to include measurement of the progress of the nuclear reaction, assembly, arming and firing, air blast studies, and the manning of microbarograph stations.[52]

The Second Lab Controversy

In the interim, the schism between Bradbury and Teller over administration of the thermonuclear program at Los Alamos widened. As Teller explained the situation: "Bradbury and I remained friends, but we differed sharply on the most effective ways to produce a hydrogen bomb at the earliest possible date.

Among the VIPs at Operation Greenhouse were AEC Chairman Gordon Dean, Senator Scoop Jackson, and Test Director General Elwood Quesada shown here standing in front of the Joint Task Force 3 VIP tent.

ERA OF THE SUPERBOMB: POLITICS TO PROOFTESTING

Alvin Graves (at left) of Los Alamos was assisted by Stan Burriss (right).

Ernest Lawrence and Edward Teller visit during testing at Greenhouse.

Glenn Fowler, Howard Austin, and Robert L. Rourke, shown here at Kwajalein, were among the participants in the Greenhouse test series.

Sandians at Greenhouse included (back row, left to right) Robert W. Henderson, Walt Treibel, Bob Krohn (LASL), Harold Poulsen, Bob Knapp, (front row, left to right) Don Cotter, Don Shuster, and Ed Udey.

Balloons used for test purposes at Greenhouse.

A B-17 drone aircraft prepares for cloud sampling.

NUCLEAR ORDNANCE ENGINEER FOR THE NATION

Field tester Carroll McCampbell takes a well-deserved break during Operation Greenhouse. Below is one of the shot towers used during the test series.

... The dissension with Bradbury crystallized in my mind the urgent need for more than one nuclear weapons laboratory."[53]

On September 11, 1951, Teller handed in his resignation to Bradbury. In the opinion of J. Carson Mark, "LLL, in a sense grew out of Edward's frustrations." Teller concluded that he could campaign more effectively for a second laboratory if he were not affiliated with Los Alamos, and on November 1, 1951, he returned to the University of Chicago. His supporters, including Commissioner Willard F. Libby, had been pressing Gordon Dean on the site issue for some time.[54]

After the Princeton meeting and a subsequent visit with Norris Bradbury, AEC Commissioner Thomas E. Murray also began to push for a separate facility for the thermonuclear program. He wrote to Dean suggesting that the "new facility might be located at Sandia" and urged that action on the matter be taken at once, rather than waiting for the results of further theoretical work.[55]

But the Teller faction faced strong opposition from GAC members who felt that a second lab would deplete Los Alamos of talent and resources, and in effect, prove detrimental to the program in the long run. Teller argued to the contrary. Herbert York, who would become the proposed Lab's first Director, recalled that Teller often expressed his beliefs in terms that made it clear that he felt that the Los Alamos leadership was "unimaginative, negative, and otherwise inadequate."[56]

The Air Force, realizing the defense potential of thermonuclear bombs adapted to their needs, provided Teller with strong support. Equally significant was a proposal from Ernest O. Lawrence that the second laboratory be established as a branch of the University of California Radiation Laboratory at Livermore, which would mean less initial expense and a cadre of personnel available to begin work on the program immediately.[57]

In early February 1952, Lawrence invited Teller to visit him on the Berkeley campus. Teller accepted, and after his arrival drove inland with Lawrence to an abandoned naval air station near the small town of Livermore. The location, Lawrence felt, would be a perfect site for the extension of Lawrence's world-famous laboratory in Berkeley—a second weapons laboratory. He urged Teller to resign from the University of Chicago and supervise setting it up. With this new development, opposition from the GAC diminished, because according to Oppenheimer, "there was an existing installation, and it could be done gradually and without harm to Los Alamos."[58]

The persistent prodding of Captain John T. "Chick" Hayward, Deputy Director of the AEC's Division of Military Application, led to a resumption of negotiations on the Laboratory issue and a commitment from Gordon Dean to include thermonuclear weapons development in the Livermore program. Historian Roger Anders summed up Dean's predicament: "Dean was forced to establish a second weapons research laboratory or watch the Department of Defense do it for him."[59]

On June 6, 1952, the Commission sanctioned the establishment of what would become known after Lawrence's death in 1958 as the Lawrence Livermore Laboratory. As Director of the new Laboratory, York was backed by a Steering Committee that included, among others, Edward Teller, John S. Foster, Jr., Arthur T. Biehl, and Harold Brown, who would take charge of the thermonuclear program.[60]

Plans were proceeding at this time for what would be the first full-scale explosion of a thermonuclear bomb, the *Mike* shot of Operation Ivy, which took place October 31, 1952, eastern standard time. Gordon Dean reported the successful event to President Dwight D. Eisenhower in now famous terms: "The island of the Atoll—Elugelab—is missing."[61] Where it stood, an underwater crater of some 1,500 yards in diameter remained. At ten megatons the detonation obliterated the island from the Pacific landscape and signalled expansion of the nuclear arsenal into the fusion field.

Designed and built by Los Alamos, the test device and all information relating to it would be kept under strict security by the AEC—a stance that prevented credit for technical contributions being given where credit was due. Teller and the Livermore Labs, because of association with the Super, received the "lion's

ERA OF THE SUPERBOMB: POLITICS TO PROOFTESTING

share" of credit from the press. This unfortunate set of circumstances, exacerbated by already strained relations and the Oppenheimer Security Hearings of 1953–1954, led to a bitter rivalry between the two nuclear labs.[62]

This confrontational relationship would have significant ramifications for the operation and eventual expansion of the Sandia Corporation. In retrospect, the destiny of the thermonuclear program, the second nuclear laboratory at Livermore, and ultimately, establishment there of a Sandia Branch laboratory to provide engineering support, were inextricably interwoven.

While politics were being played in the nation's capital, Sandia Laboratory continued to direct its research and testing activities toward pioneering studies in weapon effects, both at the Pacific Proving Grounds overseas and at the continental test site in Nevada.

Mike shot, Operation Ivy, October 31, 1952.

On April 22, 1952, Sandians (left to right) James Dempsey, Fred Smith, and Robert Lemm witnessed Operation Big Shot at Yucca Flat during an open house at the Nevada Proving Ground.

Chapter XXI.

WEAPONS TESTING ON THE CONTINENT

> *Between January 27, 1951 and May 1955, more than fifty test devices have been fired in the Pacific and Nevada. . . . In the early years, we were limited to fission type weapons of such size and shape that they could be delivered only by our biggest bomber aircraft. . . . Today, we utilize both fission and fusion types in what has been described as a whole family of weapons, meaning a variety of weapons for a variety of military purposes and delivered by many types of military vehicles.*
>
> Kenner F. Hertford, Manager,
> Albuquerque Operations Office
> Speech, Hotel Muelbach, Kansas City, March 1956.

Referred to by the AEC in its *Twelfth Semiannual Report* as "Los Alamos and Sandia Laboratory's backyard workshop," the Nevada Test Site is an area of high desert and mountainous terrain located in Nye County near Las Vegas.[1] Starting with the Ranger operation in 1951, all atmospheric nuclear weapons tests conducted within the continental United States have been staged at this 640-acre, square-mile tract adjoining the Nellis Bombing and Gunnery Range. Here tests could be set up quickly, conducted more frequently, and for less expense.[2]

Interspersed with the Ivy and Castle thermonuclear tests in the Pacific were five different continental test operations that would be directed toward studies of cratering, air pressure and blast effects, as well as the development of thermonuclear principles and design concepts. The double titles

TEST ORGANIZATION/EXERCISES DESERT ROCK I, II, AND III STRUCTURE WITHIN FEDERAL GOVERNMENT

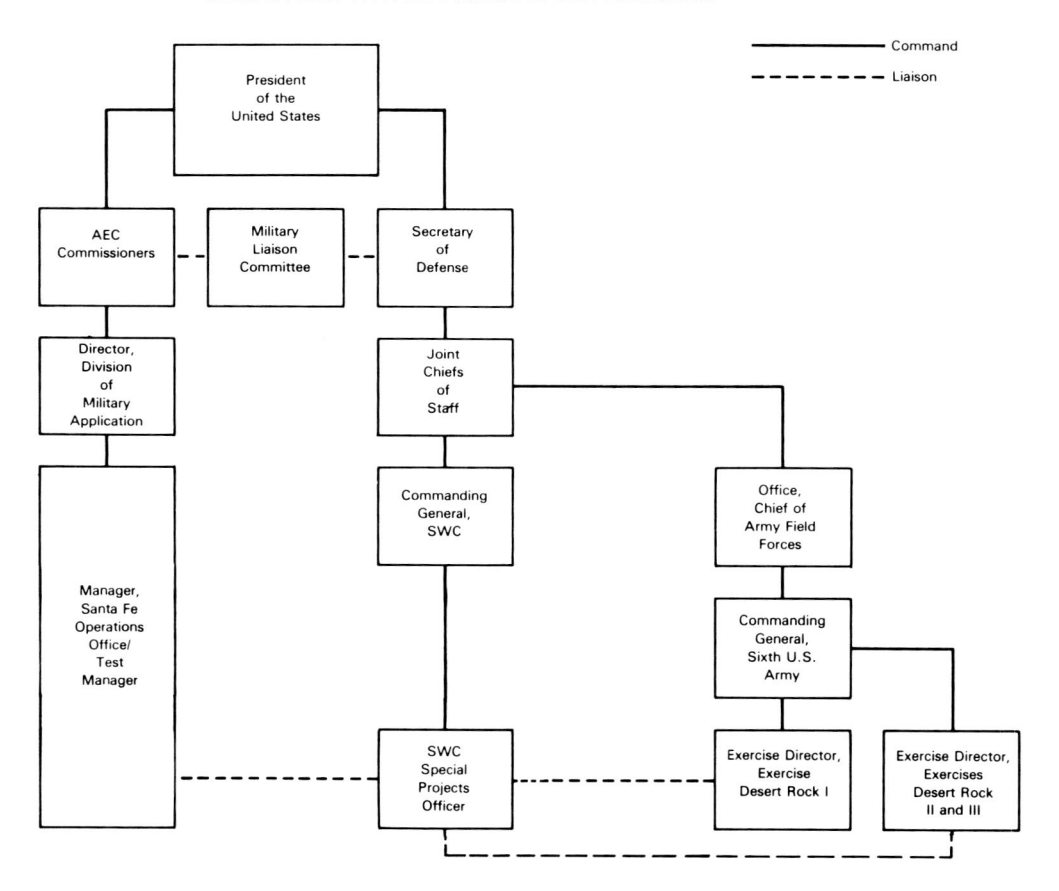

of operations Buster–Jangle, Tumbler–Snapper, and Upshot–Knothole are indicative of the dual nature of continental test administration and objectives, namely, the combination of AEC-sponsored nuclear weapon development tests with DOD-sponsored nuclear weapon effects events. In May 1955, test crews would return to the Pacific for the weapons effects test, Operation Wigwam, held off the coast of California, just as Operation Teapot was being completed in Nevada.[3]

The interest in weapons effects testing was directly related not only to the development of thermonuclear weapons with larger yields, but also to the government's interest in tactical, as well as strategic, weapons. With the Korean War at a stalemate in 1951, the idea of using tactical weapons in the field was growing in favor with military strategists and in political circles. General Omar Bradley, for example, strongly supported the concept of troop participation in nuclear weapons development and effects testing such as that planned for the Buster–Jangle series. For the first time, some 6,500 military troops participated in full-scale test maneuvers as part of the Desert Rock Exercises I, II, and III.[4]

Buster–Jangle

Historically, the combined Buster–Jangle series evolved as a result of cancellation of the *Windstorm* cratering tests in the Aleutian Islands. The military originally intended to fire both a surface and an underground burst on the remote island of Amchitka in the Aleutians, but a geological survey revealed that the alluvium and meteorological conditions were unsuitable. Meanwhile, Sandia Corporation had been designated to make air pressure measurements for the tests, so Everett Cox asked Jack Howard to fly to Amchitka to investigate conditions under which employees would have to work. Howard arrived just as the troops were pulling out, and after a quick assessment of the situation, returned with a discouraging report.[5]

NUCLEAR ORDNANCE ENGINEER FOR THE NATION

LOCATION OF NEVADA PROVING GROUND

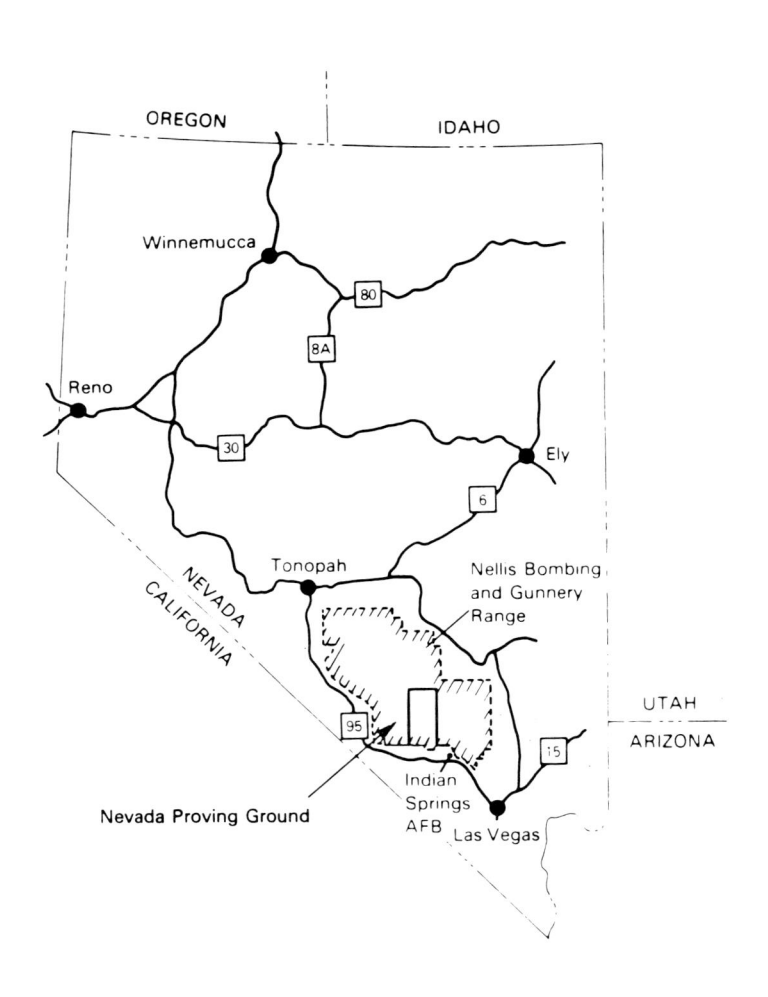

WEAPONS TESTING ON THE CONTINENT

NEVADA PROVING GROUND SHOWING GROUND ZEROS FOR OPERATION BUSTER-JANGLE

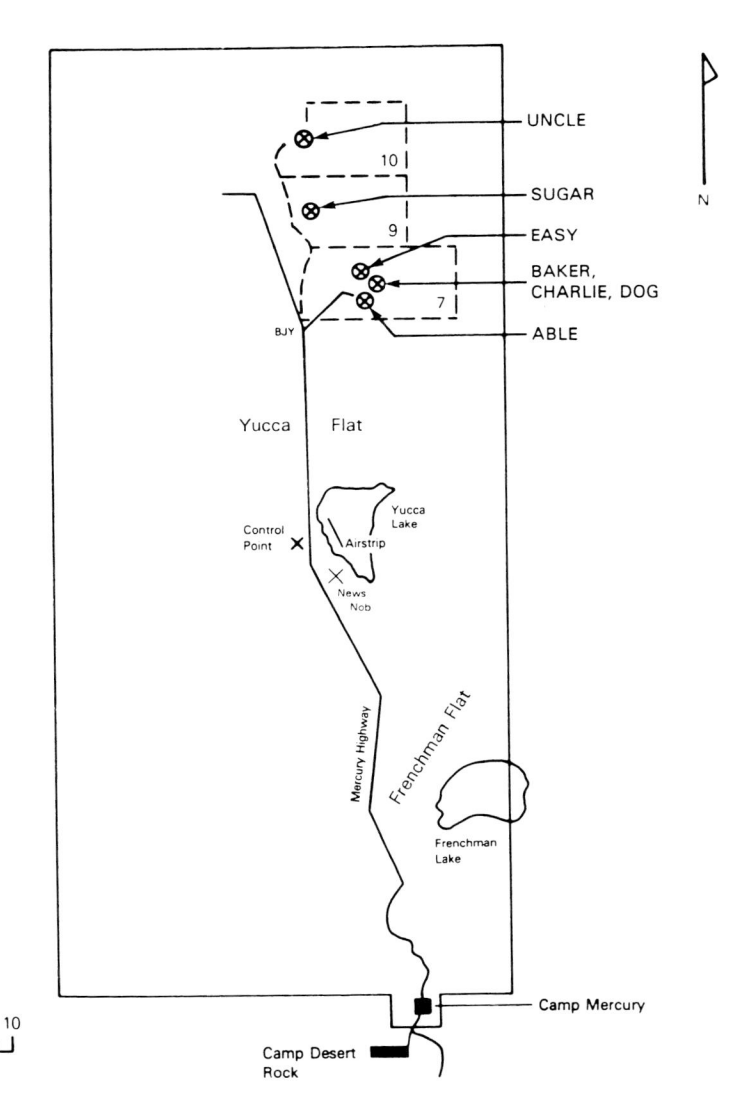

Ultimately, the AFSWP Research and Development Board made the decision to reduce the yield of the tests substantially and move the project to the Nevada Proving Grounds (as the site was known until 1955). *Windstorm*, therefore, became the second part of the Buster–Jangle series, scheduled for October 22–November 29, 1951. The original Buster nuclear weapons development tests were combined with the DOD nuclear weapon effects experiments. Accordingly, Sandia Corporation transferred its assignments to the newly merged operation.[6]

The first definitive meeting regarding Sandia's participation in Project Jangle was held on June 5, 1951. Bob Poole, Glenn Fowler, and Everett Cox of Sandia met with representatives of the military, Carroll Tyler and Admiral Kraker of the AEC, and Alvin Graves of Los Alamos. The following week, the Program Directors for Jangle visited Sandia to explain the Laboratory's tasks, which included plans to set up eight microbarographic stations. Further discussion with Test Director Alvin Graves in early August clarified the programs that Sandia would be conducting for the Los Alamos J Division.[7]

For the Buster phase of the operation, the Lab would be involved in transit time studies, blast and shock experiments, and weapons assembly. For the Jangle phase, Sandia would install instruments operated by remote control during detonation and prepare records for project teams, provide structures instrumentation, and assist with technical photography. Cox, as head of Sandia's Weapons Effects Research Program, would be assisted by H. J. Plagge, Supervisor of the Meteorological Section, and J. W. Reed, also a member of Fowler's Field Test directorate. Jack Howard, Project Leader for "Air Pressure Versus Time Studies," was placed in charge of the crew responsible for recording blast pressures at ground level stations for both the surface and underground detonations. Harlan Lenander, Project Leader for "Structures Instrumentation," supported the military structures programs. Laboratory personnel were also involved in the measurement of terrain effects, or changes caused by hills and dales, on the *Sugar* and *Uncle* shots of the Jangle series.[8]

In preparation for Sandia's assignments, Cox sent Mel Merritt out to Nevada to locate a suitable hill for conducting the pressure measurement studies. At the test site at Frenchman Flat, Merritt traded his rental car for a jeep and made his way over the rough dirt roads into the forward area. He finally selected a hill at the south end of the Banded Mountains on the east side of Yucca Valley. "The terrain was much steeper than it is in . . . cities such as San Francisco," Merritt explained, "but this was needed to exaggerate the effects we were looking for."[9] During Buster–Jangle, it would be possible for the first time to measure free air atomic blast pressures over land.

All seven shots of the Buster–Jangle series were detonated in the Yucca Flat area, a large desert valley surrounded by the mountains needed to contain overpressure and set up the weapons effects experiments. From the Control Point on the west side of Yucca Pass, observers could view the Yucca Flat test sites to the north. A network of power and timing cables connected the control building to each test area.[10]

Camp Mercury, located on the southern boundary of the Proving Grounds, served as the base of test operations. Base camp facilities offered office and living quarters for AEC and DOD participants, as well as laboratory facilities and warehouses. At Camp Desert Rock two and one-half miles to the southwest, military troops camped out in a tent city erected amid mobile trailers and quonset huts. "Water was a problem," wrote Cal Tech–trained physicist Frank Shelton. Shelton, who had grown up in the Nevada desert, recalled that the military tackled the problem "by around-the-clock runs of water tank trucks to and from Indian Springs about twenty-five miles away."[11]

With approximately 9,000 military and civilian personnel participating, Buster–Jangle was a massive operation. Seventy percent of the personnel involved took part in the Desert Rock operations. The five Buster shots (*Able, Baker, Charlie, Dog,* and *Easy*), which concentrated on AEC weapons development, had by far the majority of DOD participants. Troop maneuvers were restricted to Buster, while the *Uncle* and *Sugar* shots of the Jangle phase of the test were intended to test weapon effects.

More specifically, the objectives of Jangle were "to compare military effects and relative effectiveness of surface versus subsurface blasts, to determine the physical laws regarding shock-wave propagation, and to test military equipment and techniques."[12]

After some delay caused by weather and various operational difficulties, the Buster phase began on October 22, 1951, with a small tower detonation of less than 0.1 kiloton. There was a tense moment during the tower shot—one that remains especially memorable for Walt Treibel. Treibel had transferred to Sandia's Field Test organization in the summer of 1951, just in time to take part in an emergency situation.

On October 22, all seemed ready for the *Able* shot of the series. The device rested in its cab atop the tower, and on the ground below, countdown began: "Three, two, one, zero"—but nothing happened. The device had failed to detonate. The voice came back on the loud speaker, and with some agitation, announced the misfire: "Hold your positions! Hold your positions!" There was only one solution; someone had to disarm the device. Treibel was among the "fortunate" souls selected by the Test Director to perform the operation. "If you have ever witnessed an atmospheric detonation," Treibel recalled, "you can imagine the apprehension and fear associated with retracing the arming operation in the reverse direction."[13] To complicate matters, the elevator used to transport personnel up and down the tower had been removed. This meant that the men had to climb the ladder on the 100-foot tower to reach the device, disconnect the arming cable, and climb back down. Amazingly, the feat was accomplished without mishap.[14]

Six days later, a red-tailed B-29 from Kirtland Air Force base in Albuquerque dropped a 3.5-kiloton tactical atomic bomb—the first one of its kind—at the Yucca Flat test site. Observers at safe distances on mountains surrounding the site saw the bomb as "a burning Ping Pong ball bouncing off the black top table near Yucca Flat." The target area, cleared of all trees and shrubbery, had been blacktopped so plant matter would not interfere with the hundreds of instrument lines. According to local news reports, the mushroom cloud that followed the detonation ascended into the atmosphere "with majestic speed" and retained rigidity until it had reached an estimated 20,000 feet, where it "took on the colorful aspects of a tree, deep in the throes of autumn." Detonation of the device at such a low altitude, the report continued, "marked an important new step in the progress of atomic war games."[15]

Representative Albert Gore, a high-ranking member of the House Appropriations Subcommittee, witnessed the detonation along with three other committee members. Gore reported that "the flash of light was so intense that it outshone the sun." Upon his return to Washington, Gore told reporters that "the use of an atomic weapon is a decision for the military and the President to make, but as I witnessed the accuracy and cataclysmic effect of the explosion I felt the conviction that it [the atomic bomb] might be used in Korea if the cease-fire negotiations break down."[16] After three additional airdrop detonations ranging in yield from fourteen to thirty-one kilotons, the Jangle phase began on November 19 with the surface detonation and ended with the underground cratering shot on November 29.[17]

The problems experienced during the Ranger test series and the unexpected damage to windows in Las Vegas encouraged investigation into the cause for such an effect during Buster-Jangle. Sandia Corporation, therefore, was assigned to study what was known as the "skip-zone phenomenon." In this occurrence, blast waves leaving the point of explosion at different angles return to earth at locations varying in distance from Ground Zero. Following their initial impact with the earth, the waves bounce, carrying the sound with them in varying degrees. A related phenomenon results from sound waves created by the blast traveling up through the atmosphere to the ozonosphere. The warm layer bends them back to earth where they strike in a circle ranging from 80 to 150 miles away from Ground Zero. Thus, it is possible for a community to receive two shocks from each nuclear blast—one from sound waves skipping along the terrain, and a second one later after they have traveled thirty miles up to the ozonosphere and bent back to earth.

The "refraction–reflection" phenomenon, combined with "focussing," can cause both types of waves and can result in waves of each type being centered on a single area—hence the need for a prediction capability and a microbarographic program to help avoid atmospheric situations that could focus damaging blast waves on towns surrounding the test site.[18]

At Buster–Jangle, Sandia Laboratory was assigned still another weapons effects investigation, one that would continue through Redwing in 1956; namely, the measurement of free-field air pressure. The results from the air pressure experiments conducted by Sandians at Buster–Jangle caused consternation. The pressures turned out to be much less than anticipated and, in effect, challenged the standard reference cataloguing the "ideal" height-of-burst curves. The then-current conception of air pressures presented by Fred Reines and Francis Porzel of Los Alamos in LA-743 took for granted that the surface of the earth would be a perfect reflector and that the air pressures would have sharply rising fronts followed by decreasing pressures. Mel Merritt, who participated in these studies, explained the significance:

> For air bursts there would be increased peak pressures on the ground, particularly in the region of Mach reflection, which could mean greatly increased ranges for damaging levels of air pressure. The resulting height-of-burst curves had sharp knees leading to very strict requirements on fuzing, one of Sandia's principal businesses at the time. However, the pressures we measured on the Buster shots were anything but ideal. Instead of seeing a sharply rising shock front, we observed slowly rising waves with peak pressures much reduced from expectations.[19]

Confidence in the basic theory was so great that considerable doubt was expressed regarding the validity of the data derived from the Sandia experiments. As a result, just before Christmas a meeting was held at Princeton University to decide what should be done. Byron Murphey made a presentation, explaining what had been measured by Sandia and how. The meeting concluded with John von Neumann, whose vote was highly respected, recommending confirmation of data through additional testing.

The discrepancies caused such serious concern because the lower pressures derived from the Sandia experiments, if accurate, could dramatically impact strategic weapon planning. Since the measured peak pressures did not extend nearly as far from Ground Zero as had been assumed, it was possible that atomic weapons would not be as effective as anticipated. The military, alarmed because of what they felt to be the data's implications for effectiveness of the stockpile, wanted the test series planned for 1952 accelerated in order to confirm or deny the data. The perceived urgency of the situation encouraged the Joint Chiefs of Staff to direct AFSWP to carry out Operation Tumbler in combination with the AEC's Operation Snapper scheduled for the spring of 1951.[20]

Tumbler-Snapper

Operation Tumbler-Snapper, which would either verify or disprove Sandia's findings relating to blast phenomenology, began as planned on April 1, 1952. The entire program, involving four tower shots and four air drop detonations, extended through the fifth of June. Nuclear devices, designed and built by Los Alamos, were used for the two weapons effects shots and for the six weapons-related events. The objective of the weapons development shots was to test weapons for the nuclear stockpile and techniques to be used at the Pacific Proving Grounds for Operation Ivy.[21]

Colonel Kenner F. Hertford of the AFSWP Field Command was selected to serve as Test Director, reporting to Test Manager Carroll L. Tyler of the Santa Fe Operations Office. An energetic and talented career officer, Hertford was loaned to the AEC for the duration of the operation. Hertford explained the circumstances leading to his appointment:

> The question arose concerning the conduct of the present full-scale tests and the AEC insisted on doing the job both for the military and Los Alamos. A compromise was reached whereby the AEC assumed full control and responsibility

Colonel Kenner F. Hertford, AFSWP Field Command, was Test Director for Tumbler–Snapper.

Hertford was promoted to Major General, and after his retirement from the Army in the fall of 1955, he was appointed Manager of the AEC Albuquerque Operations Office.

with an officer on active duty as AEC Director of Test Operations. . . . Having been in this business four years and having worked with the long-hairs some, I became the people's choice and have been the ringmaster out here for this eight-ring circus. If I say so myself, it has been fairly successful and laid the basis for better mutual understanding on both sides.[22]

Hertford's broad knowledge of the atomic weapons program, combined with a strong background in engineering and military administration, made him a suitable choice to deal with the sensitive issues related to directing a joint military/civilian operation. According to Hertford, the problems were sometimes as basic as keeping military personnel from encroaching on the technical areas and disturbing the test instrumentation set up by the scientists. Once again thousands of soldiers would participate in troop maneuvers and training exercises as part of the Desert Rock IV Exercise.[23]

AEC Chairman Gordon Dean, impressed by Hertford's performance of duties, later requested his permanent assignment to the Santa Fe Operations to allow him to continue as Director, Office of Test Operations. In the fall of 1955, after retirement from the Army as a Major General, Hertford would accept the position of Manager of the AEC Albuquerque Operations Office.[24]

For Tumbler-Snapper, Scientific Test Director Alvin Graves of Los Alamos had reporting to him M. "Pete" Scoville as Deputy for Military Effects and William E. Ogle as Deputy for Weapons Development. J. C. Clark of Los Alamos served as Assistant Test Director. As at previous tests, representatives from Sandia's Research and Development groups, including Field Test, Engineering, and the Weapons Effects organizations, participated in a support capacity. Harlan Lenander, who reported to Sandia's Director of Field Test Glenn Fowler, was the Sandia representative at the site. Lenander and Everett Cox both served as members of the Test Manager's Advisory Panel. Harvey North headed the Weapons Assembly group.[25]

For the first time since Crossroads, uncleared observers, in addition to the military, AEC, and contractors, were allowed to view a nuclear detonation. On April 22, 1952, visitors watched from the News Nob observation point near the airstrip on Frenchman Flat as countdown began for the *Charlie* shot of the Snapper phase of the operation. The detonation, staged to provide the AEC's Los Alamos Laboratory with diagnostic data on nuclear weapons design, was described as an "unusually clean and spectacular atomic cloud." Rocket trails and antiaircraft smoke puffs accompanied the blinding light surrounding the fireball, which ascended against the backdrop of a deep red sky. "The cloud left a white vapor trail as it rose," one observer recalled; "there was no mushroom stem as ordinarily occurs in low level bursts."[26]

In addition to *Charlie*, five other weapons–related tests were conducted during the Tumbler–Snapper operation. Of major concern to Sandians, however, were the weapons effects tests being carried out for the DOD. During these tests, the Laboratory hoped to prove the effect of height–of–burst on blast overpressure and the existence of the precursor observed at Buster–Jangle.[27]

To verify the controversial results of the Buster–Jangle operation, two service laboratories—the Army's Ballistic-Research Laboratory and the Naval Ordnance Laboratory—made air pressure measurements in parallel with Sandia. The respective tests showed that pressures measured along the ground were indeed quite "nonideal." Photographs of the shock waves indicated that a precursor wave along the ground, presumably caused by a hot layer of air, accounted for the slow rise in pressure. For the tests, Sandia developed special gauges to observe the dynamic pressure histories and instruments ("Greg" and "Snob") to separate the air dynamic pressure and dust momentum flux of the precursor afterflow. According to Cal Tech–trained physicist John Banister, "It was found that the contribution of the dust to the effective dynamic pressures generally exceeded that of the air."[28]

Several weeks after completion of the operation, a Blast Review Meeting was held at the AFSWP Field Command headquarters at Sandia Base. At this meeting, attended by representatives from the various laboratories and AFSWP, Everett Cox and Byron Murphey gave significant presentations reporting the Lab's findings relating to blast phenomenology. Among the participants was a promising young Southerner who had just received his Ph.D. in physics from Vanderbilt University before joining Sandia Corporation in the Weapons Effects Department. Hired at age twenty-four as the youngest Ph.D. to join the Lab to date, Thomas B. Cook soon became interested in studies relating to the effects of nuclear weapons, particularly at high altitudes.[29]

At the Blast Review Meeting being conducted in 1952, Cook presented a summary of the Sandia shock-gauge evaluation tests conducted at Tumbler-Snapper. Bill Perrett, who had measured pressures in the ground on the Buster-Jangle shots, reported on earth stresses and pressures—an area of investigation that would become extremely important when full-scale testing had to be conducted underground.[30]

Despite vindication of the Laboratory's findings by the tests conducted at Tumbler-Snapper, there was a secondary result. With other research laboratories now directly involved in the area, Sandia no longer had a monopoly. Ultimately, these competent laboratories would assume the entire responsibility for the air pressure field. Sandia then would expand its research activities into related measurements such as dynamic pressures, dust loading, and particle velocities.[31]

Not long after Tumbler-Snapper and the Blast Review Meeting that followed it, Everett Cox and Byron Murphey assigned Frank Shelton to the challenging physics problem of blast-thermal interactions. Shelton was asked to learn all that he could about the abnormal blast wave forms referred to as precursors. Since Sandia's weapons effects studies were to continue during Operation Upshot-Knothole scheduled for the spring of 1953, Shelton and Carter Broyles—another talented new member of Sandia's research team—made plans "to work the Sandia blast lines at Frenchman Flat on shots *Encore* and *Grable*."[32] Carter Broyles, like Tom Cook, had recently received his Ph.D. in Physics from Vanderbilt.

Left to right: Byron Murphey, Jim Shreve, and Carter Broyles discuss results of testing activities during the Upshot–Knothole series.

As Luke Vortman observed: "An outcome of the series was that we could understand structural reponse if we understood the airblast loading. In order to improve our knowledge of the subject, the Armed Forces Special Weapons Project sponsored a program at Sandia to study blast loading of structures using high-explosive sources of blast and small-scale structures." In addition to these tests, which were conducted at the Coyote Canyon test site, Sandians made preparation for the Upshot–Knothole series planned for March through June 1953 at the Nevada Proving Grounds.[33]

Upshot-Knothole

Both fission and fusion theories were investigated at Upshot-Knothole, in addition to "a new and revolutionary method of producing deliverable thermonuclear weapons."[34] The military part of the tests included an air drop over an extensive array of instrumentation, and demonstration of the firing of a nuclear cannon shell. For the first time, the newly formed weapons laboratory in Livermore participated in a full-scale test operation.[35]

During the *Annie* event of Upshot-Knothole, as part of the Desert Rock V exercise, tactical troop maneuvers provided military personnel with experience in the battlefield use of nuclear weapons. Approximately 2,000 soldiers participated.[36]

By 1953, residents of the downwind states of Nevada, New Mexico, Arizona, and Utah were beginning to express concern over the effects of fallout from the continental tests. As historian Frederick J. Alexander recorded: "Radioactive fallout on nearby communities caused considerable public clamor."[37] Government concern relating to the effects of radiation and blast damage was reflected in active participation by the Federal Civil Defense Administration Civil Effects Group. Members of this group tested emergency preparedness plans and conducted scientific projects to assess nuclear effects on civilian structures and food supplies. According to Scientific Test Director Alvin Graves, the "nuclear diagnostic shot" incorporated civil defense experiments involving cars and houses with strategically positioned mannequins. Deputy Director Robert L. Corsbie, AEC Washington, coordinated the civil effects tests.[38]

Sandia's Weapons Effects Manager Everett Cox was assisted during the Upshot-Knothole operation by Byron Murphey and Carter Broyles, who acted as liaison for the blast program. Blast effects information, interpreted through microbarograph research, was again an important component of this test series.

On test day, March 17, 1953, microbarograph teams awaited the countdown, signalling successful completion of months of preparation. Ten people crowded into the Sandia Laboratory office at Control Point to await the first continental test shot of 1953. Those present showed a weary sort of excitement as they anticipated the climax of many long hours of work. At microbarograph stations throughout the surrounding country, other Sandians stood ready: Carl Csinnjinni, Hans Hansen, and Francis E. Thompson in Las Vegas; Price Hampson and Trevor C. Looney in Cedar City, Utah; Pete Church and Jim Valentine in Caliente, California. Other team members at stations in Goldfield, Pasadena, Indian Springs and Mercury, Nevada, and St. George, Utah, waited with eyes on sensitive equipment for the signal indicating graphically that detonation had occurred.[39]

Military participants watch an Upshot–Knothole test.

For this test, several hundred news reporters and Civil Defense representatives, eyes protected by dark glasses, also watched for the now familiar mushroom-shaped cloud. Activity at the Control Point, brain center of the operation, slowed in anticipation. Radios were silent; all was ready. Among those present, Harlan Lenander—the bill of his baseball cap turned up—looked toward Ground Zero. Everett Cox, commemorating this St. Patrick's Day event with a green shirt and socks, also gazed toward the tower. Inside the office, Norm Bollinger prepared to operate the pressure gauge recording equipment. Don Larson paused temporarily in his job of furnishing technical reports to the microbarograph stations. Albert T. Marrs and Robert A. MacArthur finished the weather computations, and Gus Gustafson and John Harding stopped monitoring the radio and telephone. Hazel Williams and Clara Koebke, their faces showing the strain of many hours spent recording communications, also waited for the blast. Secretary Dorothy Harvey learned that the device sitting on its tower some eight miles away had been "dedicated" to her by co-workers.

Harlan Lenander stands next to one of the Sandia trailers that were a part of the Lab's field offices.

In Caliente, Nevada, Pete Church and Jim Valentine heaved a sigh of relief after some anxious moments. A flat tire on a generator trailer, a double load on a pick-up truck, and some 230 miles of driving kept them from getting their microbarograph equipment set up until just twenty minutes before firing. In St. George, Utah, Robert Prichett and Robert Thompson bent over their instruments, knowing that it would still be twelve minutes after detonation before any pressures were recorded. Throughout Yucca Flat, strategically placed instruments were ready to record the massive release of energy, anticipated to be the equivalent of approximately 15,000 tons of TNT.

After the shot, when the area was deemed sufficiently safe, several Sandians would go forward to retrieve instruments and devices placed there during the past weeks. Included in this squad were Bob Bunker, Emory Whitlow, "Sid" Swartzbaugh, Dean List, and Dean Thornbrough. At the present, however, their eyes, like those of many others, were glued on the blinking light above the tower. With only two seconds to go, Henry B. Lauerson, atop a ledge on a rocky knoll, adjusted his footing and leaned forward in anticipation. He thought of the measuring devices he had placed and the results they would show. Jesse Ward, Sandia's high-explosives expert, remarked that he expected the coming spectacle to dwarf the 4,800 pounds of TNT he had placed in two holes for the pretest firing.

One second to go; little time to think; goggles adjusted. FIRE! The nuclear fireball takes shape—an awesome spectacle representing the successful efforts of the scientists and engineers who worked on it.[40]

For purposes of investigating the precursor wave phenomenon, Upshot 10 proved to be the single most important effects experiment of the operation. While it was well-known that shock waves from the explosion bounce across the surface of the earth, the trick was to predict accurately their effects. Sandia's microbarograph program helped provide the answers.[41]

WEAPONS TESTING ON THE CONTINENT

Field Testers watch an H.E. blast during Upshot-Knothole.

A recovery team prepares to enter a test area to recover data from one of the tests conducted during the Upshot-Knothole series. Left to right: Bob Bunker, Emory O. "Shorty" Whitlow, Sid Schwartzbaugh, Dean Thornbrough, John Minck, and Dean List.

Charlie Shot,
Operation Tumbler–Snapper,
April 22, 1952.

Grable Shot, Operation Upshot–Knothole, April 25, 1953.

WEAPONS TESTING ON THE CONTINENT

Met Shot Tower Burst, Operation Teapot, April 15, 1955.

Romeo Shot, Operation Castle, March 26, 1954.

Teapot

After the unfortunate *Bravo* event of the 1954 Castle operation, during which radioactive debris dusted a Japanese fishing vessel, the concern over fallout increased. The death of one of the fishermen of the ill-fated *Lucky Dragon* and the illness of other crew members reinforced the fear that "repeated atomic testing might seriously contaminate the earth's atmosphere."[42]

By the time Operation Teapot had been scheduled for February to May 1955 at the Nevada Test Site, scientists were attempting to develop operational criteria for conducting atomic tests so that fallout would be minimized. The new criteria prescribed the heights above the terrain best suited for detonation of weapons of specific yields. Meanwhile, the emphasis on weapons design was changing from almost exclusively offensive devices to those with defensive capabilities. The military was particularly interested in missile warheads for air-to-air and ground-to-air use. Small-sized, lightweight warheads with the device tailored to "a particular application without regard for interchangeability" were the result.[43]

Interest in the use of atomic warheads in antiaircraft missiles created a need for working curves showing how characteristics of parameters of atomic bursts vary with distance at different burst heights. In 1954, Tom Cook and Carter Broyles of the Sandia Weapons Effects Department, using calculations made by Klaus Fuchs of Los Alamos, produced a reference providing this information. Titled "Curves of Atomic Weapons Effects for Various Burst Altitudes (Sea Level to 100,000 feet)," this valuable document became better known as "The Cook Book."[44]

The growing interest in nuclear weapons effects coincided with investigation of Civil Defense requirements during the Teapot operation. In addition, new nuclear devices were tested for possible inclusion in the weapons arsenal, and exercises were conducted with the objective of improving military tactics, equipment, and training. The Teapot series also marked the first attempt to study the changes in blast wave coupling with increasing altitude. A device was exploded 30,000 feet above sea level after being dropped from an airplane flying at considerably higher altitude. Sandia deployed a wide variety of instruments, including bursting smoke grenades, to allow the shock afterflow history to be traced by photography. Other experiments during Teapot initiated a principal theme of field test work in underground testing; namely, examination of the response of nuclear weapons to the hostile environment generated by other nuclear bursts. The program, which started with exposure of mocked-up Mk 5's to airblast at Teapot led to more extensive testing employing the so-called "Totem Pole" in 1957. John Banister, who would participate in many such studies during his lengthy career at Sandia, observed that "these preliminary experiments were the start of the critical examination of the vulnerability of our stockpile weapons."[45]

For the Teapot series, Sandia personnel had a new Director of Field Test—Richard A. Bice. In the summer of 1954, Bice was promoted to succeed Glenn Fowler, when Fowler became Director of Electronics. During the Teapot operation, Sandia engineers under Theodore Church, Manager of the Electronic Research Department, were assigned to provide project support for the Los Alamos Laboratory's weapons development activities. Church, assisted by Barney J. Carr and Carroll B. McCampbell, were responsible for gathering information about the nuclear device at the moment of detonation. To do so, they set up complex instrumentation in shelters near the target or tower. Wireless telemetering instruments transmitted the data. Harvey North and Robert C. Spence headed other weapons projects in support of the Test Director's activities.[46]

Sandia projects relating to military effects experiments were coordinated by Roland S. Millican and Luke J. Vortman, in combination with Civil Defense Administration personnel. Luke Vortman, program director for the AEC's civil defense test program, was assigned to evaluate the various types of personnel structures exposed to an atomic explosion. This particular program also dealt with the effects of a nonideal shock wave on blast loading of a structure and with the nuclear effects on machine tools.

To carry out this assignment, Vortman directed construction of a concrete building six feet high, six feet wide, and thirty-six feet long. When the formidable structure was completed, Wiancko pressure gauges were placed on the top, back, front, and ends, and it was filled with dirt. Despite the strength of its construction, the force of the explosion completely destroyed the structure, which ended up in pieces some distance downwind. "Because of its rectangular shape," Vortman recalled, "it was properly dubbed the 'Galloping Domino.' " However, before the instrumentation cables broke, early shock wave data were obtained that proved to be of considerable interest since the structure was located in the precursor zone. During the fifties, nuclear effects studies conducted by Sandia and other agencies would expand the realm of knowledge to the extent that Samuel Glasstone's well-known 1950 publication, *The Effects of Atomic Weapons*, by 1957, would be reprinted with updated information and a new title—*The Effects of Nuclear Weapons*.[47]

Operation Teapot, the fifth in a series of atmospheric nuclear weapons tests to be conducted by the AEC within the continental United States, began on February 18, 1955, and extended through May 15, 1955. On the day of this first shot, more than 100 Sandia Corporation employees were on the scene. After having dinner at Camp Mercury, Sandia President James W. McRae and the Laboratory's Board of Directors joined other official observers to witness the test. Lewis A. Strauss, Chairman of the U. S. Atomic Energy Commission, and General John A. Dahlquist, Commanding General of the Army's Continental Command, were among the dignitaries in attendance.[48]

A short time before detonation, one or more charges of conventional explosives were set off, and readings were taken at microbarograph stations in communities around the perimeter of the test site and in neighboring states. These readings, plus meteorological data analyzed on a newly developed analog computer called Ray Pac, provided scientists with information needed to predict the intensity of the blast in populated areas. The microbarograph work, coordinated by W. A. Gustafson, provided a record of the blast pattern.[49]

The Los Angeles *Times* described this first atomic test of the 1955 winter–spring series as a "baby blast" air-dropped over Yucca Flat. The small one-kiloton detonation, so slight "that it failed to interrupt the click of dice and the hum of the roulette wheels at the gambling casinos," was the first of fourteen tests scheduled for the Teapot series. In contrast, on Washington's birthday, a tower shot exploded with a force that jarred cities 135 miles away. According to Senator Russell Long of Louisiana, one of several Congressional observers, it was "an opportunity to foresee the definite possibility of these small-sized nuclear weapons, launched either from the ground or from aircraft against invading forces."[50]

Indeed, one of the most significant results of the Teapot test series was the decision to apply new techniques to the development of smaller warheads for tactical, antisubmarine, and air defense applications. Furthermore, the analog computer system, designed by Sandia to predict distant damage effects, proved to be extremely accurate.[51]

For this test operation, Sandia also contributed a small, laboratory-developed package recording system that eliminated the need for large instrument shelters. This cost-savings measure resulted from the development of a device called WAFCOM (Wow and Flutter Compensator). WAFCOM, invented by James H. Scott and Jim Valentine, eliminated the need for heavy recording equipment previously required to record blast data on magnetic tape. Later, back in the laboratory, when the tapes were played for data analysis, WAFCOM compensated for much of the distortion introduced in the lightweight recorder used in the field.[52]

Meanwhile, in the Pacific off the coast of southern California, other Sandians were involved in preparations for a deep underwater test called Wigwam scheduled for May 14, 1955. At the request of the Navy Department, Sandia Laboratory was to do the over-water pressure instrumentation. Lenander asked Jim Scott, who by this time had been promoted to Division Supervisor in the Full Scale Test Department, to help him figure out a way to jerry-rig a skyhook device to elevate the pressure gauges to the required 500 feet above the water. As the plan developed, they conceived the idea of using an array of large, nylon-covered balloons as the means to keep the instrumentation aloft. The balloons were to be attached to a five-mile-long steel cable being towed by a seagoing tug. This cable, in turn, supported submarines positioned beneath the surface of the ocean at different distances from the detonation point. The "Zero" Barge with the nuclear device attached was secured to the end of this towed cable. The submarines served as the test vehicles. Pressure gauges and the radio transmitters for telemetering the information were mounted on the cable tethering the balloon.

Left to right, Richard Bice, Don Shuster, James Scott, and H. E. Hansen prepare to board a helicopter for an observation flight over a test site.

Development test runs had taught field testers the hazards of large inflated balloons blowing amok, towing trucks across testing fields, and landing in unpredictable places, but Operation Wigwam introduced them to still another unexpected situation. On the trial test, Scott and Lenander in a boat below, watched with dismay as two Navy planes made a sweep near the balloons. Dismay turned into real anxiety as they saw one of the airplanes clip the cable tethering the balloon. In the process, the cable nearly severed one of the wings of the plane; however, the aircraft landed safely. "Almost everything went wrong on this test," Scott and Lenander recalled. "A combination of high winds and rough seas prevented recovery of much of the test data."[53] The following year, during the Pacific test operation, Redwing, the role of Sandians continued in weapons effects studies.

The important results garnered from Pacific and continental test operations such as these promised to make the workload for Sandia Laboratory even heavier. And on the horizon, increasing concern regarding atmospheric nuclear testing would lead eventually to an international moratorium. In the interim, scientists and world leaders alike began to focus on the applications of atomic energy for both war and peace.

Left to right, James McRae, Glenn Fowler, Richard Bice, and Everett Cox stand by the Sandia Corporation tent on Ujelang Atoll, May 1956.

NUCLEAR ORDNANCE ENGINEER FOR THE NATION

FIELD TESTING
1956

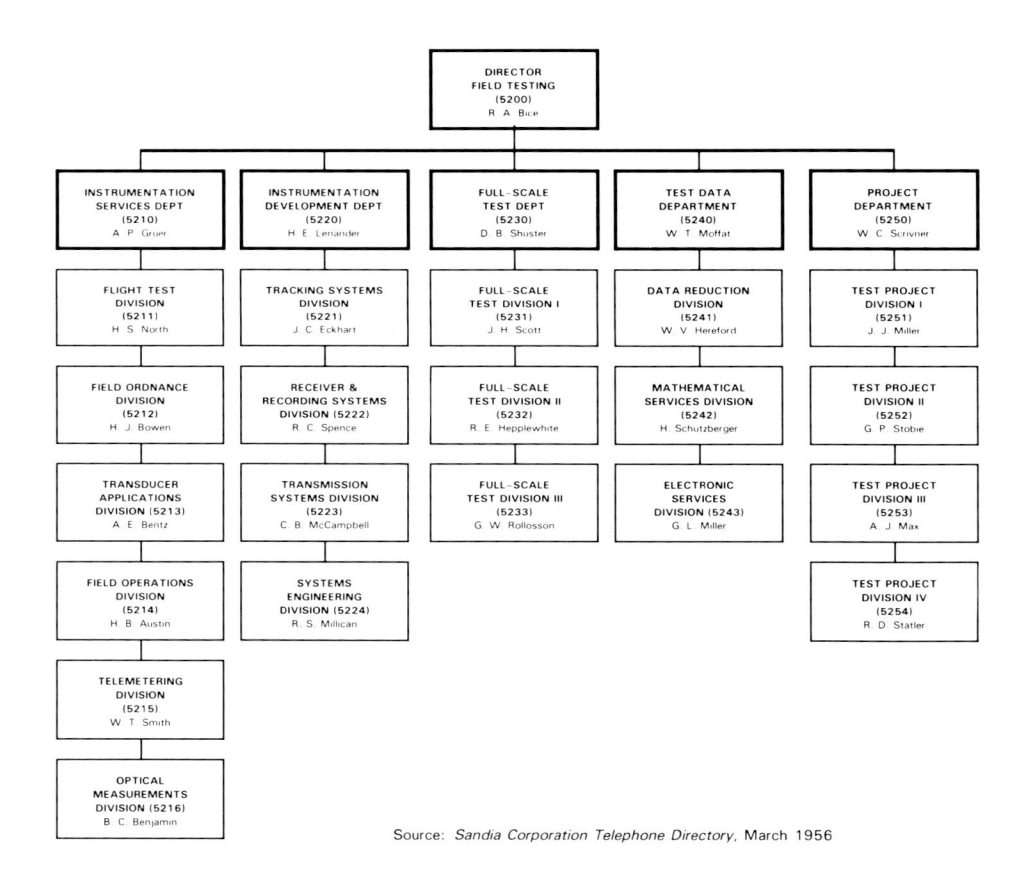

- **DIRECTOR FIELD TESTING (5200)** — R. A. Bice
  - **INSTRUMENTATION SERVICES DEPT (5210)** — A. P. Gruer
    - FLIGHT TEST DIVISION (5211) — H. S. North
    - FIELD ORDNANCE DIVISION (5212) — H. J. Bowen
    - TRANSDUCER APPLICATIONS DIVISION (5213) — A. E. Bentz
    - FIELD OPERATIONS DIVISION (5214) — H. B. Austin
    - TELEMETERING DIVISION (5215) — W. T. Smith
    - OPTICAL MEASUREMENTS DIVISION (5216) — B. C. Benjamin
  - **INSTRUMENTATION DEVELOPMENT DEPT (5220)** — H. E. Lenander
    - TRACKING SYSTEMS DIVISION (5221) — J. C. Eckhart
    - RECEIVER & RECORDING SYSTEMS DIVISION (5222) — R. C. Spence
    - TRANSMISSION SYSTEMS DIVISION (5223) — C. B. McCampbell
    - SYSTEMS ENGINEERING DIVISION (5224) — R. S. Millican
  - **FULL-SCALE TEST DEPT (5230)** — D. B. Shuster
    - FULL-SCALE TEST DIVISION I (5231) — J. H. Scott
    - FULL-SCALE TEST DIVISION II (5232) — R. E. Hepplewhite
    - FULL-SCALE TEST DIVISION III (5233) — G. W. Rollosson
  - **TEST DATA DEPARTMENT (5240)** — W. T. Moffat
    - DATA REDUCTION DIVISION (5241) — W. V. Hereford
    - MATHEMATICAL SERVICES DIVISION (5242) — H. Schutzberger
    - ELECTRONIC SERVICES DIVISION (5243) — G. L. Miller
  - **PROJECT DEPARTMENT (5250)** — W. C. Scrivner
    - TEST PROJECT DIVISION I (5251) — J. J. Miller
    - TEST PROJECT DIVISION II (5252) — G. P. Stobie
    - TEST PROJECT DIVISION III (5253) — A. J. Max
    - TEST PROJECT DIVISION IV (5254) — R. D. Statler

Source: *Sandia Corporation Telephone Directory*, March 1956

A Rad Safe monitor prepares Glenn Fowler and James McRae for entering a contaminated area to recover instrumentation records.

In this microbarograph station on the Ujelang Atoll (above), Sandia field tester W. A. Gustafson (below) records pressures resulting from detonations at the Enewetak Proving Grounds.

WEAPONS TESTING ON THE CONTINENT

Among the observers at Operation Redwing in 1956 were (beginning second from left, left to right) Edward Teller, Ernest Lawrence, Donald Cooksey, Robert Poole, and Herbert York.

James W. McRae, third President of Sandia Corporation.

Chapter XXII.

ADVANCING TECHNOLOGICAL FRONTIERS

> *President Eisenhower recently stated that the country stands ready to put all nuclear weapons permanently aside and to devote some of our huge expenditures for armament to the greater cause of mankind's welfare. . . . He noted that this nation had two tasks. One was to seek assiduously to evolve agreements with other nations that will promote trust and understanding among all peoples. The other was to make sure, in the meantime, that the quality and quantity of our military weapons command enough respect to dissuade any other nation from the temptation of aggression. Thus it is the job of Sandia to develop weapons, not to wage war, but to prevent war.*
>
> W. J. Howard, Director of
> Systems Development at Livermore
> and later ATSD(AE)
> Speech presented May 23, 1957

On a winter day in December 1953, President Dwight D. Eisenhower addressed the General Assembly of the United Nations. His introductory remarks called attention to the somber years passed, the tension of the times, and the dangers that lay ahead. As a historical frame of reference, he directed his listeners backward in time to the inception of the atomic era: "On July 16, 1945, the United States set off the world's first atomic explosion. Since that date in 1945, the United States of America has conducted forty-two test explosions."[1]

The President summarized the technological advances the nation had made and called attention to the fact that atomic bombs were now twenty-five times as powerful as the weapons "with which the atomic age dawned," while hydrogen weapons ranged in the "millions of tons of TNT equivalent." He noted that the number, size, and variety of weapons in the stockpile had increased rapidly. "But the dread secret, and the fearful engines of atomic might, are not ours alone," he warned.[2] By 1953, the U.S. monopoly of those weapons no longer existed.

Just nine months after the U.S. successfully detonated the "Mike" shot on October 31, 1952, Stalin's successor Georgi Malenkov announced triumphantly that the U.S.S.R. had exploded its own hydrogen bomb. Meanwhile, Great Britain had also joined the ranks of nuclear nations with the detonation in 1952 of an atomic bomb off the coast of Australia's Monte Bello Island. Therefore, it was time, Eisenhower urged, "to reach beyond the human degradation and destruction" to apply atomic power toward peaceful goals and "the benefit of all mankind."[3]

The President's "Atoms for Peace" plan, as the program came to be known, was an attempt to add a new dimension to the politics of the atomic age. Although the Russians would reject Eisenhower's overture for an International Atomic Energy Agency to handle allocation of fissionable materials for peaceful purposes, the atoms for peace concept encouraged new paths for scientific pursuit. Soon this expansion of atomic research and development would be reflected in government support for nuclear power reactors to generate electrical energy and for other peaceful applications. In keeping with this new line of thinking, the Atomic Energy Commission on September 26, 1953, considered an informal proposal to construct a new reactor capable of producing both plutonium and electric power. Such a dual-purpose reactor would therefore have a dual benefit, both to the defense complex and the public sector. In 1954, the new Atomic Energy Act, reflecting a similar duality, assigned the AEC to continue its weapons program while promoting the private use of atomic energy for peaceful applications and protecting the public health and safety.[4]

The mid-fifties period was indeed a time of contradictions, in ideals and in actions: The ink was hardly dry on the Korean truce of 1953 when the Civil War in Indochina erupted into a full-fledged Communist attack on Dien Bien Phu. In New York, Eisenhower spoke of atoms for peace, but people in small-town U.S.A. would soon be constructing bomb shelters in their backyards. Technologists toyed with the idea of a nuclear-powered airplane; and on the stocks in Groton, Connecticut, the world's first atomic-powered submarine, the *Nautilus*, took shape. While Eisenhower proposed to expand the parameters of the atomic shield for peaceful purposes, military and political leaders advocated "a bigger bang for the buck," ballistic missiles to span continents, the policy of massive retaliation, and a commitment to tactical nuclear weapons to counter the Russian threat in Europe. In societal terms, Eisenhower's "Atoms for Peace" epitomized the ambivalence of the age. While peaceful applications for atomic energy remained largely a political and moral ideal, the majority of defense dollars went toward military research, development, and testing.[5]

As time moved toward the tenth anniversary of the world's first atomic explosion, the volume of research and development work at the Sandia nuclear ordnance laboratory reached its highest point. Results were seen in milestone improvements in materials, fuzing systems, and long-life components. Combined, these advances led to two developments of radical conception—the laydown and the wooden bombs. In response to the national need, Sandia Laboratory, with Los Alamos, also focussed on the design of weapons for tactical as well as strategic use, programs for the development of high-yield thermonuclear weapons on a crash basis, and on fission weapons with a wider range of applications. The man with ultimate responsibility for accomplishing these tasks was the Laboratory's new president, James W. McRae.[6]

President Eisenhower signed the revised Atomic Energy Act on August 30, 1954, creating a private power industry and permitting greater atomic cooperation with American allies. Those present for the historic occasion included (seated, left to right) President Eisenhower, Representative Sterling Cole, AEC Chairman Lewis Strauss; (standing) MLC Chairman Herbert B. Loper, Senator Edwin C. Johnson, Representative Carl Hinshaw, Representative James E. VanZandt, Representative Melvin Price, Representative Carl T. Durham, and Commissioner Thomas Murray.

ADVANCING TECHNOLOGICAL FRONTIERS
McRae Becomes Sandia's Third President

When Donald A. Quarles left Sandia to become Assistant Secretary of the Department of Defense on July 29, 1953, the Board of Directors selected as his successor James McRae, a blond-haired man of medium height, kindly demeanor, and impressive credentials. A native of Vancouver, British Columbia, McRae received his Bachelor of Science degree from the University of British Columbia and his Masters and Doctorate degrees from the California Institute of Technology. In 1937, he joined the Bell System, where his early work included research on transoceanic radio transmitters and microwave techniques for both civilian and military applications. With the onset of the war, he was commissioned a major in the U.S. Army Signal Corps and coordinated development programs for airborne radar equipment and radar countermeasure devices. He became Deputy Director of the Engineering Division of the Signal Corps Engineering Laboratories before his release from the military with the rank of lieutenant colonel in 1946. He returned to BTL, and in 1947 became Director of Electronic and Television Research. In all, he held three directorate-level positions at BTL before being appointed vice-president of systems development in 1951.[7]

On September 1, 1953, McRae assumed his duties as President of Sandia Corporation. Joining him in Albuquerque were his wife Marian and their four children. For Marian, who had grown up in Oklahoma, the move was a return to the Southwest; for the children it was a new experience. McRae's closely knit family would soon be integrated into the Southwestern lifestyle and the Sandia culture. Like his predecessor, McRae and his friendly, people-oriented approach were well-received. His residence soon became a social extension of the Sandia scene. A focal point of interest for first-time visitors was an impressive train set that McRae had built for his children in the basement of their home.[8]

On the job at his office in the Administration Building inside Sandia's Tech Area I, McRae held a series of orientation

Attending the Sandia Corporation Board of Directors meeting in February 1954 were (left to right) Walter L. Brown, Vice-President and General Counsel, Western Electric Company; James E. Dingman, Vice-President and General Manager, Bell Telephone Laboratories; Frederick R. Kappel, President, Western Electric Company; Mervin J. Kelly, President, Bell Telephone Laboratories; Timothy E. Shea, Vice-President and General Manager, Sandia Corporation; Henry C. Beal, Vice-President, Manufacturing, Western Electric Company; George A. Landry, Vice-President, Purchasing and Traffic, Western Electric Company, and former President of Sandia Corporation. Standing, left to right, Frederick R. Lack, Vice-President, Radio Division, Western Electric Company; and James W. McRae, President, Sandia Corporation.

ADVANCING TECHNOLOGICAL FRONTIERS

President McRae maintained an active schedule, conducting numerous meetings with Sandia personnel and hosting visiting dignitaries. Above, McRae and General Curtis R. LeMay (middle), Commander of the U.S. Air Force Strategic Air Command, converse during a 1955 luncheon meeting. Norris Bradbury, Director of the Los Alamos Scientific Laboratory is at left. Below, President McRae listens to Edward Teller (left). Jack Howard (right) looks on.

635

meetings. The new President learned that he was in charge of a rapidly growing organization of some 5,447 employees. Based upon the age of personnel, Sandia was still a very young corporation. In the 26–35 age category were 2,490 employees, with the next largest group of 1,375 in the 36–45 category. McRae learned, too, that the number of employees assigned to Research and Development activities during the year 1953 had increased seventy percent. Throughout his five-year tenure, the emphasis on research and development and a corresponding decline in production-related activities would continue.[9]

The McRae Committee

Not long after his arrival, McRae scheduled group meetings with all Laboratory personnel. This show of interest and willingness to listen to the people and their concerns became a trademark of the McRae administration. At the managerial level, he employed a similar approach by continuing the organizational analysis begun by his predecessor. McRae formalized the Quarles charter in November 1953 by setting up a committee assigned to improve the organization's effectiveness through a methodical and meticulous scrutiny of its various components.[10]

At the first meeting held December 4, Chairman Harvey G. Mehlhouse and Committee members Frederick J. Given, J. R. Townsend, and Charles W. Campbell prepared an agenda proposing investigation into issues ranging across the spectrum of the Laboratory. The procedure followed was to invite visitors from the departments under perusal to make presentations. From the candid and spirited discussions that followed, significant decisions were made.[11]

As young as the Laboratory was, it had matured to the point by early 1954 that serious concern was being expressed about "the rigid organizational structure we have established" and "the stature attached to the word 'supervisor.'" The suggestion was made to abolish the section supervisory position and assign signatory authority to project leaders. This, they felt, would help Project Engineers to do their job and establish accountability. While all such suggestions were not implemented, others were.[12]

Among the issues addressed that did result in change was the status of the Transition Engineering, or Standards Department, within J. R. Townsend's Materials and Standards Engineering Directorate. Regarded as "the most controversial situation in the Laboratory," the interface between Design Engineering and Manufacturing had to be improved. At the crux of the controversy was whether or not Transition Engineering still served a purpose. C. W. Campbell summarized the perceived status of the group at the Committee meeting of February 11, 1954. "At one time," he said, "there was good reason for establishing a transition engineering group—to be frank, at the time we had, in a sense, two corporations. The time has come when we are one family and there should be only one channel, one face from development to manufacturing." In 1954, therefore, it was Campbell's opinion that "we have one too many organizations from a design standpoint."[13]

The consensus of the group was that the design function should be shifted from Standards Engineering to Development Engineering and that manufacturing type operations should be moved from Standards Engineering to Manufacturing Engineering. Later, when Standards Supervisor Bryan E. Arthur proposed that a better product would result if the designer were placed face-to-face with the production people, management was on the same wavelength. By 1956, the need for a formal Transition Engineering group had diminished to the extent that the decision was made to phase it out.[14]

During the Committee meeting devoted to Stu Hight's Research Directorate, discussion revealed just how far the Corporation had come in this area, and the considerable way it still had to go. Although his organization was called "Research," Hight admitted that much of the actual work was to supply specialist services in the fields of weapons effects, aerodynamics, mathematics, physics, "and other limited areas." "A research department whose major job is to supply services can do little on long-range problems," he said, adding, "it is hoped that, in time, this group will be a couple of years ahead of engineering and can establish priorities for development."[15]

Even more revealing, however, was Hight's contention that "real basic research can be very vague and abstract and, in many cases, useless." On the other hand, he said, "*our* research is aimed at useful objectives." This basic tenet—research for the betterment of engineering—would remain a distinct facet of the corporate character for years to come, even after a charter had been granted to pursue establishment of a fundamental research capability at Sandia.

During 1954, as Hight pointed out, the Research Directorate incorporated not only Weapons Effects with its analytical work, participation in full-scale testing, and in environmental testing, but also Component Development. In regard to long-range projects involving timers, spark gaps, and X-units, the Committee acknowledged that the lack of manpower contributed to the Lab's inability to respond adequately. Among other recommendations, Committee members—citing the need for a unified presentation to outsiders—proposed doing away with the fragmentation of Research activities between the different vice-presidential organizations. "Research and Exploratory Development," they suggested, "should be jointly directed into the building of advanced research products in structures *and* components." Moreover, there was a need for a "centralization of systems engineering" to achieve overall compatibility.

These concerns were serious enough to motivate reassessment and restructuring as the Laboratory anticipated the continued need to respond to priority assignments. Overall, however, the consensus of the Committee in 1954 was that the Corporation should not have "pure research as such but only applied research in the field of atomic weapons, useful to the Corporation program."[16]

Within a year's time, a lead article in the *Lab News* announced: "Reorganization Accompanies Vice-President Appointments." At age thirty-seven, Glenn A. Fowler became the Corporation's youngest Vice-President when he was promoted to head the Research organization. At the same time, Frederick J. Given was appointed Vice-President of Development Technical Services. The activities of the Electronics and Apparatus Engineering Directorates previously held by Fowler and Given were then transferred to Lou Hopkins, who became Director of Components. Hopkins' warhead engineering directorate was combined with bomb development under Bob Henderson as Director of Systems Development. These organizational moves signified recognition of component development as a discipline and its increasingly significant role in the Laboratory's quest for weapons of longer stockpile life, minimal maintenance, and unquestionable reliability.[17]

In general, the Committee's review of the Research, Development, and Manufacturing Engineering organizations revealed the need for more careful planning, the introduction of programs in an orderly pattern, and a caution not to over-commit the capacity of the Corporation by unwise scheduling. Experience had proven that approximately three years were required from inception of an idea until the first production unit was available, but reality showed that new projects, such as the TX-15 then in process, allowed only half that time. "The atmosphere in which we are working on the TX-15," the Committee concluded, "suggests we still do not have proper recognition of Sandia's place in work assignments and schedule negotiations." Laboratory leadership, therefore, should strive for "a stronger scheduling authority," which the Committee felt to be related to the need for "clarification of Corporate responsibilities in the AEC system."[18]

According to Bob Henderson, speaking as Vice-President of Development in 1958: "Relations improved greatly when General James McCormack became Chief of the DMA, and increased gradually to the high level which has prevailed since General Hertford assumed the position of Manager of ALO." "General Hertford," Henderson continued, "contributed greatly to establishing a clear cut role for Sandia Corporation in his prior post with the Army."[19]

Nevertheless, in 1954 Sandia had to work at establishing and retaining its specific role in the developing defense complex. For example, the Committee also expressed concern over the trend being shown by the military and the AEC to assign some design and development work to other organizations. While admitting that part of the reason for this might be "political," they advised that "Sandia Corporation . . . arrest this trend by firmer assertion of the leadership it has held so far."[20]

The McRae Committee also addressed the Laboratory's Quality Assurance and Inspection functions. By the mid-fifties, Sandia's interactions with the production complex had grown to the extent that the Lab's relatively small assembly shop was supported by approximately 100 vendor plants scattered nationwide. In view of the magnitude and complexity of the system, the Committee concluded that the Laboratory's Inspection Operation had received too little attention and that Quality Assurance had been overemphasized. Basically, there was the misconception that "quality assurance in itself could improve quality of the product." On the contrary, the Committee advised that the Laboratory should look "to the designer, manufacturer, and inspection for improvement."[21]

In addition to political problems and production issues, the Lab's introspective analysis touched upon topics ranging from improvement of the services of Purchasing to the value of bomb books, and the facilitation of operations in the development shops.[22] After a year of study, when the Committee presented its final reports, many of the conclusions reached formed the basis for corporate policy and philosophy, not only for the McRae administration, but for years to come. Meanwhile, the newly levied emergency capability (EC) programs for production of hydrogen bombs and warheads were among the most pressing problems confronting the Laboratory.

Emergency Capability Programs

When the decision was made to proceed full-speed ahead on thermonuclear development in preparation for Ivy and Castle, the DMA, to assure maximum progress in the field, issued a report directing that certain immediate measures be taken. The report advocated—in addition to the expansion of Los Alamos Scientific Laboratory—continuation of programs at Princeton University (the Matterhorn Project), American Car and Foundry (ACF), and at Sandia. Specifically, Sandia Corporation was to be brought "into the ordnance engineering aspects such as ballistics and fuzing."[23]

Rather than follow the pattern established on fission weapons with the theoretical work being done at Los Alamos Scientific Laboratory and the engineering work being done at Sandia, Los Alamos initially decided to bring in another firm to handle the engineering–design job on thermonuclear weapons. A letter from Bradbury to Gordon Dean, dated September 26, 1951, indicated that he had asked Marshall Holloway to coordinate the Los Alamos theoretical work and engineering design with fabrication. "Representatives of American Car and Foundry are coming to Los Alamos next Tuesday to explore the possibility of their taking the responsibility for design engineering and possibly some fabrication and procurement," he added.[24]

From the Sandia perspective, this development signalled encroachment on the Laboratory's domain. Sandia leadership watched with some apprehension as ACF set up shop in an old dance hall over by the Albuquerque Fairgrounds. For the first large thermonuclear bomb, ACF did indeed serve as the engineering design arm of Los Alamos. Meanwhile, according to Bob Henderson, "Sandia worked quietly to convince the AEC that they should allow us to continue doing their engineering for them as we had all along." As Henderson observed: "The split between Los Alamos and Sandia on design matters was a bone of contention."[25]

Sandia's involvement in the emergency capability programs began on a limited basis in May 1952. At that time, Sandia and Los Alamos jointly forwarded a letter to DMA proposing a design for the TX–14 thermonuclear weapon, soon to be categorized as an "emergency capability," meaning high-priority production for stockpile. The term *emergency capability* indicated that bombs would be stockpiled for use before completing all engineering for manufacturing, operational suitability tests, and gathering of all information on reliability. Assembly and use of these weapons required trained scientists and engineers; thus the emergency capability devices were similar to the Little Boy and Fat Man bombs of World War II vintage. The prevailing attitude was that the "emergency" situation justified the shortcuts. Under the highest secrecy and security, these big 40,000-pound monsters were assembled in a special area on Sandia Base. As Robert L. Peurifoy, Jr.—at the time an electrical engineer in Lee Hollingsworth's Systems Development Division—explained: "Because policy-makers were concerned that the Russians might do something foolish, a program was designed to put together a few thermonuclear weapons before any test had been conducted to serve as an emergency capability."[26]

The achievement of emergency capability for thermonuclear weapons depended upon close cooperation and exchange of information among the Special Weapons Center at Sandia Base, Los Alamos Scientific Laboratory, and Sandia Corporation. Consequently, early in June 1952, Major General John S. Mills, Commander, Air Force Special Weapons Center; Los Alamos Director Norris Bradbury; and Sandia President Donald A. Quarles formed a "Committee of Three Principals." The Committee was to serve as a vehicle for coordinating information on all aspects of development for the TX-14 emergency capability thermonuclear weapon. This group, joined by Major General Leland S. Stranathan, Commander of Field Command, AFSWP, in mid-1953 published quarterly a joint development progress report on the emergency capability programs for all DOD agencies concerned.[27]

Until the Upshot-Knothole test of March 24, 1953, most of the work on the TX-14 was performed under the cognizance of Los Alamos in affiliation with ACF. In July, however, the Santa Fe Operations Office suggested that the same division of responsibilities on thermonuclear weapons be made as had been previously established for fission devices. This meant that Sandia was once again in strong competition with ACF for design responsibilities. To work out the details of this division of responsibilities, an interlaboratory committee, similar in scope to the TX-N Committee, was proposed. Appropriately called the TX-Theta Committee, the group held its first meeting on October 26, 1953. By this stage, the Committee had to devote considerable time to the development of other thermonuclear devices in addition to the TX-14, which was the first thermonuclear weapon to be added to the nuclear arsenal. After only eight months in stockpile, the TX-14s were replaced by the TX-17s and retired in October 1954.[28]

In addition to the "Committee of Three Principals," which by this time had evolved into the "Quadrapartite Committee," another vehicle for interaction among the military and civilian agencies was the Joint Operational Planning Committee (JOPC). The JOPC included R. W. Henderson as Chairman and W. A. "Bill" Gardner, also of Sandia. This second cooperative joint committee had been formed in September 1953 when it became obvious that planning for the operational phase of the emergency capability programs should begin. In this time frame, Eton Draper's Engineering Department was assigned the task of producing the EC thermonuclears for the Air Force. Jim DeMontmollin, who joined Sandia in time to work on fuzing for the ECs, recalled that "Draper did an exceptional job of keeping up with a rapid pace of development."[29]

Engineering for the TX-16 was accomplished under direction of Marshall K. Holloway's Los Alamos team. Until the spring of 1954, the TX-14 and TX-16, in reality weapon prototypes, were the emergency capability weapons with which the Air Force Special Weapons Center was directed to achieve an Air Force capability. In coordination with the military, Sandia in 1953 established the TX-16 Panel to study weapon logistics such as handling, transportation, and storage. Sandia provided the afterbody, fuze, parachute, power supply, and certain test equipment. Such responsibilities necessarily required close liaison with the Air Force in order to modify carriers to carry these large thermonuclear devices.[30]

Since the Air Force had to have delivery capability the day the TX-16 hit the stockpile, external configurations and support equipment for the aircraft had to be modified with each change of the weapon's design. The maximum specifications for the TX-16 and the 17 were determined, therefore, by the dimensions of their carrier, the B-36. The huge size of these first thermonuclear weapons posed a real problem in logistics. Colonel Charles G. "Moose" Mathison, who was stationed at Kirtland Air Force Base as part of the Air Materiel Command and later placed in charge of the development directorate of the Special Weapons Center, recalled that Bill Gardner came up with the solution. Gardner proposed using a straddle-carrier, a machine used to move lumber in a lumber yard, to transport the unwieldy weapons. "We visited a lumber company and got one of their largest straddle carriers, made a cradle, and tried it," Gardner remembered. The idea worked and remained in use.[31]

Another serious problem related to the need to find a way to slow the bomb sufficiently to allow the airplane and crew to escape the effects of the detonation. To arrive at a solution, Sandians worked with the parachute laboratory at Wright-Patterson Air Development Center in Dayton, Ohio. With the assistance of Wright-Patterson personnel, Sandia engineers adapted a ribbon parachute design from the Germans and went to Pioneer Parachute Company in Hartford, Connecticut, to have the parachute made.

About this time, the B-47, the first strategic jet bomber, entered the Air Force inventory. The Strategic Air Command was interested in determining if the new bomber could be modified to carry a hydrogen bomb. To adapt the B-47, the bomb bay doors and aft fuel tank were removed and the bomb squeezed in. A horse collar device was built around the tail of the B-47. To provide the extra power needed to lift the airplane, fourteen JATO bottles were attached. At the working level, cooperation between military and civilian contingents, in contrast to interaction at higher levels, was exemplary. As Mathison put it: "The relationship between Sandia and the Air Force during this time represented the best elimination of red tape I've ever experienced."[32]

In the final quarter of 1953, the Lab's program for the manufacture of warheads for the TX-14, 16, and 17 reached the peak of design detailing. Veteran weapons engineer Jim Cocke recalled that "Sandia hand-produced a number of emergency capability TX-17 and TX-24 bombs with the bomb case and afterbody assembled; however, the fuze, power supply, parachute, and key nuclear components were stored in separate containers."[33] Both the Mk 17 and Mk 24, which were physical look-alikes, entered stockpile in 1955 with true contact fuzing and drogue parachutes to help ensure safety of the delivery aircraft crews.

The turning point that led to establishment of Sandia as the design-ordnance laboratory for the thermonuclear program occurred when Los Alamos requested emergency assistance on development of the TX-15. This device was envisioned as a small tactical hydrogen bomb to be delivered by aircraft. Later, an XW-15 warhead was approved for application in the Army's REDSTONE, and the Air Force's SNARK and NAVAHO missiles.

Although the TX-15 program was not an emergency capability in the sense that the TX-14, 16, and 17 programs were, its schedule was extremely tight and characterized by problems with ACF and contact fuze development. The schedule called for the TX-21 to follow the TX-15 by only six months. This timescale initially prevented the Laboratory from developing a satisfactory contact fuze; however, production of the fuze was achieved by 1955. Basically, the underlying source of contention associated with this program was the division of responsibilities between ACF and Sandia. As one committee member expressed it: "ACF had been told by Los Alamos that Sandia was the designated agency for the program, but they [ACF] regarded Sandia as something of an upstart." Sandia, on the other hand, regarded ACF as "a boiler-maker outfit whose work would probably have to be inspected carefully." In essence, relations were none too smooth, and Sandia felt the pressure. There was fear that if Sandia failed to produce, and produce on time, "SFO

(Santa Fe Operations) might consider ACF as the design agency."[34]

Sandia management regarded the Laboratory as being "on trial." The TX-15 program, therefore, was considered to be extremely important to Sandia's future in nuclear ordnance design and engineering. Despite the friction and the pressure, the Laboratory met the TX-15 program requirements. "It was Sandia's outstanding job on the 15 that put us into the thermonuclear weapon design business," Henderson concluded. "The result was a strengthening of relations with LASL."[35]

In December 1955, the Mk 15 improvements, including weight reduction and thermal battery, were successfully incorporated into the requirements for a new TX-39 weapon. One of these variations, the TX-39-XI, had a nose cone design made of a collapsible aluminum honeycomb material for shock absorption, a development that would be extremely important for the concept known as the laydown bomb.

In this connection, "Lone Star," an exploratory program for a very heavy weapon, was initiated. As Randy Maydew explains, "This was our first attempt to lay down a big one. The weapon, which resembled a boat more than a bomb, was dubbed appropriately 'Lone Star' after the Texas boat company of that name." Initiated to demonstrate the laydown capability of the TX-39, the bomb weighed 7,500 pounds and was mounted in a semirectangular "boat" made of metal with aluminum honeycomb on the bottom to absorb ground impact. On top, the Lone Star had a rectangular opening for the rigging of the Sandia-designed 130-foot-diameter ribbon parachute (at the time the largest fabricated). Design criteria for Lone Star included impact velocity of thirty feet per second on the ground after release at a 1,000-foot altitude from a B-47 or B-52. "This program was not carried through because the TX-39 was so fragile," Maydew explained, "but it was significant because its development demonstrated interim capability. In comparison, the B-28, developed later, was much more rugged."[36]

Meanwhile, Los Alamos Laboratory's newly established competitor, the University of California Radiation Laboratory in

Livermore (UCRL), also requested Sandia's assistance in 1953 with development of the high-yield thermonuclear bomb, designated the TX-22. Sandia provided some long-distance engineering for this program, and in so doing established a toehold as the ordnance support lab for UCRL. The unsuccessful TX-22 was superseded by the smaller Mk 27. The TX-27 missile warhead program, the first major program in which Sandia worked jointly with UCRL, had applications for the RASCAL, MATADOR, and REGULUS missiles.

The advances being made encouraged the TX-Theta Committee also to propose design of the TX-28, known as the "building block concept." Designed to be used in configurations compatible with most Air Force and Navy tactical and strategic aircraft, the Mk 28 provided the stockpile with a new nuclear posture. During this early period, John P. Cody of Sandia directed the 28 design with the assistance of Sam A. Moore, W. R. Hoagland, Jim Davis, and later Bill Denison.

Several XW-25 fission warheads were also released as emergency capabilities. For this warhead, developed for the Air Force's high-velocity rocket, code-named DING DONG, Sandia was responsible for the X-unit, fuzing switch, warhead case, and associated assembly, test, and handling equipment. In connection with the interest in the effects of high altitude on weapon operation as well as safety issues, the satisfactory results of the sealed nuclear system were significant. Bill Wells was responsible for the XW-25 in 1954; later, Walt Treibel took the project. At the Redwing series, held primarily to test high-yield thermonuclear devices, technology in the field had advanced to the stage that the device could be dropped from a B-52 bomber. Redwing demonstrated to the world the deliverability of thermonuclear weapons.[37]

As early as April 1954, plans were being made to begin phaseout of the emergency capability programs. It was at this time that the Sandia and ACF positions were further clarified. As Bradbury wrote to General D. J. Leehy: "With the development of new . . . weapons following the SC-LASL pattern . . . , it is not anticipated that LASL will ask the ACF to undertake any major new development programs." Los Alamos and Sandia did plan to continue using ACF for some manufacturing, notably "retrofit operations."[38] Bradbury, General Mills, General Stranathan, and McRae agreed, therefore, that after termination of the EC program, further technical development in the field would follow "the normal organizational pattern" between Los Alamos Scientific Laboratory and Sandia Corporation. "We look forward in the future to the same vigor and enthusiasm as the EC Program has shown in our relations with both DOD organizations and SC [Sandia Corporation] in the pursuit of subsequent developments," Bradbury concluded.[39]

Advances made during development of the high-yield thermonuclear bombs of the emergency capability program led to other radical weapon concepts. Among these was the "laydown bomb."

The Laydown Bomb

"In the late summer of 1958," read an article in *The Western Electric Engineer*, "an Air Force fighter flying fast and low over an abandoned airstrip near Dalhart, Texas, dropped an inert nuclear test weapon of unusual shape and, as it developed, of unusual significance. As the delivery plane swept upward, the spike-tipped experimental 'shape' struck the seven-inch-thick concrete runway and embedded itself like a dart thrown into a cork board. Telemetry indicated that the components within the casing had survived the shock . . . and were ready to trigger the detonators at the preset time after impact."[40]

This successful test represented the culmination of a five- to six-year effort by Sandia engineers and scientists. The task assigned had been to "design for delivery to a hard target by a low-flying airplane—a nuclear weapon containing ordnance systems that would keep the weapon from ricocheting off the target, from failing to detonate because of impact damage, or from detonating before the delivery aircraft was a safe distance away."[41]

Whereas nuclear bombs traditionally were designed for delivery by aircraft flying at high altitudes, which permitted escape before detonation, the new laydown bomb with its delayed-action ground burst provided a capability that the military had been seeking for some time. Refinements in radar defense and improvements in antiaircraft missiles made high-altitude bombing less and less attractive; therefore, both the Air Force and the Navy wanted a technology that would permit release of atomic weapons in the altitude range of 50 to 200 feet above the ground. To reach this milestone, a number of delivery and shock-mitigating techniques—including spikes, parachutes, and cushioning materials—had to be developed.[42]

Sandia's involvement in development of new weapon concepts, such as the laydown bomb, began under Walter A. MacNair during his tenure as Vice-President of Research. Basic to this focus was MacNair's belief that research should be involved in the weapon's evolution. Whereas the Lab's earliest years were characterized by "nuts and bolts physics," Sandia's scientific endeavors, within a short timespan, required an expanded research orientation.

Such a trend was already being seen in George Hansche's Weapons Component Department, where George Anderson and Frank Neilson were making impressive inroads on the development of firesets and detonators. And in 1954, Richard S. "Dick" Claassen, Supervisor of the Electro-Mechanical Division, invented the concept of an initiator driven by a ferroelectric power supply.[43] Breakthroughs such as these provided the antecedent for weapons in years to come.

In addition to the Component Department, investigation leading to the laydown concept involved the efforts of an exploratory development group called "Weapons Analysis." Department Manager Kenneth W. Erickson, described as "a powerhouse of a man with piercing blue eyes and chiseled features," assigned Sheldon Dike and Walt Wood to perform the initial systems analysis.[44] The feasibility of both the concept and the materials to be used had to be proven.

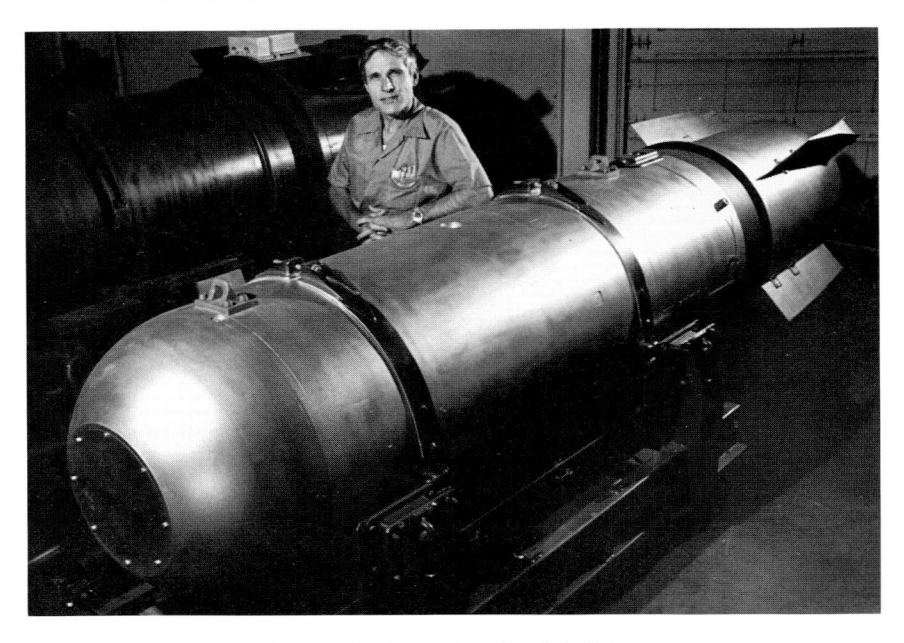

Grover Hughes with the Mk 27.

Ray Brin with the Mk 15.

ADVANCING TECHNOLOGICAL FRONTIERS

Del Olson with the Mk 39.

Bill Hoaglund and Walt Treibel with the Mk 25/Genie.

NUCLEAR ORDNANCE ENGINEER FOR THE NATION

TX-28/34 LAYDOWN BOMB

Above, an illustration of a parachute-retarded W-34 warhead in a Mk 28 casing was an example of the early laydown bomb.

Luke Stravasnik and Vic Engel with the Mk 43.

ADVANCING TECHNOLOGICAL FRONTIERS

The Mk 7 (above) and the Mk 28 (below): As the bomb program evolved, less equipment was required to support the bombs being developed.

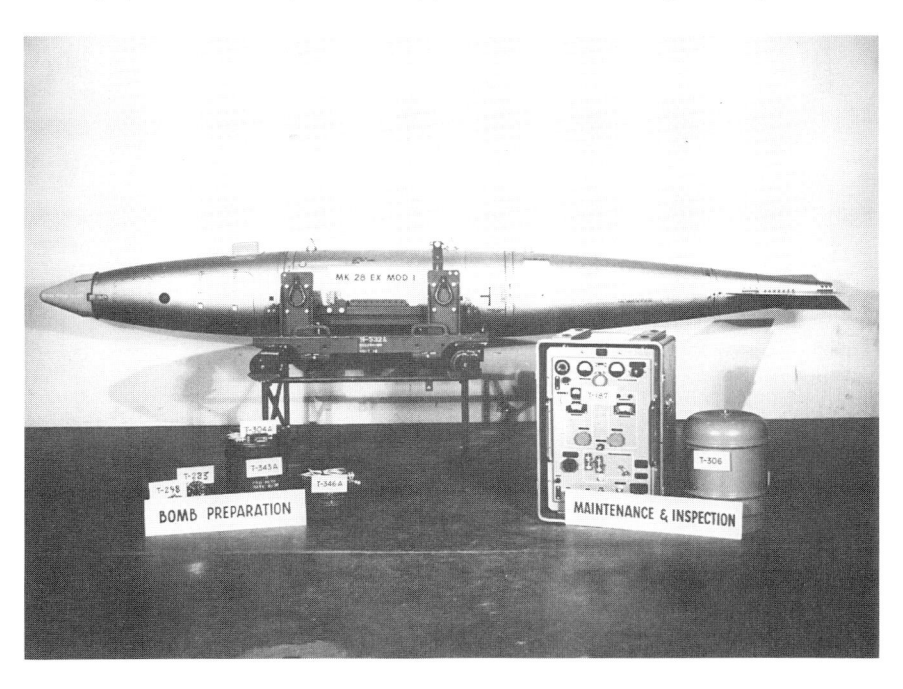

The Mk 28, known as the "Building Block" bomb, was configured with different nose caps and tail fins, making it compatible with most aircraft.

In late 1953, Robert F. Brodsky, head of the Analytical Aerodynamics Division, stopped by Randy Maydew's office. "Randy," he said, "I've got a new half-time project for you. I'd like for you to look at the feasibility of delivering a tactical, externally carried, nuclear weapon from low altitudes."[45] Brodsky encouraged Maydew to investigate the state-of-the-art parachute technology to see if parachute deceleration of the bomb could provide enough additional time so the aircraft and pilot could escape the effects of the nuclear explosion upon bomb-ground impact. The objective of the project, which preceded the laydown bomb concept, was to deliver the weapon from 2,000 feet at speeds up to Mach 1.0. Maydew, a new staff member recruited from the NASA Ames Laboratory in California, had spent three years in aerodynamics research before coming to Sandia in 1952. Brodsky's request would channel the course of Maydew's career at the Labs for the next thirty-five years.[46]

Maydew started out by conducting a literature survey and found that the Germans had excelled in parachute development prior to and during World War II. Among these outstanding German parachute scientists were two in particular: Dr. Helmut Heinrich and Theodore Knacke. Both Knacke, who invented the FIST ribbon parachute, and Heinrich, inventor of the guide surface parachute, came to the United States as part of Operation Paperclip after World War II and brought with them some of the best parachute technology known at the time.

Maydew presented the findings of his survey in a memo titled "Feasibility of Parachutes for Low Level Delivery," in January 1954. The results were initially discouraging because Maydew's research indicated that a parachute could be deployed reliably only at Mach number up to 0.5, whereas Sandia hoped to deliver nuclear weapons at Mach numbers up to 1.0. However, when parachute experts around the country were presented with specifications and results, the response was that it could be done. (Compared to previous parachute requirements, i.e., heavy objects [cargo] at low speeds and light objects [pilots] at high speed, the weapons parachute requirements were for heavy objects [bombs] at high speeds.) The next step was for Maydew to write a memo for Director of Research Stu Hight's signature. The memo discussed the feasibility of transonic parachutes and recommended a parachute low-level delivery (LLD) flight test program for Sandia Laboratory. The mission of the LLD would be to demonstrate that heavy-duty parachutes could be designed to operate up to Mach 1.0.[47]

MacNair formalized the initiative in the Research organization with the "New Weapons Program" and assigned Dick Claassen to launch it. "I had the job of proving the feasibility of such a project," Claassen recalled, adding that the laydown program required a three-pronged approach. "The first," Claassen said, "was to determine whether you could have a retarding device that would work at sea level at Mach 1." The second challenge was to develop some kind of an impact absorption system once the bomb hit the ground. "One idea," Claassen added, "was to use a nose spike that could penetrate concrete or dirt to absorb the energy and stop tumbling. Later, however, shock-absorbing aluminum honeycomb came into use." The third element, Claassen said, "was to make the physics package itself rugged enough to withstand the parachute shock and the actual impact of laydown. That task was accomplished by the nuclear labs, Los Alamos and Lawrence Livermore."[48]

Claassen assigned his division, including physicist Al Bridges, to investigate all types of crushable materials to be used for the nose, such as aluminum and stainless steel honeycomb. Bridges, who became president of the Kaman Nuclear Division of Kaman Aircraft after leaving Sandia, was so obsessed with finding a solution, that he reportedly dreamed one night that corn flakes packed into the bomb nose would do the job. To conduct the impact tests, Ken Erickson persuaded management that a 300-foot drop tower should be built in Area III. Despite initial opposition to the idea, the tower was built and used extensively.[49]

The low-level delivery parachute program began in 1954 as a joint research and development program with the Parachute

Branch at Wright Air Development Center (WADC). Maydew recalled that he would meet about once a month with WADC engineer Frank Vlasic. Although the original ribbon-type parachute design must be credited to German technology. Sandia designed the parachute test vehicle (PTV) with spike nose tip. The ribbon-type parachute also had to be redesigned by Sandia and WADC for this very high-speed application. Dick Browne's division, which was a part of Bob Henderson's Weapons group, conducted the mechanical and electrical design for the PTV. Considering that the parachute program was being carried out at the height of thermonuclear weapons development, it is not surprising that Ken Erickson had to do some real persuading to convince Browne's department manager, John Cody, to allow Browne's group to allot the time and manpower necessary to design the parachute test vehicle.[50]

Claassen recalled that there was enough cynicism about parachute-aided laydown at the inception of the program that Kaman Aircraft was asked to develop a propeller-shaped "rotochute" that would open up as it entered the airstream. Sandia engineer Carroll Osborne later developed a similar device at Sandia. Company founder Charles H. Kaman described the rotochute, saying: "The best way to visualize a rotochute is to imagine a maple seed pod autogyrating as it spins to the earth. The function of the rotoblades is to control the rate of descent." As Maydew explained: "Subsequent flight tests demonstrated that a ribbon parachute is more efficient than a rotochute (based on weight and volume of each) in rapidly decelerating a bomb for laydown delivery."[51]

Later when Don Robinson, Bob Strieby, and a group from Kaman visited Sandia to discuss the application of rotochutes to the laydown bomb, they met Ken Erickson for the first time. Apparently impressed by Erickson's forceful personality and knowledge, they reported back to Charles H. Kaman. When Erickson met with Kaman at the Connecticut plant, the association resulted in Erickson joining the company and setting up the Kaman Nuclear division in Albuquerque.[52]

Sandia's parachute test vehicle, designed by George Norris, was first tested in October of 1955 at Mach 0.6 using an F7U aircraft at the Los Lunas test range. A twenty-foot diameter parachute was used and the test vehicle was fully equipped with state-of-the-art telemetry and camera pods. Pluto cameras, developed by Ben Benjamin, were positioned so that the parachute-opening phenomena could be photographed. "By the summer of 1956," Maydew recalled, "we had conducted thirty-three drop tests at Mach numbers from 0.6 to 1.0 with parachutes of sixteen-, twenty-, twenty-four-, twenty-eight-, and forty-foot diameter, with test vehicles weighing from 1,750 to 2,600 pounds, and established that you could reliably deploy a parachute at low altitudes at release speeds up to Mach 1."[53]

Two courageous pilots, Navy Lieutenant Duck and Air Force Lieutenant Forrest McCartney, flew the test planes. "Thirty years ago," Maydew said, "we weren't nearly as safety conscious as we are today about mounting an untested new shape store on an aircraft and asking a pilot to fly the aircraft with full afterburner (up to Mach 1) and release the store at low altitude."[54]

Based upon Claassen's exploratory program, the DOD in October 1955 officially requested a Phase 2 study of laydown weapons. The DOD specifically proposed a joint study involving the staff of DNA, the AEC, the Navy, Air Force, and Sandia to investigate the feasibility of a laydown weapon system for an eighteen-inch-diameter tactical weapon.

The "Tableleg Committee," as the group was known, met for the first time in November. Proposals were made for design of a weapon called "Hotpoint," subsequently stockpiled as the Mk 34; a weapon based on the TX-28 design, which became the Mk 28, Mod 1 bomb; and a design to combine the best features of the 34 and the 28, with multiple options of airburst, contact, or delayed laydown. This became the Mk 43 bomb—the result of a long-term development program that would include advances in the state of the art in component and weapon design. The development phase for the first laydown weapon, the B-43, began in November 1956. Members of Dick Browne's division represented the development organization on the Tableleg Committee. Design responsibility was assigned to Sandia supervisor J. W. Jones, who was assisted by section heads A. A. Lieber, L. A. Dillingham, and G. W. Randle.[55]

The laydown concept represented a vast improvement in the delivery and flexibility of nuclear weapons, and one of Sandia's first and most successful exploratory development programs. Being able to ensure that a weapon gets to the target has always been a prime element in deterrence. The combined efforts of Sandia's aerodynamicists and mechanical and electrical engineers resulted in a capability that was superior to any existing in the military at the time. In 1958, based upon a proposal by Randy Maydew, Sandia set up its own parachute laboratory. Another result of this effort, therefore, was establishment of parachute technology as one of the Laboratory's centers of excellence.

In addition to the laydown concept, other notable advances in technology began to bring the possibility of the "wooden bomb" closer to reality.

The Quest for the Wooden Bomb

The Laboratory had made considerable progress since the early days of the nuclear weapons era when weapons were stored as components and assembled for use. During these formative years, test equipment design and procurement personnel such as Lee Deeter, Jim Jones, Ben Bright, and Jerry Hinman were among the twenty to thirty testers associated with each program. A significant portion of the Laboratory's effort went into checking the basic component, the cartridge assembly, or the electrical system. Separate storage of nuclear materials provided a safety factor. With introduction of thermonuclear weapons, however, Military Characteristics for new weapons became more stringent.[56]

The mid–fifties, therefore, marked a new focus in the nuclear weapons field. The emphasis on deploying a credible stockpile as quickly as possible shifted toward technologies that would prolong stockpile life and improve weapons assembly, testing, and reliability. Over the years a variety of steady evolutionary developments working together, often in symbiotic relationship, contributed toward development of a nuclear bomb that would be as easy to maintain, "as trouble-free as a block of pine," hence the name of the design goal—the "wooden bomb." Bob Peurifoy, one of the Sandia pioneers who developed the concept, along with Eton Draper, Lee Hollingsworth, and others, explains that the term implies that "a properly designed nuclear weapon should be like an inert log in a forest—able to stay put for years while requiring no attention."[57]

This ultimate goal was rigorous, as specified in an August 1954 memo detailing the basic maintenance philosophy of the Defense Atomic Support Agency (DASA). "The design goal," read the memorandum, "is a wooden bomb which will require no test or maintenance during its stockpile life. However, the design agency must give attention to reasonable accessibility in order to permit the replacement of limited life components . . . and to permit modernization, and exchange of components for which functional integrity cannot be predicted."[58]

With this as the stated goal, an initial challenge was how to prolong the useful life of tritium-bearing components. Tritium, the rare hydrogen isotope with a half-life of 12.3 years, is a common ingredient in most nuclear weapons; therefore, decay to less than required levels for initiation was to be expected. The need existed, therefore, for limited maintenance to replace neutron generators, the critical tritium-containing components that act as triggers to initiate the nuclear reaction. In response, the Laboratory focussed efforts on development of the external initiator, or timed neutron source. A version of the external initiator was tested successfully during the 1955 Teapot test series in Nevada.[59]

At the Teapot operation, the sealed pit concept, developed mainly by Los Alamos, was also successfully tested and represented a major breakthrough in the quest for the wooden bomb. The ability to seal off or isolate various weapon components from other explosive materials and the environment facilitated the assembly process and improved weapon performance and reliability. On the other hand, the new sealed pit raised certain safety concerns that would lead eventually to increased involvement of the Laboratory in the safety aspect of the nuclear weapons program. The sealed pit, along with the development of thermal battery technology, were key developments leading to an inert bomb that could be shelved indefinitely.[60]

According to Peurifoy, the first wooden bomb to enter the nation's stockpile "was either the Mark 15 or the Mark 28, depending on how rigorous a definition is used." Other early candidates would be the Mk 11 bomb, which became known as the Mk 91 when design responsibility was transferred to the Navy. W. J. Cocke points out that this bomb employed reversible nuclear safing and, therefore, could be stored in the fully assembled state. This weapon system, which entered the stockpile in 1957, replaced the Mk 8 on a one-for-one basis. Cocke observed that "the Mk 17, Mod 1 has also been suggested for the honor; however, the fuze, power supply, parachute, and key nuclear components for this weapon were stored in separate containers," and, therefore, this Mark does not fit the definition. Jim DeMontmollin recalls that "all of the strategic bombs later than the 21, and the tactical weapons after the 28, were of the one-shot, wooden bomb variety, as opposed to the earlier bombs that required retesting and storage of components separately."[61]

While the sealed pit has been recognized as a driving force in development of the wooden bomb, advances made in battery development were equally significant. Because of short life and the need for constant maintenance, the battery has been considered the weakest link in the component chain. By 1955, however, the evolution of battery technology had run the gamut from lead–acid, nickel cadmium, and silver–zinc to thermal. The first atomic bombs used lead–acid until 1953 when the nickel-cadmium and silver zinc batteries were introduced. These early batteries suffered similar limitations: They were bulky, contained corrosive liquid electrolytes, and had to be recharged frequently. In contrast to these aqueous electrolyte batteries—the typical automobile type—the thermally activated battery solved the need for frequent recharging.

ADVANCING TECHNOLOGICAL FRONTIERS

Some considered the Mk 15 to be one of the first weapons referred to as the wooden bomb.

Lee Hollingsworth stands next to a Mk 17/24. This bomb could be carried only by the B-36 aircraft, which had ten engines.

German scientist Otto Erb, who planned to incorporate the thermal battery into the V-1 and V-2 rockets during World War II, is credited for introduction of thermal battery technology into the United States. After the war, during interrogation by British Intelligence, Erb provided information on the subject; and in this manner, the thermal battery concept reached the United States, where it underwent further development. The advantages of the dry-thermal battery were soon recognized: its ability to stand inactive for many years at normal ambient temperatures, and its ability to be activated simply and remotely by an external electrical or mechanical signal relayed to solid, heat-producing chemicals within the battery. The high temperatures that are produced, in the 550-degree Celsius range, melt the solid electrolyte, generating the electricity. These one-shot primary electrochemical devices have the added advantage of extreme reliability in the range of 0.99995.[62]

The story of thermal battery development necessarily includes the contributions of Roger Curtis and the Ordnance Development Division of the National Bureau of Standards. Curtis combined his work with that of the Catalyst Research Corporation to develop the first self-contained thermal battery and to demonstrate the use of thermal cells for ordnance applications. After representatives from the AEC heard the Curtis report on thermal batteries, both the military and Sandia Corporation adopted the battery for use in the wooden bomb program.[63]

Bob Wehrle, who worked in thermal battery development for many years, recalled a memorable first-time demonstration. The occasion was introduction of the new device to Frederick J. Given upon his promotion to Director of Apparatus Engineering in 1952. "When the signal to go was announced," Wehrle said, "the igniter in the battery did its job of igniting the pyrotechnic heat source; and as the battery warmed up, the metal case of the battery developed a crack and a stream of molten electrolyte at 500° Celsius squirted from the battery. Luckily, the visiting dignitaries were spared and the only real damage was a charred floor," Wehrle reported.[64] Despite this inauspicious beginning, the demonstration failed to squelch the enthusiasm for thermal batteries, and Sandia embarked on a full-scale development program."[65]

Under the aegis of Sandia Corporation, subcontractors built components to specifications. Clarence Simpson, a test engineer at the Crab Orchard, Illinois, plant (later to become Unidynamics of Phoenix, Arizona) commented on the symbiotic working relationship. Cooperation was so effective that at one time Sandia gave serious consideration to making Unidynamics an integrated contractor.[66]

By 1955, order-of-magnitude improvements for weight, ruggedness, and effectiveness had been made in all four of the principal types of Sandia Laboratory components, namely, radar altimeter fuzes, external neutron initiation subsystems, firing subsystems, and low-voltage power supplies. It was in this year that Frank Neilson and George Anderson first demonstrated explosive to electric transducers—a major technological breakthrough. Jack Marron, also a contributor in the field, notes that these were ultimately developed into weapon components by the Engineering side of the house. "They were significantly smaller than conventional devices of the day," Marron said, "and provided bonus benefits such as high radiation resistance, simplicity, and reliability."[67]

Concurrent with research and development for transducers, the supporting activity in ferroelectric ceramics research made the ferroelectric explosive transducer possible. Advances such as these, largely Sandia-generated, contributed toward establishment of technologies such as shock dynamics, explosives, hydrodynamic codes, and lower-energy exploding-bridgewire detonators, among others.[68]

George Anderson and Frank Neilson, who joined the Lab in 1954, personified in many respects the creative and dedicated scientific pioneers of this era. In the area of exploding bridgewires, studies made by Neilson and Anderson contributed to a better understanding of the process involved to allow for a reduction by a factor of ten in the electrical energy required for a detonator to be properly initiated. Neilson, personally and collegially, published over twenty papers on energy-conversion,

Personnel in the Component group include D. A. Whitcomb at drafting table; center row, front to back: Mel Chaney, Aggie Mahinske, Jerry Rynders, Don Schultz, Bob Quinlan, Bob Pinkham, Bob Stromberg, Ken Gillespie; and Paul Field at foreground desk.

specifically explosive-to-electrical or shock-wave-induced energy. The second decade in the life of the Laboratory would mark the coming of age of component development.[69]

Toward Fundamental Research

Another promising young scientist, Frank Hudson, arrived at Sandia in the summer of 1953. Intellectually precocious, he had entered the fourth grade at age seven and college at sixteen. He joined Sandia with a Ph.D. in physical chemistry. Frank's academic training, combined with industrial and engineering experience that included an apprenticeship in ship construction, various mechanical and electrical design jobs, industrial planning, and two years in the U.S. Navy electronics program during World War II, provided a balance of science and engineering that would prove meaningful to Sandia's future. As the Lab's first physical chemist, his only assignment was to explore the uses of his field at the Laboratory. In retrospect, he views this as an advantage. "I had to figure out what to do," he said, "W. E. Boyes tutored me and, in effect, became my mentor and my guide. . . . A good share of my first two weeks was spent meeting people and visiting the shops, labs, and offices—an experience that provided a good perspective on the Laboratory and its goals."[70]

In that first year, Hudson participated in a variety of programs, including the wooden bomb systems study, the quality assessment of the Mk 7, and initiation of a quality assurance group in research. He also worked with the battery development division on storage problems of the nickel-cadmium batteries, and on exploratory studies of thermal batteries. This rapid and broad-spectrum exposure gave Hudson the lingering impression that Sandia had "high quality, cooperative, and conscientious people. . . . I concluded that Sandia was a first-class engineering operation," he said, "but Research did not have a real orientation."[71]

A little more than six months after the McRae Committee agreed that the Laboratory should continue to focus on Applied

Research, Hudson wrote a memo to Ken Erickson suggesting "A Systematic Program for Component Studies."[72] The context of the memorandum was much more significant than the title implied and served as an impetus for the expansion of Research at Sandia. Using as a frame of reference a previous discussion with Ken Erickson regarding the possibility of a research program to study the basic functional parts of weapons, Hudson formally proposed incorporation of fundamental research into the Laboratory's program. He explained the basis for this philosophy accordingly: First, he said, "there are fundamental differences between the engineering and research approaches to a problem and in the personnel required for each." Second, "there is great value in the research approach as the experience of industry has indicated for thirty years." And, finally, "there are research problems in connection with the ordnance phases of weapons."[73]

As a case study Hudson summarized the history of the Laboratory's experience with the nickel–cadmium battery. To prove its worth, the Laboratory had ordered fifty Ni–Cads from a contractor to conduct a comparison study with the lead–acid batteries then in use. When the nickel–cadmium battery proved to be superior, a million–dollar contract was let to a company for development of the battery according to Sandia specifications. Sandia, however, had no control of the research developmental work, despite the price. Problems occurred during the development period, Hudson recalled, and two years after initiation of the contract, there were still serious problems. Furthermore, Hudson added, "We found it necessary to institute a $20,000 research contract with Bell Telephone Laboratories to gain the fundamental knowledge of the battery required to assist in solving our problems."[74] He concluded by proposing incorporation of organized, systematic, and continuing research studies (as opposed to engineering studies) and the personnel to conduct them in order to solve problems internally or prevent them from occurring in the future.[75]

Hudson's initial attempt at selling a Basic Research program at Sandia didn't get anywhere, but he continued his campaign through Ken Erickson. The first positive sign backed by action was permission to hire two additional Ph.D.s into the department for fundamental work in radiation effects. Erickson sent Hudson a copy of the personnel status report with a victorious scrawl across the bottom of the page: "Frank: Get the Hell to work! Ken."[76]

About this time, the interest in nuclear-powered aircraft and the question of delivery systems and speculations regarding the effects of nuclear radiation on components generated interest in the acquisition of a nuclear reactor. Tom Cook, who would be promoted to Department Manager of the Nuclear Burst Department, elaborated on this aspect of the evolution that was occurring:

> Sandia was becoming concerned with the development of technology and testing activities having to do with the vulnerability of nuclear warheads. The need to destroy incoming nuclear warheads and to design warheads that could not be "killed" by other nuclear warheads, therefore, marked a branch point that led to the whole developmental pulsed power technology. As a complementary activity related to the nuclear-powered aircraft evolved, and based upon our involvement, we justified a reactor at Sandia. That was our first entry into nuclear radiation effects.[77]

Supported by Bob Townsend's Materials and Standards Directorate, enthusiasm for a Sandia Engineering Reactor Facility (SERF) grew. While there was some disagreement over its real benefit to the Laboratory, President McRae saw in the situation an opportunity for the Laboratory to expand its Research activities, and he advised capitalizing on it. With McRae's encouragement, Erickson set up a committee of Ph.D.s to discuss characteristics of a Basic Research program.

By this time, Richard S. Claassen, Supervisor of the Experimental Weapons Division, had reviewed the 1954 Hudson memo and brought it again to the attention of management in 1956. Although Erickson agreed that it had considerable pertinence in view of recent discussions on the subject with Vice-President of Research Glenn Fowler, and President McRae, the timing was

PHYSICAL SCIENCES RESEARCH

	SOLID STATE	RADIATION EFFECTS	COMBUSTION PROCESSES	PHYSICAL ELECTRONICS	HYDRO-MAGNETICS
PROGRAMS	SEMI-CONDUCTORS FERRO-ELECTRICS FERRO-MAGNETICS	SEMI-CONDUCTORS, POLYMERS, IONIC CRYSTALS	PROPELLANTS HEAT POWDERS EXPLOSIVES	GASEOUS DISCHARGE ION PRODUCTION EXPLODING WIRES	IMPULSE TUBE, RF MOVING FIELD MACHINE
TYPICAL PROBLEMS	ELECTRONIC VS IONIC NATURE OF FERRO-ELECTRICITY RELATION OF SURFACE CONDITION TO ELECTRICAL PROPERTIES. HIGH ENERGY DENSITY PROPERTIES OF FERRO-ELECTRICS AND FERRO-MAGNETICS	DAMAGE MECHANISMS IN SEMI-CONDUCTOR STRUCTURAL CHANGES INDUCED IN POLYMERS COLOR CENTERS IN ALKALI HALIDES RATE EFFECTS IN NEUTRON DOSES	HEAT POWDER BURNING PARAMETERS ENERGY TRANSFER INTO EXPLOSIVES. STUDY OF TRANSIENT GAS PRODUCTS IN EXPLOSIVES RELATIONSHIP OF STRUCTURE TO PROPELLANT BURNING	ENERGY PROPAGATION IN GASEOUS DISCHARGE MECHANISMS OF ION PRODUCTION ENERGY PRODUCTION AND TRANSFER RELATED TO EXPLODING WIRES	MAGNETIC COUPLING SEPARATION OF OHMIC HEATING AND MAGNETIC ACCELERATION MAGNETIC STRUCTURE OF SHOCKS.
SANDIA INTEREST	TRANSISTORS DIODES VOLTAGE SOURCES	VULNERABILITY RADIATION ENVIRONMENT	EXPLODING SWITCHES THERMAL BATTERIES DETONATORS PROPELLANT ACTIVATED DEVICES	SPARK GAPS SWITCHING TUBES DETONATORS ZIPPER SHERWOOD	SHERWOOD, FIREBALL RE-ENTRY HIGH TEMPERATURE SOURCES

Source: Richard S. Claassen Collection and Frank Hudson Collection—Derived from

HIGH TEMPERATURE	THEORETICAL MECHANICS	GEOPHYSICS	THEORETICAL	ANALYTICAL
THERMAL SHOCK ENERGY TRANSFER MATERIAL PROPERTIES ABLATION	DYNAMIC SIMILITUDE STRESS-STRAIN-TIME RELATIONSHIP	UPPER ATMOSPHERE PHYSICS		
EFFECT OF THERMAL TRANSIENTS ON SURFACES GRAIN BOUNDARIES NEAR MELTING. ABLATION IN CONTROLLED ATMOSPHERE.	SCALING LAWS IN DYNAMIC STUDIES. MOLECULAR STRUCTURE-MECHANICAL PROPERTY CORRELATION. STRAIN PRODUCED BY SHORT DURATION STRESS.	ENERGY TRANSFER AT HIGH ALTITUDES. ALTITUDE DEPENDENCE OF VARIOUS PARAMETERS	THEORETICAL PHYSICS SUPPORT OF OTHER'S RESEARCH PROGRAMS INDEPENDENT PROGRAMS	DEVELOPMENT OF SPECIAL TECHNIQUES FOR DETECTION OF PHYSICAL CHANGES AND INTERPRETATION OF SECONDARY INDICATIONS OF SUCH CHANGES IN SUPPORT OF OTHER RESEARCH PROGRAMS.
FIREBALL RE-ENTRY SHERWOOD ROVER	SHOCK AND VIBRATION MODEL STUDIES. WARHEAD DESIGN.	HIGH ALTITUDE SYSTEMS	RESEARCH SUPPORT	RESEARCH SUPPORT

Hudson's initial plan for Fundamental Research at Sandia.

still not right. Claassen returned the letter to Hudson with the humorous but optimistic comment: "Pls file for use in '58."[78] Nevertheless, the management attitude was changing. George Anderson, after a report from a Small Staff meeting, made a prophetic entry in his journal for June 1, 1956: "Apparently a new philosophy is going to take over with much greater scope of research at Sandia, including an expansion of research efforts into other than weapons lines."[79]

Meanwhile, Erickson and Claassen persisted in their efforts, and on August 7, 1956, Erickson decided to send Fowler a copy of Hudson's carefully detailed chart outlining a research program. "Dick and I think that you might be interested in the attached papers, . . ." he wrote. "While I believe that we all came away from the McRae meeting feeling that we had won every battle but lost the war (shades of the guided missile scrap), we nevertheless derived some cheer out of statements made by both you and McRae which indicated that the Corporation might be ready to give some support to a research effort."[80]

Fowler was definitely interested and receptive to Hudson's initiative to create a balanced research program. In addition, he realized that the advent of the nuclear burst with its tremendous release of energy in various forms had left the engineer with a shortage of some of the essential information needed for nuclear ordnance design. As Fowler explained:

> During my tour as Director of Electronics, I became convinced that solid state devices, with their small size, low power consumption, and inherent resistance to shock would replace vacuum and gaseous discharge devices in weapons fuzing and firing systems and in the instrumentation required during the development testing phase. It also was apparent to me that the weapons fuzing and firing systems would be required to operate safely and reliably at some level of the fierce environment of other nuclear weapon bursts. There would be no handbook data available about these new devices and their operation in this radiation and shock environment. In order for Sandia engineers to have the information required to produce valid designs, a fundamental understanding of solid state physics and of the

ADVANCING TECHNOLOGICAL FRONTIERS

Richard Claassen (left) and Frank Hudson (right) were instrumental in the establishment of fundamental research at Sandia.

Tom Cook, who would become an Executive Vice-President of the Laboratories, spent his early career in the field of weapons effects.

effects of weapon-produced radiation and shock on various solid state devices would have to be developed in-house.[81]

Subsequently, when a new research building (806) was proposed in 1956, Fowler took the position that it was important to know what would be done in the building to develop a sensible design. To formulate a plan for the research effort, Fowler organized the "Four O'Clock Group" early the next year. FOG, as the committee was known, consisted at one time or another of Fowler, Director Stu Hight, Dick Claassen, Tom Cook, Mel Merritt, W. W. Bledsoe, Alan Pope, M. J. Norris, and Frank Hudson. During the month of February, Fowler's Four O'Clock Group met nearly every working day at 4:00 P.M. to come up with ideas for establishment of a fundamental research program.[82]

Discussion during the meeting held February 11, 1957, laid the ground rules. Fundamental studies—that is, the study of phenomena—were to be called "Physical Science or Physical Research," while the application of fundamental knowledge to Sandia components was to be called "Applied Research." A two-pronged approach was discussed, with one branch oriented toward weapons and the incorporation of Weapons Effects—already in place at the Laboratory—and including "Systems Analysis," "New Weapons," and "Aerodynamics"; the other branch incorporating "Fundamental Research" activities. The discussion was interrupted by a call for Fowler from McRae with news that DMA had approved the Van de Graaff accelerator for the radiation effects program.[83]

Toward the end of February, Fowler met with Small Staff to present the Committee's justification for the expansion and formalization of a fundamental research program. He emphasized that, in addition to research aimed at exploring selected areas in physical science, the research organization should actively exploit the results of fundamental research to determine feasibility "of the new knowledge being applied effectively to the ordnance problems facing the company in the future."[84] Among other key points, he stressed the benefit of fundamental research in analyzing "the role of atomic weapons to guide research and

RESEARCH 1956

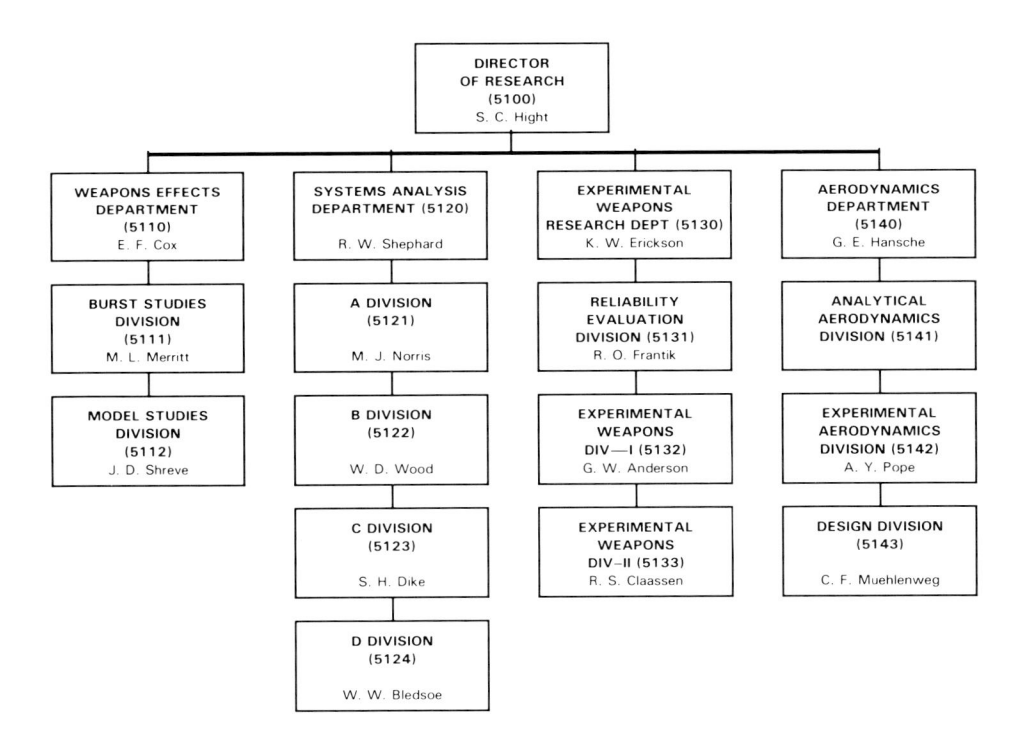

- **DIRECTOR OF RESEARCH (5100)** — S. C. Hight
  - **WEAPONS EFFECTS DEPARTMENT (5110)** — E. F. Cox
    - BURST STUDIES DIVISION (5111) — M. L. Merritt
    - MODEL STUDIES DIVISION (5112) — J. D. Shreve
  - **SYSTEMS ANALYSIS DEPARTMENT (5120)** — R. W. Shephard
    - A DIVISION (5121) — M. J. Norris
    - B DIVISION (5122) — W. D. Wood
    - C DIVISION (5123) — S. H. Dike
    - D DIVISION (5124) — W. W. Bledsoe
  - **EXPERIMENTAL WEAPONS RESEARCH DEPT (5130)** — K. W. Erickson
    - RELIABILITY EVALUATION DIVISION (5131) — R. O. Frantik
    - EXPERIMENTAL WEAPONS DIV—I (5132) — G. W. Anderson
    - EXPERIMENTAL WEAPONS DIV—II (5133) — R. S. Claassen
  - **AERODYNAMICS DEPARTMENT (5140)** — G. E. Hansche
    - ANALYTICAL AERODYNAMICS DIVISION (5141)
    - EXPERIMENTAL AERODYNAMICS DIVISION (5142) — A. Y. Pope
    - DESIGN DIVISION (5143) — C. F. Muehlenweg

Source: *Sandia Corporation Telephone Directory*, March 1956

development at Sandia, and to provide information to the AEC and DOD."[85]

After Fowler's successful bid to gain management support for the proposed program, the next step was to present the plan to Mervin J. Kelly of BTL. Dick Claassen was selected to make the presentation. Claassen began by summarizing the Corporation's evolution, which he related to the present need for fundamental research. "Originally," Claassen began,

> Sandia was split from Los Alamos Laboratories, fostered mainly by geographic considerations. The primary function to be served was that of matching into the military system, and as time went on, to this was added the problem of production of engineering designs which had been partially completed elsewhere. Still later the increased scope and variety of weapon designs has created a logical need for development of a variety of components and systems at Sandia.[86]

Claassen observed that these developments had been somewhat dictated by "giant strides in warhead advancement made by the weapons warhead Laboratories." However, Sandia had now reached the point, he said, where achievements in development and design work at the Laboratory had led to the "obligation to make original contributions in the nuclear weapons field." "By this, I do not mean new warhead designs," he added, "but rather new weapons systems designs which may be sparked by specialized components originated at Sandia."[87]

Claassen acknowledged that during this period of establishment of a new laboratory, there had been a research activity at Sandia Corporation. However, a large share of the research effort was the result of the development of weapons programs. "In a mature organization," Claassen said, "the research effort should be expended in advance of the development programs." To maintain superiority in the weapons race, he maintained that the Laboratory needed to find a way "to form a very close coupling between advances in fundamental knowledge and engineering designs of new weapons."[88]

ADVANCING TECHNOLOGICAL FRONTIERS

SANDIA CORPORATION
1956

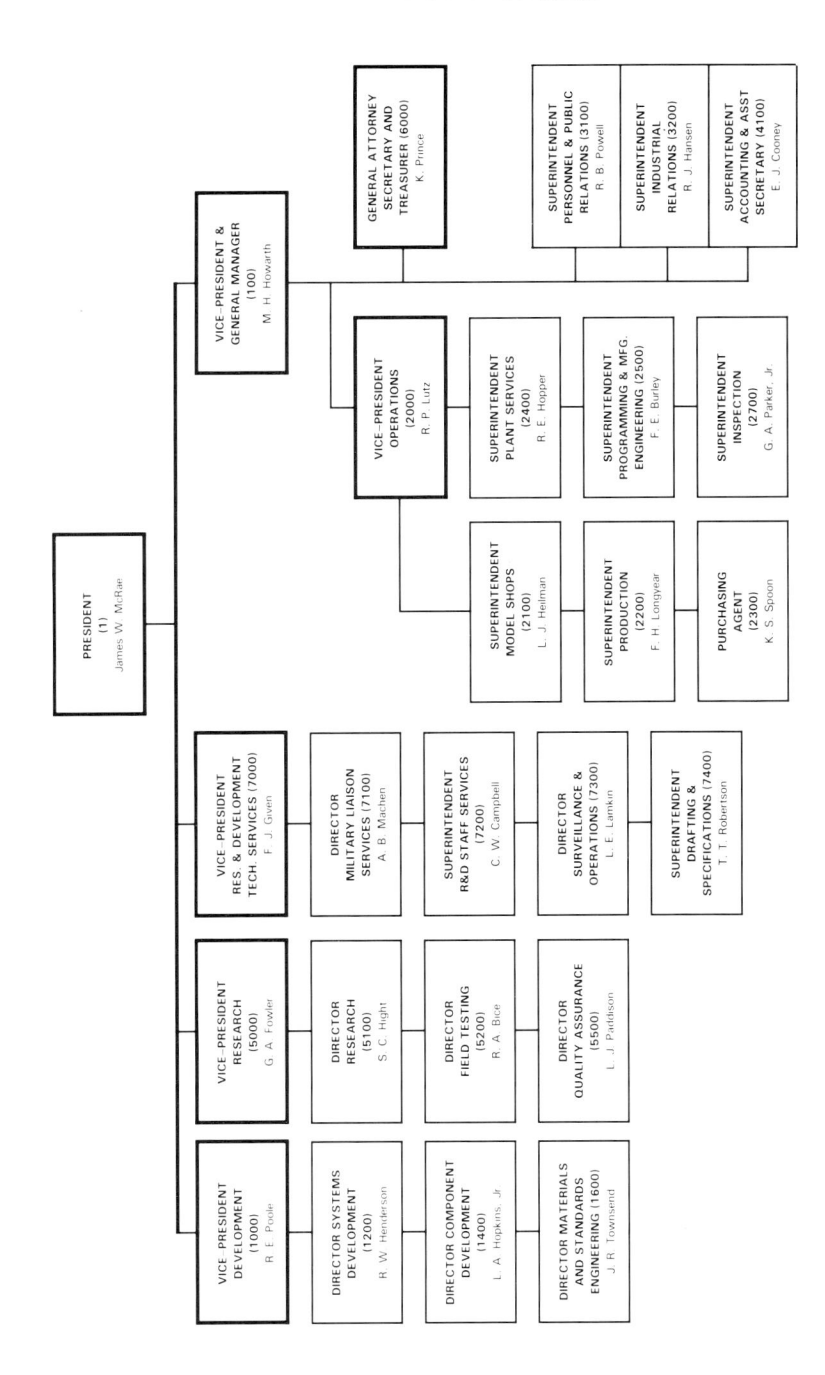

PRESIDENT (1) — James W. McRae

VICE-PRESIDENT & GENERAL MANAGER (100) — M. H. Howarth

- GENERAL ATTORNEY SECRETARY AND TREASURER (6000) — K. Prince
- SUPERINTENDENT PERSONNEL & PUBLIC RELATIONS (3100) — R. B. Powell
- SUPERINTENDENT INDUSTRIAL RELATIONS (3200) — R. J. Hansen
- SUPERINTENDENT ACCOUNTING & ASST SECRETARY (4100) — E. J. Cooney

VICE-PRESIDENT OPERATIONS (2000) — R. P. Lutz

- SUPERINTENDENT PLANT SERVICES (2400) — R. E. Hopper
- SUPERINTENDENT PROGRAMMING & MFG. ENGINEERING (2500) — F. E. Burley
- SUPERINTENDENT INSPECTION (2700) — G. A. Parker, Jr.
- SUPERINTENDENT MODEL SHOPS (2100) — L. J. Heilman
- SUPERINTENDENT PRODUCTION (2200) — F. H. Longyear
- PURCHASING AGENT (2300) — K. S. Spoon

VICE-PRESIDENT RES. & DEVELOPMENT TECH. SERVICES (7000) — F. J. Given

- DIRECTOR MILITARY LIAISON SERVICES (7100) — A. B. Machen
- SUPERINTENDENT R&D STAFF SERVICES (7200) — C. W. Campbell
- DIRECTOR SURVEILLANCE & OPERATIONS (7300) — L. E. Lamkin
- SUPERINTENDENT DRAFTING & SPECIFICATIONS (7400) — T. T. Robertson

VICE-PRESIDENT RESEARCH (5000) — G. A. Fowler

- DIRECTOR RESEARCH (5100) — S. C. Hight
- DIRECTOR FIELD TESTING (5200) — R. A. Bice
- DIRECTOR QUALITY ASSURANCE (5500) — L. J. Paddison

VICE-PRESIDENT DEVELOPMENT (1000) — R. E. Poole

- DIRECTOR SYSTEMS DEVELOPMENT (1200) — R. W. Henderson
- DIRECTOR COMPONENT DEVELOPMENT (1400) — L. A. Hopkins, Jr.
- DIRECTOR MATERIALS AND STANDARDS ENGINEERING (1600) — J. R. Townsend

Source: *Sandia Corporation Telephone Directory*, March 1956

Claassen then introduced some of the research programs being considered, including continuation of the study of radiation effects, already underway in Everett Cox's department. In this regard, Claassen mentioned that General Hertford in a letter to AFSWP had already formalized Sandia's role. The Hertford letter stated that the Laboratory would have "primary responsibility for the AEC in the study of radiation effects on weapons and their components." As Claassen acknowledged, "the radiation effects program was the one full-fledged fundamental research program we have started." Other possibilities suggested for investigation were: polymers and their interaction with radiation, alkali halides and their potential for use in dosimetry, and the effect of radiation on ferroelectric materials, among others.[89]

Claassen's presentation convinced Mervin Kelly of BTL that the program for fundamental research at Sandia should be supported. Meanwhile, Glenn Fowler continued his efforts in coordinating the program with Sandia management and the AEC. "It was Fowler," Hudson said, "who was responsible for the actual implementation of the program. He listened to us, let us develop a coherent program, and sold it to Small Staff and the AEC."[90]

In September 1957, the new Physical Sciences organization, directed toward fundamental research, became a part of the Laboratory's expanded research program. Thus, within the span of three years, the management philosophy of the Laboratory had broadened from one of a focus on Engineering and Applied Physics to the inclusion of an expanded role for fundamental research in the Laboratory infrastructure. Claassen was named as the new Department Manager.[91]

The establishment of fundamental research at Sandia marked a milestone in the maturation and evolution of the Laboratory. Within the short span of a decade, Sandia had grown from a small nuclear weapons design, assembly, and field testing facility to a well-respected research and development organization, recognized as the nuclear ordnance engineer for the nation.

ADVANCING TECHNOLOGICAL FRONTIERS

The Westward Migration

The end of the postwar decade signalled not only the expansion of Sandia Laboratory into new technical fields, but also its physical and geographic expansion. As the University of California Radiation Laboratory (UCRL) grew, so did its need for an engineering organization in close proximity. Sandia had been providing design and development engineering support for UCRL, first from headquarters in Albuquerque. Then in the fall of 1952, Herb York, Director of UCRL, invited Bob Henderson to California to work out an arrangement for a group of Sandia personnel to work on site.[92]

Subsequently, on October 27, Henderson, accompanied by Ralph Wilson and Richard Bice, boarded a plane for San Francisco. From the Claremont Hotel at Berkeley, they traveled inland by car to the small town of Livermore, California, nestled east of the Oakland hills among family owned vineyards. Cooled by fogs and breezes from the bay, the area was well suited to the growing of premium grape varieties, and it was this locale—at the time isolated from urban areas—that Ernest Lawrence and Edward Teller had selected as the site for the Radiation Lab.[93]

In Livermore, the men met with York and together made plans for Sandia to provide expanded weaponization support for the Rad Lab. A focus of their efforts would be Upshot-Knothole test operation scheduled for 1953. But this type of support soon became inadequate as well; and in 1955, a pioneer group of fifteen Sandians moved to Livermore to work on the XW-27. The following year, in March 1956, Sandia Corporation formally established a second weapons laboratory with headquarters in an old WAVE's barracks building across the street from UCRL in Livermore, California.[94]

At the outset, there was only one department, Engineering, headed by W. Jack Howard and supported by two project divisions under supervisors C. R. "Barnie" Barncord and R. L. "Ray" Brin. The population of Livermore increased by nine when Leo Gutierrez, his wife, and seven children arrived. Gutierrez also assumed duties as a Project Division supervisor. Soon the

Engineering Department had grown to such a size that it was split into two departments, with Jack Howard heading one and L. E. "Lee" Hollingsworth heading the new department. In November 1956, Howard was promoted to Director of Systems Development for the Laboratory; and in 1957, Bob Poole became the Lab's first Vice-President.[95]

The Livermore Laboratory's charter was to provide engineering support for UCRL in much the same manner as the Sandia Laboratory in Albuquerque supported Los Alamos. From these beginnings, Sandia, Livermore, would develop its own distinct capabilities as a research and development organization during the Laboratory's second decade.

Vice-President Robert E. Poole visits with Department Managers Jack Howard and Charlie Campbell at Livermore.

ADVANCING TECHNOLOGICAL FRONTIERS

Charles Gump hangs a sign outside the WWII WAVE barracks that would serve as headquarters for the Livermore Branch of Sandia Corporation.

The first transfer of personnel to Livermore included (left to right) C. E. Barncord, Charles A. Gump, Clifford O. Erickson, Benjamin F. Fisher, Jr., Robert L. Siglock, S. Gayle Cain, Vernon M. Field, Charlie Winter, Wayne A. Grimshaw, Frank J. Thomas, Orval W. Wallen, James McMinn, Mary A. VanBrocklin, William B. Marsh, and Nora-Bell Byrd.

ADVANCING TECHNOLOGICAL FRONTIERS

In Retrospect

The genesis and growth of Sandia National Laboratories epitomizes the complex, but effective interaction of government, industry, cooperative research, and sound engineering. From a small nuclear ordnance and assembly operation established to meet a temporary need, the organization gained permanence as an integral facet of the nation's developing defense complex.

The disintegration of U.S.-Soviet solidarity and the collapse of attempts to establish international control of atomic energy led to a buildup of the nation's nuclear arsenal, refinement of the early atomic devices, emergence of the thermonuclear era, and full-scale testing. Such factors also led to the continued existence and rapid growth of the nuclear ordnance facility located at Sandia Base. Despite adversity, the Laboratory showed a surprising resiliency and responsiveness—through the turmoil of organizational change, crash projects, and programs to establish emergency capabilities. It benefited from strong advocacy in Washington, and especially from the industrial management expertise of its AT&T sponsor.

For Sandia, therefore, the postwar decade was a time of technical accomplishments and transition—from a small offshoot of Los Alamos to an independent Laboratory of nearly 6,000 people; from academic to industrial management; from a production orientation to systems engineering based upon solid research. But above all, the story of Sandia National Laboratories during its formative years is the story of the mission-oriented Manhattan Project pioneers, the founders of the organization, and those who came soon after. The Sandia image was one of technical competence, youth, exuberance, and dedication—an employee population imbued with the pragmatic, "can-do" ethos that enabled the Laboratory to meet the requirements of the Cold War era.

NOTES

PROLOGUE:

1. N.A., "The Atomic Bomb," *Yank: The Army Weekly*, 7 September 1945 [p. 15]. See also "Truman Bares Deadly New U.S. Discovery," *LA Herald Express*, 6 August 1945 and "Hiroshima Devastation Still Not Known," *The Montreal Daily Star*, 7 August 1945.

2. Otto Hahn in "The Discovery of Fission," *The Scientific American* (February 1958), pp. 82-83, recalled that: "In January, 1939, we published an account of these 'experiments that are at variance with all previous experiences in nuclear physics.' "

Leo Szilard to Frederick Joliot-Curie, 2 February 1939, quoted in William Manchester, *The Glory and the Dream; A Narrative History of America* 1932-1972, Vol. I (Boston: Little, Brown and Company, 1973), p. 255.

3. "Woman Jewish Atom Expert Fled Germany," newsclipping in scrapbook, dated 7 August, New York (newspaper title unavailable). This scrapbook was donated by the National Atomic Museum, Kirtland Air Force Base, Albuquerque, New Mexico. Other clippings of interest include "Lise Meitner Says Bomb Should Be Controlled," 10 August, New York, and "Mystery Town Cradled the Bomb . . . . the work of many minds," n.d., n.p.; Peter Wyden, *Day One, Before Hiroshima and After* (New York: Simon and Schuster, 1984), pp. 22-23. Richard Rhodes in *The Making of the Atomic Bomb* (New York: Simon and Schuster, 1986), pp. 168-97, contains an excellent account of the exodus of these scientists to the United States and England.

See also Otto Hahn, "The Discovery of Fission," *The Scientific American* (February 1958), pp. 76-78, 80. On pp. 82, 84, he states: "Immediately after our paper appeared, [Lise] Meitner and Otto R. Frisch came out independently with their historic publication showing how Niels Bohr's model of the atom could explain the cleavage of a heavy nucleus into two nuclei of medium size. Meitner and Frisch named the process 'fission.' " The Meitner and Frisch account was published as "Physical Evidence for the Division of Heavy Nuclei under Neutron Bombardment," *Nature*, 18 February 1939, Vol. 143, p. 276. See also Manchester, *The Glory and the Dream*, pp. 254-57.

NOTES

Accounts of these developments may also be found in Richard G. Hewlett and Oscar E. Anderson, Jr., *The New World*, 1939/1946, *A History of the United States Atomic Energy Commission*, Vol. I (University Park: Pennsylvania State University Press, 1962), pp. 10-11 (hereinafter referred to as AEC *History*, Vol. I). Vannevar Bush in his "Statement to the National Policy Committee," 28 May 1945, also gives a good summary, S-1 (Uranium Committee) Folder 36 "Bush V. 1944," Record Group No. 227, National Archives.

4. Manchester, *The Glory and the Dream*, pp. 257, 259. See also F. B. Jewett, President National Academy of Sciences to J. B. Conant, Chairman NDRC, 17 July 1941, mentioning a report submitted by Fermi, entitled "Some Remarks on the Production of Energy by a Chain Reaction in Uranium," both in Folder 5: "Historical File," Record Group No. 227, National Archives.

5. Manchester, *The Glory and the Dream*, pp. 260-61, and James MacGregor Burns, *Roosevelt: The Soldier of Freedom* 1940-1945 (New York: Harcourt, Brace, Jovanovich, 1970), pp. 249-50, record this conversation. The AEC *History*, Vol. I, pp. 16-17 provides a slightly different version.

Original documents include Albert Einstein to Franklin D. Roosevelt, 2 August 1939, the Szilard memorandum, and Sach's covering letter, all of which may be found in the *Papers of Franklin D. Roosevelt*, Franklin D. Roosevelt Library, Hyde Park, New York (hereinafter referred to as FDR). Sach's version of the story is told in a typescript manuscript, "Early History Atomic Project in Relation to President Roosevelt 1939-40," to be found in *Records of the Manhattan Engineer District*, 1942-1948, World War II Records Division, National Archives and Records Service, Alexandria, Virginia.

6. James F. Byrnes, *Speaking Frankly* (New York: Harper and Brothers, 1947), p. 3. Byrnes' commentary on the Roosevelt years gives a good flavor of this trying time in history from the presidential perspective, although similar information may be found in numerous secondary sources. For a good discussion of public debate over Lend-Lease, see Charles A. Beard, *President Roosevelt and the Coming of the War 1941*, (New Haven: Yale University Press, 1948), pp. 14-68.

7. Lauritson to Bush, 11 July 1941, Draft Report by MAUD, Technical Committee on the Release of Atomic Energy for Uranium (hereinafter referred to as Maud Report), enclosure, F. L. Hovde to C. L. Wilson, 4 July 1941, Office of Scientific Research and Development (hereinafter referred to as OSRD). See also Henry L. Stimson, "The Decision to Use the Atomic Bomb," *Harper's Magazine* (February 1947), p. 97.

Anthony Cave-Brown and Charles B. MacDonald, eds., *The Secret History of the Atomic Bomb* (New York: Dell Publishing Co., 1977), pp. 31-34; AEC *History*, Vol. I, pp. 41, 43, 49-52, which includes photocopies of original

NOTES

documents Bush to Conant, 22 December 1941, OSRD, Headquarters, U.S. Atomic Energy Commission, Washington, D.C.

8. Bush, "Statement to the National Policy Committee," 28 May 1945, pp. 7, 14, 29 (S-1 Uranium Committee), Folder 36, "Bush V. 1944," Record Group No. 227, National Archives; AEC *History*, Vol. I, p. 52; Brown and MacDonald, *Secret History*, pp. 50, 51, 55, 56.

9. Brown and MacDonald, *Secret History*, pp. xix, xx, 33, 86–87.

10. Ibid., p. xiii.

11. For comprehensive coverage of Soviet espionage activities during this period, see Justin Atholl, *How Stalin Knows*, (London: Jarrold and Sons Limited, no date), pp. 1–182. See also *Secret History*, pp. xiii–xvi, and Barton J. Bernstein, *The Atomic Bomb* (Boston: Little, Brown and Company, 1976), pp. vii–viii.

12. AEC *History*, Vol. I, pp. 44, 55–62; Donald M. Kerr, Director, LASL, "Forty Years of Service to the Nation," LASL Annual Report, 1982, p. 6; Manchester, *The Glory and the Dream*, pp. 379–81.

13. AEC *History*, Vol. I, pp. 115–19, 227–28.

14. John H. Dudley, "Ranch School to Secret City" in *Reminiscences of Los Alamos*, 1943–1945, ed. by Lawrence Badash, Joseph O. Kirschfelder, and Herbert P. Broida (Boston: D. Reidel Publishing Co., 1980), pp. 4–5. See also Leslie R. Groves, *Now It Can Be Told* (New York: Harper and Brothers, 1962), pp. 65–67; AEC *History*, Vol. I, p. 41; and Public Relations Staff, *The First 20 Years at Los Alamos January 1943–January 1963* (Los Alamos: LASL News, 1963), pp. 8–16.

15. Stimson, "The Decision to Use the Atomic Bomb," p. 102. See also Henry L. Stimson and McGeorge Bundy, *On Active Service In Peace and War* (New York: Harper and Brothers, 1947), pp. 615–18, and Henry L. Stimson, "The Official Explanation: Statement and Challenge" including "Memorandum for the President" in Barton J. Bernstein, ed., *The Atomic Bomb: The Critical Issues* (Boston: Little, Brown and Company, 1976), pp. 10–11. See also Manchester, *The Glory and the Dream*, pp. 452–54. While recent scholarship shows that some American leaders and military planners estimated that an invasion of Japan would cost no more than 46,000 deaths—those leaders making the decisions such as Stimson were obviously informed otherwise.

16. AEC *History*, Vol. I, pp. 383–84; Manchester, *The Glory and the Dream*, p. 463.

17. AEC *History*, Vol. I, pp. 393–94; Brown and MacDonald, *Secret History*, pp. xv–xvi. See also Truman's *Years of Decision*, p. 416, which gives a firsthand account of the conversation. Churchill's *Triumph and Tragedy* (Boston: Houghton Mifflin Co., 1953), pp. 669–70, provides a witness's account of the conversation.

NOTES

18. Stimson, "The Decision to Use the Atomic Bomb," p. 99; Churchill, *Triumph and Tragedy*, pp. 638–39; and Stimson, "The Official Explanation," p. 2; and Stimson and Bundy, *On Active Service*, p. 634.

19. *The Conference of Berlin* (The Potsdam Conference), 1945 (Washington, D.C., 1961), I, p. 884, and *The Conference of Berlin*, II, p. 1293. See also AEC *History*, Vol. I, pp. 395–97, and Stimson, *On Active Service*, pp. 620–25; Groves, *Now It Can Be Told*, pp. 327–28. Original statements quoted in *Manhattan District History*, Book I, *General*, Vol. IV, Auxiliary Activities, Ch. 8. Press releases in New York *Times*, 7 August 1945.

See also Harry Truman to James Lea Cate, Washington, D.C., 12 January 1953, quoted in *The Army Air Forces in World War II*, Vol. 5, ed. by Wesley Frank Craven and James Lea Cate (Chicago: University of Chicago Press, 1953), pp. 1–2, which contains a reprint of the letter in which Truman explains that the original directive for delivery of the bomb was dated July 25, in order "to set the military wheels in motion"; however, the final decision was not made until "we were returning from Potsdam."

20. Harry Truman to Leroy A. Wilson, Washington, D.C., 13 May 1949 (copy), SNL Archives.

CHAPTER 1:

1. *New York Times*, 16 July 1945; Sandia *Lab News*, 15 July 1955, p. 1; Albuquerque *Morning Journal and Tribune*, 16 July 1945. See Lansing Lamont, *Day of Trinity* (New York: Atheneum, 1965), p. 186. The Lamont book is excellent for capturing the spirit of Trinity.

2. Interview, Leo M. "Jerry" Jercinovic, Sandia National Laboratories, Albuquerque Archives (hereinafter referred to as SNL Archives), n.d.; Leslie R. Groves, *Now It Can Be Told: The Story of the Manhattan Project* (New York: Harper and Brothers, 1962), pp. 292–93.

3. Interview, Charles R. Barncord, and Ray Brin, n.d., SNL Archives. Barncord and Brin helped build the site and develop loading techniques for the Fat Man. Later, they went to Wendover to assist in training. Interview, Jerry Jercinovic, n.d.; Lamont, *Day of Trinity*, p. 169; Lawrence Badash, Joseph O. Hirschfelder, and Herbert P. Broida, eds., *Reminiscences of Los Alamos 1943–1945* (Boston: D. Reidel Publishing Co., 1980), p. 58. Kistiakowsky claimed that he insisted upon Friday 13 because he believed in "unorthodox luck."

According to Arthur A. Machen, "What came down from Los Alamos was the H.E. sphere assembly." A complete atomic bomb, i.e., nuclear assembly ballistic case, proximity fuzing, and ancillary firing hardware, was

NOTES

not assembled until the Tinian event. See Arthur B. Machen to Furman, Rogue River, Oregon, 4 March 1984, SNL Archives.

4. Arthur Machen generously lent the author a copy of a memorandum from Norris E. Bradbury titled "TR Hot Run" that provides a daily log with times and actions taking place. Enclosure, Machen to Furman, Rogue River, Oregon, 4 March 1984, SNL Archives.

5. People said of Slotin: "He was a brave man but brave in an odd sort of way. Slotin had a positive hankering for danger." See Charles Neider, ed., "The Strange Death of Louis Slotin," *Man Against Nature: Tales of Adventure and Exploration* (New York: Harper and Brothers, 1954), pp. 10–11.

6. Among the euphemisms used by the scientists, the "core," which could refer to the plutonium only, or to the initiator and the plutonium, was also called the "globe." The core, in turn, could be divided into two "hemispheres." The initiator was also called the "urchin," and the "reflector" was called the "tamper."

Machen to Furman, Rogue River, Oregon, 4 March 1984, SNL Archives.

7. Machen to Furman, Rogue River, Oregon, 4 March 1984, SNL Archives; Taped interview, Robert W. Henderson, 23 January 1984, SNL Archives; Lamont, *Day of Trinity*, pp. 173–75; Interview, Jerry Jercinovic, n.d., SNL Archives; Interview, Robert Bacher, 3 March 1986, File TR–86–032, Los Alamos Archives.

8. Interview, Robert W. Henderson, 23 January 1984, SNL Archives.

9. Interview, Jerry Jercinovic, n.d., SNL Archives. See also Ferenc Szasz' comprehensive treatment of Trinity, *The Day the Sun Rose Twice: The Story of the Trinity Site Nuclear Explosion, July 16, 1945* (Albuquerque: University of New Mexico Press, 1984).

Telephone interview, William Caldes, 15 December 1983. Caldes tells a story about asking Oppenheimer for a raise. Others had received theirs, so Caldes walked into the director's office and presented the problem: " 'I didn't get a raise. How come?' 'You didn't?' Oppie responded. 'Well, I'll write you one.' " And he did.

10. Selected to head the Manhattan Engineer District September 17, 1942, Groves previously had helped to build the Pentagon. See Groves, *Now It Can Be Told*, p. 290, and Szasz, "The Day the Sun Rose Twice," p. 18. Walt Treibel to Beryl Hefley, 29 March 1982, SNL Archives.

See also Lamont, *Day of Trinity*, pp. 179, 195–96, and James W. Kunetka, *City of Fire, Los Alamos and the Birth of the Atomic Age 1943–1945* (Englewood Cliffs, N.J.: Prentice–Hall, Inc., 1978), p. 159. The Kunetka book on the detonation of the atomic bomb is based on primary sources from the Los Alamos Archives.

NOTES

11. Interview, Robert W. Henderson, 23 January 1984, SNL Archives; K. T. Bainbridge to Bradbury, Memo re: "Jumbo," 11 July 1945, Los Alamos National Laboratory Records Center and Archives (hereinafter referred to as Los Alamos Archives).

12. Groves, *Now It Can Be Told*, pp. 291-94 and p. 436, Appendix VIII, "Memorandum for the Secretary of War" from General Farrell, 18 July 1945. See also Leona Marshall Libby, *The Uranium People* (New York: Crane Rusak and Co., 1979), pp. 104-5.

13. Letter of transmittal, Dr. Jack Hubbard to Los Alamos Archives, 20 May 1976, and p. i in "Los Alamos 1945, Journal of J. Hubbard, Meteorologist," Los Alamos Archives (hereinafter referred to as Hubbard *Journal*).

14. Bob Henderson to F. H. McCollum, Los Alamos, New Mexico, 7 September 1945, (copy) SNL Archives; Groves, *Now It Can Be Told*, p. 289; K. T. Bainbridge, "All in Our Time: Prelude to Trinity," *Bulletin of Atomic Scientists*, Vol. 31 (April 1975), p. 43 (hereinafter referred to as "Prelude to Trinity"), reprinted in *All in Our Time: The Reminiscences of Twelve Nuclear Pioneers*, ed. by Jane Wilson, under the title "Orchestrating the Test," pp. 202-31 (Chicago: University of Chicago Press, 1975); *American Men & Women of Science*, 14th Ed., Vol. I, (New York: R. R. Bowker Company, 1979), p. 192.

15. Barbara Storms, *Reach to the Unknown*, Los Alamos publication, 1965, pp. 23-25; Kunetka, *City of Fire*, p. 168.

16. Interview, Glenn A. Fowler, 17 May 1983, SNL Archives; reprinted in Sandia *Lab News*, 12 June 1983, p. 5.

17. Descriptions of the explosion are taken from the combined testimony of Sandians who witnessed the Trinity test: Robert W. Henderson, Director of Engineering and later Vice-President; Leo M. Jercinovic, Supervisor of Model Materials Division; Arthur B. Machen, Director of Military Liaison Services; Glenn A. Fowler, Director of Field Testing and later Vice-President; William Caldes, Supervisor of Field Simulation Department; and Louis F. Jacot, Technical Illustrator, all in SNL Archives. See also Groves account in *Now It Can Be Told*, pp. 295-96.

18. Louis Jacot, interview in Sandia *Lab News*, 15 July 1955, p. 5.

19. William L. Laurence, *Dawn Over Zero: The Story of the Atomic Bomb* (New York: Alfred A. Knopf, 1947), p. 193, quoting a report to the War Department written by General Farrell, assistant to Groves, giving his impressions of those last seconds before zero. Farrell's memorandum is also reprinted in full in Groves, *Now It Can Be Told*, pp. 433-38.

20. George Kistiakowsky, quoting himself in "Reminiscences of Wartime Los Alamos," in book of same title, ed. by Badash et al., p. 60.

21. Robert W. Henderson, quoted in Sandia *Lab News*, 12 June 1983, p. 5, and Glenn A. Fowler, quoted in same.

CHAPTER 2:

1. Both Franklin D. Roosevelt and Winston Churchill saw the Trinity test as signalling a hasty end to World War II. In his book *Triumph and Tragedy*, p. 638, Churchill recounts how an attack on the mainland would have resulted in the loss of a million American lives and many British as well. With the successful detonation of the first atomic bomb, however, "all this nightmare picture vanished," he said. "In its place was the vision—fair and bright indeed it seemed—of the end of the whole war in one or two violent shocks."

2. Interview, Robert W. Henderson, 23 January 1984, SNL Archives; also told in Szasz, *The Day the Sun Rose Twice*, p. 40.

3. Lamont, *Day of Trinity*, pp. 69–70. Kunetka, *City of Fire*, p. 146, claims that the suggestion was made by Oppenheimer in a meeting at his office. See also Nuel Pharr Davis, *Lawrence and Oppenheimer* (New York: Simon and Schuster, 1968), pp. 224-25; and Peter Goodchild, *J. Robert Oppenheimer: Shatterer of Worlds* (Boston: Houghton Mifflin Co., 1981), p. 130; Szasz, *The Day the Sun Rose Twice*, pp. 40–41.

4. Szasz, *The Day the Sun Rose Twice*, pp. 40–41; and Marjorie Bell Chambers, "Technically Sweet Los Alamos. The Development of a Federally Sponsored Scientific Community," Ph.D. dissertation, University of New Mexico, 1974.

5. Szasz, *The Day the Sun Rose Twice*, p. 41, suggests that the name be best defined geographically at latitude 33° 28'–33° 50', longitude 106° 22'–106° 41'.

6. Arthur Holly Compton, *Atomic Quest: A Personal Narrative* (New York: Oxford University Press, 1956), p. 212.

7. Quoted in Storms, *Reach to the Unknown*, p. 3.

8. Groves, *Now It Can Be Told*, p. 63. Biographer Nuel Pharr Davis in *Lawrence and Oppenheimer*, p. 148, indicates that "since the beginning of 1942, Oppenheimer had been under investigation. FBI and Army intelligence agents had swarmed about him compiling an enormous dossier."

9. John Manley, "Experiment Results and Description of Available Equipment," Los Alamos Archives, quoted in *Los Alamos Science*, Winter/Spring 1983, p. 12. James B. Conant and Leslie R. Groves to Dr. J. R. Oppenheimer, OSRD, Washington, D.C., 25 February 1943, Los Alamos Archives. Specifically, the letter read: "The Laboratory will be concerned with

NOTES

10. The actual contract, No. W-7405, was signed 20 April 1943 and made retroactive to 1 January 1943. See Memo to Oppenheimer, "Transmittal of Contract No. W-7405 Eng. 36, 27 July 1943," Los Alamos Archives. Los Alamos would not become known as Los Alamos Scientific Laboratory until January 1947.

11. David Hawkins, *Manhattan District History, Project Y*, Vol. I, Washington, D.C.: Office of Technical Services, 1945-1947, pp. 38-39. This is an official history of Project Y written by a participant in the project and reprinted in 1983 as a commemorative hardbound volume. This volume has been reprinted as #2 of *The History of Modern Physics* 1800-1950 series. Volume II is titled *Project Y: Los Alamos Story* with Part I: *Toward Trinity* by David Hawkins and Part II: *Beyond Trinity* by Edith C. Truslow and Ralph Carlisle Smith (Los Angeles: Tomash Publishers, 1983). See also *Los Alamos, The Beginning of an Era 1943-1945*, pp. 11-18. The latter is a booklet compiled by the staff of the Los Alamos Public Relations Office; Interview, Ray B. Powell, n.d., SNL Archives.

12. For problems of assembly, see AEC *History*, Vol. I, pp. 245-46; and Storms, *Reach to the Unknown*, p. 45.

13. Groves, *Now It Can Be Told*, p. 61.

14. Ibid., pp. 159-60. See also Kunetka, *City of Fire*, pp. 81-82.

15. Hawkins, *Project Y*, Vol. I, pp. 124-25; Groves, *Now It Can Be Told*, pp. 159-61; and Kunetka, *City of Fire*, p. 81.

16. At the Tuesday Ordnance meeting, this change in the organization was announced. See Los Alamos *Daily Log*, 13 October 1943 to 4 July 1944 (hereinafter referred to as Daily Log #1), pp. 68, 69, for March 22, 23, 1944. See also Hawkins, *Project Y*, Vol. I, pp. 124-29; and Bainbridge, "Prelude to Trinity," p. 44.

For references to Ayers and Dreesen, see AEC *History*, Vol. II, p. 138, and Daily Log #1, pp. 9, 15, 25. See also "Robert Brode File," SNL Archives.

17. Norman F. Ramsey, "History of Project A," Los Alamos Archives, quoted in *Los Alamos Science*, Winter/Spring, 1983, p. 23.

18. Ibid. See also Frederick C. Alexander, Jr., *History of Sandia Corporation: Through Fiscal Year 1963* (Albuquerque: Sandia Corporation, 1963), p. 2; and Lee Bowen, "Project Silverplate, 1943-1946," in "A History of the Air Force Atomic Energy Program 1943-1953," Vol. 1 of 5 (unpublished manuscript) USAF History Division, Maxwell Air Force Base Archives, Alabama (copy in SNL Archives; source hereinafter designated as Bowen, "Silverplate"). Microfilm and manuscript copies can also be found at the Atomic Museum and the Technical Library, Sandia National Laboratories, Albuquerque.

NOTES

19. Interview, Robert W. Henderson, 23 January 1984, SNL Archives; Hawkins, *Project Y*, Vol. I, p. 145. The first two versions of the Thin Man were numbered as Mk 1 and Mk 2. The Fat Man was the Mk 3.

20. For information on modifications of the B-29s, see the book written by Henry H. Arnold, the Air Force General in charge of the program (code name "Silverplate"): *Global Mission* (New York, 1949), p. 492.

21. Hawkins, *Project Y*, Vol. I, pp. 145-46. Daily Log #1, p. 18, indicates that models had also been dropped in San Francisco Bay. Mention of the accident can be found in this same source, pp. 65, 67. A Fat Man and "several slim men" were delivered to Muroc in early February. See p. 43.

22. Hawkins, *Project Y*, Vol. I, p. 145 and pp. 132-35. See also Los Alamos *Daily Log*, 5 July 1944 to 13 September 1945 (hereinafter referred to as Daily Log #2) for TWX to Parsons, "attn. Brode," re: disappointment with barometric fuzes, p. 94; and Ordnance Meeting, 8 August 1944, p. 127.

23. Ramsey, "History of Project A" and Minutes of the Governing Board, 17 June 1943, pp. 3-4, Los Alamos Archives. See also AEC *History*, Vol. I, p. 251.

24. Ramsey, "History of Project A," Los Alamos Archives.

25. Daily Log #1, pp. 67, 74. Log #2, p. 116, details continued Silverplate problems with the design of Fat Man tails. An entry for 24 August 1944 concerns problems between "people in Detroit," who wanted to use the "Buffalo design on the Fat Man tail" rather than the design created at Los Alamos. The Buffalo design would have cut down on the clearance needed; therefore, Parsons made the decision "to tell the Detroit people that our drawings would be modified to include some of their minor changes, and that the tails were to be made from the drawings prepared here." The writer went on to explain that the situation was "somewhat of a sore-spot anyhow since nearly everyone here is of the opinion that the Buffalo drawings were a duplicate of effort." See also Ramsey, "History of Project A," and Hawkins, *Project Y*, Vol. I, p. 146. Captain David Semple was later killed in a B-29 crash in New Mexico.

26. To facilitate this development, an impressive contingent of scientists from Great Britain and other European countries arrived. Daily Log #1, entry for 29 December 1943, p. 24, notes Niels Bohr and son, Aage, were expected to arrive at the site the following day and adds that "Dr. Bohr will be known as Nicholas Baker and son as Jim Baker." Bohr's presence was top secret information. In fact, the entry noted that "All E-3 group members had to sign a paper which required them to keep this information secret." AEC *History*, Vol. I, pp. 310-11.

27. AEC *History*, Vol. I, p. 252. For excellent coverage of events and personalities contributing to development of the bomb and postwar research

named Poole, who wrote a brief paper on the subject early in the war. (See p. 9 of interview typescript.) However, Henry Linschitz, a young section leader, according to Kistiakowsky, was the first to use the lens on the atomic bomb. policy, see Daniel J. Kevles, *The Physicists, The History of a Scientific Community in Modern America* (Cambridge, Massachusetts: Harvard University Press, 1971), pp. 324–48.

28. Minutes of the Governing Board, 23 September 1943, pp. 2–3; 4 November 1943, pp. 2–3, Los Alamos Archives; Kunetka, *City of Fire*, pp. 84–85.

29. Hawkins, *Project Y*, Vol. I, p. 138.

30. "Outline of Present Knowledge," presented by J. Robert Oppenheimer at the orientation meetings 15–24 April 1943, Los Alamos Archives. See also Kunetka, *City of Fire*, pp. 47–55.

31. "Status of Ordnance Work at Y" (as of March 1, 1943), 21 pp., Los Alamos Archives.

32. Ibid., pp. 2, 7–9, 12, 20. Referenced also were reports by John Manley, Oppenheimer, and Serber, "Use of Materials in a Fission Bomb" and "Effects of Impurities on Detonation" by Edward Teller. George Bogdan Kistiakowsky, among others, had previously experimented with the idea of using high explosives in a spherical configuration and, in fact, had written about the subject in a National Academy of Science report for President Franklin Roosevelt in "either '41 or '42." See Interview, "George Bogdan Kistiakowsky" (typescript), Los Alamos Archives, p. 11.

33. Goodchild, *J. Robert Oppenheimer*, pp. 82, 112; Kistiakowsky, "Reminiscences of Los Alamos," pp. 49–50; Kunetka, *City of Fire*, pp. 52–55.

34. Goodchild, *J. Robert Oppenheimer*, pp. 106–7; Minutes of the Governing Board, 23 September 1943, pp. 2–3.

35. Hawkins, *Project Y*, Vol. I, p. 139; Stephane Groueff, *Manhattan Project: The Untold Story of the Making of the Atomic Bomb* (Boston: Little, Brown and Company, 1976), pp. 321–22 (hereinafter cited as *Manhattan Project*).

36. Hawkins, op. cit.; Kunetka, *City of Fire*, p. 84.

37. Kunetka, op. cit., 86; Minutes of the Governing Board, 4 November 1943, p. 2.

38. Minutes of the Governing Board, 4 November 1943, p. 3.

39. Groueff, *Manhattan Project*, pp. 327–29.

In an interview conducted by Los Alamos personnel with George B. Kistiakowsky, he maintains that "several people had the idea of using explosive lenses" and that the original idea, he felt, came from an Englishman

NOTES

the development and final manufacture of an instrument of war, which we may designate as Projectile S-1-T."

40. Memo, George Kistiakowsky to Robert Oppenheimer and William Parsons, 13 June 1944, Los Alamos Archives; Goodchild, *J. Robert Oppenheimer*, p. 112, says that Kistiakowsky came in January 1944, for the first time. On page 116, Goodchild quotes Neddermeyer: "Oppenheimer lit into me . . . . From my point of view, he was an intellectual snob. He could cut you cold and humiliate you . . . . On the other hand, I could irritate him." AEC *History*, Vol. I, pp. 246-47; Kunetka, *City of Fire*, pp. 84-86; Kistiakowsky, "Reminiscences of Los Alamos," p. 50.

41. Kistiakowsky, "Reminiscences of Los Alamos," p. 49; Interview, Kistiakowsky (typescript), Los Alamos Archives, p. 8.

42. Kistiakowsky, "Reminiscences of Los Alamos," p. 50; Oppenheimer to Seth Neddermeyer, 15 June 1944, Library of Congress, *Oppenheimer Papers*, Government File Supplement.

43. Quotation from "August 1944 Reorganization" printed in *Los Alamos Science*, Winter/Spring, 1983, p. 15; See also Hawkins, *Project Y*, Vol. I, Chapter IV.

44. Hawkins, *Project Y*, Vol. I, pp. 173-74.

45. Ibid.; Oppenheimer to R. F. Bacher, "Organization of Gadget Division," 14 August 1944, Los Alamos Archives; Oppenheimer to G. B. Kistiakowsky, "Organization of Explosives Division," 14 August 1944, Los Alamos Archives; printed in *Los Alamos Science*, Winter/Spring, 1983, p. 15; Telephone interview, John Manley, 10 July 1989.

46. AEC *History*, Vol. I, p. 248, quoting "Secret Report on Status of Ordnance Work at Y" (as of March 1, 1944), Los Alamos Archives. Kistiakowsky's schedule is in the Manhattan Engineer District files.

47. Kistiakowsky, "Reminiscences of Los Alamos," pp. 54-55.

48. Ibid., p. 51. See also Interview, Kistiakowsky (typescript), Los Alamos Archives, p. 3.

49. Interview, Ray Brin, SNL Archives; Hawkins, *Project Y*, Vol. I, p. 141; Kunetka, *City of Fire*, pp. 79, 124-25, 135; and Neider, "The Strange Death of Louis Slotin," pp. 12-13; "Louis Slotin, Radiation Victim, Mourned by Hill," Los Alamos *Times*, 7 June 1946, p. 1. See also Richard L. Miller, *Under the Cloud, The Decade of Nuclear Testing* (New York: The Free Press, 1986), pp. 66-71, which covers the accidents of Slotin and Daghlian in detail.

50. Kistiakowsky, "Reminiscences of Los Alamos," pp. 55-56.

51. Ibid.; Groves, *Now It Can Be Told*, pp. 288-89; Hawkins, *Project Y*, Vol. I, pp. 142-43, 266.

NOTES

52. Daily Log #1, 10 June 1944, p. 107; Hawkins, *Project Y*, Vol. I, p. 270; Oppenheimer to Bainbridge, "Gadget Testing Using Water for Recovery and Control," 22 December 1944, Los Alamos Archives.

53. Oppenheimer, "Memo on Test of Implosion Gadget," 16 February 1944, and Oppenheimer to Groves, Washington, D.C., 10 March 1944, Los Alamos Archives. See Daily Log #1, 3 May 1944, p. 91, re: Jumbo.

54. Kistiakowsky to Parsons, Memo, Subject: "Engineering Activities," 18 July 1944, p. 1, Los Alamos Archives.

55. Hawkins, *Project Y*, Vol. I, p. 137. According to Henderson, Lawrence had a very clear perception of the complementary role played by physicists and engineers. As he said, "I've heard him make speeches to his top staff there, time and time again: I want your physicists to stay out of the hair of the engineers and 'vicie versi'—I want you engineers to let the physicists do their physics; now you work together; I'm not going to tolerate second guessing by physicists of engineer or vicie versi." From interview, Robert W. Henderson (by Hank Willis), n.d., SNL Archives.

56. Kistiakowsky to Parsons, Memo, Subject: "Engineering Activities," 18 July 1944, Los Alamos Archives.

57. Jim Mitchell, News Release, 16 January 1974, Public Information Division SNL Archives; Interview, Robert W. Henderson; "Bob Henderson to Take Early Retirement," Sandia *Lab News*, 18 January 1974, pp. 2, 4.

58. Interview, Robert W. Henderson, 23 January, 1984, SNL Archives.

59. Ibid.

60. R. W. Carlson, "Study of a Modified Jumbo Utilizing Ductility and Inertial Metal," n.d., Los Alamos Archives; R. W. Carlson, "Confinement of an Explosion By a Steel Vessel," LA Report-390, 14 September 1945, Los Alamos Archives.

61. Ibid.; Daily Log #1, entry for 26 April 1944, p. 86, gives first mention of Jumbo and indicates that it would take six months to produce the first units.

62. Interview, Robert W. Henderson, 23 January 1984, SNL Archives.

63. Bainbridge to Hirschfelder, "Fragment Sizes, Velocity, Ranges," 29 August 1944, p. 2, Los Alamos Archives. See also Hubbard, *Journal*, p. 19.

Jack Hubbard, meteorologist and weatherman for the Trinity test, described Hirschfelder as "an indescribably unassuming" man with a "sunburned nose and a nasal twang," who referred to the other scientists as "long hairs." See p. 19.

NOTES

64. Telephone Interview, Roy W. Carlson, 19 September 1982, quoted in Szasz, *The Day the Sun Rose Twice*, p. 35. Apparently the association here is to both size and the 19th-century elephant Jumbo.

65. Interview, Robert W. Henderson, 23 January, 1984, SNL Archives; Hawkins, *Project Y*, Vol. I, p. 248. See also, Roy Carlson to Furman, San Jose, California, 2 July 1984, SNL Archives.

66. Interview, Robert W. Henderson, 23 January, 1984, SNL Archives; Bainbridge, "Orchestrating the Test," p. 222; Szasz, *The Day the Sun Rose Twice*, pp. 35-36; Storms, *Reach to the Unknown*, p. 5.

67. K. Bainbridge to N. E. Bradbury, 11 July 1945, Los Alamos Archives, quoted in *Los Alamos Science*, Winter/Spring, 1983, p. 21.

68. Mr. Penney to Mr. Carlson and Mr. Mack, "The Heat of Combustion of Jumbo," 15 December 1944, Los Alamos Archives; Interview, Robert W. Henderson, 23 January, 1984, SNL Archives.

69. Interview, Robert W. Henderson, ibid.

70. Bainbridge, "Orchestrating the Test," p. 222.

71. Ibid., pp. 222-23; Interview, Robert W. Henderson, 23 January, 1984, SNL Archives; Carlson, "Confinement of an Explosion By a Steel Vessel," LA Report-390, 14 September 1945, Los Alamos Archives (report provided courtesy of R. W. Henderson); Groves, *Now It Can Be Told*, pp. 288-89, 297. A final irony concerning Jumbo: When the City Fathers of Socorro visited the site, they found a chunk of Jumbo that had been blown out by the explosives test and thinking that it was a piece of the atomic bomb, took it home, and placed it on display with a sign indicating that it was a piece of the atomic device.

72. Hawkins, *Project Y*, Vol. I, p. 267; Bainbridge, "Prelude to Trinity," pp. 44-45; Bainbridge, "Orchestrating the Test," pp. 209-12; Robert W. Henderson, Commentary, 12 December 1988, SNL Archives; Szasz, *The Day the Sun Rose Twice*, pp. 27-31.

73. Interview, Robert W. Henderson, 23 January, 1984, SNL Archives, in which he tells of Groves' reaction and the site selection process.

74. Ibid.; Groves, *Now It Can Be Told*, p. 289; Bainbridge, "Orchestrating the Test," pp. 209-15; Bainbridge, "Test Site," p. 1 (typescript in Trinity Site Construction Folder), Los Alamos Archives; Bainbridge, "Prelude to Trinity," pp. 44-46; Interview, Robert W. Henderson, 23 January 1984, SNL Archives.

75. Interview, Robert W. Henderson, 23 January, 1984, SNL Archives; Bainbridge, "Prelude to Trinity," pp. 45-47.

76. Interview, Robert W. Henderson, ibid. Bainbridge's official report of the search for a test site tells of a similar incident. Bainbridge recalled that on

NOTES

September 6, General Uzal Ent had arranged with Colonel Wriston to cease all air-to-air gunnery and bombing from 7:00 A.M. to 7:00 P.M., Sunday, September 17. But as Bainbridge wryly reported: "Col. Wriston slipped up badly here. Just as we approached target B-4 in the vicinity of site position #1 . . . , 5 to 7 B-17s approached target B-4 with bomb bay doors open and at least one and probably more dropped practice bombs . . . . As we saw the bombers approaching B-4 with bomb bays open, we stopped the cars and watched the practice. If we had arrived 10 minutes earlier, we would have been at B-4 and it might have been more exciting." See Bainbridge, "Test Site," p. 2 (typescript in Trinity Construction Site Folder), Los Alamos Archives.

77. *Los Alamos, The Beginning of an Era 1943-1945*, pp. 34-35; "Construction and Equipment Requirements for Proposed Test Site, Trinity," Manhattan Engineer District files, quoted in Szasz, "The Day the Sun Rose Twice," p. 79; Telephone interview, John Manley, 10 July 1989. On more than one occasion, according to Manley, antelope would come racing across the desert, tearing down the wires within their range of height.

78. AEC *History*, Vol. I, pp. 318-19; Hawkins, *Project Y*, Vol. I, pp. 91, 143, 175.

79. Hawkins, *Project Y*, Vol. 1, p. 177. See references to Weapons Committee and Alberta in Daily Log #2, pp. 28, 30. Taped interview, Glenn A. Fowler, SNL Archives.

80. Hawkins, ibid. See also *Los Alamos, The Beginning of an Era 1943-1945*, pp. 39-42 and 43, which shows a photo of Ben Benjamin, later of Sandia, and George Economu preparing charges for shock velocity determinations.

81. Hawkins, *Project Y*, Vol. 1, p. 273; Ramsey, "The History of Project A," Los Alamos Archives; *Los Alamos, The Beginning of an Era 1943-1945*, p. 36.

82. Interview, Robert W. Henderson, 23 January, 1984, SNL Archives; *Los Alamos, The Beginning of an Era 1943-1945*, p. 42.

83. Telephone Interview, Ben Benjamin, 22 November 1983; *Los Alamos, The Beginning of an Era 1943-1945*, p. 42.

84. Interview, Robert W. Henderson, 23 January, 1984, SNL Archives; Bob Campbell, et al., "Field Testing, the Physical Proof of Design Principles," *Los Alamos Science*, Winter/Spring, 1983, pp. 167-70; Telephone interview, John Manley, 10 July 1989.

85. Campbell, op. cit.

86. Interview, Robert W. Henderson, 23 January, 1984, SNL Archives; quoted also in Szasz, "The Day the Sun Rose Twice" (manuscript), p. 71.

NOTES

CHAPTER 3:

1. Hawkins, *Project Y*, Vol. I, pp. 177, 221, 281-82; Interview, Robert W. Henderson by Hank Willis, n.d., SNL Archives.

2. For an eyewitness account by a participant, see *Nuclear Hostages* (Boston: Houghton Mifflin Co., 1983), by Bernard O'Keefe, Chairman of the Board of Edgerton, Germeshausen, and Grier. O'Keefe was a Navy ensign assigned to Los Alamos and Tinian and later participated in other test shots. See also Wesley Frank Craven and James Lea Cate, ed.; *The Army Air Forces in World War II*, Vol. V (Chicago: University of Chicago Press, 1953), p. 707.

3. Interview, Leon Smith, 19 January 1984, SNL Archives. Smith is Director of Monitoring Systems at Sandia.

4. Hawkins, *Project Y*, Vol. I, p. 177. For good primary coverage of Project A at Tinian, see *Nuclear Weapons Excerpt*, SNL Archives.

Secondary source coverage includes William L. Laurence, *Dawn Over Zero: The Story of the Atomic Bomb* (New York: Alfred A. Knopf, 1947), pp. 196-211; Michael Amrine, *The Great Decision: The Secret History of the Atomic Bomb* (New York: G. P. Putnam's Sons, 1959), pp. 72-73, 152-55, 192-98; and Fletcher Knebel and Charles W. Bailey, *No High Ground* (New York: Harper and Brothers, 1960), pp. 87-94.

5. Hawkins, *Project Y*, Vol. I, p. 285; Captain F. L. Ashworth, U.S.N., Address at Ft. Belvoir, Va., "The Atomic Bombing of Nagasaki," 23 September 1946. Lieutenant General M. F. Harmon, Commanding General of the Army Air Forces, strongly recommended to Ashworth that Tinian be selected (see p. 109, Bowen, "Silverplate"). Parsons, "Paraphrased Teletype Reference TA-2050 dated 16 July," Los Alamos Archives. See also Appendix "Destination."

For a listing of personnel assigned to Destination, see N. F. Ramsey to Major Peer de Silva, 2 April 1945, pp. 1-8, SNL Archives. See also an interesting aside in R. S. Warner, Jr., to Commander N. E. Bradbury, Subject: Group II - "Destination." Warner says: "If this group should be changed or augmented, I should be amiss not to bring to your attention such qualifications for the job as emotional stability and an experimental approach. Therefore, if such circumstances arose, and given the status of his job on this site, I should recommend R. W. Henderson."

6. Hawkins, *Project Y*, Vol. I, p. 179.

7. Ibid., pp. 178-79, and entries in Daily Log #2, for 29 September 1944 and 30 September 1944, p. 91.

8. Interview, G. C. Hollowwa, 15 December 1983, SNL Archives; *Nuclear Weapons Excerpt*, SNL Archives. Note that this reference is to the introductory article in this report.

NOTES

9. Interview, G. C. Hollowwa, 15 December 1983, and Memo, Hollowwa to Furman, 23 June 1986, with attached military orders dated 13 August 1945, both in SNL Archives; Interview, Leon Smith, 19 January 1984.

10. Interview, G. C. Hollowwa, 15 December 1983.

11. Interview, Leon Smith, 19 January 1984.

12. 509th *Pictorial Album* (lent by the Interservice Nuclear Weapons School, Kirtland Air Force Base), p. 7.

13. Robert W. Henderson, p. 1 of Errata to *Enola Gay* by Gordon Thomas and Max Morgan Witts (New York: Pocket Books, 1978).

14. Interview, G. C. Hollowwa, 15 December 1983, and Memo, Hollowwa to Furman, 23 June 1986, both in SNL Archives.

15. Ibid.

16. Ibid. On one of these trips into Salt Lake, they returned to find that the Officers' Club had burned on Saturday night.

17. V. A. Miller, "Assembly of the 1561 Fat Man," in *Nuclear Weapons Excerpt*, SNL Archives. See also James Les Rowe, *Project W-47* (Livermore, California: Ja A Ro Publishing, 1978), pp. 16, 28-33, 39. A controversial book, *Project W-47* contains a good description of loading operations at Wendover. Memo, G. C. Hollowwa to Necah S. Furman, Subject: History Manuscript, 19 July 1986, SNL Archives.

18. Rowe, *Project W-47*, pp. 85-89, 177-78; Hawkins, *Project Y*, p. 284.

19. Bowen, "Silverplate," pp. 110-11. Robert Brode had charge of the small electronics laboratory at Wendover.

20. Phillip R. Owens, "A History of the Salton Sea Test Base" (typescript in SNL Archives); see Chapter I, pp. 1-8.

21. TWX, Parsons to Groves, 25 October 1944, Daily Log #2, p. 75; Ashworth and Brode to Parsons, 6 November 1944, Daily Log #2, p. 70; and Admiral Purnell to Parsons, 9 November 1944, Daily Log #2, p. 68.

22. Purnell to Parsons, 9 November 1944, Daily Log #2, p. 68; Interview, Leon Smith, 19 January 1984, SNL Archives.

23. Ramsey, "History of Project A," *Nuclear Weapons Excerpt*, SNL Archives; Bowen, "Silverplate," pp. 95, 102. See also "Brief Narrative and Administrative History of the 509th Composite Group," Vol. 1, December 44-April 46, copy SNL Archives, p. 2 (hereinafter cited as "509th Composite Group").

24. 509th *Pictorial Album*, p. 2; "509th Composite Group," pp. 45-50, 56-60; Bowen, "Silverplate," p. 132.

25. "Silverplate," pp. 103, 112; "509th Composite Group," p. 4.

NOTES

26. "509th Composite Group," pp. 4-5, 56, 59; Amrine, *The Great Decision*, pp. 178-79.

27. Interview, Leon Smith, 19 January 1984, SNL Archives.

28. Ramsey, "History of Project A," *Nuclear Weapons Excerpt*, SNL Archives; O'Keefe, *Nuclear Hostages*, p. 83.

29. Sheldon H. Dike, "Atomic Bomb Project Aircraft," *Nuclear Weapons Excerpt*, SNL Archives; Bowen, "Silverplate," pp. 96-100. On page 136 Bowen records that "through long work with the practice bombs, Jeppson was technically competent to serve as weaponeer despite his lack of years and rank."

30. Dike, "Atomic Bomb Project Aircraft," *Nuclear Weapons Excerpt*, SNL Archives. Dike, who received his B.S. in electrical engineering from the University of New Mexico and his Ph.D. from Johns Hopkins, became Supervisor of the Systems Analysis Division at Sandia and later founded Dikewood Corporation. See also "Silverplate," p. 98.

31. Dike, "Atomic Bomb Project Aircraft," pp. 127-28.

32. Ibid., pp. 127-30; "Silverplate," p. 112; Interview, Ray Brin, SNL Archives. Milo Bolstad also became a member of Z Division.

33. 509th *Pictorial Album*, p. 45.

34. Bowen, "Silverplate," p. 133.

35. Ibid., "Nobody Looked," *Newsweek*, 27 August 1945, in *Atomic Age Scrapbook*, SNL Archives.

36. Bowen, "Silverplate," p. 134.

37. Ibid.; Amrine, *The Great Decision*, pp. 195-96; Commentary, Robert W. Henderson, 18 November 1988.

38. Bowen, "Silverplate," p. 134; 509th *Pictorial Album*, p. 34.

39. Interview, Leon Smith, SNL Archives; Lecture, Henry Monteith, "Nuclear Principles," 13 February 1984, SNL Archives.

40. Interviews, Leon Smith and George H. "Howie" Mauldin, 18 December 1984, SNL Archives.

41. Bowen, "Silverplate," p. 137; Ramsey, "History of Project A," *Nuclear Weapons Excerpt*, SNL Archives; Hawkins, *Project Y*, p. 288.

42. Interview, Robert W. Henderson, 23 January 1984, SNL Archives.

43. Ramsey, "History of Project A at Tinian," *Nuclear Weapons Excerpt*, SNL Archives.

44. "509th Composite Group," pp. 72-73; Bowen, "Silverplate," pp. 138-39.

NOTES

45. M. G. Holloway and R. E. Schreiber, "Design and Assembly of Pit for Model 1561 Fat Man," Nuclear Weapons Excerpt; A. B. Machen, "Mechanical Design of Model 1561 Fat Man, Nuclear Weapons Excerpt; and V. A. Miller, "Assembly of the 1561 Fat Man," Nuclear Weapons Excerpt. See also R. W. Henderson, R. Bice, and P. J. Larsen, Final Evaluation Report MK3 Mod 0 Fat Man Bomb Excerpt, SNL Archives.

46. Hawkins, *Project Y*, Vol. 1, p. 284; Interview, G. C. Hollowwa, who tells of Machen's participation; O'Keefe, *Nuclear Hostages*, p. 98; Commentary, Robert W. Henderson, 18 November 1988.

47. O'Keefe, *Nuclear Hostages*, p. 98.

48. Ibid., pp. 100-101.

49. Ibid., pp. 101-2; Interview, Ray Brin, SNL Archives.

50. Bowen, "Silverplate," pp. 139-41. On p. 139, Bowen explains in a footnote the "inexplicable reluctance to admit that the bomb was dropped by *Bock's Car*." "The mistake," he says, "may have originated in press releases prepared as soon as Sweeney was selected to pilot the bomber for Strike Two." "History of Project A at Tinian," Nuclear Weapons Excerpt; Hawkins, *Project Y*, p. 290.

51. Peter Wyden, *Day One, Before Hiroshima and After* (New York: Simon and Schuster, 1984), pp. 253-361; Interview, Leon Smith, 11 May 1984, SNL Archives.

52. Bush, "Statement to the National Policy Committee," 28 May 1945, pp. 31-35, S-1 (Uranium Committee), Folder 36, "Bush V. 1944," Record Group No. 227, National Archives.

CHAPTER 4:

1. Robert Oppenheimer to F. C. Alexander, Jr., Princeton, New Jersey, 25 May 1961, SNL Archives.

2. See photo caption, Goodchild, *J. Robert Oppenheimer*, p. 170; and *Los Alamos 1943-1945, The Beginning of an Era*, pp. 61-62.

3. Oppenheimer to Alexander, Princeton, New Jersey, 25 May 1961, SNL Archives.

4. Leslie R. Groves to F. C. Alexander, Jr., Stamford, Connecticut, 6 June 1961, p. 1, SNL Archives.

5. F. C. Alexander, Jr., to Hartley Rowe, Boston, Massachusetts, 21 November 1961, SNL Archives; Interview, Norris E. Bradbury, Los Alamos, New Mexico, 8 August 1984, SNL Archives.

NOTES

6. Oppenheimer to Alexander, Princeton, New Jersey, 25 May 1961, SNL Archives.

7. Hartley Rowe to F. C. Alexander, Jr., Newton Centre, Massachusetts, 24 November 1961, SNL Archives.

8. In his *History of Sandia Corporation Through Fiscal Year 1963*, Alexander has a good section on "Sandia Origins, 1942–1945." See p. 4 in particular.

9. Ibid., pp. viii–xiv; Telephone interview, Caroline Meade (daughter of James G. Oxnard), May 1989; Signed statements, John B. Dalbey (son of Dalbey), 25 May 1989, and Mary MacGregor (daughter of Dalbey), 26 May 1989, and Mrs. J. G. Oxnard, 28 May 1989; Certificate of Incorporation, State of New Mexico, for *Aircraft Holdings, Inc.*, by Directors James G. Oxnard, T. Thornton Oxnard, and D. E. Dalbey, 12 June 1929, SNL Archives; and Summary of Documentation by Tom Lane, 31 May 1989. My appreciation to Tom Lane for researching this topic. See p. xii especially. See also Don E. Alberts, *Balloons to Bombers, Aviation in Albuquerque 1882–1945* (Albuquerque: Albuquerque Museum, 1987), pp. 26–43.

10. Ibid., p. xii.

11. License between the Secretary of War and the Defense Plant Corporation, signed by the Secretary of War, 18 May 1945, and accepted 5 June 1945, pp. 1–4, SNL Archives.

12. Memo, A. C. Johnson, Lieutenant Colonel Corps of Engineers, to Chief of Engineers, Attn. Major Wolf, Subject: Assignment of Buildings, 14 July 1945, SNL Archives. See also memo transferring "Albuquerque Army Air Field, Albuquerque, New Mexico from Jurisdiction of U.S. Engineers, Manhattan District," 21 July 1945, SNL Archives; and Colonel L. E. Seeman to Colonel Johnson, Santa Fe, New Mexico, 2 July 1945, SNL Archives, which encloses list of buildings.

13. Alexander, *History of Sandia*, p. 3.

14. Don E. Alberts and Allan E. Putnam, *A History of Kirtland Air Force Base 1928–1982*, p. 39; Rowe, *W-47*, pp. 172–73.

15. Taped interview, Ray B. Powell, 23 February 1984, SNL Archives. Powell was Vice-President, Administration, SNL.

16. Memo for record, "Z Division – Ordnance Engineering," N. A., July 1945, SNL Archives.

17. Memo, J. R. Zacharias to J. R. Oppenheimer, Subject: "In Reference to Organization of Z Division . . . ," 6 August 1945, p. 1, Los Alamos Archives.

18. Memo, N. E. Bradbury to J. R. Oppenheimer and J. R. Zacharias, Subject: "Z Division," 2 August 1945, p. 1, Los Alamos Archives. A log book

NOTES

kept by Glenn Fowler from 18 June 1945 to June 1948 shows the evolution of Z-Division Field Test activities and the involvement of Weapons Committee personnel, SNL Archives (hereinafter cited as Fowler Logbook).

19. Memo, Captain Larkin to Commodore Parsons, Subject: "Organization of Z Division," 7 September 1945, p. 1, SNL Archives.

20. Memo, R. B. Brode to Group O-3, Subject: "Immediate Program," 17 August 1945, SNL Archives.

21. Memo, Captain Larkin to Commodore Parsons, Subject: "Organization of Z Division," 7 September 1945, pp. 2-3, SNL Archives.

22. Ibid., p. 3.

23. Memo, Frank Oppenheimer to J. R. Zacharias, Subject: "Z-4 Organization," 5 September 1945, p. 4, SNL Archives. Note: The B-29s that crashed were not a part of Tibbets' 509th. See also A. Timothy Warnock, "Statistics on Crashes of B-29s at Takeoff," 17 June 1982, File VFA-350, Los Alamos Archives.

24. Memo, Robert W. Lockridge to All Concerned, Subject: "Transfer of Group Z-2A to Sandia," 26 September 1945, Los Alamos Archives. See also Alexander, *History of Sandia*, p. 12.

25. J. R. Oppenheimer to J. R. Zacharias, Subject: "In Reference to Organization of Z Division . . . ," 6 August 1945, p. 1, Los Alamos Archives. See also Alexander, *History of Sandia*, p. 12.

26. Memo, J. R. Oppenheimer to All Concerned, Subject: "Additions to Organization of the Lab," 12 September 1945, and Memo, W. T. Theis, Subject: "Function and Responsibility of Group Z-6," 1 October 1945, Los Alamos Archives.

27. Memo, W. T. Theis, Subject: "Function and Responsibility of Group Z-6," 1 October 1945, Los Alamos Archives; Memo, W. T. Theis to J. R. Zacharias, Subject: "Personnel Assigned to Group Z-6," 8 October 1945, Los Alamos Archives.

28. Memo, N. E. Bradbury to All Division and Group Leaders, Subject: "Formal Notification," 17 October 1945, SNL Archives. AEC *History*, Vol. I, p. 625; Interview, Richard A. Bice, 21 June 1984, SNL Archives; Glenn A. Fowler Logbook, entry for 17 October 1945, p. 18, records: "R. S. Warner replaces J. R. Zacharias as Division Z Chief." See also entries for 17 October 1945 and 12 December 1945, SNL Archives.

29. Interviews, Norris E. Bradbury, 8 August 1984, and Robert W. Henderson, 23 January 1984, SNL Archives.

30. Ibid. See also Z-Division Leadership chart and Alexander, *History of Sandia*, p. 9. Memo for record, 4 December 1945, Los Alamos Archives, gives

the exact date as 4 December 1945. See also M. F. Roy to Technical Personnel Office, 6 December 1945, Los Alamos Archives.

31. Rowe, *W-47*, pp. 156-62, and Interview, James Les Rowe, Livermore, California, 18 September 1984, SNL Archives.

32. Rowe, *W-47*, pp. 171-72; Appendix, Z-Division Leadership chart; Interviews, James Les Rowe, 18 September 1984, and Interview, James Les Rowe by Sam Johnson, 7 February 1985, both in SNL Archives.

Rowe said that a pit was dug at Building T-941 to facilitate filling test drop units called "pumpkins" with concrete. When the concrete pouring project was completed, they didn't know what to do with the pit, so they filled it with concrete and covered it with dirt. A number of years later, the contractor pouring the footings for Building 840 unexpectedly ran into this large block of concrete. It was almost impossible to remove.

33. "Z-Division Leadership Chart." Note that Z-12 and Z-13 are unnamed. See also Wilbur F. Shaffer to Frederick C. Alexander, Jr., Knoxville, Tennessee, 10 May 1962, SNL Archives; Interview Ray Schultz by Bill Stevens; "LAMS-635" (copy provided by Art Eiffert), SNL Archives; "Working Draft of Production and Manufacturing," SNL Archives; and Interview, G. C. Hollowwa, 10 September 1984, SNL Archives.

34. Memo, W. S. Parsons to N. E. Bradbury, Subject: "Paraphrased Teletype Reference TA-2461 dated 28 August [1945]," Los Alamos Archives.

35. Untitled memo for record, 15 September 1945, Los Alamos Archives.

36. Interview, G. C. Hollowwa, 23 January 1984, SNL Archives.

37. Untitled memo for record, 15 September 1945, Los Alamos Archives.

38. Interviews, Richard A. Bice, 21 June 1984, and Bill Kraft, 24 August 1984, SNL Archives. Kraft recalled that a Mr. Peden from Denver purchased the surplus aircraft and after collecting and selling all the gas and propellers, he had cleared his debt—"All the rest was gravy."

39. Ibid. See also Alexander, *History of Sandia*, p. 12.

40. Memos, Lyle E. Seeman to Colonel G. R. Tyler, Subject: "Construction at Sandia," 2 September 1945; District Engineer to Colonel L. E. Seeman, Subject: "Appointment," n.d.; and L. R. Groves to Commanding Officer, Subject: "Construction at Sandia Project," 17 October 1945, all in Los Alamos Archives. See also Teletype memo, Clear Creek to Washington Liaison Office, 27 October 1945, Los Alamos Archives.

41. Memo, A. J. Frolich to Commanding Officer, Kirtland AFB, Subject: "Use of Facilities During Closing of Kirtland Field," 11 December 1945; Teletype, L. E. Seeman to Colonel Nichols and Colonel Derry, 8 February 1946; and Memo to All Concerned from A. J. Frolich, Subject: "Housing at

NOTES

42. Teletypes, A. W. Betts to Washington Liaison Office, 19 January 1946, and Washington Liaison Office to Commanding Officer, Clear Creek, 23 January 1946, Los Alamos Archives.

43. Interview, Ray B. Powell, 23 February 1984, SNL Archives.

44. Memo for record, 27 October 1947, Los Alamos Archives.

45. Memo, R. L. Colby to R. S. Warner, Subject: "Plans for Z-3 and Z-5 Merger," 2 March 1946, Brode File, SNL Archives. See also Z-Division Leadership chart.

46. Alexander, *History of Sandia*, pp. 10-11.

CHAPTER 5:

1. Interview, Glenn A. Fowler, 20 December 1983, SNL Archives; see also interview conducted by Bruce Hawkinson for Sandia *Lab News*, 12 June 1983, SNL Archives; "Commendation for Meritorious Civilian Service," Army Air Forces to Glenn A. Fowler for "his exceptional ability to initiate radar research and development programs and to advise and make recommendations with respect to the proper utilization of such equipment," SNL Archives. The citation also commended Fowler's contribution to the war effort through development of airborne radar equipment and techniques.

2. Ibid.

3. Memo, N. E. Bradbury to Mr. Galloway, Subject: "Testing Laboratory at Sandia," 23 August 1945, Los Alamos Archives; Interview, Glenn A. Fowler, Sandia *Lab News*, 12 June 1983.

4. Fowler Logbook, p. 15, SNL Archives; Memo, Commander N. E. Bradbury to Glenn A. Fowler and Dale R. Corson, Subject: "Organization and Functions of Group Z-1A," 12 September 1945, Los Alamos Archives.

5. "Los Lunas Bombing Range, New Mexico," manuscript in SNL Archives, pp. 1-2.

6. *Manhattan District History*, Book VIII, Los Alamos Project (Y), Vol. 3, Auxiliary Activities, Chapter 6, Sandia, p. 3, microfilm, Sandia Technical Library; Interview, Robert Hepplewhite, 1 May 1987, SNL Archives. Hepplewhite, who spent approximately twenty-five years in Field Test, was a member of the Z-1 Los Lunas test group.

7. Memo, B. H. Schaffer to Commanding Officer, W. S. Engineer Office, Subject: "Sanitation of Los Lunas Range," 20 November 1946, copy in "Los Lunas Bombing Range, New Mexico," pp. 3-4, SNL Archives.

8. Fowler Logbook, pp. 15-21, SNL Archives.

NOTES

9. Interview, Tom Pace and Howard Austin, 19 October 1983, SNL Archives. Pace, now retired, was formerly head of the Field Instrumentation Department. Austin, also retired, was Supervisor of the Simulation Instrumentation Department.

10. Interview, Howard Austin, 19 October 1983, SNL Archives.

11. Memo, Bradbury to Galloway, Subject: "Testing Laboratory at Sandia," 23 August 1945, Los Alamos Archives; Memo, Fowler and Corson to Bradbury, Subject: "Organization and Functions of Group Z-1A," p. 2, Los Alamos Archives.

12. Interview, Glenn A. Fowler, 31 January 1984, SNL Archives.

13. Ibid.; Memo, Parsons to Major General T. F. Farrell, Subject: "Bombing Range at Salton Sea," SNL Archives; Memo, Fowler to Roger Warner, 15 November 1945, SNL Archives; and John K. Merillat, "Geological Investigation of the Salton Sea Test Base Area, California," SAND Report No. SC-1408 (TR), p. 6, SNL Archives.

Fowler's Logbook entry of 6 November 1945 (p. 21) reads: "Warner: Think we should plan flight to Salton Sea to look over facilities at Thermal and Palm Springs." The entry for 13 and 14 November 1945 reads: "Trip to Salton Sea area, Semple, DeSelm, Johnson."

14. Letter, Parsons to Bradbury and Zacharias, and Memo, Parsons to Farrell, Subject: "Bombing Range at Salton Sea," n.d., SNL Archives.

15. Phillip R. Owens, "History of Salton Sea Test Base," p. 12, SNL Archives.

16. Ibid., p. 13; Interview, Howard Austin, 19 October 1983, SNL Archives.

17. Interview, Howard Austin and Tom Pace, 19 October 1983, SNL Archives.

18. Carroll Tyler, Manager, SFOO, and Daniel F. Worth, Jr., Field Manager AEC-Sandia, "Three Years Report for Sandia Field Office," 12 October 1953, Department of Energy Archives, Albuquerque (hereinafter cited as DOE, ALO): Folder—"Old Sandia Field Office File, Policy and Historical Documents." The transfer was made under provision of Executive Order 9816.

19. Interview, Howard Austin and Tom Pace, 19 October 1983, SNL Archives.

20. Owens, "History of Salton Sea Test Base," p. 16, SNL Archives.

21. Interview, Howard Austin and Tom Pace, 19 October 1983, SNL Archives.

NOTES

Kirtland for Sandia Personnel," 21 December 1945, all in Los Alamos Archives. See also Alexander, *History of Sandia*, p. 10.

22. Interview, Hank Willis, 17 February 1985, SNL Archives; Letter, Guy Willis to Hank Willis, 9 May 1985, SNL Archives. Willis said that Bill Chown and Irv Lenz had a hand in the latter project.

23. Guy Willis, ibid.

24. Interview, Howard Austin, 19 October 1983, SNL Archives.

25. Interview, Gordon Miller, 11 January 1983, SNL Archives; Interview, Orville Howard, 1 May 1987, SNL Archives.

26. Interview, Howard Austin and Tom Pace, 19 October 1983, SNL Archives.

27. Interview, Tom Pace, 19 October 1983, SNL Archives.

28. W. T. Moffat, Staff Assistant SLT to R. W. Henderson, Associate Director, "Report on 'The Case of the Mysterious Niland Bomb,'" 6 June 1949, SNL Archives; Owens, "History of Salton Sea Test Base," pp. 30–31, SNL Archives.

29. Owens, "History of Salton Sea Test Base," p. 31, SNL Archives.

30. Ibid., pp. 31–32; Sandia *Lab News*, 17 December 1954, p. 6; Interview, Robert Hepplewhite, 1 May 1987, SNL Archives.

31. Ibid., p. 17.

32. Ibid., pp. 17–23.

33. Ibid., pp. 27–29.

34. Ibid., pp. 23–24; Interviews, Glenn A. Fowler, 20 December 1983, and Howard Austin, 19 October 1983, SNL Archives.

35. John Banister, "History of Field Engineering Directorate," p. 3, SNL Archives; Interview, Glenn A. Fowler, 20 December 1983, SNL Archives; and Interview, Glenn Fowler, Sandia *Lab News*, 12 June 1983.

36. Interviews, Richard A. Bice, 21 June 1984, and Hank Willis, 5 March 1985, SNL Archives. During his career Fowler hired eight men who reached the director level: Carter Broyles, William A. Gardner, Harlan Lenander, Bill Myre, Don Shuster, Jim Scott, W. C. Scrivner, and Hank Willis.

37. Interview, Glenn A. Fowler, 20 December 1983, SNL Archives.

38. Interview, W. C. Myre, 29 August 1984, SNL Archives; Interview, Gene Harling, 18 February 1988, SNL Archives; Interview, Orville Howard, 1 May 1987, SNL Archives; Interview, Lyle C. Hake, 27 April 1987, SNL Archives.

39. Interview, W. C. Myre, 29 August 1984, SNL Archives; Interview, Gene Harling, 18 February 1988, SNL Archives.

NOTES

40. Owens, "History of Salton Sea Test Base," p. 26, SNL Archives. There were sixty full-time resident employees at Salton Sea by 1951.

CHAPTER 6:

1. Memo, Roger S. Warner, Jr., to Captain R. A. Larkin, Subject: "Inspection Party to Crossroads Site," 12 February 1946, Los Alamos Archives; Interview, Richard A. Bice, 21 June 1984, SNL Archives; David Bradley, *No Place to Hide* (Boston: Little, Brown and Company, 1984), p. 11. (This is a log kept by a doctor assigned to Operation Crossroads.) Letter, H. M. Lehr to Marshall Holloway, Kwajalein, Marshall Islands, 27 March 1946, Los Alamos Archives.

2. Neal O. Hines, *Proving Ground, An Account of the Radiobiological Studies in the Pacific, 1946–1961* (Seattle: University of Washington Press, 1962), pp. 20–21; William A. Shurcliff, Historian of Joint Task Force One, *Bombs at Bikini: The Official Report of Operations Crossroads* (New York: William H. Wise and Company, 1947), pp. 10–13, 27.

3. Lloyd J. Graybar, "Bikini Revisited," *Military Affairs* 44 (October 1980), p. 118. See also Lloyd J. Graybar, "The 1946 Atomic Bomb Tests:Atomic Diplomacy or Bureaucratic Infighting?" *Journal of American History* 72 (March 1986), pp. 888–89, and 892, in which the author posits that Operation Crossroads "not only gave the appearance of atomic diplomacy," but also was a "central event" of the immediate postwar years.

4. "94 From Lab En route to A-Bomb Tests," Los Alamos *Times*, 19 April 1946 (quotes President Truman), pp. 1, 4.

5. Shurcliff, *Bombs at Bikini*, pp. 10–13, 27. See also W. S. Parsons, "Atomic Energy Whither Bound," *United States Naval Proceedings* 73 (August 1947) p. 904. Blandy also earned the nickname "The Buck Rogers of the Navy." See Sidney Shalett, "The Buck Rogers of the Navy," New York *Times*, 10 January 1946, pp. 44–45.

6. "Final Report for Tests Able and Baker," Joint Task Force One Operation Crossroads, 18 October 1946, SNL Archives.

7. Memo, N. E. Bradbury to All Division Leaders, Subject: "B Division," 23 January 1946, Los Alamos Archives; *Manhattan District History*, Book VIII, pp. 8.20, 8.22, 8.8–8.9; "History of Los Alamos B Division," 30 April 1946, Los Alamos Archives; Minutes of Group Leader's Meeting, 2 January 1946, Los Alamos Archives.

The Bradbury memo indicated that B-Division personnel would receive "premium pay" of $200 per month, partly as "incentive pay" and partly as "hazardous duty pay."

709

NOTES

See also Shurcliff, *Bombs at Bikini*, p. 28; Memo, N. E. Bradbury to Dale R. Corson, Subject: "Organization of Z Division While Naval Test is in Progress," 27 March 1946, Los Alamos Archives; Hawkins, *Project Y*, pp. 20–21, 98–99; and Memo, Dale R. Corson to N. E. Bradbury, Subject: "Future Z Division Program," 11 April 1946, Los Alamos Archives.

8. Memo, Dale R. Corson to N. E. Bradbury, Subject: "Future Z Division Program," 11 April 1946, Los Alamos Archives.

9. Memo, Commander Joint Task Force One to Chief of Naval Operations, et al., Subject: "General Information on Atomic Bomb Tests," 28 January 1946, Los Alamos Archives; *Manhattan District History*; Book VIII, pp. 8.24–8.26; Interview, R. W. Henderson, 23 January 1984, SNL Archives.

10. Memo, Roger Warner to E. B. Doll, Subject: "Plans for Survey Trip to Marshall Islands," 8 March 1946, and Memo, Roger Warner to R. A. Larkin, Subject: "Inspection Party to Crossroads Site," 12 February 1946, both in Los Alamos Archives; Interview, R. A. Bice, 21 June 1984, SNL Archives.

11. "History of Los Alamos B Division," 30 April 1946, Los Alamos Archives; Interviews, R. W. Henderson, 23 January 1984, and Ira Hamilton, 16 November 1984, SNL Archives. Shurcliff, *Bombs at Bikini*, pp. 21–22, 50; *Manhattan District History*, Book VIII, pp. 8.23–8.24.

Passage of House Resolution 307 on 14 June 1946 indicated Congressional approval of the operation and authorized the use of 33 U.S. combat vessels in the target arrays for the test. See *Congressional Quarterly*, Vol II, 1946, pp. 340–41; *Manhattan District History*, Book VIII, p. 8.6, and J. H. Williams, "Trip Report to Bikini Area," 29 March 1946, Los Alamos Archives.

Another reason for the postponement of the test was ostensibly to improve the atmosphere at the United Nations where talks were in progress concerning international control of atomic energy. See Graybar, "Bikini Revisited," p. 120 and footnote 9.

12. Graybar, "Bikini Revisited," p. 120; "Baruch Presents 'Control' Plan to Senate Group," Los Alamos *Times*, 14 June 1946, p. 1.

13. Shurcliff, *Bombs at Bikini*, pp. 38–39; Memo, Commander of Joint Task Force One to Chief of Naval Operations, Subject: "General Information on Atomic Bomb Test," 28 January 1946, p. 6, Los Alamos Archives. See also Hines, *Proving Ground*, p. 38.

14. Memo, Roger S. Warner, Jr., to Mr. E. J. Demson, Subject: "Responsibilities of the Personnel Office to B Division," 30 January 1946, p. 1, Los Alamos Archives.

15. Memo, E. J. Demson, R. A. Larkin, and J. A. Muncy to Roger S. Warner, untitled draft response to Memo: "Responsibilities of the Personnel Office to B Division," 30 January 1946, Los Alamos Archives.

NOTES

16. Shurcliff, *Bombs at Bikini*, pp. 40–41, 57, 84–85; N. Bradbury, "Los Alamos Activity," Report prepared for Technical Director, JTF–I, 8 July 1946, Los Alamos Archives. For information on the biomedical program, see "Joint U.S. Atomic Energy Commission and Department of Defense Press Conference," Washington, D.C., 13 June 1951, A–83–1005, p. 7, Los Alamos Archives.

17. Interview, Leon Smith, 19 January 1984, SNL Archives; Memo, R. A. Larkin to Colonel Seeman, Subject: "Separation of Army Personnel to Take Part in Crossroads Operation," 11 February 1946, and Memo, Lt. Col. E. E. Wilhoyt to W. O. McCord, Jr., Subject: "Reclassification of Military Personnel, B Division," 12 March 1946, both in Los Alamos Archives.

18. Shurcliff, *Bombs at Bikini*, pp. 57–58.

19. Film, "Crossroads," made by Arthur Machen, SNL Archives; "And That Means Los Alamos," Los Alamos *Times*, 26 July 1946. The movie made by Machen exhibits the techniques of the Assembly crew aboard the U.S.S. *Albemarle*; *Manhattan District History*, Book VIII, p. 8.31; Bradley, *No Place to Hide*, p. 17; "B Division—Crossroads," Personnel Chart, Los Alamos Archives.

20. William L. Laurence, *Dawn Over Zero, The Story of the Atomic Bomb* (New York: Alfred A. Knopf, 1947), p. 279; Interview, Howard Austin, 19 October 1983, SNL Archives.

21. Shurcliff, *Bombs at Bikini*, pp. 104–5; "Rehearsal of Pacific Drop is Successful," Los Alamos *Times*, 22 March 1946.

22. *Operation Crossroads Electronics Crosstalk* (hereinafter cited as *Crosstalk*), 2 July 1946, p. 7. *Crosstalk* is a newspaper printed by members of the electronics crew aboard the *Avery Island*. Copies were donated by David L. Anderson, former ensign aboard the *Avery Island*.

23. Interview, Leon Smith, 19 January 1984, SNL Archives; "And That Means Los Alamos," Los Alamos *Times*, 26 July 1946; Commentary, Robert W. Henderson, 18 November 1988. The author appreciates the donation of copies of the Los Alamos *Times* by Mike Michnovicz.

24. Shurcliff, *Bombs at Bikini*, pp. 104–5. See footnote, bottom of p. 104.

25. Interview, Leon Smith, 19 January 1984, SNL Archives.

26. Bradley, *No Place to Hide*, pp. 12–13.

27. *Crosstalk*, 2 July 1946, p. 7.

28. Ibid.; *Operation Crossroads*, commemorative album (unpaginated), also loaned by David L. Anderson.

29. *Crosstalk*, 2 July 1946, p. 3.

30. Ibid.

NOTES

31. Interview, Leon Smith, 19 January 1984, SNL Archives.

32. Ibid; Shurcliff, *Bombs at Bikini*, pp. 104–5; "Hill Man on Bomb Plane Unimpressed," Los Alamos *Times*, 12 July 1946, p. 3.

33. Interview, Leon Smith, 19 January 1984, SNL Archives. See also *Crosstalk*, 2 July 1946, p. 4.

34. *Crosstalk*, 2 July 1946, p. 4.

35. "Hill Man on Bomb Plane Unimpressed," Los Alamos *Times*, 12 July 1946, p. 3.

36. Report to President Truman from Carl A. Hatch, Chairman of the President's Evaluation Commission, included in Press Release, dated 11 July 1946, Los Alamos Archives; Bradley, *No Place to Hide*, pp. 72–73, 81.

37. Mailgram, Commander Joint Task Force One to Public, 19 July 1946, Los Alamos Archives; Interview, Ira Hamilton, 16 November 1984, SNL Archives. The device detonated approximately ninety feet beneath the surface of the lagoon.

38. One of the best unclassified sources recording the radiological effects of the tests on the lagoon environment is the previously cited Neal O. Hines, *Proving Ground*; see particularly pp. 40–49. See also Shurcliff, *Bombs at Bikini*, pp. 154–64; Carl Hatch to President Truman, "Report of the Second Bikini Atom Test," included in Press Release dated 2 August 1946; "Early Report Shows 11 of 87 Target Ships Sunk," Los Alamos *Times*, p. 1; Comments, Glenn Fowler to Furman, 18 November 1988.

39. Joint U.S. Atomic Energy Commission and Department of Defense Press Conference, 13 June 1951, Washington, D.C., pp. 33–34, Los Alamos Archives; Bradley, *No Place to Hide*, p. 131.

40. Laurence, *Dawn Over Zero*, pp. 284–85.

41. Walmer Elton Strope, "The Navy and the Atomic Bomb," U.S. *Naval Institute Proceedings* (October 1947), p. 1222; Graybar, "Bikini Revisited," p. 122 quoting Gregg Herken, "American Diplomacy and the Atomic Bomb, 1945–1947" (Ph.D. dissertation, Princeton University, 1974), p. 363, footnote.

42. "This Week's Atomic Age News Brief," Los Alamos *Times*, 6 September 1946; Graybar, "Bikini Revisited," p. 122.

43. *Manhattan District History*, Book VIII, p. 8.48, quoting the "Report of the President's Evaluation Board," SNL Archives.

CHAPTER 7:

1. Interview, Richard A. Bice, 21 June 1984, SNL Archives.

NOTES

2. AEC *History*, Vol. I, pp. 1–2; M. C. Latta to Lilienthal, The White House, Washington, D.C., 31 December 1946, with enclosure: Harry S. Truman, Executive Order [9816] "Providing For The Transfer of Properties and Personnel to the Atomic Energy Commission," The White House, 31 December 1946; and Memo, Carroll L. Wilson to the Commissioners, Subject: "Transfer Orders," United States Atomic Energy Commission, Washington, D.C., 17 December 1946, both in the National Archives, Record Group 326, Records of the Atomic Energy Commission, Office Files of David E. Lilienthal, Folder: "Manhattan District, Transfer of."

3. "Statement of Secretary of War Patterson on Transfer of Manhattan District," War Department, 31 December 1946, National Archives, Record Group No. 326, "Records of the Atomic Energy Commission Office Files of David E. Lilienthal," Folder: "Manhattan District, Transfer of."

4. The most comprehensive treatment of the effort to establish national and international control of atomic energy remains the previously cited AEC history, *The New World*, by Richard G. Hewlett and Oscar E. Anderson, Jr. See pp. 411–619. Secondary sources providing synthesized versions of the legislative battles include Bernard J. O'Keefe, *Nuclear Hostages* (Boston: Houghton Mifflin Company, 1983), pp. 127–34; Gregg Herken, *The Winning Weapon, The Atomic Bomb in the Cold War 1945–1950* (New York: Vintage Books, 1982), pp. 151–91; Alice L. Buck, *A History of the Atomic Energy Commission* (Washington, D.C.: U.S. Department of Energy, 1982), pp. 1–26; James R. Newman and Byron S. Miller, *The Control of Atomic Energy* (New York: McGraw-Hill Book Company, 1948); Richard J. Barnett, *Who Wants Disarmament?* (Boston: Beacon Press, 1960), pp. 11–13; and Joseph L. Nogee, *Soviet Policy towards International Control of Atomic Energy* (Notre Dame: University of Notre Dame Press, 1961), pp. 22–32.

5. United Nations Atomic Energy Commission, "Official Record of the First Meeting, Remarks by the Provisional Chairman Bernard Baruch," Hunter College, New York, 14 June 1946, p. 4, National Archives, Record Group 326, Records of the Atomic Energy Commission, Miscellaneous, Relating to the U.N. Atomic Energy Commission 1946–1949. See also Herken, *The Winning Weapon*, p. 172; James F. Byrnes, *Speaking Frankly* (New York: Harper and Brothers, 1947), p. 270; Bernhard G. Bechhoefer, *Postwar Negotiations for Arms Control* (Washington, D.C.: Brookings Institute, 1961), pp. 41–82; and Nogee, *Soviet Policy Towards International Control of Atomic Energy*, pp. 34–48.

6. Herken, *The Winning Weapon*, pp. 174–75; Barnett, *Who Wants Disarmament?*, pp. 21–22.

7. On March 5, 1946, during a speech at Westminster College at Fulton, Missouri, British Prime Minister Winston Churchill focussed attention on what

NOTES

he called "an Iron Curtain running from Stettin in the Baltic to Trieste in the Adriatic," symbolizing the beginning of the Cold War and the heightening of tensions between the United States and Russia. Actually, however, Churchill's "Iron Curtain" phrase had been a favorite of his for some time and, in fact, had been used shortly after the German surrender in a prophetic cable to Washington. See Godfrey Hodgson, *America In Our Time* (New York: Vintage Books, 1976), p. 27.

8. See S. 1463, 3 October 1945 (copy) in "Atomic Energy Bills" File, SNL Archives, and *Congressional Quarterly*, Vol. I, 1945, pp. 674–75, and Vol. II., 1946, pp. 505–14, Albuquerque Public Library. See also Alice Kimball Smith, *A Peril and a Hope, The Scientists' Movement in America*: 1945–47 (Chicago: University of Chicago Press, 1965), pp. 261–75 for an excellent analysis of the Bill.

9. *Congressional Quarterly*, Vol. II., 1946, pp. 505–14, carries an account of the 79th Congress and reviews the introduction and passage of the McMahon Bill. See also Newman and Miller, *The Control of Atomic Energy*, pp. 8–10, 42.

10. AEC *History*, Vol. I, pp. 490–91, 499–502.

11. Ibid., pp. 506–8; Buck, *A History of the Atomic Energy Commission*, p. 1; Newman and Miller, *The Control of Atomic Energy*, p. 42.

12. *Atomic Energy Legislation Through 95th Congress*, 2nd Session, Committee on Armed Services, U.S. House of Representatives, USGPO, 1979, p. 40. The Act was amended in 1954. See also Byron S. Miller," A Law Is Passed—The Atomic Energy Act of 1946," *University of Chicago Law Review* 15 (Summer 1948), pp. 799–821; and Newman and Miller, *The Control of Atomic Energy*, pp. 3–4, 45.

13. The National Security Act of 25 July 1947 created the Department of Defense as a single unit and established the Air Force as a service separate from the Army Air Corps. The National Security Act also set up the National Security Resources Board, the Central Intelligence Agency, and the National Security Council. James V. Forrestal succeeded Patterson as the first Secretary of Defense. See AEC *History*, Vol. II, p. 128.

14. A memo to Chief of Staff, United States Army, Chief of Naval Operations, Chief of Staff, United States Air Force, dated 21 October 1947, subject: "Armed Forces Special Weapons Project" served the AFSWP in lieu of a charter. See RG 218, Records of the US Joint Chiefs of Staff, CCS 471.6 (8/16/45) Section 7, Control of Atomic Weapons and Test, National Archives, Washington, D.C.

A memo dated 12 July 1951 to the Chief of Armed Forces Special Weapons Project superseded the memo of 8 July 1947, subject: "Mission and Responsibilities of the Armed Forces Special Weapons Project," RG 218,

NOTES

Records of the US Joint Chiefs of Staff, CCS 471.6 (8/15/45) Section 27-A, Control of Atomic Weapons and Test, National Archives, Washington, D.C.

AEC *History*, Vol. II, pp. 130-32, 157; Newman and Miller, *The Control of Atomic Energy*, pp. 39-40; Groves, *Now It Can Be Told*, pp. 399-400. AFSWP was established 29 January 1947.

15. In 1959, AFSWP was disestablished in favor of the Defense Atomic Support Agency (DASA), the first agency to be formed under an amendment to the DOD Reorganization Act of 1958. This amendment was alleged to be a Congressional acknowledgment that integration and coordination of any common supply or service would not be done on the initiative of the military departments within their structures. See Theodore W. Bauer and Eston T. White, *Defense Organization and Management*, National Security Management Series (Washington, D.C.: Industrial College of the Armed Forces, VI Defense Agencies, 1975), pp. 93-95, 98-101; and AEC *History*, Vol. II, p. 131.

16. *Science and Technology Act of 1958, Analysis and Summary*, Prepared by the Staff and Submitted to the Senate Government Operations Committee, 85th Congress, 2nd Session (1958), S. Doc. 90, p. 38. See also Harold Green and Alan Rosenthal, *Government of the Atom* (New York: Atherton Press, 1963), p. 266.

17. Green and Rosenthal, *Government of the Atom*, p. 105.

18. Quote taken from "Thinking Ahead with . . . Senator Clinton P. Anderson," an interview in *International Science and Technology* (April 1964), p. 62, recorded in Orlans, *Contracting for Atoms*, p. 157.

19. Orlans, *Contracting for Atoms*, pp. 160-62.

20. Ibid., pp. 183-91.

21. *Atomic Energy Legislation Through the Ninety-Fifth Congress*, p. 40; *The Government and Atomic Energy* (Albuquerque, New Mexico: Kirtland AFB, 1958), p. 16.

22. AEC *History*, Vol. II, p. 132; *Story of Albuquerque Operations*, DOE (1978), U.S. Department of Energy, pp. 1-2; *The Government and Atomic Energy*, p. 42.

Sandia School, as it was known, was owned by Albert Simms and leased by the Army during the war for use as a hospital and modified with government funds. The property was returned to Simms in 1946 and, subsequently, leased—and later purchased—by the New Mexico School of Mines. By August 5, 1948, the Sandia Branch of Los Alamos Laboratory was seriously considering acquisition of the property. See "Atomic Energy Commission, Sandia Branch of the Los Alamos Laboratory, Note by the Secretary," pp. 4-5. U.S. DOE Archives, Record Group 326, U.S. Atomic Energy Commission, Collection: Secretariat, Box 1235, Folder 601 (8-5-48) LASL.

NOTES

23. AEC *History*, Vol. II, p. 132.

24. Luther J. Heilman to Furman, 22 June 1986, SNL Archives; Sam Johnson, "Building Construction—1946 through 1956," in Plant Engineering History, SNL Archives; Interview, Luther J. Heilman, 10 January 1985, SNL Archives.

25. Luther J. Heilman to Furman, 22 June 1986, SNL Archives; Johnson, "Building Construction," SNL Archives.

26. Luther J. Heilman to Furman, 22 June 1986, SNL Archives; Johnson, "Building Construction," SNL Archives; Interview, Virgil Harris by Frederick Alexander, 26 September 1960, SNL Archives; AEC *History*, Vol. II, p. 133; Alexander, *History of Sandia*, p. 10.

27. "Report by the Manager, Office of Santa Fe Directed Operations and Chief of Engineering," Sandia Branch of the Los Alamos Laboratory, 5 August 1948, p. 4, U.S. DOE Archives, Record Group 326, Collection: Secretariat, Box: 1235, Folder 601 (8-5-48) LASL; Alexander, *History of Sandia*, pp. 13, 27.

28. Alexander, *History of Sandia*, p. 13; Memo, A. W. Betts to Major Gen. L. R. Groves, Subject: "Housing at Sandia," 11 June 1946, Los Alamos Archives.

29. Memo, Dale R. Corson to N. E. Bradbury, Subject: "Future Z Division Program," 11 April 1946, Los Alamos Archives.

30. Alexander, *History of Sandia*, pp. 12-13; Interview, Robert W. Henderson, 7 March 1985, SNL Archives.

CHAPTER 8:

1. J. H. Manley, "Report on Z-Division (Prepared for the General Advisory Committee)," 10 November 1947, pp. 1-2, Los Alamos Archives. A cover memo indicates that "the absence of Mr. Henderson, Z-Division Leader, did not make it possible to consult with him subsequent to . . . preparation of the report."

2. AEC *History*, Vol. I, pp. 633-42; David E. Lilienthal, *The Journals of David E. Lilienthal*, Vol. II (New York: Harper and Row, 1964), p. 106, entry dated 16 November 1946; Telephone interview, John Manley, 10 July 1989.

3. "Conference Notes Re: Genesis of Sandia Corporation," Kimball Prince with General James McCormack and Mr. Carroll Wilson (at McCormack's office, MIT, Cambridge, Massachusetts), 29 September 1959, SNL Archives; AEC *History*, Vol. II, pp. 15, 30-32, 45-46, 58.

NOTES

4. AEC *History*, Vol. II, pp. 58–60; N. E. Bradbury, Weapon Program (Excerpt), 27 January 1947, Los Alamos Archives.

5. AEC *History*, Vol. II, pp. 60–61.

6. David E. Lilienthal, "The Kind of Nation We Want," *Colliers*, 14 June 1952, p. 49, SNL Archives.

7. "Report to the President of the United States from the Atomic Energy Commission January 1–April 1, 1947," and "Appendix C, 2 April 1947," U.S. DOE Archives, Collection: Energy History, Germantown, Maryland. The report to the President had been drafted previously with blanks provided for the actual number of bombs in stockpile. On April 16, Lilienthal met with the President and orally supplied the stockpile figures. See AEC *History*, Vol. II, p. 53, and Bradbury, Weapon Program (Excerpts), 27 January 1947, Los Alamos Archives. See also Interview, Norris E. Bradbury by Arthur Lawrence Norberg, Albuquerque, New Mexico, 11 February 1976. Copies of this interview are housed at the Bancroft Library, University of California, Berkeley; the American Institute of Physics, New York, New York; and at SNL Archives, Albuquerque. Permission to quote generously provided by Dr. Bradbury and the director of the Bancroft Library, UC, Berkeley.

8. Manley, "Report on Z Division," p. 1.

9. Ibid., pp. 1–12.

10. Ibid., pp. 6–7.

11. A. B. Machen to Manual Control Group Z-3, 14 February 1947, Los Alamos Archives.

12. Manley, "Report on Z-Division," p. 2; Memo for Information, Roy B. Snapp to James McCormack, 18 December 1947, and Extracts from Minutes of the 22nd AEC-MLC Meeting, 17 December 1947, both in Legal Files, SNL Archives.

13. Interview, R. W. Henderson, 23 January 1984, SNL Archives.

14. Interview, N. E. Bradbury, 8 August 1984, SNL Archives; AEC *History*, Vol. II, p. 139.

15. Memo, N. E. Bradbury to Paul J. Larsen, 10 November 1947, p. 3, Los Alamos Archives.

16. AEC *History*, Vol. II, p. 139.

17. Ibid., pp. 150–51.

18. Ibid., p. 155.

19. A cover memo to Groves' letters reads: "*Note By the Secretaries to the Joint Chiefs of Staff on The Position of the Military Establishment in the Development of Atomic Weapons*: At the request of the Chief of Staff, U.S.

NOTES

Army, the Chief of Naval Operations, and the Chief of Staff, U.S. Air Force, the views of the retiring chairman of the Armed Forces Special Weapons Project on military participation in Atomic Energy development contained in Enclosures A, B, and C, are circulated for information." The enclosures are addressed to Armed Forces Special Weapons Project with A going to General Omar N. Bradley, B to Admiral Louis E. Denfeld, and C to General Carl Spaatz, all dated 28 February 1948, National Archives, Record Group 218, Records of the United States Joint Chiefs of Staff, CCS 471.6 (8/15/45) (Section 9), Washington, D.C. See particularly Enclosure C, p. 7.

20. AEC *History*, Vol. II, pp. 157–58.

21. Ibid., p. 161. Appendix to JCS 1848/3, K. A. Nichols to D. F. Carpenter, 29 June 1948, pp. 25–26, enclosed in memo. William Burke to JCS, Subject: "Custody of Atomic Weapons," 13 July 1948, National Archives, Record Group 218, CCS 471.6 (8/15/45) (Section 11), Washington, D.C.

22. AEC *History*, Vol. II, p. 165.

23. Ibid., pp. 166–67.

24. Memo, Chief of Staff, U.S. Air Force to the Joint Chiefs of Staff, Subject: "Atomic Bomb Assembly Teams," 27 July 1948, with Enclosure, James Forrestal to Joint Chiefs of Staff, Subject: "Custody of Atomic Weapons," 28 July 1948, pp. 28–29, quoting Truman's announcement, National Archives, Record Group 218, CCS 471.6 (8/15/45) (Section 11), Washington, D.C.

25. Lilienthal, *The Journals of David E. Lilienthal*, pp. 350–51; also mentioned in Gregg Herken, *The Winning Weapon, The Atomic Bomb in the Cold War* 1945–1950 (New York: Vintage Books, 1982), p. 257.

26. Herken, ibid., p. 263 footnote; AEC *History*, Vol. II, pp. 537–39.

27. Herken, *The Winning Weapon*, p. 263 footnote.

28. Good secondary source material on this period includes the aforementioned Herken, *The Winning Weapon*, pp. 245–46; AEC *History*, Vol. II, p. 158; and William Manchester, *The Glory and the Dream*, Vol. I (Boston: Little Brown and Company, 1973), pp. 532–37; p. 652 quotes MacArthur.

29. David E. Lilienthal, "The Kind of Nation We Want," *Colliers*, 14 June 1952, p. 49, extract in SNL Archives.

CHAPTER 9:

1. Memo, Bradbury to Distribution, Subject: "Z Organization," 4 December 1947, SNL Archives; Paul Larsen, Engineer, "Atomic Research Chief," loose news clipping, SNL Archives.

NOTES

2. Interview, R. W. Henderson, 23 January 1984, SNL Archives; Alexander, *History of Sandia*, p. 15. Bradbury's memo indicated that Larsen's office would be in Room 16, Building T-911.

3. Memo, Bradbury to Distribution, Subject: "Z Oganization," 4 December 1947, SNL Archives.

4. Ibid.; Memo, Larsen to All Concerned, Subject: "Division of Responsibilities Between the Director and Associate Director, Sandia Laboratory," 23 July 1948, Los Alamos Archives; "Note," R. W. Henderson to N. Furman, 8 May 1989, SNL Archives.

5. Memo, Bradbury to Larsen, 10 November 1947, pp. 2-3, Los Alamos Archives.

6. Memo, B. H. Schaffer to All Concerned, Subject: "Minutes of the Meeting on the Mk 4," 16 April 1947, SNL Archives.

7. The new title, effective 1 April 1948, became Sandia Laboratory, Branch Los Alamos Scientific Laboratory. See Memo, Larsen to All Concerned, 1 April 1948; File 310.1, Z Organization, 6 April 1948; Bradbury to All Group and Division Leaders, 26 July 1948; and Bradbury to Sproul, University of California, File 322, all in SNL Archives. Of these individuals on the preliminary branch lab roster, three department heads, Bice, Fowler, and Campbell, would reach the level of corporate Vice-President; Paddison would be named a Director.

8. Taped recollections, provided by Charles W. Campbell for SNL History Project, 31 August 1984, SNL Archives. See also May 1948 organizational chart.

9. Interview, G. Corry McDonald (regarding Powell), 11 February 1986, SNL Archives; Alexander, *History of Sandia*, p. 19; Memo, L. G. Hawkins, Business Manager-Sandia to A. E. Dyhre, Project Business Manager-Los Alamos, Subject: "History Sandia Business Office, July 1-December 31, 1948," 30 March 1949, pp. 1-4, Los Alamos Archives.

10. Alexander, *History of Sandia*, p. 19; G. T. Kupper, "Purchasing History," SNL Archives.

11. Memo, Sylvan Harris to L. G. Hawkins, Subject: "Sandia Laboratory News," 25 August 1948, Los Alamos Archives: 322-Sandia Corporation Correspondence, Bay 9, Drawer 58.

12. Ibid.

13. Ibid.

14. L. G. Hawkins, Business Manager, Sandia Laboratory, to A. E. Dyhre, Project Business Manager, Los Alamos Laboratory, Subject: "Sandia Laboratory News," Los Alamos Archives.

NOTES

15. *Sandia Lab Weekly Bulletin*, 26 November 1948, SNL Archives.

16. D. M. Olson, "A History of the Coronado Club," manuscript in SNL Archives; Alexander, *History of Sandia*, p. 19; "5 Years History Told By Coronado Club Presidents," Sandia *Lab News*, 17 June 1955, SNL Archives. Other members of the Club's first Board of Directors were: T. G. O'Hara, Harold W. Sharp, George W. Landry, Geneva Bishop, Robert E. Roy, Robert W. Henderson, R. J. Hansen, J. Les Rowe, D. F. Worth, Dorothy Youel, and L. J. Heilman. See Sandia *Lab News*, 17 June 1955, p. 3, SNL Archives.

17. Olson, "A History of the Coronado Club," manuscript in SNL Archives.

18. Alexander, *History of Sandia*, pp. 19, 46, SNL Archives.

19. Memo, Warner to Bradbury, 29 November 1946, SNL Archives.

20. Memo, E. L. Cheeseman to All Groups Concerned, Subject: "Nomenclature Conference," n.d., SNL Archives; Interview, Richard Bice, 21 June 1984, SNL Archives.

21. W. L. Stevens, "History of Nuclear Weapons Activities," manuscript, SNL Archives; Commentary, Robert W. Henderson, 18 November 1988.

22. W. L. Stevens, "Component Development History," manuscript, SNL Archives; Leon Smith, Commentary, 11 November 1989, SNL Archives; N. S. Furman, "Historical Background for the Evolution of Fuzing and Firing Technology," special project draft, SNL Archives.

23. Leon Smith, Commentary, 11 November 1989, SNL Archives; Furman, "Historical Background . . . ," op. cit.; Philip R. Owens, "A History of Component Development at Sandia" manuscript, 7 October 1964, SNL Archives.

24. Leon Smith, Commentary, 11 November 1989, SNL Archives; Memo, Henderson to Bradbury, Subject: "Congressional Report" (Microfilm), 25 November 1947, SNL Archives.

25. Stevens, "History of Nuclear Weapons Activities," manuscript, SNL Archives.

26. Memo, Henderson to Bradbury, Subject: "Congressional Report," (Microfilm), 25 November 1947, SNL Archives.

27. "Sandia Branch of the Los Alamos Laboratory," Report by the Manager, Office of Santa Fe, Director of Operations and Chief of Engineering, 5 August 1948, p. 6, DOE Archives, Record Group 326, Atomic Energy Commission, Collection: Secretariat, Box 1235, Folder 601 LASL, Germantown, Maryland; Alexander, *History of Sandia*, p. 16.

NOTES

28. Stevens, "History of Nuclear Weapons Activities," manuscript, SNL Archives. Hansche joined Sandia on 17 August 1948; on 2 June 1948 Everett Cox became the first Ph.D. actually recruited; Byron Murphey was hired 15 June 1949, while Richard Claassen on 21 March 1951 and Tom Cook on 29 August 1951, respectively, became the fourth and fifth Sandia Ph.D.s.

29. Howard G. Bunker, USAF Chairman of the Tactical and Liaison Committee, KAFB to Chief of Staff, USAF, Washington, D.C., Folder: "Documents on the Air Force Tactical and Liaison Committee, 1946–48," KAFB, N.M.

30. Interview, G. Corry McDonald, 11 February 1985, SNL Archives.

31. Memo, P. J. Larsen, Associate Director, to N. E. Bradbury, Director, Subject: "Joint Sandia Cooperation," 2 April 1948, Los Alamos Archives; Memo, P. J. Larsen to C. L. Tyler through N. E. Bradbury, Subject: "Participation by the Armed Forces in the Design and Development of Atomic Weapons," 22 December 1947, SNL Archives.

32. Memo, N. E. Bradbury, Subject: "Proposed Research and Development Board," 21 January 1943, SNL Archives.

33. "Minutes of the First Meeting," Joint Research and Development Board, Sandia Base, Albuquerque, 2 March 1948, SNL Archives. Mr. C. W. Campbell of Sandia was appointed recording secretary at the Board's second meeting held March 12, 1948.

34. Robert LeBaron, Chairman MLC, to George A. Landry, 3 May 1951, Washington, D.C., SNL Archives.

35. Ibid.; Memo, P. J. Larsen to J. H. Manley, Subject: "Los Alamos Representation on Sandia Boards," 22 April 1948, Los Alamos Archives; Interview, Richard A. Bice, 21 June 1984, SNL Archives.

36. Interview, Richard A. Bice, 21 June 1984, SNL Archives.

37. Memo, P. J. Larsen to J. H. Manley, Subject: "Los Alamos Representation on Sandia Boards," 22 April 1948, Los Alamos Archives.

38. Interview, Richard A. Bice, 21 June 1984, SNL Archives.

39. *'Weaponizing,' Sandia Corporation and the Nuclear Weapons Program*, n. d. (brochure), p. 8, SNL Archives.

40. Ibid., pp. 8–9.

41. Interview, Lee Schulz by Fred Duimstra, 24 February 1985, SNL Archives.

42. Ibid.

43. "Z Division Progress Reports," Los Alamos, 20 May 1947, 20 October 1947, SNL Archives.

NOTES

CHAPTER 10:

1. AEC, *Second Semiannual Report*, USGPO, July 1947, p. 7, AEC *History*, Vol. II, p. 84; *Atomic Weapons Tests, Operation Sandstone, 1948*, Vol. I, Report of the Commander, Joint Task Force Seven to the Joint Chiefs of Staff, 16 June 1948, pp. vi, 6 (hereinafter cited as *Operation Sandstone*, Report to JCS), SNL Archives.

2. AEC *History*, Vol. II, pp. 84–85, quoting Lilienthal to Brereton, 25 April 1947. Carroll L. Wilson, General Manager of the Atomic Energy Commission, prepared on 24 April 1947, a memorandum for the Commissioners in which he summarized the objective of the Los Alamos Scientific Laboratory in requesting a field test. See *Operation Sandstone*, Report to JCS, p. v, SNL Archives.

3. AEC *History*, Vol. II, p. 85; *Sandstone Report 41, Scientific Director's Report of Atomic Weapon Tests*, Annex 17, Parts II and III: Sandia Laboratory Group and Forward Area Administration, p. 5 (hereinafter cited as *Sandstone Report 41: SNL Group*), SNL Archives.

4. *Operation Sandstone*, Report to JCS, pp. xix, xxi; *Operation Sandstone* : 1948, DNA Technical Report, p. 1, SNL Archives.

5. *Sandstone Report 41: SNL Group*, pp. 6–7. Lilienthal wrote to Brereton, Chairman of the MLC: "Only site 1 (Enewetak) will permit full realization of the objectives of the tests"; in DOE Archives, Collection: Secretariat, Box 4942, Folder 471.6 (4/21/47), Sec. 1, Germantown, Maryland.

6. *Operation Sandstone*: 1948, DNA Technical Report, pp. 18, 26, SNL Archives.

7. Ibid., p. 30. AEC, *Fourth Semiannual Report*, USGPO, 1948, p. 2.

8. *Operation Sandstone*, Report to the JCS, pp. 17–18, and Part I: Introduction, n.p., SNL Archives.

9. Ibid., Part 1: Introduction, n.p., SNL Archives.

10. Ibid., chart on p. 31 and p. 89; *Sandstone Report 41: SNL Group*, pp. 5–6, 12–13.

11. Interview, Leo "Jerry" Jercinovic, 21 January 1982, SNL Archives.

12. *Sandstone Report 41: SNL Group*, pp. 7–8, 13–15, 20, 26.

13. Ibid., 16, 28.

14. *Operation Sandstone* : 1948, DNA Technical Report, pp. 35–36, 39, 48–63.

NOTES

15. Interviews, Ira Hamilton, 16 November 1984, and Leo M. Jercinovic, 21 January 1982, SNL Archives. See also *Operation Sandstone 41: SNL Group*, p. 20.

16. Art Machen to Furman, 19 January 1985, Rogue River Oregon, SNL Archives.

17. Ibid.; Interview, Jercinovic, 21 January 1982, SNL Archives.

18. *Sandstone Report 41: SNL Group*, pp. 27-28; *Operation Sandstone*: 1948, DNA Technical Report, p. 17; Bernard O'Keefe, *Nuclear Hostages*, p. 139.

19. O'Keefe, *Nuclear Hostages*, pp. 140-41.

20. *Operation Sandstone: 1948*, DNA Technical Report, pp. 17, 89; Robert W. Henderson, "The Development of the Hydrogen Bomb," Speech presented at the National Atomic Museum, 20 February 1985, SNL Archives.

21. *Operation Sandstone*, Report to the JCS, p. xxi; *Operation Sandstone*, DNA Technical Report, p. 18; DOE, Bulletin WNP-46A: "Declassification of Early Stockpile Data," 19 February 1982, SNL Archives.

22. AEC *History*, Vol. II, p. 175.

23. Ibid., pp. 175-76.

24. Ibid., p. 176.

25. Interview, Richard A. Bice, 21 June 1984, SNL Archives; *Sandstone Report 41: SNL Group*, p. 11.

26. *Sandstone Report 41: SNL Group*, p. 11.

27. Interview, Richard A. Bice, 21 June 1984, SNL Archives.

28. *Sandstone Report 41: SNL Group*, p. 11.

29. Sandia *Bulletin*, 19 November 1948, SNL Archives.

30. Memo, "Sandia Laboratory Training Requirements," to Attn: A. B. Bonds, AEC, Washington, D.C., 30 June 1948, SNL Archives. See also Alexander, *History of Sandia*, p. 17.

31. AEC *History*, Vol. II, pp. 175-76.

CHAPTER 11:

1. Excerpt, "Minutes of the 6th Meeting of the Sandia Research and Development Board," 6 May 1948, SNL Archives.

2. "Report to the President by the Special Committee of the National Security Council on 'The Proposed Acceleration of the Atomic Energy

NOTES

Program,'" 10 October 1949, and Report to the Joint Chiefs of Staff by the Joint Strategic Survey Committee in "Annual Report of the Military Liaison Committee, Atomic Energy Commission," 11 October 1948, both in Record Group 218, Records of the United States Joint Chiefs of Staff, National Archives.

3. Alexander, *History of Sandia*, p. 16.

4. W. F. Shaffer, Jr., to Frederick Alexander, Knoxville, Tennessee, 10 May 1962, SNL Archives; Alexander, *History of Sandia*, p. 16.

5. Memo, P. J. Larsen, Associate Director, to N. E. Bradbury, Director, Subject: "Joint Sandia Cooperation," 2 April 1948, Los Alamos Archives.

6. According to the *Ninth Semiannual Report* of the AEC, January 1951: "To carry out necessary production and research operations, the Commission has followed the Manhattan Engineer District's practice of using contractor operators. Contracts are made with industrial concerns, universities, and other scientific organizations to carry out the principal research operations." See p. 5.

7. The author would like to express appreciation to Robert "Bob" Duff (organization 3180) for his generous assistance with this chapter during classification review. Interview, John Sundberg, Nuclear Weapons Lab, Sandia National Laboratories, Albuquerque, New Mexico, 30 April 1985; "An Historical Perspective on the Nuclear Weapons Production System" (hereinafter cited as "Nuclear Weapons Production") DOE AL, August 1985, p. 3.

8. "Nuclear Weapons Production," August 1985, p. 3; Alexander, *History of Sandia*, p. 19; Jim Babcock, "A bomb research triggered birth of Miamisburg plant," n. p., news clipping service, SNL Archives.

9. Alexander, *History of Sandia*, p. 19.

10. N. E. Bradbury to President Robert G. Sproul, University of California, Berkeley, 18 November 1948, SNL Archives.

11. Memo, DMA to Circulation, AEC Files, Series 199, ALO (copy), SNL Archives.

12. Paul J. Larsen to C. L. Tyler, Manager Santa Fe Operations, 11 January 1949, Los Alamos Archives.

13. Ibid.; Memo, N. E. Bradbury to P. J. Larsen, Subject: "Sandia Branch of Los Alamos Scientific Laboratory," 7 October 1948, Los Alamos Archives.

14. Shaffer to Alexander, 10 May 1962, SNL Archives.

15. Interview, Richard A. Bice, 28 June 1984, SNL Archives; Wilbur F. Shaffer, Jr., to Frederick C. Alexander, Jr., Knoxville, Tennessee, 10 May 1962, SNL Archives.

NOTES

16. Shaffer to Alexander, 10 May 1962, SNL Archives; "History of Design Engineering Services," SNL Archives.

17. No author, "Sandia Corporation's Job in the Nuclear Weapons Program," n.d., manuscript, SNL Archives.

18. D. E. Hurt, "History of Quality Assurance at Sandia National Laboratories, Albuquerque," manuscript, SNL Archives.

19. Sandia Laboratory, Branch of Los Alamos Scientific Laboratory, University of California, Organizational Chart, 5 May 1948, SNL Archives. Memo, G. M. Dorland to R. W. Henderson, Subject: "Road Canning and Nitrogenizing Program," 29 April 1947, SNL Archives; H. P. Kelsey and L. Jean LaPaz, "Quality Assurance Program for Atomic Weapons," April 1967, p. 4, SNL Archives.

20. Interview, Louis Paddison, 16 April 1985, SNL Archives.

21. Ibid.

22. Interview, G. "Corry" McDonald, 11 February 1985, SNL Archives.

23. Kelsey and Paz, "Quality Assurance Program for Atomic Weapons," pp. 3–4.

24. A. J. Clark, Jr., "History of Field Engineering, Test, and Manuals," manuscript, SNL Archives.

25. Interview, Louis Paddison, 16 April 1985, SNL Archives.

26. Herken, *The Winning Weapon*, pp. 219–20.

27. *Memorandum By the Chief of Staff, U.S. Air Force* to the Joint Chiefs of Staff on *Atomic Bomb Assembly Teams*, Ref.: JCS 1745/15 and decision dated 2 September 1948, Record Group 218, Records of the US JCS, National Archives.

28. Memo, US AEC to General James McCormack, Subject: "Action By the Armed Forces and Atomic Energy Commission For Transfer of Atomic Weapons," 16 July 1948, Record Group 218, Records of the US JCS, National Archives.

29. Interview, Louis Paddison, 16 April 1985, SNL Archives; Bruce de Silva, "Military Bases: Are They Untouchable?" *Albuquerque Journal*, 28 April 1985, p. 1.

30. Interview, Larry Neibel and William O'Neill, 11 February 1985, SNL Archives; D. E. Hurt, "History of Quality Assurance at Sandia National Laboratories, Albuquerque," pp. 2–3, SNL Archives.

31. Interview, Louis Paddison, 10 April 1985, SNL Archives.

32. Interview, Larry Neibel and William O'Neill, 11 February 1985, SNL Archives.

NOTES

33. Interview, Sam Johnson, 2 October 1982, SNL Archives.

34. Interview, Jerry Jercinovic, 21 January 1982, SNL Archives. See also "Atomic Energy Commission Future Storage Requirements, Report by the Director of Military Applications," p. 1, including Appendix A, "Conclusions and Recommendations of Board on Future Storage," pp. 6-9, 12 May 1950, DOE Archives, Record Group 326 AEC, Collection: Secretariat, Box 4943, Folder 633, Germantown, Maryland.

35. Interview, Larry Neibel and William O'Neill, 21 January 1985, SNL Archives.

36. Ibid. See also "Atomic Energy Commission Future Storage Requirements, Note by the Secretary [Roy B. Snapp]," 5 January 1950, pp. 4-9, U.S. DOE Archives, Record Group 326, Collection: Secretariat, Folder 633, Germantown, Maryland, and Atomic Energy Commission "Status of Underground Storage Sites, Report by the Director of Military Applications," p. 2, 15 January 1949, DOE Archives, Record Group 326 AEC, Collection: Secretariat, Folder 633, Germantown, Maryland.

37. Interview, Larry Neibel and William O'Neill, 11 February 1985, SNL Archives; Memo, Carroll L. Tyler to George P. Kraker, Subject: "Report on Flood at Igloo Area," 10 August 1950, SNL Archives.

38. Interview, Larry Neibel and William O'Neill, 11 February 1985, SNL Archives.

39. Ibid.

40. Sandia Corporation Telephone Directory, October 1950, SNL Archives; "Atomic Energy Commission Future Storage Requirements, Report by the Director of Military Applications," pp. 1-4, 12 May 1950, DOE Archives, Record Group 326 AEC, Collection: Secretariat, Box 4943, Folder 633, Germantown, Maryland.

CHAPTER 12:

1. Interview, Norris E. Bradbury, Los Alamos, New Mexico, 8 August 1984, SNL Archives; "Dr. Kelly to Retire from BTL; Is Named Chairman of the Board"; "We Wish Him Well" (Editorial); and "Fellow Scientists Praise Dr. Kelly for Contribution to Program," all in Sandia *Lab News*, 6 February 1959, pp. 1, 3.

2. Interview, Bradbury, 8 August 1984, SNL Archives; "We Wish Him Well" (Editorial), and "Fellow Scientists Praise Dr. Kelly for Contribution to Program," Sandia *Lab News*, 6 February 1959, p. 3.

NOTES

3. James McCormack to Norris Bradbury, 23 February 1949, SNL Archives.

4. R. M. Underhill to Norris Bradbury, Berkeley, California, 24 June 1947, SNL Archives.

5. Ibid.

6. Interview, Norris E. Bradbury by Arthur Lawrence Norberg, Albuquerque, New Mexico, 11 February 1976. Copies of this interview are housed at the Bancroft Library, University of California, Berkeley; the American Institute of Physics, Center for the History of Physics, New York, New York; and at SNL Archives, Albuquerque. This was a joint interview, focussing on Bradbury with interjections by Robert Underhill. Permission to quote generously provided by Dr. Bradbury and the director of the Bancroft Library, UC, Berkeley. Robert Underhill to Carroll L. Tyler, Berkeley, California, 31 December 1948, SNL Archives.

7. Interview, Kimball Prince, SNL Attorney, with James McCormack and Carroll Wilson, 29 September 1959, recorded in Kimball Prince, "History of Formation," p. 3, Manuscript in SNL Archives.

8. Ibid., pp. 3–4. James McCormack to Ted Alexander, MIT, Cambridge, Massachusetts, 5 December 1961, SNL Archives.

9. Underhill to Tyler, 31 December 1948, SNL Archives.

10. Interview, Norris Bradbury, 8 August 1984, SNL Archives; Bradbury to Robert G. Sproul, President UC, Berkeley, 18 November 1948, SNL Archives.

11. Alexander, *History of Sandia*, pp. 19–20.

12. Memo, Howard P. Bunker, Brigadier General USAF, to Major General Schlatter, Subject: Sandia Lab Proposed Contractor Changes, KAFB Files, General Correspondence, Office of Atomic Energy, 12 August 1948, pp. 27–49, SNL Archives.

13. Ibid.

14. Ibid.

15. Paul J. Larsen, "Information Presented to Sandia Laboratory Department Managers, Division Leaders and Staff Assistants at Sandia Base Theater on May 12, 1949," SNL Archives.

16. Michael F. Wolff, "Mervin J. Kelly: Manager and Motivator," IEEE *Spectrum*, December 1983, pp. 71–75.

17. In a personal letter with Attachment, "Comments on the Sandia History by Frederick C. Alexander, Jr." to D. P. Severance of MIT, Mervin Kelly indicates that he had "read with interest the Prologue and Chapter 1 of

NOTES

the *History of Sandia* by Frederick C. Alexander, Jr., and although the history "is in accord with my memory and generally accurate . . . one item needs correction." Kelly goes on to record that he was not *assigned* to make the review; rather that Buckley advised Lilienthal that Kelly's "time would be made available"; however, arrangements would have to be made directly with Kelly. Kelly to Severance, MIT, 25 July 1961, SNL Archives.

18. Ibid. Buckley was also in charge of the Signal Corps Laboratory in Paris in World War I. See W. H. Doherty, "The Spirit of Research," Ch. 10 of *A History of Engineering and Science in the Bell System: The Early Years (1875-1925)* (Bell Telephone Laboratories, 1975), p. 972.

19. Kelly to Severance, MIT, 25 July 1961, SNL Archives; Kimball Prince, "History of Formation," p. 4, quoting interviews with McCormack and M. J. Kelly, conducted jointly 20 December 1958, SNL Archives.

20. Kelly to Severance, MIT, 25 July 1961, Attachment, SNL Archives; Report of 264th AEC Meeting, 4 May 1949, SNL Archives; AEC, "Administration of Sandia Laboratory–Report of Mr. Mervin J. Kelly," Notes by the Secretary (Roy B. Snapp), SNL Archives.

21. Report of Mervin J. Kelly, ibid.

22. Ibid.

23. Ibid.; Testimony of Mervin J. Kelly, "In the Matter of J. Robert Oppenheimer," Transcript of Hearing Before Personnel Security Board, Washington, D. C., April 12, 1954, through May 6, 1954.

24. "Report of Mervin J. Kelly," SNL Archives.

25. Ibid.

26. Ibid.

27. Memo, Roy B. Snapp to Brigadier General James McCormack, Jr., Director of Military Application, "Commission Action on AEC 199/1, Prospective Change in Management of the Sandia Laboratory," 12 May 1949, SNL Archives.

28. John Brooks, *Telephone, the First Hundred Years* (New York: Harper and Row, 1975), p. 3.

29. Ibid., pp. 43-44.

30. "Report of Mervin J. Kelly," p. 2; Kimball Prince, "Sandia Corporation, a Science–Industry–Government Approach to Management of a Special Project," excerpt from *Federal Bar Journal*, Vol. 17, 1957, SNL Archives.

31. Brooks, *Telephone*, p. 13.

32. Ibid., pp. 10-11.

NOTES

33. Ibid.

34. Ibid., pp. 11–12.

35. This agreement was reached at AEC Meeting 266. See "Chronology of Events Leading to Execution of Contract With Western Electric For Operation of Sandia," AEC 199/6, copy in SNL Archives.

36. Memo, David E. Lilienthal to the President, AEC, Washington, D.C., 13 May 1949, copy in SNL Archives.

37. Drafts, President Truman to Leroy Wilson and Truman to Dr. O. E. Buckley, Washington, D.C., 13 May 1949, copies in SNL Archives. See also "Prospective Change in the Management of Sandia Laboratory," AEC 199/3, 24 May 1949, copy in SNL Archives.

38. Wilson to Truman, Washington, D.C., 17 May 1949, and Buckley to Truman, Washington, D.C., 17 May 1949, AEC Files, Series 199, ALO, copies in SNL Archives.

39. [McCormack] to AEC, SFO, Memo Re: "Change of Laboratory," SNL Archives; Leroy A. Wilson to David E. Lilienthal, New York, 1 July 1949, copy, SNL Archives.

40. Interview, Kimball Prince with James McCormack, noted in SNL "History of Formation," p. 7. See Section 201, First War Powers Act, and relative to it, House Report No. 1507, 77th Congress 1st Session as discussed in H. C. Anderson to Walter E. Brown, Re: Atomic Energy Commission Contracts, 7 September 1949, SNL Archives.

David Lilienthal also recalled the conversation with Wilson as indicated in Lilienthal, "The Kind of Nation We Want," *Colliers*, 14 June 1952, p. 49.

The act resolved the issue by incorporating the cost–plus–fixed–fee concept, employed by DuPont when constructing the nuclear plants at Clinton (Oak Ridge), Tennessee, and Hanford, Washington.

41. [McCormack] AEC, Washington to AEC, SFO, Memo re: "Change of Laboratory," SNL Archives.

42. "Chronology of Events Leading to Execution of Contract . . . ," SNL Archives; "Prospective Change in the Management of Sandia Laboratory," 15 June 1949, SNL Archives, and Memo, James McCormack to Carroll L. Wilson, Subject: "Sandia Laboratory Forms Attachment to AEC," SNL Archives.

43. The meeting took place on 14 June 1949. Memo, James McCormack to Carroll L. Wilson, AEC, Washington, D.C., 15 June 1949, SNL Archives.

44. Ibid.

NOTES

45. Memo for Record, Files of Leroy Wilson, AT&T, 15 August 1949, 4:50 P.M., copy in SNL Archives. The timing of these "implied" assurances from Tom Clark is interesting, considering that only a year earlier, the Attorney had been persuaded to sue to separate legally Western Electric from the Bell System.

46. Leroy A. Wilson to David E. Lilienthal, New York, 1 July 1949, SNL Archives.

47. Ibid., John Brooks in *Telephone*, pp. 234–38, discusses the negotiations between the government and AT&T (which he refers to as "a game of cat and mouse") in some detail. He questions the Bell System's motives in accepting the management responsibility and concludes also that both the antitrust suit and service in the national interest were factors in AT&T's decision.

48. Brooks, *Telephone*, pp. 253–54.

49. Wilson to Lilienthal, New York, July 1949, SNL Archives.

50. Quoted in Brooks, *Telephone*, p. 253. See also p. 234.

51. Ibid., p. 237.

52. See W. Brooke Tunstall; *Disconnecting Parties, Managing the Bell System Break-Up: An Inside View* (New York: W. McGraw-Hill, 1985), pp. 12–13, and 17 for quote.

53. Brooks, *Telephone*, p. 238.

CHAPTER 13:

1. Paul J. Larsen, Director, Sandia Laboratory, "Information to Sandia Laboratory Department Managers, Division Leaders and Staff Assistants," Sandia Base Theater, 12 May 1949, p. 8, SNL Archives.

2. Ibid., p. 2.

3. Ibid., pp. 2–3.

4. Ibid., p. 3. For comparison, see AEC, "Administration of Sandia Laboratory—Report of Mr. Mervin J. Kelly." Notes by the Secretary [Roy B. Snapp], AEC 199/11, copy, SNL Archives.

5. "Information Presented to Sandia Laboratory . . . ," p. 3, SNL Archives.

6. Ibid., pp. 3–5.

7. Ibid., pp. 6–7.

730

NOTES

8. Ibid., p. 6.

9. Ibid., p. 7.

10. Ibid., pp. 8–9.

11. Ibid., pp. 10–12.

12. Ibid., pp. 13–14.

13. "AT&T Subsidiaries to Operate AEC Sandia Laboratory," (AEC press release) Washington, D.C., 11 July 1949; "News Briefs for Western Electric Management," New York, 13 July 1949; "Private Firms to Run AEC Plant at Sandia," New York *Times*, 12 July 1949, all in SNL Archives.

14. David E. Lilienthal to Mr. Wilson, New York, 6 July 1949, Re: arrangements for the trip; "Western Electric Heads Plan Lab Change," and "Western Electric, Bell Telephone Have Key Role at Sandia A–Base," miscellaneous clippings, SNL Archives.

15. David E. Lilienthal to Brien McMahon, Washington, D.C., 8 July 1949, copy, SNL Archives; John Brooks, *Telephone, The First Hundred Years* (New York: Harper and Row, 1975), p. 236.

16. Photograph with caption, *Albuquerque Journal*, 14 July 1949, in Kimball Prince, "History of Formation," n.p., "Western Electric, Bell Telephone Have Key Role at Sandia A–Base," and "News Brief for Western Electric Management," New York, 13 July 1949, all in SNL Archives; Bradbury to Underhill, Los Alamos, N. M., 18 July 1945, SNL Archives.

17. "Notes of Conference of Kimball Prince with General James McCormack and Mr. Carroll Wilson . . . at General McCormack's Office—MIT," 29 September 1959, Re: "Genesis of Sandia Corporation," p. 4, SNL Archives.

18. Ibid. See also Kimball Prince, "Sandia Corporation, A Science–Industry–Government Approach to Management of a Special Project," *Federal Bar Journal*, Vol. 17, 1957, p. 433.

19. Frank L. Yates, Acting Comptroller General of the United States to David Lilienthal, Washington, D.C., 14 January 1948, pp. 1–7, copy, Prince, "History of Formation," SNL Archives.

20. Ibid.

21. Prince, "History of Formation," p. 10, SNL Archives. See also "Contract AT–(29–1)–789 between United States Atomic Energy Commission and Western Electric Company, Incorporated (including as Appendix A, copy of Contract between Sandia Corporation and Western Electric Company, Incorporated dated October 6, 1949)," p. 11, SNL Archives.

22. Memo, Director, Los Alamos Scientific Laboratory to Distribution, Subject: "Comments and Suggestions on Transfer of the Sandia Laboratory

from the University of California to a New Contractor," Los Alamos, 8 August 1949, SNL Archives.

23. Memo of Record giving "Schedule of Visit" [12 August 1949], Fred R. Lack to Distribution; "Takeover Agreement" including "Minutes of Conference," Los Alamos, New Mexico, 29 August 1949, signatories: C. L. Tyler, R. M. Underhill, Fred R. Lack, SNL Archives.

24. B. Edelman, Record of Conference, Subject: "Proposed Draft of Contract on Sandia Project," 16 September 1949, in Prince, "History of Formation," SNL Archives; Memo, P. D. Wesson to J. A. Dempsey, Subject: "Certificate of Incorporation," New York, 11 November 1949, p. 1.

25. P. D. Wesson to John E. Hall, 19 September 1949, Albuquerque, New Mexico, SNL Archives. For secondary coverage, see Brooks, *Telephone*, pp. 236–37; and C. A. Warren, J. A. Hornbeck, and Morgan Sparks, "Special Projects—Sandia and Bellcom" in *A History of Engineering and Science in the Bell System—National Service in War and Peace* (1925–1975), ed. by M. D. Fagen (Bell Labs, Inc., 1978), pp. 655–56.

26. "Unclassified Minutes of the First Meeting of the Board of Directors of Sandia Corporation," 6 October 1949, SNL Archives, and Contract AT–(29–1)–789, p. 14, and Appendix A, p. 6, SNL Archives. See also Alexander, *History*, p. 24.

27. Contract AT–(29–1)–789, p. 1.

28. Alexander, *History of Sandia*, p. 33.

29. SNL *Annual Reports*, 1950, p. 41; 1953, n.p.; Interview, R. W. Henderson, 23 September 1986, SNL Archives.

30. Contract AT–(29–1)–789 between United States Atomic Energy Commission and Western Electric Company, Incorporated (including as Appendix A, Copy of Contract between Sandia Corporation and Western Electric Company, Incorporated, dated October 6, 1949). See Article VI, p. 5. See also *Atomic Energy Act of 1946*, Public Law 585 [S1717] in Laws of 79th Congress–Second Session, 1 August, pp. 722–41.

31. Memo, P. D. Wesson To W. L. Brown, Subject: "Sandia Contract—Unlimited Indemnity Agreement," 26 September 1949, SNL Archives.

32. Ibid., Article XI, Security, p. 10.

33. Ibid., XII, General, p. 12, Sec. 6.

34. Ibid., X, Patents and Inventions, pp. 8–9, Sec. 1; Interview and Notes, Kurt Olsen, SNL Albuquerque, New Mexico, 7 October 1986.

35. After almost four decades, effective with the October 1, 1983 contract, that limitation was deleted. AT&T can sell its products and services to Sandia. Thus, AT&T and Sandia can now do business, but only

NOTES

with government approval. See Contract DE-AC04-76DP00789 between United States Department of Energy and Western Electric Company, Incorporated, Mod. No. M086, 24 October 1983 (effective 1 October 1983) including as Appendix A the contract between Sandia Corporation and Western Electric Company, Incorporated, 24 October 1983 (effective 1 October 1983). See specifically Article IX, Sec. 1. Interviews, Larry Greher and Kurt Olsen, SNL, Albuquerque, N. M., 7 October 1986.

36. Interview, Leonard Jacobvitz, Albuquerque, N. M., 1 October 1985, SNL Archives; Contract DE-AC04-76 DP00789, (1983) Article VI, Property, p. 9; Article III, Payment of Costs, pp. 4-5; Article XX, Status and Authority of Sandia Corporation as an Agent of the DOE, pp. 40-42; Interview, Harold Folley, Albuquerque, N. M., 20 November 1986; Interviews, Greher and Olsen, 7 October 1986.

CHAPTER 14:

1. James S. Lay, Executive Secretary, "NSC-68: A Report to the National Security Council," Washington, D.C., 14 April 1950 (includes letter, Harry S. Truman to Mr. Lay, Washington, D.C., 12 April 1950), copies, SNL Archives.

2. Ibid., Truman appointed the committee on 18 November 1949; NSC-68 was approved in April 1950. See also Paul Johnson, *Modern Times; The World From the Twenties to the Eighties* (New York: Harper and Row, 1983), pp. 442-43.

3. Johnson, *Modern Times; The World From the Twenties to the Eighties*, pp. 443-44; AEC, *Seventh Annual Report of the Atomic Energy Commission*, January 1950 (Washington, D.C.: USGPO), p. viii; Peter Pringle and James Spigelman, *The Nuclear Barons* (New York: Holt, Rinehart, and Winston, 1981), pp. 86-87; Eric F. Goldman, *The Crucial Decade, America, 1945-1955* (New York: Alfred A. Knopf, 1956), p. 112. Robert Chadwell Williams in *Klaus Fuchs, Atom Spy* (Cambridge, Massachusetts: Harvard University Press, 1987), pp. 70-88, tells the story of the Fuchs, Greenglass, Gold, and Rosenberg spy connections of the time.

4. Lewis L. Strauss, *Men and Decisions* (New York: Doubleday and Company, 1948), pp. 204-5; Ross is quoted by Goldman in *The Crucial Decade* on p. 99.

5. Strauss, *Men and Decisions*, p. 205; Pringle and Spigelman, *The Nuclear Barons*, p. 88.

6. "Who Was Who in America," Vol. 4, 1961-68, photocopy, SNL Archives; Arch Napier, "Sandia Corporation: On the Frontier of Engineering," *Sandia Corporation*, n.d., p. 3, SNL Archives.

NOTES

7. Interview, Ray Powell, 23 August 1984, SNL Archives.

8. George A. Landry, "The Responsibilities of Sandia Corporation," p. 14, Speech presented at Skytop Management Conference, late 1950, SNL Archives; Alexander, *History of Sandia*, p. 25; Press release, "AEC Signs Contract with Western Electric Company for Operation of Sandia Corporation," 7 October 1949, SNL Archives.

9. George A. Landry, "The Responsibilities of Sandia Corporation," p. 14, Speech presented at Skytop Management Conference, late 1950, SNL Archives;"Personnel History," B-1-2, SNL Archives; *Sandia Laboratories*, February 1952, p. 4, SNL Archives; Alexander, *History of Sandia*, pp. 25-26.

10. "Personnel History," B-1-2, SNL Archives.

11. H. H. Benedict to Mr. Underhill, University of California, Berkeley, 25 August 1949, Los Alamos Archives.

12. Interview, Corry McDonald, 11 February 1985, SNL Archives.

13. Interview, Ray Powell, 23 August 1984, SNL Archives.

14. Ibid.; Interview, Robert W. Henderson, 30 October 1985, SNL Archives.

15. Landry to R. E. Gibson, Director Applied Physics Laboratory, Johns Hopkins University, Silver Springs, Maryland, 17 November 1949, SNL Archives.

16. Alexander, *History of Sandia*, p. 25.

17. Sandia Corporation Telephone Directory, July 1950, SNL Archives.

18. Alexander, *History of Sandia*, pp. 26, 30.

19. AEC, *Eighth Semi-Annual Report* (USGPO), July 1950, pp. ix-xi, 3.

20. Ibid., p. xi.

21. Speech, Carl T. Durham to House of Representatives, 24 June 1952, photocopy, National Archives, Washington, D.C., RG 128, Joint Committee on Atomic Energy, Box 222.

22. Ibid.

23. "Nuclear Weapons Production," pp. 60-61; 13 and Appendix A-1, Section 148, SNL Archives.

24. The principle of civilian control has been given continuing consideration. In 1975, for example, the "Transfer Study" (ERDA 97), examined the matter; and most recently, the issue was raised again as part of an investigation conducted by the President's Blue Ribbon Task Force (1985). (See "Report of the President's Blue Ribbon Task Group on Nuclear Weapons Program Management," July 1985, SNL Archives.) However, legislation such

as the 1954 version of the Atomic Energy Act and various studies have failed to change the policy of civilian control of atomic energy.

25. Originally, ALO was responsible for nuclear explosives test operations as well as the operation of the production complex; however, in 1962, management of test operations was transferred to the Nevada Operations Office (NV) in Las Vegas, Nevada. See "Nuclear Weapons Production," pp. 3-4.

The Economy Act of 1932 established the original legal guidelines and restraints pertaining to the cooperation between government and industry, although provisions of the act were not converted into formal policy by Bureau of the Budget Bulletin #60-2 until 1959. See Science Policy Research Division. *A Case Study of the Utilization of Federal Laboratory Resources* (Washington, D.C.: United States Government Print Office, 1966), pp. 34, 38-39; and "Nuclear Weapons Production," Appendix A-1, Section 148.

26. Sandia Corporation, *Annual Report of the Board of Directors to the Stockholders*, 31 December 1950, SNL Archives (hereinafter cited as Sandia Corporation, *Annual Report*); W. L. Stevens, "A Brief History of Early Nuclear Weapons Production and Manufacturing Engineering Activities at Sandia National Laboratories 1947-1951," manuscript, SNL Archives.

27. William P. Thomas, "Birth and Demise of the Manufacturing Development Engineering Technical Organization at Sandia," pp. 10-15, in History of Design Engineering Services Collection, SNL Archives.

28. *A Summary of 2100: 1949 to 1951*, Sandia Corporation, 1951, p. 6, SNL Archives.

29. Ibid., pp. 3, 16.

30. Sandia Corporation Telephone Directory, July 1950, SNL Archives.

31. *A Summary of 2100: 1949 to 1951*, pp. 3-16; Stevens, "A Brief History of Production and Manufacturing Engineering Activities at Sandia National Laboratories 1947-1951," manuscript, SNL Archives.

32. *A Summary of 2100: 1949 to 1951*, p. 3; Sandia Corporation, *Annual Report*, 31 December 1950, SNL Archives.

33. Interview, G. Corry McDonald, 11 February 1985, SNL Archives.

34. *A Summary of 2100: 1949 to 1951*, pp. 3, 38-43; Sandia *Bulletin*, 19 January 1951, pp. 1-3, SNL Archives.

35. Ibid.; Interview, Robert W. Henderson, 26 April 1988, SNL Archives; Stevens, "A Brief History of Production and Manufacturing Engineering Activities at Sandia National Laboratories, 1947-1951," manuscript, SNL Archives; Interview, Robert E. Hopper, 4 May 1987, SNL Archives; "Hopper

NOTES

Named Superintendent in Sandia Organization Revision," Sandia *Lab News*, 24 February 1956, R. E. Hopper Collection, SNL Archives.

Among those accepting staff positions at Sandia following completion of the US Navy's Reserve Officer's Training course at the University of New Mexico were Charles L. Hines, Harry Kinney, Robert D. Statler, Henry W. Willis, Jr., and George Hildebrandt. Hines would later serve as a New Mexico State Representative; Kinney would serve as a City Commissioner, and after retirement, as Mayor of Albuquerque. Willis would become President of the Albuquerque School Board. Statler became a member of the Field Test organization at Sandia, and Hildebrandt would begin his career at Sandia in Production Engineering.

36. "History of Design Engineering Services," SNL Archives, and Sandia Corporation Telephone Directory, 1949, SNL Archives; Interview, Charlie Winter, 16 April 1985, SNL Archives; Interview, Robert W. Henderson, 26 April 1988, SNL Archives.

37. Memo, "Sandia Laboratory Training Requirements," to Attn: A. B. Bonds, AEC, Washington, D.C., 30 June 1948, and Minutes: "Division Leaders Meeting," 16 November 1948 (report dated 23 November 1948), SNL Archives.

38. Interview, Robert W. Henderson, 26 April 1988, SNL Archives.

39. Interview, G. Corry McDonald, 11 February 1985, SNL Archives.

40. Ibid.

41. Ibid.

42. Interview, Louis Paddison, 16 April 1985, SNL Archives.

43. Ibid.

44. Ibid.

45. Interview, Bernard S. Biggs, 20 September 1984, SNL Archives.

46. Interview, George Hildebrandt, 5 May 1984, SNL Archives. While attending the University of New Mexico, Hildebrandt was a member of the UNM football team.

47. Ibid.

48. Sandia Corporation Telephone Directory, July 1950, SNL Archives.

49. SNL Budget History, p. 29, and accompanying documents: Temporary SCI 3036, 10 November 1950, SNL Archives; SCI 3506 notes that "In reviewing some of my SCI's, it seemed that SCI 3506 was in need of . . . elimination. I discussed it with Cost Division and we agreed that in view of the fact that we no longer do any manufacturing, this SCI should be eliminated—ERW 9/2/60."

NOTES

CHAPTER 15:

1. "25 Lab Employees Called to Colors," Sandia *Bulletin*, 22 December 1950, p. 4, SNL Archives. Those listed were: Benny Anaya, James M. Bedeaux, John G. Boyes, Lyle E. Bonn, Fred J. Clark, Jr., Robert W. Copeland, Paul L. Cruze, Jesse A. Floyd, Johnny B. Freelove, Claudio J. Gonzales, Roger M. Herring, Roy H. Kendall, Vernon E. Kerr, Ray H. Lee, Alfonso Lujan, Albert J. Mallet, Adolfo Martinez, Arsenio P. Montoya, Zachary R. Ortiz, Eden A. Raney, Myron D. Roepke, Donald L. Russell, Tom Vigil, Billy D. Wiley, and Guy C. Willis.

2. James McCormack, Jr., Director DMA, to Carroll L. Tyler, Manager SFO, Washington, D. C., 31 July 1950, p. 1.

3. Ibid., p. 2.

4. Interview, G. Howard Mauldin, 18 December 1984, SNL Archives.

5. Memo, AEC. "Preliminary Examination of the Work Load at Santa Fe Operations," 7 December 1951, Secretariat Files 1951–58, MR and A9–1 Weapons Development, DOE Archives, Albuquerque, New Mexico; Carroll L. Tyler to George Landry, Sandia Base, 18 November 1949, SNL Archives; "Notes for Meeting at Los Alamos with Dr. Bradbury," 15 June 1949, SNL Archives.

6. "Notes for Meeting at Los Alamos with Dr. Bradbury," on June 1949 (pertaining to responsibilities between Sandia and Los Alamos), SNL Archives; Memo, P. J. Larsen to N. E. Bradbury, Subject: "Division of Responsibilities Between Sandia Laboratory and Los Alamos Scientific Laboratory," 15 August 1949, SNL Archives; Norris E. Bradbury, "Scope of the Inter–Related Work of the Sandia Laboratory and the Los Alamos Scientific Laboratory," 29 August 1949, SNL Archives. On May 1, 1955, Lamkin was promoted to Director of Surveillance and Operations with L. R. Neibel and H. E. Viney reporting to him as Department Managers. See "Lessel E. Lamkin Appointed Sup't," Sandia *Lab News*, 22 April 1955; Elias Klein to L. E. Lamkin, Office of the Secretary of Defense, Washington, D.C., 8 December 1953 (with Henderson note attached), and L. E. Lamkin, "Packaging Criteria," paper presented to the Twenty–First Shock and Vibration Symposium, Los Angeles, California, 17–18 November 1953, published in *Material Handling Flow*, March 1955, all in Lamkin Collection, SNL Archives.

7. Memo, Rainbow Team, Joint Strategic Plans Group to Director, Joint Staff, Col. C. V. Clifton, 2 May 1952, p. 2, RG 218, Records of the USJCS, CCS 471.6, Section 30, Control of Atomic Weapons and Test, National Archives, Washington, D. C.

8. Sandia Corporation, Annual Report Excerpt, 31 December 1950, p. 81, SNL Archives; "An Analysis of the Strategic Uses of the Air–Burst Atomic

NOTES

Bomb with Special Reference to the Fuzing Problem," 24 May 1951, Excerpt in SNL Archives. The author gratefully acknowledges the assistance of Gene Harling and Jim King in locating this document.

9. Sandia Corporation Telephone Directory, 1951, SNL Archives. Lillian Hoddeson in "The Emergence of Basic Research in the Bell Telephone System, 1875–1915," *Technology and Culture* 22 (1981), pp. 512–44, provides excellent documentation of the development of basic research at Sandia's parent company.

10. "An Analysis of the Strategic Uses of the Air-Burst Atomic Bomb with Special Reference to the Fuzing Problem," 24 May 1951, and the Newhouse Committee, "Technical and Cost Aspects of the Radar Fuze Countermeasures, An Appendix to a Systems Study of Strategic Bombing with Atomic Weapons," 13 July 1950, both in SNL Archives; W. L. Stevens, "Research History," manuscript, SNL Archives.

11. SWDB, Weapons Reliability Committee Excerpt, 4 December 1952, including reference to Bradbury's memorandum for file; N. E. Bradbury to R. E. Poole, 10 October 1950, SNL Archives.

12. W. L. Stevens, "Research History," SNL Archives; Sandia Corporation Telephone Directories, 1951, 1952, SNL Archives.

13. Interview, Randy Maydew, 22 August 1985, and Don Shuster, 13 July 1986, both in SNL Archives; Chief AFSWP to Chairman AEC, 4 December 1948, Mk 4 Files, SNL Archives.

14. Ibid.; 21 October 1948, Mk 4 Files, SNL Archives.

15. Memo for the Chairman, Sandia Research and Development Board, 13 February 1950, Microfilm File SC-35, SNL Archives; Chief, Field Office AEC to Col. A. A. Fickel, Subject: "Military Characteristics for FM Type Weapons." SWC Historical Files, Kirtland Air Force Base, copy in SNL Archives. See also Gregg Herken, *The Winning Weapon, The Atomic Bomb in the Cold War*, 1945–1950 (New York: Vintage Books, 1982).

16. Sandia Corporation Field Testing Summary (Excerpt), 1946–1954, 13 May 1955, SNL Archives.

17. Interview, G. Howard Mauldin, 18 December 1984, SNL Archives.

18. Ibid.

19. Ibid.

20. Ibid.

21. Ibid.

NOTES

22. Memo, Landry to McCormack, Subject: The "Mk 41" [4N] Program, Microfilm File SC-35, Sandia Technical Library. It is noteworthy that the quantity of bombs provided by the 4-N program is precisely that specified in the Joint Chiefs of Staff's first theoretical plan for war with Russia, code name "Pincher," to destroy twenty cities. Several updated war plans, however, had been considered by 1950. See Herken, *The Winning Weapon, The Atomic Bomb in the Cold Way, 1945-1950* (New York: Vintage Books, 1982).

23. Memo, Bice to All Concerned. Re: "Stamp Code Designation for New Mk 4 Program." Microfilm File, SC-35, Sandia Technical Library. Memo, McCormack to US Air Force, Subject: "Drop Test Program," Microfilm File, SC-29, Sandia Technical Library. Information Research Division, Mk 4 Excerpt, SNL Archives.

24. W. L. Stevens, "A Brief History of Early Nuclear Weapons Activities at Sandia National Laboratories 1947-1951," manuscript, SNL Archives.

25. Ibid.; Telephone Interview, Robert P. Stromberg, 26 January 1988, SNL Archives.

26. Stevens, "A Brief History of Early Nuclear Weapons Activities at Sandia National Laboratories 1947-1951," manuscript, SNL Archives.

27. R. N. Brodie, US Nuclear Weapon Safety (Excerpt), 1945-1986, SNL Archives.

28. Military Liaison Committee to Chairman AEC, 3 July 1952, Central Technical File, Folder 5-4, Sandia Technical Library; Interview, Richard S. Claassen, 21 August 1987, SNL Archives.

29. Sandia Corporation, Annual Report Excerpt, 1950, p. 73, SNL Archives.

30. Ibid.

31. Stevens, "A Brief History of Early Nuclear Weapons Activities," manuscript, SNL Archives.

32. Ibid.; Sandia Corporation, Annual Report Excerpt, 1950, p. 73; "5000 Weapons History, Mk 5 Bomb," manuscript, SNL Archives.

33. Memo, Joint Strategic Plans Group to Colonel C. V. Clifton, Washington, D. C., 2 May 1952, p. 2, RG 218, Records of the Joint Chiefs of Staff, National Archives, Washington, D. C.

34. Stevens, "A Brief History of Early Nuclear Weapons Activities," manuscript, SNL Archives.

35. "History of the Mk 7 Bomb," manuscript, SNL Archives.

NOTES

36. Ibid.; Sandia Corporation, Annual Report Excerpt, 1950, pp. 74-78, SNL Archives; John Wendell Bailey, "History of Sandia Base," Field Command Historian AFSWP, manuscript excerpt in SNL Archives.

37. Information Research Division, "History of the TX-9," SNL Archives; Stevens, "Brief History of Early Nuclear Weapons Activities," manuscript, SNL Archives; Interview, Douglas A. Ballard, 15 October 1985, SNL Archives. Before joining Sandia, Ballard was hired by Los Alamos Scientific Laboratory in October 1945 to work in the plutonium metallurgy group.

38. Sandia Corporation, Annual Report Excerpt, 1950, p. 78, SNL Archives; "5100 Weapons Development," manuscript, SNL Archives.

39. Stevens, "Brief History of Early Nuclear Weapons Activities," manuscript, SNL Archives.

40. Ibid.

41. Alexander, *History of Sandia*, p. 48.

42. Stevens, "History of Components," manuscript, SNL Archives.

43. Alexander, *History of Sandia*, p. 48; Robert L. Peurifoy, Jr., Interview, 16 October 1985, SNL Archives.

44. Sandia Corporation, Annual Report Excerpt, 1951, SNL Archives.

CHAPTER 16:

1. "Minutes of the Board of Directors of Sandia Corporation," 13 December 1949 and 21 February 1950, Sandia Corporate File. At the meeting on 21 February 1950, the number of directors was increased from four to six. Stanley Bracken and Donald A. Quarles were added. For a good description of the atmosphere of 195 Broadway, see John Brooks, *Telephone, the First Hundred Years* (New York: Harper and Row, 1975), pp. 19-20.

2. "Minutes of the Board of Directors of Sandia Corporation," 13 December 1949 and 21 February 1950, Sandia Corporate File; Ray B. Powell to Necah Furman, 27 February 1984, Sandia National Laboratories, Albuquerque, New Mexico; Alexander, *History of Sandia*, p. 31.

3. Memo, George A. Landry to All Employees, Subject: [Revised Vacation Plan], 28 December 1949, SNL Archives; Powell to Furman, 27 February 1984, SNL Archives; Telephone Interview, Luther J. Heilman, February 1989.

4. Ernest Petersen, "History of Labor Unions at Sandia," 1958, excerpt in SNL Archives.

NOTES

5. Ibid. See also D. Joy Deininger, "Unions—Yesterday and Today," 24 April 1986, manuscript in SNL Archives.

6. Petersen, "History of Labor Unions," 1958, SNL Archives; Alexander, *History of Sandia*, p. 38. F. B. Smith, Bulletin, "Significant Labor Relations Developments at Sandia Laboratory, American Federation of Labor," 15 February 1950, SNL Archives.

7. Petersen, "History of Labor Unions at Sandia," 1958, SNL Archives; F. B. Smith to All Supervisors, Subject: "Union Activities at Sandia Corporation," 21 March 1950, SNL Archives.

8. Smith, Treasurer, Personnel and Public Relations, Sandia Corporation, Bulletin: "Significant Labor Relations Developments at Sandia Laboratory, American Federation of Labor," 15 February 1950, SNL Archives; "Notice to All Supervisors from the Management of Sandia Corporation," 14 November 1949, SNL Archives.

9. Smith, Bulletin: "Significant Labor Relations Developments," 30 March 1950, SNL Archives.

10. Petersen, "History of Labor Unions at Sandia," 1958, SNL Archives; Sandia Corporation, Annual Report Excerpt, 1951, pp. 39–40, SNL Archives; Smith, Bulletin: "Significant Labor Relations Developments," 30 March 1950, SNL Archives.

11. "3200, Labor Relations Report," 31 August 1950, SNL Archives; Sandia Corporation, Annual Report Excerpt, 1950, p. 39, SNL Archives.

12. Petersen, "History of Labor Unions at Sandia," excerpt in SNL Archives; Sandia Corporation, Annual Report Excerpt, 1950, SNL Archives.

13. Ibid.

14. "Sandia Corporation Bulletin to All Employees," 1 June 1951; "3200, Labor Relations Report," 31 August 1950, SNL Archives; Alexander, *History of Sandia*, p. 39; Sandia Corporation, Annual Report Excerpt, 1950, SNL Archives; Sandia Corporation, Annual Report Excerpt, 1951, SNL Archives; Deininger, "Unions—Yesterday and Today," p. 18, SNL Archives.

15. Petersen, "History of Labor Unions at Sandia," 1958, SNL Archives; Alexander, *History of Sandia*, p. 39; Sandia Corporation, Annual Report Excerpt, 1950, p. 41, SNL Archives.

16. Photo Caption, Sandia *Bulletin*, 27 April 1951, p. 1, SNL Archives; Petersen, "History of Labor Unions at Sandia," excerpt in SNL Archives; Alexander, *History of Sandia*, pp. 39–40, notes that "this jockeying for position between the two unions carried on until 1955, when joint bargaining by both unions with the Corporation was started."

NOTES

17. Sandia Corporation, Annual Report Excerpt, 1953, p. 7, SNL Archives; Sandia Corporation, Annual Report Excerpt, 1955, p. 14, SNL Archives; Petersen, "History of Labor Unions at Sandia," SNL Archives.

18. "Minutes of the Annual Meeting of Stockholders," 24 April 1951, Sandia Corporate File, SNL Archives; Sandia Corporation, Annual Report Excerpt, 1951, SNL Archives.

19. "Sandia Corporation Bulletin to All Employees," 1 June 1951, SNL Archives.

20. Sandia Corporation, Annual Report Excerpt, 1950, SNL Archives; Interview, James Hinson, 7 March 1988, SNL Archives.

21. Interview, Hal F. Gunn, for Sandia *Lab News*, n.d., SNL Archives; Interview, James Hinson, 7 March 1988, SNL Archives; Telephone Interview, Luther J. Heilman, February 1989. Heilman recalls that Weaver also played professional football for the Chicago Bears.

22. Interview, Hal F. Gunn for Sandia *Lab News*, n.d., SNL Archives.

23. Ibid.

24. Ibid.

25. Interview, Robert L. Stewart, for Sandia *Lab News*, n.d., SNL Archives.

26. "Changing of the Guard," Sandia *Lab News*, 3 March 1987; Interview, Hal F. Gunn, Sandia *Lab News*, n.d., SNL Archives; Interview, James Hinson, 7 March 1988, SNL Archives. See also Norman Moss, *The Man Who Stole the Atom Bomb* (New York: St. Martin's Press, 1987), pp. 170–76, which discusses the series of events, including the Fuch's case, that contributed to the temper of the fifties era.

27. Telephone Interview, Robert Byrd, February 1989.

28. Ann Hogan, "History of Directorate 3400," manuscript, with contributions from Hank Willis, Jerry Jercinovic, J. D. Martin, and Mel Mefford, SNL Archives; "President Landry Presents Annual Report to Directors," Sandia *Bulletin*, 11 May 1951, SNL Archives.

29. Interview, Hal F. Gunn, for Sandia *Lab News*, n.d., SNL Archives.

30. Interview, James Hinson, 7 March 1988, SNL Archives; Interview, Verne Honeyfield, Sandia *Lab News*, n.d., SNL Archives.

31. "Changing of the Guard," Sandia *Lab News*, 3 March 1987, quoted from story written by Sharon Ball.

32. Interview, Ed Sims with Verne Honeyfield, Ted Varoz, and Robert Stewart, for Sandia *Lab News*, n.d., SNL Archives; Interview, William O.

NOTES

Bramlett, Marvin H. Brown, Robert L. Stewart, George Davies, and Albert Joe Angel, 24 January 1986 (hereinafter referred to as Interview, Union Members), 24 January 1986, SNL Archives.

33. Marvin H. Brown, "A History of Local No. 27, International Guards Union of America," p. 1, manuscript, SNL Archives.

34. Brown, "A History of Local No. 27," p. 3, SNL Archives.

35. Telephone Interview, Luther J. Heilman, February 1989.

36. Alexander, *History of Sandia*, p. 40; Interview, Union Members, 24 January 1986, SNL Archives. See also William Bramlett, ed., *International Guards Union of America*, August–September 1958, Vol. 1, No. 10, p. 7, SNL Archives.

37. Interview, James Hinson, 7 March 1988, SNL Archives; Sandia Corporation Telephone Directory, June 1952, SNL Archives. In 1952, the Plant Security Department under Manager E. P. Hutson reported to Superintendent of Plant Services Luther Heilman.

38. Contract, "Agreement Between Sandia Corporation and the Albuquerque Guards Union, Local 27, International Guards Union of America," 22 April 1953, SNL Archives; Telephone Interview, Luther Heilman, February 1989; Petersen, "History of Labor Unions at Sandia," SNL Archives; Interview, Union Members, 24 January 1986, SNL Archives; William Bramlett, ed., *International Guards Union of America*, June–July 1958, Vol. 5, No. 1, which also enumerated union gains, SNL Archives; Interview, James Hinson, 7 March 1988, SNL Archives.

39. Interview, Union Members, 24 January 1986, SNL Archives.

40. President George A. Landry, "The Responsibilities of Sandia Corporation," p. 14, Speech presented at Skytop Management Conference, late 1950, SNL Archives.

CHAPTER 17:

1. Sandia Corporation, *Annual Report of the Board of Directors to the Stockholders*, Excerpts, 1950, 1951, 1952, 1955, SNL Archives.

2. Alexander, *History of Sandia*, p. 32.

3. Sandia Corporation, Annual Report Excerpt, 1952; Kimball Prince, "Sandia Corporation . . . ," A Science–Industry–Government Approach to Management of a Special Project," *Federal Bar Journal*, Vol. 17, No. 3, 1957, pp. 438–39, SNL Archives.

NOTES

4. Sandia Corporation, Annual Reports Excerpts, 1950, 1951, 1953 (which includes 1952 because of change in fiscal year), 1954, 1955, SNL Archives.

5. Sandia Corporation, Annual Report Excerpt, 1950.

6. Ibid., Sandia Corporation Telephone Directories, 1950, 1951, 1955; Sandia Corporation, Annual Report Excerpt, 1950.

7. "Personnel History," B-1, B-2, SNL Archives.

8. Sandia Corporation, Annual Report Excerpt, 1951, SNL Archives.

9. Ibid., Alexander, *History of Sandia*, p. 37.

10. Sandia Corporation, Annual Report Excerpt, 1950, SNL Archives; Alexander, *History of Sandia*, p. 31.

11. Alexander, *History of Sandia*, pp. 32-33.

12. Ibid., p. 33.

13. Interview, Lloyd Fuller, 6 October 1961, SNL Archives; Alexander, *History of Sandia*, p. 35.

14. Interview, Fuller, 6 October 1961, SNL Archives; Alexander, *History of Sandia*, pp. 40-41.

15. Interview, Fuller, 6 October 1961, SNL Archives; Sandia Corporation Telephone Directories, July 1951, June 1952.

16. Interview, Fuller, 6 October 1961, SNL Archives; Alexander, *History of Sandia*, p. 41.

17. Interview, Fuller, 6 October 1961, SNL Archives; Alexander, *History of Sandia*, p. 42.

18. Interview, Fuller, 6 October 1961, SNL Archives; Alexander, op. cit.

19. Interview, Fuller, 6 October 1961, SNL Archives; Alexander, op. cit.; "Personnel History," B-2, SNL Archives.

20. George Landry to W. L. Brown, 7 November 1949, Sandia Corporation, Sandia Base, Albuquerque, New Mexico, SNL Archives.

21. Alexander, *History of Sandia*, p. 33; Kimball Prince, Sandia Corporation . . . ," *Federal Bar Journal*, Vol. 17, No. 3, 1957, SNL Archives.

22. Sandia Corporation, Annual Reports Excerpts, 1950, 1951, 1952; "Phillip D. Wesson Retires After 30 Years of Service With W. E.," Sandia *Lab News*, 7 October 1955, p. 1, SNL Archives.

23. Report by the Manager, Santa Fe Operations and the Director of Military Applications, "Atomic Energy Commission Legislative Jurisdiction Over AEC Properties at Sandia Base," Note by the Secretary Roy B. Snapp,

NOTES

25 September 1951, AEC 161/1, DOE Archives, Record Group 326, AEC, Collection: Secretariat, Germantown, Maryland; Sandia Corporation, Annual Report Excerpt, 1950, SNL Archives; Prince, "Sandia Corporation . . . ," *Federal Bar Journal*, Vol. 17, No. 3, 1957, SNL Archives: "How Can You Be Taxed if You Can't Vote? And Why You Can't Vote?" Sandia *Lab News*, 12 March 1954, p. 3, SNL Archives.

24. Interview, Phillip D. Wesson, 17 October 1961, SNL Archives; Prince, "Sandia Corporation . . . ," *Federal Bar Journal*, Vol. 17, No. 3, 1957, SNL Archives; Sandia Corporation, Annual Report Excerpt, 1951, SNL Archives.

25. Sandia Corporation, Annual Report Excerpt, 1953, SNL Archives.

26. Phillip D. Wesson Retires November 1 After 30 Years Service with W. E.," Sandia *Lab News*, 7 October 1955, p. 1, SNL Archives.

27. Interview, F. G. Hirsch, n. d., January 1962, SNL Archives.

28. Ibid.

29. Sandia Corporation Telephone Directory, July 1951, SNL Archives.

30. Interview, Hirsch, n. d., January 1962, SNL Archives; Sandia Corporation, Annual Report Excerpt, 1950, SNL Archives.

31. Interview, Hirsch, n. d., January 1962, SNL Archives; Sandia Corporation, Annual Report Excerpt, 1950, SNL Archives.

32. Alexander, *History of Sandia*, pp. 43–44.

33. Ibid., pp. 44–45; Sandia Corporation, Annual Report Excerpt, 1950, SNL Archives; Interview, Hirsch, n. d., January 1962, SNL Archives.

34. Sandia Corporation Telephone Directory, June 1952, SNL Archives; Alexander, *History of Sandia*, p. 45; Sandia Corporation, Annual Report Excerpt, 1953, SNL Archives.

35. Interview, Henry Moeding, 18 September 1986, SNL Archives; "In 1951 Sandia's Purchasers Did A Multi–Million Dollar Buying Job," Sandia *Bulletin*, 6 June 1952, p. 1, SNL Archives.

36. Ibid.; Sandia Corporation Telephone Directories, 1950, 1952.

37. H. R. Crumley, "Purchasing at Sandia," Purchasing History, SNL Archives.

38. Interviews, Larry Neibel, 28 July 1986, and Jay Hughes, 2 December 1986, SNL Archives.

39. Interview, Waylon Ferguson, 18 July 1986, SNL Archives; Interview, Jay Hughes, 2 December 1986, SNL Archives.

40. Interview, Frank Duggan, 12 December 1986, SNL Archives. Duggan became Supervisor of Purchasing for Sandia, Livermore.

NOTES

41. Ibid.

42. Carroll L. Tyler, Manager SFOO and Daniel F. Worth, Jr., Field Manager, AEC–Sandia, "Three Year Report for Sandia Field Office," 12 October 1953, Folder: Old Sandia Field Office Files: Policy and Historical Documents, DOE AL Archives, Kirtland Air Force Base, Albuquerque, New Mexico.

43. Interview, Ed Herrity and Larry Neibel, 23 July 1986, SNL Archives.

44. Ibid.; Interview, Frank Duggan, 18 July 1986, SNL Archives.

45. Interview, Lyle Whelchel, 30 September 1986; Ed Herrity and Larry Neibel, 28 July 1986; Waylon Ferguson, 18 July 1986, all in SNL Archives.

46. Interview, Frank Duggan, 18 July 1986, SNL Archives.

47. Ibid.

48. Interview, Lyle Whelchel, 30 September 1986, SNL Archives; Sandia Corporation Telephone Directories, 1954, 1955; and Leroy Crumley and George Kupper, "Purchasing History," SNL Archives.

49. Donald A. Quarles, "The Public Relations Aspect of Our Job," Speech delivered to Sandia supervisors at annual dinner meeting, 17 February 1953, SNL Archives.

50. Ibid.

51. Interview, Ted B. Sherwin, conducted by Hank Willis with Bill Carstens, Jim Mitchell, and Sherwin [hereinafter cited according to speaker], 13 February 1985, SNL Archives.

52. Memo, P. F. Meigs to S. Harris, Department Manager SLD, Subject: "Sandia Laboratory Public Relations," 16 June 1949; Memo, T. B. Sherwin to P. F. Meigs, Division Leader, Subject: "Public Relations," 13 June 1949, both in SNL Archives.

53. Interview, T. B. Sherwin, 13 February 1985, SNL Archives.

54. Ibid.; Sandia Corporation Telephone Directory, July 1951, SNL Archives.

55. Sandia Corporation Telephone Directory, July 1951, SNL Archives; Interview, T. B. Sherwin, 13 February 1985, SNL Archives.; Interview, T. B. Sherwin, 13 February 1985, SNL Archives.

56. Sandia Corporation Telephone Directories, July 1951, June 1952, SNL Archives; Interview, T. B. Sherwin, 13 February 1985, SNL Archives; "A Newspaper Goes to Press," Sandia *Lab News*," 18 June 1954, SNL Archives.

57. Interview, T. B. Sherwin, 13 February 1985, SNL Archives; "Editorial Policy—Sandia *Lab News*," 2 February 1955, Sherwin Collection, SNL Archives.

NOTES

58. Interview, T. B. Sherwin, 13 February 1985, SNL Archives.

59. Ibid.

60. Fred Smith, "Industrial and Public Relations Program, Sandia Corporation," Speech presented at New Mexico Public Relations Conference," 24 January 1953, SNL Archives.

61. Ibid.

62. "Guide for AEC Contractors," Part 9, Public Information, SNL Archives; Memo, R. B. Powell to Distribution, Subject: "Proposed Public Relations Guide," December 1955, p. 2, SNL Archives.

63. Ted Sherwin, "Notes for File Re: Public Relations Program," 15 October 1954, SNL Archives; Interview, T. B. Sherwin, 13 February 1985, SNL Archives.

64. Ibid.

65. Sandia Corporation Telephone Directory, June 1952, SNL Archives; Interview, T. B. Sherwin, 13 February 1985, SNL Archives.

66. Ibid.

67. Ibid.

68. Interview, Hank Willis, 13 February 1985, SNL Archives.

69. Sandia Corporation, Annual Report Excerpt, 1950, SNL Archives.

70. Interview, Hank Willis and T. B. Sherwin, 13 February 1985, SNL Archives.

71. "Service Recognition Plan for Employees Is Announced," Sandia *Lab News*, 20 May 1955; "Service Pin Suggestion Deadline Friday, June 24," Sandia *Lab News*, 3 June 1955; "Sandia Service Emblem Design Submitted to Vote," Sandia *Lab News*, 12 August 1955; "Thunderbird Pin Is Winner of Contest," Sandia *Lab News*, 26 August 1955, all in SNL Archives.

72. Interview, T. B. Sherwin, 13 February 1985, SNL Archives.

73. "Public Relations Highlights," 3100 Information Services Collection, SNL Archives. For excellent coverage of full-scale test operations and the effects of nuclear testing on residents of the downwind states of Utah, Nevada, and Arizona, see Howard Ball, *Justice Downwind, America's Testing Program in the 1950's* (New York: Oxford University Press, 1986), and Richard L. Miller, *Under the Cloud, The Decades of Nuclear Testing* (New York: The Free Press, 1986). The third volume of the AEC histories, *A History of the United States Atomic Energy Commission*, 1952–1960 by Richard G. Hewlett and Jack M. Holl with Essay on Sources by Roger M. Anders just released through the Office of Scientific and Technical Information, Oak Ridge, Tennessee, addresses the serious issue of fallout in considerable and candid

detail. On p. x-5, for example, the authors point out that "the enormous fallout pattern from *Bravo* . . . indicated that thermonuclear weapons were far more deadly as a radiation device than as an explosive."

74. Ibid.

CHAPTER 18:

1. "Summary of C. N. Hickman's Participation in Developing Warhead Installations for Guided Missiles," 27 March 1953, pp. 1-15, SNL Archives.

2. Ibid. Because of the difficult terrain and missions, Clark Carr (Carco) flew in support of Sandia and Los Alamos for years. He employed only bush pilots, who were exceptionally suited to the job.

3. Tom Bower, *The Paperclip Conspiracy, The Hunt for the Nazi Scientists* (Boston: Little, Brown and Company, 1987). Bower's book maintains that many of the scientists and doctors brought over from Peenemünde were indeed guilty of war crimes. Despite their wartime activities, according to Bowers, interrogating military officers decided that their knowledge and expertise were essential to America's national interest. The files of those individuals selected to be sent to America were identified by a simple paperclip—hence the title of the book. See p. 4.

After the V-1 launch sites were overrun by the Allies, hundreds of V-1s were carried and launched by Heinkel He-III bombers. Thus, the ALCM was "scooped" by thirty years, as Andrew Lieber of Sandia National Laboratories pointed out to the author.

Herbert York, *Race to Oblivion: A Participant's View of the Arms Race* (New York: Simon and Schuster, 1970), pp. 76-77; Milton Lehman, *This High Man, The Life of Robert H. Goddard*, with a preface by Charles A. Lindbergh (New York: Farrar, Strauss and Company, 1963), p. 379; Eugene M. Emmé, ed., *The History of Rocket Technology* (Detroit: Wayne State University Press, 1964), which contains excellent and concise coverage of the work of Goddard in "Pioneer Rocket Development in the United States" by G. Edward Pendray, pp. 19-23; and Heintz Gartmann, *The Man Behind the Space Rockets* (London: Weidenfeld and Nicolson, 1955), pp. 41-44 and 35-37.

4. Andrew Lieber to N. Furman, Sandia National Laboratories, 31 August 1988, SNL Archives; Lehman, *This High Man*, pp. 379-80; and York, *Race to Oblivion*, pp. 76-80.

5. Bower, *The Paperclip Conspiracy*, pp. 59-107, and Gartmann, *The Men Behind the Space Rockets*, pp. 41-44 and 35-37.

6. York, *Race to Oblivion*, pp. 79-80.

NOTES

7. Wernher von Braun and Frederick I. Ordway, III, *History of Rocketry and Space Travel* (New York: Thomas Y. Crowell Company, 1975), pp. 122-23.

8. Michael H. Armacost, *The Politics of Weapons Innovation, the Thor-Jupiter Controversy* (New York: Columbia University Press, 1969), pp. 25-27.

9. Ibid.

10. Ibid., p. 26; Emme, *The History of Rocket Technology*, pp. 108-9.; von Braun and Ordway, *History of Rocketry and Space Travel*, p. 127.

11. Herbert F. York and G. Allen Greb, "Military Research and Development: A Postwar History," *Bulletin of the Atomic Scientists* (January 1977), p. 19.

12. Ibid.

13. Ibid.

14. Ibid.

15. York, *Race to Oblivion*, pp. 80-84.

16. Memo, Groves to Bradbury, Subject: "Atomic Warhead in a Guided Missile," 12 December 1946, SNL Archives; Alexander, Mk 5 Warhead Excerpt, SNL Archives.

17. Alexander, Mk 5 Warhead Excerpt, p. 10, SNL Archives.

18. DMA to Sandia Laboratory, 29 June 1949, AEC Files, MRA-5, Vol. V, 1949-50, cited in ibid.; Memo, Sandia Research to Engineering, Subject: "Proposed Conference on Guided Bomb," 26 August 1949, Central Technical File, Microfilm Reel 745, Sandia Technical Library.

19. Memo, Research and Development Board to MLC, Subject: "Guided Missiles With Atomic Warheads," 27 January 1950, AEC Files, MRA-5, 4/49-6/50, excerpt in SNL Archives; Minutes, Sandia Weapons Development Board to Distribution, Subject: "Minutes of 39th Meeting," Central Technical File, Sandia Technical Library.

20. Sandia Corporation Telephone Directories, October 1950 and June 1952, SNL Archives; "L. A. Hopkins is Appointed Director of Engineering II," Sandia *Bulletin*, 9 November 1951, p. 1, SNL Archives.

21. "Summary of C. N. Hickman's Participation in Developing Warhead Installations for Guided Missiles," p. 1, SNL Archives. See also Robert H. Goddard, *Rocket Development, Liquid-Fuel Rocket Research, 1929-1941*, ed. by Esther C. Goddard and G. Edward Pendray (New York: Prentice-Hall, 1948), p. xv.; and von Braun and Ordway III, *History of Rocketry and Space Travel*, pp. 84, 93-95.

NOTES

22. Memo, Sandia Weapons Development Board to Distribution, Subject: "Minutes of the 45th Meeting," 18 October 1950, SNL Archives; Glenn Fowler, Logbook, 27 December 1950, SNL Archives.

23. "Summary of C. N. Hickman's Participation in Developing Warhead Installations for Guided Missiles," p. 7, SNL Archives.

24. Ibid.; Sandia Corporation, Annual Report Excerpt, 1951, SNL Archives.

25. "Summary of C. N. Hickman's Participation in Developing Warhead Installations for Guided Missiles," p. 7, SNL Archives; Sandia Corporation Telephone Directory, July 1952, SNL Archives.

26. Sandia Corporation to Distribution, 15 February 1951, Microfilm Reel MF-SF-SC-134, Sandia Technical Library. The historical groundwork laid by Sandia's first historian, Frederick C. Alexander, Jr., has been invaluable in providing a synthesis of weapons information and reference to pertinent source materials. While it is not possible to give full credit, the author wishes to acknowledge Alexander's excellent scholarship.

27. Memo, Santa Fe Operation Office to Los Alamos and Sandia Corporation, Subject: "Implosion Type Weapon Program," Excerpt, 19 January 1951, SNL Archives.

28. Memo, Military Liaison Committee to Division of Military Application, Subject: "Definition of Responsibility in the Marriage Program of Guided Missiles to Atomic Warheads," AEC Files, MRA-5, 7/50-6/51, excerpt in SNL Archives, Alexander Collection.

29. Memo, SC to File, Subject: "Proposed Nuclear Arming System for Atomic Warheads Used With Guided Missiles," Excerpt, SNL Archives.

30. "Report of Second Meeting of the Guided Missiles Committee of the Sandia Weapons Development Board," 26 February 1951, SNL Archives; Minutes, Sandia Weapons Development Board, 22 November 1950, SNL Archives; Telephone Interview, W. E. Boyes, 29 June 1988, SNL Archives.

Redundancy applies to both safety and "reliability" (i.e., probability of detonation). Safety is enhanced by use of components in series; i.e., two components *must* function to complete an action. Probability of detonation is increased by components in parallel; i.e., either one can perform the function.

Reliability is defined simply as the probability of success. Success can be defined for any action or operation of a component or of an entire system, for either a safety (premature prevention) or mission success definition.

31. Second Meeting Excerpt, Guided Missile Committee of the SWDB, 26 February 1951, SNL Archives; Memo, Sandia Corporation to Santa Fe

NOTES

Operations Office, Subject: Universal XW-7 Warhead Excerpt, 13 April 1953, Central Technical File, Sandia Technical Library; Memo, Sandia Corporation to Distribution, Subject: Arming and Fuzing, Corporal TX-7 Excerpt, 11 July 1951, Central Technical File, Sandia Technical Library; Excerpt, Minutes of Corporal TX-7 Working Group, 12-13 June 1951, Central Technical File, Sandia Technical Library.

32. Memo, Sandia Corporation to Distribution, Subject: "Arming and Fuzing for the CORPORAL/TX-7 Marriage," 11 July 1951, Central Technical File, Sandia Technical Library.

33. Interview, Andrew Lieber, 8 June 1988, SNL Archives.

During World War II, the most practical of the various fuels used during the war was aniline, and the most common oxidizer was red fuming nitric acid. Temperamental and risky to handle, these fuels had the advantage of being storable at room temperature and of igniting on contact. The fuels would be pumped into a combustion chamber where they would mix and burn, producing the necessary pressure and temperature to force their combustion products to exhaust through the tail of the rocket, creating thrust.

34. Interview, Louis A. Hopkins, 13 April 1987, SNL Archives; Telephone Interview, 6 February 1989.

35. Interview, Andrew Lieber, 8 June 1988, SNL Archives; "Summary of C. N. Hickman's Participation in Developing Warhead Installations for Guided Missiles," p. 10, SNL Archives.

36. Interview, Andrew Lieber, 8 June 1988, SNL Archives.

37. "A Summary of C. N. Hickman's Participation in Developing Warhead Installations for Guided Missiles," p. 4, SNL Archives; Lieber to Furman, 31 August 1988, SNL Archives.

38. Memo, Guided Missiles Committee to Special Weapons Development Board, Excerpt, Subject: "Initial Recommendations on REGULUS/TX-5 Arming and Fuzing," Addendum to the 55th Meeting of the Board, 26 August 1951, SNL Archives; Interview, Del Olson, 19 August 1985, SNL Archives.

39. Excerpt of Minutes, Special Weapons Development Board, 88th Meeting, 1 December 1954, XW-8/REGULUS, Part I, SNL Archives; Memo, Chief of Naval Operations to Distribution, Subject: "Penetrating Atomic Warhead Program for the Regulus I Guided Missile, Termination of," AEC Files, MRA-5, Regulus I-55, excerpt in SNL Archives, Excerpt in Alexander Collection.

40. Excerpt, Military Liaison Committee to Division of Military Application, Subject: "Requirements in Connection with the Regulus Assault Missile (Project Ram)," 18 August 1952, SNL Archives; Memo, Chief of Naval Operations to Bureau of Aeronautics, Subject: "Regulus XW-5 Warhead

NOTES

Fuzing Systems, Change of," March 1956, AEC Files, MRA-5, Regulus, excerpt in SNL Archives.

41. York, *Race to Oblivion*, pp. 80-81. Minutes, XW-5/Matador Ad Hoc Working Group to Distribution, 6th Meeting, AEC Files, MRA-5, Matador, Vol. I; TWX, Santa Fe Operations Office to Sandia Field Office, AEC, 9 April 1953, AEC Files, MRA-5, Matador, Vol. II, excerpts in SNL Archives, Alexander Collection.

42. Memo, Field Command to Santa Fe Operations Office, Subject: "Early Emergency Capability XW-5/Matador," 17 March 1954, AEC Files, MRA-5, Matador, Vol. III, Alexander Collection; Excerpt of Minutes, SWBDB to Distribution, Minutes of the 94th Meeting, Part I, SNL Archives.

43. Memo, Department 1270 to File, Subject: "Report of Trip to Pentagon, Washington, D.C.," SNL Archives; Interview, Jim Scott, 9 April 1987, SNL Archives; York, *Race to Oblivion*, pp. 80-81.

44. TWX, AEC, Washington, D.C. to AEC, Sandia Base, 12 August 1951, SNL Archives; Telephone Interview, Ray Schultz, 9 February 1989.

45. "A Summary of C. N. Hickman's Participation in Developing Warhead Installations for Guided Missiles," pp. 5, 8, 9, SNL Archives; Excerpt of Minutes, TX-N Steering Committee, 40th Meeting, 18 April 1952, Central Technical File, Sandia Technical Library.

46. "Commentary on Father John," provided to SNL Archives by Andrew Lieber, September 1988.

47. Ibid.

48. Interview, W. A. Little, n. d., SNL Archives; Telephone Interview, Ray Schultz, 9 February 1989.

49. "A Summary of C. N. Hickman's Participation in Developing Warhead Installations for Guided Missiles," p. 9; Dave Bickel, "Rocket Powered Centrifuge," SNL Archives, Environmental Test Collection; Interview, Paul Adams, 14 November 1983, SNL Archives.

50. "A Summary of C. N. Hickman's Participation in Developing Warhead Installations for Guided Missiles," pp. 7-9; "Commentary on Father John," Andrew A. Lieber, provided to SNL Archives, September 1988.

51. Excerpt of Minutes, 63rd Meeting, Special Weapons Development Board, 25 June 1952, Central Technical File, Sandia Technical Library.

52. Ibid.; Minutes, 3rd Meeting, Honest John Ad Hoc Working Group to Distribution, 12 June 1952, AEC Files, MRA-5, excerpt in SNL Archives, Alexander Collection; Memo, Kenneth E. Fields to Carroll Tyler, Subject: "Development of Honest John and Corporal," AEC Files, MRA-5, 7-53 to 8-53, excerpt in SNL Archives, Alexander Collection.

NOTES

53. Sandia Corporation to DMA, Subject: "Proposed Ordnance Characteristics of the XW-7/HJ-XI Honest John Atomic Warhead Installation," AEC Files, MRA-5, Honest John, Vol. I; Memo SFO to Sandia Corporation, Subject: XW-7/HJ-XI Program, AEC Files, MRA-5, 5-54 to 6-54; Memo, Santa Fe Operations Office to Sandia Corporation, Subject: "Development of an Adaption Kit for *Demi-John*," AEC Files, MRA-5, Demi-John, 7-54, excerpts in SNL Archives, Alexander Collection; Lieber to Furman, 31 August 1988, SNL Archives.

54. Memo, Research and Development Board to Military Liaison Committee, Subject: "Subsurface Atomic Weapons of Implosion Type," 6 July 1951, AEC Files, MRA-5, Alias, Vol. I, 6-51 to 6-53, excerpts in SNL Archives, Alexander Collection.

55. Memo, R. E. Poole to Worth, Subject: "Project Alias/Betty Complete Design Release," 16 August 1954; AEC Files, MRA-5, *Lulu*, 7-54 to 12-54, SNL Archives; Minutes, Ad Hoc Committee for Alias Betty to Distribution, 11 June 1952, AEC Files, MRA-5, Alias, Vol. I, 6-51 to 6-53, excerpt in SNL Archives, Alexander Collection.

56. Memo, Sandia Corporation to Sandia Field Office, AEC, Subject: "Project Alias Betty Complete Design Release," 16 August 1954, AEC Files, MRA-5, *Lulu*, 7-54 to 12-54, excerpt in SNL Archives, Alexander Collection; "Summary of the Participation of C. N. Hickman in Developing Warhead Installations for Guided Missiles," pp. 9-10, SNL Archives.

57. Douglas Aircraft Corporation to Sandia Corporation, Subject: "TX-7 Rocket Propelled," 29 June 1951, Central Technical File, Sandia Corporation to Distribution, Subject: "First Meeting of the BOAR Committee on July 22 and 23, 1952," XW-7/BOAR, 1-, 2-, 1952-7, Central Technical File; Military Liaison Committee to AEC, 21 October 1952, AEC Files, MRA-5, BOAR, SNL Archives, Alexander Collection; BOAR Ad Hoc Committee to Distribution, "Minutes of Meeting, 13 to 14 July 1955, Forward of," AEC Files, MRA-5, BOAR Minutes, 7-55, excerpts in SNL Archives, Alexander Collection.

58. Excerpt of Minutes, RASCAL/XW-5 Ad Hoc Working Group to Distribution, 7 November 1951, Central Technical File, Sandia Technical Library.

59. Santa Fe Operations Office to Sandia Field Office, AEC, and Sandia Corporation, Subject: "XW-5/Rascal Fuzing Development," 19 January 1953, AEC Files, MRA-5, January-February 1953; Memo, Sandia Corporation to DMA, "Sandia Corporation Proposal for XW-5/Rascal Fuzing System," 23 March 1953, AEC Files, MRA-5, Rascal, excerpts in SNL Archives, Alexander Collection.

60. Excerpt of Minutes, Special Weapons Development Board, 89th Meeting, 12 January 1955, Part I, SNL Archives; Memo, A. W. Fite to

NOTES

Distribution, Subject: "Nike Meeting Held at Sandia Corporation 22 April 1953," 22 April 1953, SNL Archives; Report on the Nike B, Prepared by the Bell Telephone Laboratories, 1 February 1954, AEC Files, MRA-5, excerpt in SNL Archives, Alexander Collection.

61. Telephone Interviews, Louis A. Hopkins, 3 February 1989, and Leon Smith, 8 February 1989; Sandia Corporation Telephone Directory, March 1956, SNL Archives; "Reorganization Accompanies Vice-President Appointments," Sandia *Lab News*, 16 December 1955, p. 3. Glenn A. Fowler was promoted to Vice-President of Research (at age thirty-seven), and Frederick J. Given to Vice-President of Research and Development Technical Services, after which Lou Hopkins took over Given's previous position as Director of Components.

62. Sandia Corporation Telephone Directory, March 1956, SNL Archives.

63. York, *Race to Oblivion*, pp. 83, 102.

64. Ibid., pp. 84-88.

65. Telephone Interview, Leon Smith, 8 February 1989.

66. Telephone Interview, Jack Howard, 9 February 1989; York, *Race to Oblivion*, pp. 94-97. The ATLAS, THOR, and TITAN were all under development by 1955 and fell in the category of "highest national priority." The program for the MINUTEMAN, destined to become the most widely deployed U.S. strategic missile of the time, began in 1957.

CHAPTER 19:

1. Speech to Sandia Supervisors, G. A. Landry, Santa Fe Rotary Club, 10 January 1952, SNL Archives.

2. "Weapon Program—Organization of Sandia Corporation," 647th AEC Meeting, 11 January 1952, RG 326 AEC, Coll.: Secretariat, Box 4930, Folder: Weapons Development, DOE Archives, Germantown, Maryland; "Walter A. MacNair," 1 February 1952, Sandia *Bulletin*, p. 3, SNL Archives.

3. Ibid.; "Sandia Corporation Expands," Albuquerque *Tribune*, 1 February 1952, SNL Archives. An interesting biographical note on Stuart Hight: Before entering graduate school at the University of California in Berkeley, Hight took two years off to earn money as a song and dance man in Hollywood; Commentary, W. L. Stevens, 9 November 1988.

4. Interview, R. W. Henderson, 26 April 1988, SNL Archives.

5. "D. A. Quarles Succeeds Mr. Landry As President of Sandia Corporation," Sandia *Bulletin*, 29 February 1952, p. 1, SNL Archives.

NOTES

6. Interview, R. W. Henderson, 26 April 1988, SNL Archives.

7. "Mr. Quarles Will Speak at Meeting of Engineer Group," Sandia *Bulletin*, 9 May 1952, p. 1, SNL Archives.

8. Interview, R. W. Henderson, 26 April 1988, SNL Archives; Interview, Rosalie Franey Crawford, 2 July 1984, SNL Archives.

9. Special credit for much of the basic research, writing, and analysis for this chapter goes to William L. Stevens, whose internal lab studies "On the Division of Responsibilities Between the AEC and DOD for Fuzing of Nuclear Warheads Used on Guided Missiles and Rockets 1950-1953" and "On the Evolution of the AEC/DOD Agreement on Development, Production, and Standardization of Atomic Weapons of March, 1953" have been liberally incorporated (hereinafter cited as "On the Division of Responsibilities").

Atomic Energy Act of 1946 and Amendments, USGPO, Washington, D.C., 1975, p. 10.

10. Ibid., 43; Memo: Gordon Dean to Chairman, MLC, Subject: *AEC Missile and Rocket Responsibilities* 564/1, 22 January 1953, p. 3, copy in SNL Archives. See also Lewis Strauss, Chairman AEC, to Mr. President, the White House, 1 August 1956, and Dwight D. Eisenhower to Mr. Chairman, Washington, D.C., 8 August 1956, copies in SNL Archives.

11. Memo: H. K. Roper, Colonel, U.S. Army, to Roy B. Snapp, Secretary, Subject: "Military Guidance to the Atomic Energy Commission in the Atomic Weapons Field," 19 July 1951, p. 2, with attachment, "Procedure for Providing Military Guidance to the Atomic Energy Commission in the Atomic Weapons Field, AEC 453, Box 1264, Folder: Design and Development; DOE Archives, Germantown, Maryland.

12. Gordon Dean, Chairman, to [R. L.] Gilpatrick, Washington, D.C., 15 May 1952, AEC 485/16, Box 1264, Folder MRA 9-1 Design and Development, DOE Archives, Germantown, Maryland.

13. R. L. Gilpatrick, Undersecretary, Department of the Air Force, to Gordon E. Dean, Chairman, AEC, Washington, D.C., 3 June 1952, AEC 485, Box 1264, Folder MRA 9-1 Design and Development, DOE Archives, Germantown, Maryland.

14. *Agreement Between the AEC and DOD for the Development, Production, and Standardization of Atomic Weapons*, drafts include Appendix C: "Sandia Corporation Comments on Agreement . . . ," 18 November 1952, AEC 485/20, Box 1264, Folder MRA 9-1 Design and Development, DOE Archives, Germantown, Maryland.

15. "Agreement Between the AEC and DOD . . . ," 21 January 1953, AEC 485/24, Box 1264, Folder MRA 9-1 Design and Development, DOE Archives,

NOTES

Germantown, Maryland. Included also is letter, Gordon Dean to Robert LeBaron, Washington, D.C., 6 January 1953 (circulated as AEC 485/23).

16. Memo, R. T. Coiner, Jr., Colonel, USAF, Acting AEC/DMA, to C. L. Tyler, Manager AEC/SFO, Subject: "Guided Missile Fuzing," 14 June 1950, Reel MF-SF-SC-101, SNL Archives; Minutes, Special Weapons Development Board, 39th Meeting, 21 June 1950, Excerpts in SNL Archives. The term "fuzing" generally referred to the component parts of a weapon system which signal the warhead to detonate at a particular altitude above the target area.

17. Excerpt of Minutes, Special Weapons Development Board, 21 June 1950, SNL Archives; Stevens, ibid., p. 8.

18. Excerpt of Minutes, Special Weapons Development Board, 42nd Meeting, 9 August 1950, SNL Archives; Memo, G. A. Landry to James McCormack, Jr., USAF and AEC/DMA, Subject: "Guided Missiles with Atomic Warheads," 1 August 1950, SNL Archives.

19. Memo, G. A. Landry to James McCormack, Jr., USAF and AEC/DMA, Subject: "Guided Missiles with Atomic Warheads," 1 August 1950, SNL Archives.

20. Ibid.

21. Memo, Captain J. S. Russell, USN, Acting Director AEC/DMA to C. L. Tyler, Manager, AEC/SFO, Subject: "Guided Missiles with Atomic Warheads," 13 November 1950, SNL Archives; Minutes, Special Weapons Development Board, 56th Meeting, 12 October 1951, SNL Archives.

22. Ibid.

23. Memo, D. A. Quarles to Colonel K. E. Fields, Subject: "Reevaluation of Arming and Fuzing Responsibilities for Atomic Warheads," 4 March 1952, SNL Archives.

24. Ibid.

25. Ibid.

26. Ibid.

27. "Responsibility for Development and Procurement of Adaption Kits for Missiles," Captain R. P. Hunter, USN, Executive Secretary DOE/MLC to Director AEC/DMA, 7 August 1952, SNL Archives; Stevens, "On the Division of Responsibilities," p. 19, SNL Archives.

28. Memo, Gordon Dean to Chairman, MLC, Subject: *Missile and Rocket Responsibilities*, AEC 564/1," 22 January 1953, copy, SNL Archives; Stevens, "On the Division of Responsibilities," pp. 20-22, SNL Archives.

29. Stevens, "On the Division of Responsibilities," p. 22, SNL Archives.

NOTES

30. Ibid., pp. 5-6.

31. Ibid., p. 24.

32. Excerpt of Memo: E. H. Draper to J. P. Molnar, Subject: "History and Status of the Ballistic Missile Program," 7 July 1959, p. 6, Central Technical File, SNL.

33. Ibid., Cover letter; Stevens, "On the Division of Responsibilities," pp. 23, 26-27, SNL Archives.

34. *Information on AEC Interface with Concept Formulation and Contract Definition* (Washington, D.C.: USGPO, 1970), pp. 9-12, citing DOD Instructions 5030.10 (July 27, 1965), and 3200.9 (Appendix C), hereinafter cited as *Information on AEC Interface*, SNL Archives.

35. "Minutes of the Fourteenth Meeting of the McRae Committee" (which includes presentation by R. W. Henderson), 2 February 1954, *McRae Collection*, SNL Archives.

36. Ibid., p. 2.

37. Ibid.

38. Ibid.

39. *Information on AEC Interface*, p. 13, SNL Archives.

40. "Minutes of the Fourteenth Meeting of the McRae Committee," p. 2, SNL Archives.

41. Ibid., p. 3.

42. Ibid.

43. G. C. McDonald, "Weapon: Birth – Life – Death, Sandia Corporation Controlled Items," 16 November 1959, p. 5, SC-TM-358-59, SNL Archives.

44. Ibid., p. 6.

45. Ibid.

46. Ibid., pp. 8-9; "Minutes of the Fourteenth Meeting of the McRae Committee," p. 3.

47. McDonald, "Weapon: Birth-Life-Death, Sandia Corporation Controlled Items," 16 November 1959, p. 5, SC-TM-358-59, SNL Archives.

48. Ibid., p. 10.

49. Ibid.

50. Ibid., pp. 11-13; AEC-Albuquerque Operations Major Assembly Release and Hold Order Systems," May 1958 (copy provided by Edwin L. Johnson), SNL Archives. The system outlined in this document has been in effect from the early Mk 6's to the present day. Interview, Douglas A. Ballard, 15 October 1985, SNL Archives.

NOTES

51. Memo, Donald Quarles to T. E. Shea, R. E. Poole, F. Schmidt, W. A. MacNair, Subject: "Transition—Manufacturing Engineering," 29 November 1952, p. 1, *McRae Committee Collection*, SNL Archives.

52. R. W. Henderson (signing for R. E. Poole) to D. A. Quarles, 17 November 1952, *McRae Committee Collection*, SNL Archives.

53. T. E. Shea to D. A. Quarles, 26 November 1952, *McRae Committee Collection*, SNL Archives.

54. Memo, D. A. Quarles to Shea, Poole, Schmidt, MacNair, 29 November 1952, *McRae Committee Collection*, SNL Archives.

55. Ibid.

56. The internal effort to improve the organization's effectiveness begun in 1952 by Quarles was formalized into a committee study on December 4, 1953, by his successor James McRae. See also York and Greb, "Military Research and Development: A Postwar History," *Bulletin of the Atomic Scientists* (January 1977), p. 20.

57. Ibid., p. 20.

58. Ibid., pp. 20–21. See also "James W. McRae Heads Corp.; Mr. Quarles Takes Defense Post," Sandia *Bulletin*, 14 August 1953, p. 1, SNL Archives.

CHAPTER 20:

1. Memo, Chief of Staff, USAF for the Joint Chiefs of Staff, Subject: "Review of Military Participation in the Atomic Energy Program," 25 November 1949, File CCS 471.6, National Archives, Washington, D.C.

2. Interview, J. Carson Mark by Arthur Norberg, University of California, Berkeley, 28 March 1979, Bancroft Library. This interview may be also found in the oral history collection of the American Institute of Physics, Center for the History of Physics, New York, New York. For a good summary of other factors influencing the public attitude, see Norman Moss, *Klaus Fuchs, The Man Who Stole the Atom Bomb* (New York: St. Martin's Press, 1987), pp. 170–71.

3. Lewis L. Strauss, quoting a personal memo to the AEC, in *Men and Decisions* (New York: Doubleday and Company, 1948), p. 217.

4. Ibid.

5. *Zeit f. Phys.* 54, 656 (1928). Interview, Charles Critchfield by Lillian Hoddeson, 5 August 1980, Los Alamos Archives. See also Edward Teller with Allen Brown, *Legacy of Hiroshima* (Garden City, N. Y.: Doubleday and Company, 1962), pp. 34–36; and *A History of the Air Force Atomic Energy Program, 1942–1953*, Lee Bowen and Robert D. Little, et al., 1959, Vol. IV, p.

NOTES

175. See also Chuck Hansen's chapter on "Thermonuclear Weapons Development, 1942–1962," in *U.S. Nuclear Weapons* (New York: Orion Books, 1988), pp. 43–103.

6. Stanley A. Blumberg and Gwinn Owens, *The Life and Times of Edward Teller* (New York: G. P. Putnam's Sons, 1976), pp. 108–9.

7. Ibid., pp. 114–19. See also Excerpt of Memo: Edward Teller to Distribution, Subject: "Relations Between Various Activities of the Laboratory," by Samuel K. Allison, Central Technical Library, Sandia National Laboratories; and Hewlett and Anderson, Jr., *AEC History*, Vol. I, p. 104, and Frederick C. Alexander, Jr., Excerpt of Early Thermonuclear Weapons Development: The Origins of the Hydrogen Bomb, May 1969, SNL Archives.

8. Frederick C. Alexander, Jr., Excerpt of "Early Thermonuclear Weapons Development," pp. 9–12, SNL Archives; "Memorandum of the History of the Thermonuclear Program by Hans A. Bethe, 28 May 1952, RG 326, AEC, Coll: Secretariat, Box 4930 Weapons Development, DOE Archives, Germantown, Maryland, published as Hans A. Bethe, "Comments on the History of the H–Bomb," *Los Alamos Science* (Fall 1982), pp. 46–47; and J. Carson Mark, "A Short Account of Los Alamos Theoretical Work on Thermonuclear Weapons, 1946–1950," Los Alamos Archives; Don Shuster, "Notes on H–Bomb Development, Pre–Greenhouse and Ranger," Taped Commentary, December 1988, SNL Archives.

9. Strauss, *Men and Decisions*, p. 217.

10. Hewlett and Duncan, *The Atomic Shield*, pp. 375–85; Herbert York, *The Advisors, Oppenheimer, Teller, and the Superbomb*, which includes as an appendix "The GAC Report of October 30, 1949," pp. 150–59.

11. Strauss, *Men and Decisions*, p. 219. See also p. 222 in which he quotes from the memorandum sent to the President: "In sum, I believe that the President should direct the Atomic Energy Commission to proceed with all possible expedition to develop the thermonuclear weapon."

12. AEC, "Eighth Semiannual Report of the Atomic Energy Commission," July 1950 (Washington, D.C.: USGPO), p. ix. See also "Executive Secretary on United States Objectives and Programs for National Security," *A Report to the National Security Council*, 14 April 1950, Washington, D.C., copy #26 in SNL Archives. See also Williams, *Klaus Fuchs, Atom Spy*, pp. 78–88. For an in–depth and well–written narrative of the temper of the times, including the activities of Fuchs, Greenglass, Gold, and the Julius Rosenbergs, see Ronald Radosh and Joyce Milton, *The Rosenberg Files, A Search for the Truth* (New York: Holt, Rinehart, and Winston, 1983).

13. Strauss quotes Bacher in *Men and Decisions*, p. 227.

14. Strauss quotes Urey in *Men and Decisions*, pp. 227–28.

NOTES

15. Gordon Dean to Mr. LeBaron, Chairman MLC, 13 July 1950, RG 326 AEC, Coll.: Secretariat, Box 1228, Folder: Greenhouse, DOE Archives, Germantown, Maryland. See also Roger Anders, ed., *Forging the Atomic Shield, Excerpts from the Office Diary of Gordon E. Dean* (Chapel Hill: University of North Carolina Press, 1987), which also provides interesting commentary on this stage of Dean's career.

16. Interview, J. Carson Mark by Arthur Norberg, 28 March 1979, Bancroft Library.

17. Stanislaw Ulam, *Adventures of a Mathematician* (New York: Charles Scribners' Sons, 1976), pp. 217-20; Hewlett and Duncan, *AEC History*, Volume II., pp. 376-77; Bethe, "Memorandum on the History of the Thermonuclear Program," 28 May 1952, p. 4, RG 326 AEC, Coll.: Secretariat, Box 4930, Folder, "Weapons Research and Development," DOE Archives, Germantown, Maryland; Shuster, "Notes on H-Bomb Development," 13 December 1988, maintains that Ulam has not received the credit he deserves for his contributions to development of the Super.

18. Ulam, *Adventures of a Mathematician*, p. 220. See also Bethe, op. cit.

19. Edward Teller with Allen Brown, *The Legacy of Hiroshima* (New York: Doubleday and Company, 1962), p. 50. See also Chuck Hansen, *U.S. Nuclear Weapons* (New York: Orion Books, 1988), pp. 49-50.

20. Alvin C. Graves, Press Release, AEC, Washington, D.C., 13 June 1951, Los Alamos Archives.

21. *Announced United States Nuclear Tests, July 1945 through December 1982* (DOE, January 1983), p. 2; Memo Edward Teller to Norris E. Bradbury, Subject: "Plan for Setting Up a Separate Thermonuclear Division," 24 March 1951, RG 326 AEC, Coll.: Secretariat, Box 1235, Folder: "Los Alamos," Germantown, Maryland; Bethe, "Memorandum on the History of the Thermonuclear Program," 28 May 1952; Roger Anders, ed., *Forging the Atomic Shield*, pp. 68-69.

22. Memo, Teller to Bradbury, Subject: "Plan for Setting Up a Separate Thermonuclear Division," 24 March 1951, RG 326 AEC, Coll.: Secretariat, Box 1235, Folder: "Los Alamos," Germantown, Maryland; Gordon Dean Diary, entry for 18 September 1950, n. p., excerpt, RG 326, Coll.: "Dean Diary" (hereinafter cited as "Dean Diary") DOE Archives, Germantown, Maryland; "Frenchman's Flat Designated as U.S. Alternate Atom Site," *Las Vegas Review*, 1 May 1951, SNL Archives.

23. *Operation Ranger, Shots Able, Baker, Easy, Baker-2, Fox* January-6 February 1951, DNA 6022F, pp. 18-29; *Announced United States Nuclear Tests, July 1945 through December 1982*, NVO-209, January 1983, pp. 2-5; Sandia Corporation Telephone Directories, 1950-1954; Glenn A. Fowler Logbook, entries for 25-28 January 1951, SNL Archives.

NOTES

24. *Operation Ranger, Shots Able, Baker, Easy, Baker-2, Fox* January–6 February 1951, DNA 6022F, pp. 18–29; Glenn A. Fowler Logbook, entry for 28 January–3 February 1951, SNL Archives; Interview, Jim Scott, 9 April 1987, SNL Archives; Shuster, "Notes on H–Bomb Development," 13 November 1988, SNL Archives.

25. Shuster, "Notes on H–Bomb Development," 13 November 1988, SNL Archives.

26. Ibid.

27. Ibid.

28. Ibid.

29. Ibid.; Mel Merritt, "Weapons Test Experience," 3 January 1985, SNL Archives; "Sandia's Part in Nevada Tests Told: Project Ranger—We Were There," Sandia *Bulletin*, 16 March 1951, SNL Archives; *Operation Ranger*, DNA 6022F, pp. 30, 37, 41.

30. *Operation Ranger*, DNA 6022F, pp. 86–87.

31. Bethe, "Memorandum on the History of the Thermonuclear Program," 28 May 1952, pp. 7–8.

32. Memo, Teller to Bradbury, Subject: "Plan for Setting Up a Separate Thermonuclear Division," 24 March 1951, RG 326 AEC, Coll.: Secretariat, Box 1235, Folder: "Los Alamos," Germantown, Maryland.

33. Excerpts from the Gordon Dean Diary, entry for 18 September 1950, n. p., DOE Archives, Germantown, Maryland.

34. "Memorandum of Conversation with Lewis Strauss," Dean Diary, entry for 12 February 1951, DOE Archives, Germantown, Maryland; *Twelfth Annual Report of the Atomic Energy Commission*, USGPO (July 1952), p. 81.

35. *Operation Greenhouse*, DNA 6034F, p. 21.

36. Ibid., p. 228; *AEC History*, Vol. II, pp. 539–41; Ralph Carlisle Smith to Ted Sherwin, Proposed Release, 19 June 1951, Los Alamos Archives.

37. "Las Vegas Rocked by Greatest Atomic Blast," Los Angeles *Evening Herald Express*, 2 February 1951, p. 1, carries subtitle "Hits Like Quake, Windows Broken"; M. L. Merritt, "Weapons Effects Research at Sandia," p. 5, manuscript in SNL Archives; Banister, "History of the Field Engineering Directorate," p. 8, SNL Archives.

38. Merritt, "Weapons Effects Research at Sandia," p. 5; Banister, "History of the Field Engineering Directorate," p. 5, SNL Archives.

39. Ibid., p. 1; Interview, Harlan Lenander, 19 August 1986, SNL Archives.

NOTES

40. Lt. General Elwood R. Quesada, *History of Operation Greenhouse 1948-1951*, Joint Task Force Three, n.d., pp. 49, 177; Interview, Harlan Lenander, 19 August 1986, SNL Archives.

41. Quesada, *History of Operation Greenhouse*, p. 177; Interview, Harlan Lenander, 19 August 1986, SNL Archives; Interview, Jim Scott, 9 April 1987, SNL Archives.

42. *Operation Greenhouse*, DNA 6034F, p. 157; Interview, Harlan Lenander, 19 August 1986, SNL Archives.

43. Interview, Harlan Lenander, 19 August 1986, SNL Archives.

44. Ibid.; Quesada, *History of Operation Greenhouse*, p. 82, Central Technical File, SNL Technical Library.

45. Interview, Harlan Lenander, 19 August 1986, SNL Archives.

46. *Operation Greenhouse, 1951*, DNA 6034F, pp. 98-99, 103; AEC History, Vol. II, p. 542.

47. *Operation Greenhouse, 1951*, DNA 6034F, pp. 111, 119-20.

48. Mel Merritt, "Weapons Effects Research at Sandia," p. 3, manuscript in SNL Archives; Interview, Don Shuster, 13 July 1984, SNL Archives; *Operation Greenhouse, 1951*, DNA 6034F, p. 2 (showing Sandstone and Greenhouse detonation sites), SNL Archives.

49. Ibid., *Operation Greenhouse, 1951*, DNA 6034F, p. 121.

50. York, *The Advisors*, p. 81, quoting Oppenheimer.

51. See Teller with Allen Brown, *The Legacy of Hiroshima* (Garden City, N. Y.: 1962), pp. 52-53; Stanley A. Blumberg and Guinn Owens, *Energy and Conflict, The Life and Times of Edward Teller* (New York: G. Putnam's Sons, 1976), pp. 277-78; "In the Matter of J. Robert Oppenheimer," Transcript of Hearings, Personnel Security Board, Washington, D.C., 12 April 1954-6 May 1954, USGPO, 1954, p. 305, gives Dean's testimony.

52. York, *The Advisors*, pp. 81-82, in which he quotes from the official planning document. See also *IVY*, DNA 6036F, p. 252, which indicates that Sandia staff at Ivy numbered 67, and *Castle Series, 1954*, DNA 6035F, pp. 404, 169, 172-76.

53. Teller, *Legacy of Hiroshima*, pp. 53-55. See also Roger Anders, ed., *Forging the Atomic Shield, Excerpts from the Office Diary of Gordon Dean*, pp. 146-47, which note Dean's attendance at the Princeton conference, and pp. 154-55, which provide a concise summary of Teller's reasons for leaving Los Alamos and his quest for a second laboratory.

54. Teller, *Legacy of Hiroshima*, pp. 54, 58. Boyer "Diary," entry for 11 September 1951 notes Teller's resignation, U.S. DOE Archives, copy provided

NOTES

by Roger Anders, Historian DOE. See also Interview, J. Carson Mark by Arthur L. Norberg, 28 March 1979, University of California, Berkeley, Bancroft Library; Atomic Energy Commission Thermonuclear Research and Development," Note by the Acting Secretary, 5 April 1951, AEC 425, with enclosure A: Letter from Dr. Libby to Chairman of the AEC, 28 March 1951, AEC, RG 326, Coll.: Secretariat, Box 1235, Folder: 635.12 Los Alamos, DOE Archives, Germantown, Maryland.

55. Memo, Thomas E. Murray to Gordon Dean, Subject: "Thermonuclear Program," 21 June 1951, RG 326 AEC, Coll.: Secretariat, Box 4930, Folder: "Weapons Research and Development," DOE Archives, Germantown, Maryland; Memo, H. D. Smyth, Thomas E. Murray, and T. Keith Glennan, Subject: "Weapons Research and Development," 15 August 1951, AEC, RG 326, Coll.: Secretariat, Box 4930, Folder: "Weapons Development," DOE Archives, Germantown, Maryland. Dean subsequently took action by setting up a subcommittee to study the need for a new weapons research laboratory.

56. Herbert York, "The Origins of Lawrence Livermore Laboratory," *Bulletin of the Atomic Scientists* (September 1975), p. 11.

57. Ibid., p. 12.

58. *LLL, Twenty-Five Years*, preface, n.p. 1981; "In the Matter of J. Robert Oppenheimer" (Cambridge, Massachusetts: MIT Press, 1971), p. 248.

59. Anders, *Forging the Atomic Shield*, p. 201.

60. York, "The Origins of Lawrence Livermore Laboratory," *Bulletin of the Atomic Scientists* (September 1975), p. 13.

61. Gordon Dean to General Eisenhower, Augusta, Georgia, November 1952, AEC, Historical Document Number 356, DOE Archives, Germantown, Maryland.

62. Ibid., pp. 13–14; York, "The Origins of Lawrence Livermore Laboratory," *Bulletin of the Atomic Scientists* (September 1975), pp. 13–14.

CHAPTER 21:

1. *Twelfth Semiannual Report of the Atomic Energy Commission*, July 1952, pp. 79–81.

2. Ibid.; *Operation Ranger*, DNA 6022F, pp. 20–21; A. Constandina Titus, *Bombs in the Backyard, Atomic Testing and American Politics* (Reno and Las Vegas: University of Nevada Press, 1986), pp. 56–57. Titus, along with authors John G. Fuller (*The Day We Bombed Utah*) and Howard Ball (*Justice Downwind*) present thought-provoking coverage of the impact of

NOTES

continental testing on residents of the "downwind" states: Utah, Arizona, and Nevada. See especially Howard Ball, *Justice Downwind, America's Atomic Testing Program in the 1950s* (New York: Oxford University Press, 1986), pp. 59-83, for coverage of the continental test series.

3. *Announced United States Nuclear Tests, July 1945 Through December 1982*, NVO-209 (January 1983), p. 5. See also Miller, *Under the Cloud*, pp. 80-248.

4. See *Operation Buster-Jangle 1951*, DNA 6023F. Frank H. Shelton, who is in the process of finalizing his forthcoming manuscript "Reflections of a Weaponeer," discusses the Buster-Jangle operation on pages (5-14) through (5-25), of his draft script. Shelton, a Cal Tech physicist and former Sandian, provides an interesting firsthand perspective into the weapons effects experiments conducted during this period.

5. Mel Merritt, "Weapons Effects Research at Sandia," p. 6, manuscript in SNL Archives.

6. Ibid.; *Announced United States Nuclear Tests*, pp. 2-3; Shelton, "Reflections of a Weaponeer," p. (5-14).

7. Fowler *Logbook*, 5 June 1951, SNL Archives.

8. *Operation Buster-Jangle 1951*, DNA 6023F, pp. 93, 95. Interviews, Jack Howard, 5 September 1984, and Harlan E. Lenander, 19 August 1986, both in SNL Archives. See also W. J. Howard and R. D. Jones, "Free Air Pressure Measurements, Jangle Project 1.4" in *Operation Jangle, Blast and Shock Measurements II*, WT-367, 231 pp., Sandia Corporation, Washington, D.C.: AFSWP, WT-306, February 1952; H. E. Lenander, "Structure Instrumentation, Jangle Project 3.28," Sandia Corporation, Washington, D.C.: AFSWP, WT-406, October 1952.

9. Interview, Mel Merritt, 2 March 1988, SNL Archives.

10. *Operation Buster-Jangle* 1951, DNA 6023F, p. 22.

11. Ibid., pp. 25-27; Shelton, "Reflections of a Weaponeer," p. (5-14).

12. *Operation Buster-Jangle 1951*, DNA 6023F, pp. 27-28.

13. Ibid., p. 25; Memo, W. E. Treibel to Beryl Hefley, Subject: "Sandia Laboratories History," 29 March 1982, pp. 4-5, SNL Archives.

14. Memo, Treibel to Hefley, op. cit.

15. William L. Laurence, "Desert Blast Held 'Baby Bombs' Debut, Atomic Explosion . . . Termed Start of 'New Era' for Tactical Weapons," *New York Times* special (dateline 23 October 1951, Chicago), Los Alamos Archives; "Lawmaker Hails New A-Bomb, Gore Cites Possible Use in Far East," Albuquerque *Tribune*, 31 October 1951, "Medium Size A-Bomb Dropped in

NOTES

Nevada Test," Albuquerque *Journal*, 29 October 1951; "Medium Sized A-Bomb Heard 200 Miles Away," Boulder *Daily Camera*, 29 October 1951, SNL Archives.

16. "Lawmaker Hails New A-Bomb, Gore Cites Possible Use in Far East," Albuquerque *Tribune*, 31 October 1951, SNL Archives.

17. *Announced United States Nuclear Tests*, pp. 2-3.

18. "Sandians Study Wave Effects of H. E. and Nuclear Blast at Nevada Tests," Sandia *Bulletin*, 23 November 1951, SNL Archives; Alan Y. Pope, "Some Experiences With Atomic Weapons," 26 March 1956, pp. 6, 13-15, Manuscript, SNL Archives.

19. Merritt, "Weapons Effects Research at Sandia," pp. 6-7; Banister, "History of the Field Engineering Directorate," p. 8, SNL Archives.

20. Merritt, op. cit, 7. See also Shelton, "Reflections of a Weaponeer, pp. (5-13)-(5-22).

21. *Announced United States Nuclear Tests*, p. 3; *Shots Able, Baker, Charlie, and Dog, The First Tests of the Tumbler-Snapper Series, 1 April-1 May 1952*, DNA 6020F, pp. 11-13, and *Shots Easy, Fox, George, and How, The Final Tests of the Tumbler-Snapper Series, 7 May-5 June 1952*, DNA 6021F, pp. 10-13.

22. Kenner to Pat, Camp Mercury, Nevada (Las Vegas), 15 May 1952, SNL Archives.

23. Interview, Kenner F. Hertford, Albuquerque, New Mexico, 9 February 1989; Gordon Dean, Chairman AEC to the Honorable Robert A. Lovett, Secretary of Defense, Washington, D.C., 6 May 1952. Dean commended Hertford highly saying: "Colonel Hertford has already contributed much to the success of our test operations and, specifically, has been highly successful in organizing the current *Tumbler-Snapper* series of tests at the Nevada Proving Grounds." Dean concluded by classifying Hertford as "almost indispensable to our atomic weapons test program."

24. Interview, Kenner F. Hertford, op. cit.; Dean to Lovett, op. cit.; Memo, K. D. Nichols, Major General, U.S.A., Chief AFSWP to Chief of Staff, U.S. Army, Subject: "Recommendation for Promotion [Colonel Kenner F. Hertford]," Washington, D.C., 2 January 1951. Nichols' recommendation for promotion cited Hertford's "sound judgment and superior personal characteristics . . . [that] have enabled him to further develop cooperation between the officers and men of the three Services who make up the population of Sandia Base and with the civilian officials of the Atomic Energy Commission at Sandia Base and nearby Los Alamos."

25. Shelton, "Reflections of a Weaponeer," p. (5-24).

NOTES

26. *The First Tests of the Tumbler-Snapper Series, 1 April-1 May 1952*, DNA 6020F, pp. 37, 55, 93-94,159, quoting observer, *Announced United States Nuclear Tests*, NVO-209, p. 3.

27. *The First Tests of the Tumbler Snapper Series*, 1 April-1 May 1952, DNA 6020F, p. 37.

28. Banister, "History of the Field Engineering Directorate," p. 9, SNL Archives. Merritt, "Weapons Effects Research at Sandia," pp. 8-9.

29. Shelton, "Reflections of a Weaponeer," p. (5-25). Frank Shelton, who in 1955 was named Deputy Technical Director for AFSWP, was present as an observer at this meeting. See in addition to the previous reference, "F. H. Shelton to Take AFSWP Post in Washington," Sandia *Lab News*, 26 August 1955.

30. Shelton, op. cit.; Merritt, "Weapons Effects Research at Sandia," pp. 10-11.

31. Merritt, op. cit., 10.

32. Shelton, op. cit.

33. Upper Management Biographical File, SNL Archives; *Announced United States Nuclear Tests*, NVO-209, pp. 3-4; Interview, Luke Vortman, 14 March 1984, SNL Archives.

34. Alexander, *History of Sandia*, p. 51.

35. Ibid.

36. *Shots Annie to Ray, The First Five Tests of the Upshot-Knothole Series, 17 March-11 April 1953*, DNA 6017F, p. 17, and *Shots Encore to Climax, The Final Four Tests of the Upshot-Knothole Series*, 8 May-4 June 1953, DNA 6018F, pp. 11-16.

37. Alexander, op. cit.

38. "Outcasts of Yucca Flat; Mannequins are Martyrs for Science in 'Nuclear Diagnostic Shot,' " *Life*, 30 March 1953, pp. 24-25, quoting Alvin Graves, SNL Archives.

39. "It's H-Hour at the AEC's Nevada Proving Ground," Sandia *Bulletin*, 27 March 1953, pp. 1, 3, SNL Archives.

40. Ibid.

41. "Operation Upshot-Knothole," film taken by Bryan E. Arthur at Lookout Mountain Laboratory, 1952, Management Staff Files, SNL. See Frank Shelton, "The Precursor, Its Formation, Prediction, and Effects," Sandia Corporation, 27 July 1953, referred to on p. (5-25) of Shelton's *Reflections of a Weaponeer.*

42. Alexander, *History of Sandia*, p. 51.

NOTES

43. Ibid.; *The Final Teapot Tests 23 March 1955–15 May 1955*, DNA 6013F, pp. 4, 14–18.

44. Thomas B. Cook and Carter D. Broyles, Excerpt, "Curves of Atomic Weapons Effects for Various Burst Altitudes (Sea Level to 100,000 feet)," 9 March 1954, Central Technical File, Sandia Technical Library. Interviews, Tom Cook, 1 August 1984, and Carter Broyles, by Beryl Hefley, 18 March 1982, both in SNL Archives.

45. Banister, "History of the Field Engineering Directorate," pp. 10, 13, SNL Archives.

46. "Sandia Personnel Preparing for AEC Nevada Atomic Tests," Sandia *Lab News*, 28 January 1955, SNL Archives; "Sandia Corporation Role in A-Tests Big One," Albuquerque *Tribune*, 15 March 1955, SNL Archives.

47. Interview, Luke J. Vortman, 14 March 1984, SNL Archives; "Sandia Personnel Preparing for AEC Nevada Atomic Tests," Sandia *Lab News*, 28 January 1955, SNL Archives; Robert L. Corsbie (Director, Civil Effects Test Group) *Operation Teapot: Project Summaries of Civil Effects Tests*, including Program 34 "Shelters for Civilian Populations" by L. J. Vortman (Program Director), p. 21, Central Technical File, Sandia Technical Library. See also Program 34.2 "Investigation of Rise Time and Duration of Pressures in Certain Regions" (R. S. Millican, Project Officer), p. 24 and L. J. Vortman, "Evaluation of Various Types of Personnel Shelters Exposed to an Atomic Explosion" (May 1956), Central Technical File, Sandia Technical Library; L. J. Vortman, "Effects of a Non-Ideal Shock Wave on Blast Loading of a Structure," May 1956, pp. 1–46, Central Technical File, Sandia Technical Library. See also Banister, "History of the Field Engineering Directorate," pp. 6–7, SNL Archives; Samuel Glasstone and Philip J. Dolan, eds., *The Effects of Nuclear Weapons*, (Washington: DOD and ERDA, 1977).

48. "Sandians View First Shot of AEC's Operation Teapot," Sandia *Lab News*, 25 February 1955, SNL Archives.

49. Ibid.; Alexander, *History of Sandia*, p. 51.

50. "Baby Atom Blast Set Off in Nevada," Los Angeles *Times*, 20 February 1955, SNL Archives; "Atomic Blast Jolts Cities for 135 Miles," Norfolk-Virginia *Pilot*, 23 February 1955.

51. Sandia Corporation, Annual Report Excerpt, 1955, SNL Archives.

52. Ibid., pp. 9–10; James H. Scott to N. Furman, Albuquerque, New Mexico, 12 April 1989, SNL Archives.

53. Interviews, Harlan E. Lenander, 19 August 1986, and Jim Scott, 9 April 1987, SNL Archives.

NOTES

CHAPTER 22:

1. "Address Before the General Assembly of the United Nations on Peaceful Uses of Atomic Energy" (hereinafter referred to as Atoms for Peace), New York City, 8 December 1953, in *Public Papers of the Presidents of the United States: Dwight D. Eisenhower* (20 January–31 December 1953), pp. 813–22.

2. Ibid.

3. Paul Johnson, *Modern Times, The World from the Twenties to the Eighties* (New York: Harper and Row Publishers, 1983), p. 467; William Manchester, *The Glory and the Dream, A Narrative History of America 1932–1972*, Vol. II (Boston: Little, Brown and Company, 1973), p. 703. The U.S.S.R. announced detonation of a thermonuclear device on August 28, 1953.

4. For a concise analysis of this "dual attitude" toward atomic energy, see George T. Mazuzan and Samuel J. Walker, "A Short History of Nuclear Regulation 1946–1985," NRC: July 1985, pp. 1–3 (manuscript published by History Staff, NRC). Mazuzan and Walker point out that these "competing responsibilities and the precedence that the AEC gave to its military and promotional duties gradually damaged the agency's credibility on regulatory issues and undermined public confidence in its safety program." See p. 2. The first chapter of this overview is drawn from *Controlling the Atom: The Beginnings of Nuclear Regulation, 1946–1962* (Berkeley: University of California Press, 1984) by Mazuzan and Walker.

To implement a civilian nuclear power policy, Eisenhower supported amendment of the Atomic Energy Act of 1946. The Act as amended in 1954 "created a private power industry and permitted greater atomic cooperation with American allies." See *The United States Nuclear Power Policy 1954–1984: A Summary History* by Jack M. Holl, Roger Anders, and Alice Buck, History Division: U.S. DOE (October 1985), p. 1; and "Atomic Energy Commission History of Expansion of AEC Production Facilities," AEC 1140, 16 August 1963, p. 70, DOE Archives, RG 326, Coll: Secretariat, Box 1435, Folder: landP 14 History, DOE Archives, Germantown, Maryland. Senator Clinton P. Anderson, Chairman of the Joint Committee, again raised the possibility of a dual-purpose reactor in May 1956 while hearings were in progress on the Gore–Holifield bill to accelerate the civilian power reactor program. See p. 71. Allan A. Needell in "Nuclear Reactors and the founding of Brookhaven National Laboratory," *Historical Studies in the Physical Sciences* 14: 93–122 sees the establishment of BNL as an attempt by the government to control the technology of nuclear reactors and its first reactor project as an example of the tensions "between notions of good science and good management and between scientific independence and public accountability."

NOTES

5. Hewlett, AEC *History*, Vol. II, p. 593; Peter Pringle and James Spigelman, *The Nuclear Barons* (New York: Holt, Rinehart, and Winston, 1981), pp. 218-19; Eric F. Goldman, *The Crucial Decade* (New York: Alfred A. Knopf, 1956), pp. 260-62. In 1955 expenditures of the Atomic Energy Commission would reach the highest point since its establishment. See Warner R. Schilling, Paul Y. Hammond, and Glenn H. Snyder, *Strategy, Politics, and Defense Budgets* (New York: Columbia University Press, 1962) for an excellent analysis of the impact of Eisenhower's "New Look" on defense budgets of the 1954-1955 period.

Robert W. Seidel in "A Home for Big Science: The Atomic Energy Commission's Laboratory System," *Historical Studies in the Physical and Biological Sciences* 16:135-75, posits that Atoms for Peace had a negative impact on weapons laboratories because demands for AEC funding of "university" accelerators, rather than solely for the AEC laboratories, threatened "the scientific rationale" for their survival. See p. 156. For excellent coverage of the attempt to provide "Nuclear Power for the Marketplace," see Hewlett and Holl, *A History of the United States Atomic Energy Commission, 1952-1960* (Oak Ridge, Tennessee: Office of Science and Technical Information, 1989), pp. vii-1 through vii-19. Hewlett and Holl conclude that even by 1955 "nuclear power was not yet ready for the marketplace."

6. Sandia Corporation, *Annual Report* 1955, pp. 3-4.

7. For his work in developing radar countermeasure devices, McRae received the Legion of Merit. See "James W. McRae, Bell Telephone Laboratories, Inc." (biographical data) July 1953, AT&T Archives; "J. W. McRae, Former W. E. Vice-President and Sandia President, Dies Suddenly," *News Briefs for Western Electric Management* 13:5 (February 2, 1960), p. 2, AT&T Archives; "Dr. James W. McRae Transferred to Atomic Development Project," Madison *Eagle*, 6 August 1953, p. 2; "Defense Post to D. Quarles; J. McRae New Sandia Head," *Kearnygram*, August 1953, AT&T Archives (copies in SNL Archives).

8. "James W. McRae Takes Up New Duties as Laboratory President," Sandia *Bulletin*, 25 September 1953; Interview, Rosalie Crawford, 24 February 1989; Interview, Leon Smith, 7 January 1986, SNL Archives.

9. Sandia Corporation, Annual Report Excerpt, 1953, SNL Archives.

10. Telephone interview, Leon Smith, 17 March 1989; "First Report of Committee for the Review of Sandia Corporation Practices," 9 February 1954, with letter of transmittal dated 24 May 1954, contained in McRae Committee Collection (hereinafter cited as McRae Collection), SNL Archives. The transmittal letter gives date of establishment of the committee as 24 November 1953.

NOTES

11. "Minutes of the First Meeting of the McRae Committee," 9 December 1953, p. 1, McRae Committee Collection, SNL Archives.

12. "Minutes of the Fifteenth Meeting of the McRae Committee," 4 February 1954, p. 3, McRae Committee Collection, SNL Archives.

13. "Minutes of the Sixteenth Meeting of the McRae Committee," 11 February 1954, pp. 3-4, McRae Committee Collection, SNL Archives. See also "Second Report of Committee for Review of Sandia Corporation Practices," 17 May 1954, p. 14, McRae Committee Collection, SNL Archives.

14. "Minutes of the Sixteenth Meeting of the McRae Committee," 11 February 1954, p. 4, McRae Committee Collection, SNL Archives; Interview Bryan E. Arthur, 7 April 1987, SNL Archives; *Sandia Corporation Telephone Directories*, August 1954 and March 1956, SNL Archives. The Materials Standards Directorate retained the Environmental Test Laboratory Department, which in 1956 was headed by William A. "Bill" Gardner, who had succeeded T. B. Morse.

15. "Memorandum of Record," Subject: "Committee for Review of Corporation Practices, Minutes of the Seventeenth Meeting," 18 February 1954, p. 1, McRae Committee Collection, SNL Archives.

16. Ibid., p. 2.

17. "Reorganization Accompanies Vice-President Appointments," Sandia *Lab News*, 16 December 1955; Telephone Interview, Lou Hopkins, 3 February 1989; Sandia Corporation *Telephone Directories*, August 1954, July 1955, and March 1956, SNL Archives. At the time of this reorganization, the Board of Directors also approved appointment of R. P. Lutz to succeed Fred Schmidt as Vice-President of Operations. Ray Powell, who became Superintendent of Personnel and Public Relations at age 38, was the second youngest to reach the rank of Vice-President. Albert Naratn became the third when he was to promoted to Vice-President of Research at age 40.

18. "First Report of Committee for Review of Sandia Corporation Practices," 9 February 1954, p. 1, McRae Committee Collection, SNL Archives.

19. "Special Planning Meeting of Small Staff," 24-25 September 1958, SNL Archives.

20. Ibid.

21. Ibid.

22. "Minutes of the Nineteenth Meeting of the McRae Committee," 8 March 1954; "Minutes of the Fifteenth Meeting of the McRae Committee," 4 February 1954, both in McRae Collection, SNL Archives. A bomb book can be defined as the single document that lists the serial numbers of every component contained in that particular weapon so that problems arising later can be pinpointed in the production process and, therefore, suspect items in

NOTES

other weapons can be pulled and examined—a technique that has saved the AEC millions of dollars over the years.

23. AEC, "Thermonuclear Research at the University of California Radiation Laboratory," Report by the Director of Military Application, 13 June 1952, DOE Archives, RG 326, Collection: Secretariat, DOE Archives, Germantown, Maryland.

24. Teletype, N. E. Bradbury to Gordon Dean, Chairman AEC, Washington, D.C., 26 September 1951, AEC, RG 326, Collection: Secretariat, Box 4930, Folder: Weapon Development, DOE Archives, Germantown, Maryland.

25. "Special Planning Meeting of Small Staff," 24–25 September 1958 (Scopes), Corporate File, Sandia National Laboratories.

26. Interview, R. L. Peurifoy, Jr., 17 October 1985, SNL Archives; Information Research Division, Excerpt of History of Early Thermonuclear Weapons," SNL Archives. While there is no author listed on this report, it was more than likely compiled by Frederick C. Alexander, Jr., Sandia's first historian, or a member of his staff.

27. "The Thermonuclear Weapons Program at the Air Force Special Weapons Center," July 1952–July 1954, pp. 3–5, excerpt in Alexander Collection, SNL Archives.

28. "The Thermonuclear Weapons Program at the Air Force Special Weapons Center," July 1952–July 1954, Excerpt in Alexander Collection, SNL Archives; Information Research Division, "History of Early Thermonuclear Weapons," SNL Archives; Hansen, *U.S. Nuclear Weapons, The Secret History*, p. 145.

Special credit goes to Jim Cocke for compiling the Weapons Directorate History Collection. These concise summaries of the different weapon systems have been most helpful. See Cocke, "Weapons Directorate History," (Introduction) pp. 1–2, SNL Archives.

29. "The Thermonuclear Weapons Program at the Air Force Special Weapons Center," July 1952–July 1954, Excerpt in Alexander Collection, SNL Archives; Interview, Jim DeMontmollin, 14 August 1988, SNL Archives.

30. Interview, Colonel Charles G. Mathison and William A. Gardner, 7 December 1984, SNL Archives. Colonel Mathison was Base Commander at Kirtland Air Force Base during the years 1971–1973. In addition to his work with the Special Weapons Center, he has served as Director of Ballistic Missile Testing at Patrick Air Force Base, the Rocket Propulsion Lab, and the Air Force Systems Command; "The Thermonuclear Weapons Program at the Air Force Special Weapons Center," July 1952–July 1954, Excerpt in Alexander Collection, SNL Archives; Information Research Division, "History of Early

NOTES

Thermonuclear Weapons," p. 24, SNL Archives; Hansen, *U.S. Nuclear Weapons, The Secret History*, p. 145; Cocke, "Weapons Directorate History" (Thermonuclear), p. 4, SNL Archives.

31. Interview, Mathison and Gardner, 7 December 1984, SNL Archives.

32. Ibid.; Information Research Division, Excerpt of History of Early Thermonuclear Weapons, SNL Archives.

33. Hansen, *U.S. Nuclear Weapons, The Secret History*, p. 147; Memo, J. W. McRae to D. J. Leehy, SFO, Excerpt: "Contact Fuzing for Two-Stage Weapons," Central Technical File, Folder F-4, 1954, Sandia Technical Library; Excerpt, Development Progress Report No. 4, "Emergency Capability of Thermonuclear Weapons," Alexander Collection, SNL Archives; Cocke, "Weapons Directorate History" (Thermonuclear), p. 5, SNL Archives.

34. Hansen, op. cit., pp. 145-46; "Minutes of Twenty-Third Meeting of the McRae Committee," 12 March 1954, pp. 1-2, McRae Committee Collection, SNL Archives; Cocke, "Weapons Directorate History" (Thermonuclear), p. 3, SNL Archives.

35. "Minutes of the Twenty-Third Meeting of the McRae Committee," op. cit, p. 2; Interview, R. W. Henderson, 30 October 1985, SNL Archives.

36. Cocke, "Weapons Directorate History" (Thermonuclear), pp. 3-4, SNL Archives; Hansen, op. cit, p. 146; Information provided by Randy Maydew, 5 May 1989, SNL Archives.

37. Ibid., pp. 148-54; Interview, R. W. Henderson, 29 May 1987, SNL Archives; Cocke, "Weapons Directorate History" (Introduction), p. 1; (History of the Mk 25 Warhead), pp. 3-4; (Mk 28 Weapon), pp. 1-2, SNL Archives; *Operation Redwing* 1956, DNA 6037F, p. 2; Sandia Corporation, Annual Report Excerpt, 1956, p. 9, SNL Archives.

38. Excerpt, Bradbury to Leehy, Archives TX-14, 1-52 through 54, Alexander Collection, and Bradbury to Fields, 21 May 1954, Archives TX-14, 1-52 through 54, Alexander Collection, SNL Archives; Cocke, "Weapons Directorate History," p. 3, SNL Archives.

39. Excerpt, Bradbury to Fields, 21 May 1954, Archives TX-14, 1-52 through 54, Alexander Collection, SNL Archives.

40. John A. Hornbeck, "*SANDIA,*" *The Western Electric Engineer*, April 1967, p. 2, reprint in SNL Archives.

41. Ibid.; Information Research Division, Excerpt, Mk 43 Bomb, January 1948, SNL Archives.

42. R. S. Claassen, *New Weapon Proposal: Towed Laydown Weapon*, Technical Memorandum, Sandia Corporation, 21 June 1957; Interview, Don

NOTES

Cotter by Frederick Alexander, 22 December 1961, SNL Archives; Information Research Division, "History of the Mk 43 Bomb," January 1948, SNL Archives.

43. Harry Saxton, Component History Collection, SNL Archives. Under the organizational expertise of Jack Marron and Sandra Barber, directorate representatives have put together an excellent compilation of materials on the history of components, which will be most valuable for use in volume II of the history of Sandia. See J. J. Marron to Furman, 7 January 1985, Saxton Component History Collection, SNL Archives. Interview, Richard S. Claassen, 25 March 1985, SNL Archives.

44. Charles H. Kaman, *Kaman, Our Early Years* (Indianapolis: Curtis Publishing Company, 1985), pp. 110, 113.

45. Interview, Randy Maydew, 22 August 1985.

46. Ibid.

47. Ibid.; R. C. Maydew, "Feasibility of Parachutes for Low-Level Delivery," 19 January 1954, pp. 3-5, Central Technical File, SNL Technical Library.

48. Interview, Richard S. Claassen, op. cit.; Interview, R. S. Claassen by Bruce Hawkinson for Sandia *Lab News*, 17 March 1987, and "Sandia and the Nuclear Deterrent: Extra Dimensions of the 'Can-Do' Ethos," Sandia *Lab News*, pp. 12-13.

49. Interview, Randy Maydew, op. cit.

50. Ibid.; R. C. Maydew, "Chronology of Parachute R&D at SNL," draft report, Aerodynamics Department Files, Sandia National Laboratories.

51. Kaman, *Kaman, Our Early Years*, pp. 109-10; Randy Maydew to N. Furman, 26 April 1989, SNL Archives.

52. Kaman, *Kaman, Our Early Years*, pp. 109-11.

53. Interview, Randy Maydew, op. cit. See also T. Blanchard, Jr., "Bimonthly Progress Report of LLD Parachute Research Drop Test Program (September-October 1956)," Central Technical File, Sandia Technical Library. The abstract for this report notes that "results of the thirty-three drop tests covering the period October 1955 through August 1956 are summarized in the Appendix."

54. Interview, Randy Maydew, op. cit. As a three-star general, Forrest McCartney would become director of NASA's Cape Kennedy . . . responsible for space shuttle launches.

55. Ibid.; Maydew, "Chronology of Parachute R&D at SNL," draft report, Aerodynamics Department Files, Sandia National Laboratories; Cocke, "Weapons Directorate History," "Mk 43 Bomb," pp. 1-3, SNL Archives;

NOTES

Information Research Division, Excerpt, Mk 43 Bomb, January 1948, SNL Archives.

56. W. J. "Jim" Cocke, "Quest for the Wooden Bomb," 29 September 1988, Summary Paper, SNL Archives. See also W. L. "Bill" Stevens, "On the Characteristics of Nuclear Weapons," another valuable compilation of historical data put together for use by the Sandia History Project.

See also "Weapon Longevity is Engineered," Sandia *Lab News*, 10 October 1986, p. 7, quoting Herman Mauney and R. L. Peurifoy, Jr.

57. Cocke, "Quest for the Wooden Bomb," op. cit.; "Working Toward the Maintenance-Free Weapon," Sandia *Lab News*, 10 October 1986, p. 8, quoting R. L. Peurifoy, Jr.; Interview, R. L. Peurifoy, Jr., October 16, 1985, SNL Archives.

58. Memo to Chief, DASA, Washington, D.C., Subject: "Basic Maintenance Philosophy of Defense Atomic Support Agency," 5 August 1954, Central Technical File, SNL Technical Library.

59. "Working Toward the Maintenance-Free Weapon," op. cit.

60. Interview, Jim DeMontmollin, 14 August 1988, SNL Archives; Sandia Corporation, Annual Report Excerpt, 1955, SNL Archives. Donald R. Cotter, *Managing Nuclear Operations* (Washington, D.C.: Brookings Institute, 1987), p. 28, explains concisely and clearly how the Los Alamos development of "boosting" (the technique in which a mix of deuterium and tritium gas introduced into the nuclear system) resulted in a more efficient burn of the fissile material and a tripling of the number of weapons within the "inventory of special nuclear materials." He notes that it was this development that led to need for permanent installation of the nuclear material in high explosives and the "sealed-pit warhead."

61. Interview, R. L. Peurifoy, Jr., 16 October 1985, SNL Archives; W. J. "Jim" Cocke, "Quest for the Wooden Bomb," cp. cit.; "Working Toward the Maintenance Free Weapon," op, cit.; Interview, Jim DeMontmollin, 14 August 1988, SNL Archives.

62. B. H. VanDomelen and R. D. Wehrle, "A Review of Thermal Battery Technology," pp. 665–70, offprint in SNL Archives. See also "Overview of Component History," Manuscript draft prepared for Sandia History Project by Arthur Wright, Summer 1988. Arthur Wright, who worked for the History Project during the summers of 1987 and 1988 also provided valuable assistance with compilation of the bibliography and the collection and organization of news clippings relating to testing activities during the fifties.

63. Van Domelen and Wehrle, op. cit.; Wright, op. cit.; R. D. Wehrle and R. K. Quinn, "Long-Life Thermal Battery Development," *Sandia Technical Review*, Vol. 9 (January 1979), p. 1.

NOTES

64. R. D. Wehrle, "History of Power Sources at Sandia, Part I," p. 1, SNL Archives.

65. Ibid., pp. 1-2.

66. Ibid.; W. L. Stevens, "A Brief History of Early Component Development Activities," draft working paper, SNL Archives; Wright, "Overview of Component History," op. cit.

67. J. J. Marron to Furman, 7 January 1985, Harry Saxton Component History Collection, SNL Archives.

68. Ibid.

69. R. S. Claassen, "Presentation to M. J. Kelly," 16 March 1957, p. 14, Claassen Collection, SNL Archives.

70. Interview, Frank Hudson, 31 March 1989, SNL Archives; "Biographical Material," Frank Hudson Collection, SNL Archives.

71. Interview, Frank Hudson, 31 March 1989, SNL Archives; "Biographical Material," op. cit.

72. Memo, Frank Hudson to K. W. Erickson, Subject: "A Systematic Program for Component Studies," 26 August 1954, Frank Hudson Collection, SNL Archives.

73. Ibid.

74. Ibid.

75. Ibid.

76. "Personnel Status Report, Department 5110 and 5130," 16 January 1956, Frank Hudson Collection, SNL Archives.

77. Interview, Tom Cook, 1 August 1984, SNL Archives.

78. Memo, K. W. Erickson to R. S. Claassen, Subject: "Future Research Program," 7 August 1956, Frank Hudson Collection, SNL Archives.

79. George Anderson, Diary, entries May–June 1956, Saxton Component Collection, SNL Archives.

80. Memo, K. W. Erickson to G. A. Fowler through S. C. Hight, Subject: "Future Research Program," 7 August 1956, op. cit.; Memo, K. W. Erickson to R. S. Claassen through S. C. Hight, Subject: "Future Research Program," 7 August 1956, op. cit. See also Dick Claassen to Frank Hudson, Shirley, New York, 19 September 1987, Frank Hudson Collection, SNL Archives, in which he commends him for the "central contributions which you made at a critical time in Sandia's evolution."

81. Glenn A. Fowler to Necah S. Furman, 7 May 1989, SNL Archives.

82. Dick Claassen to Necah Furman, 1 May 1989, SNL Archives.

NOTES

83. "Minutes of FOG Meeting of 11 February 1957," Frank Hudson Collection, SNL Archives. See also FOG Minutes of 12 February 1957; and Memo, Hudson to Fowler, Subject: "Research Planning, Phase II," 18 February 1957, Frank Hudson Collection, SNL Archives; Fowler to Distribution, 18 February 1957, including draft of Memo to McRae, Subject: "The Role of Research at Sandia," 4 February 1957, Frank Hudson Collection, SNL Archives; Dick Claassen to N. Furman, 1 May 1989, SNL Archives.

84. Fowler to Distribution, 18 February 1957, includes draft of presentation to Small Staff; see p. 3, Frank Hudson Collection, SNL Archives.

85. Ibid.

86. Memo, Claassen to Fowler, Subject: ",Meeting with Dr. Kelly on March 16, 1957," 20 March 1957, with Claassen presentation attached, R. S. Claassen Collection, SNL Archives.

87. Ibid.

88. Ibid.

89. Ibid.

90. Interview, Frank Hudson, 31 March 1989, SNL Archives. Memo, Claassen to Fowler, Subject: "Meeting with Dr. Kelly on March 16, 1957," 20 March 1957, R. S. Claassen Collection, SNL Archives.

91. "New 5100 Department Organized for Fundamental Research at Sandia," Sandia *Lab News*, 6 September 1957, Frank Hudson Collection, SNL Archives.

92. Interview, Robert W. Henderson, 29 May 1987, SNL Archives.

93. Ibid.; Interview, Barry Schrader, 15 September 1988, SNL Archives.

94. *History 8100 Directorate*: "The First Thirty Years . . . ," compiled by Donald Bohr, Barbara Kerr, Captain (U.S. Army) Allen Strouphauer, and Donald Gregson, SNL Archives, provides excellent coverage of the early Livermore Lab.

95. Ibid., pp. 1–4. See also Sandia *Lab News*, 23 March 1956, SNL Archives.

ARCHIVAL DEPOSITORIES

American Institute of Physics. Center for History of Physics. New York, New York.

AT&T/Bell Laboratories Archives and Records Center. Warren, New Jersey.

Bancroft Library. University of California, Berkeley. Berkeley, California.

Department of Energy, Albuquerque Operations. Records Management Center. Albuquerque, New Mexico.

Department of Energy, Headquarters. Historical Archives. Germantown, Maryland.

Franklin D. Roosevelt Library. Hyde Park, New York.

General Services Administration. National Archives and Records Administration (National Archives). Washington, D.C.

Harry S. Truman Library. Independence, Missouri.

Lawrence Livermore National Laboratory Archives. Livermore, California.

Los Alamos National Laboratory Records Center and Archives (L.A. Archives). Los Alamos, New Mexico.

Maxwell Air Force Base Archives. Alabama USAF History Division.

Sandia National Laboratories. Central Technical File. Albuquerque, New Mexico.

Sandia National Laboratories. Corporate History Archives (SNL Archives). Albuquerque, New Mexico.

Sandia National Laboratories. Legal Files. Albuquerque, New Mexico.

Sandia National Laboratories. Public Relations Collection. Livermore, California.

BIBLIOGRAPHY

Interviews

Arthur, Bryan E. Interview with author. 7 April 1987. SNL Archives.

Austin, Howard. Interview with author. 19 October 1983. SNL Archives.

Bacher, Robert. Interview with Lillian Hoddeson. 3 March 1986. Los Alamos Archives.

Barncord, Charles R., and Ray Brin. Interview with author. No date. SNL Archives.

Benjamin, Ben. Telephone interview with author. 22 November 1983.

Bice, Richard A. Interview with author. 21 June 1984. SNL Archives.

──────. Interview with author. 28 June 1984. SNL Archives.

Biggs, Bernard S. Interview with author. 20 September 1984. SNL Archives.

Boyes, W. E. Telephone Interview with author. 29 June 1988. SNL Archives.

Bradbury, Norris. E. Interview with Arthur Lawrence Norberg. 11 February 1976. Bancroft Library, AIP Center for History of Physics, SNL Archives.

──────. Interview with author. 8 August 1984. SNL Archives.

Bramlett, William O., Marvin H. Brown, Robert L. Stewart, George Davies, and Albert Joe Angel. Interview with author. 24 January 1986. SNL Archives.

Bramlett, William O. Telephone interview with author. 7 March 1988. SNL Archives.

Brin, Ray. Interview with author. No date. SNL Archives.

Broyles, Carter D. Interview with Beryl Hefley. 18 March 1982. SNL Archives.

Caldes, William. Telephone interview with author. 15 December 1983, SNL Archives.

BIBLIOGRAPHY

Campbell, Charles W. Taped Recollections for the Sandia History Project. 31 August 1984. SNL Archives.

Claassen, Richard S. Interview with author. 25 March 1985. SNL Archives.

———. Interview with Bruce Hawkinson for Sandia *Lab News*. 17 March 1987. SNL Archives.

———. Interview with author. 21 August 1987. SNL Archives.

Cook, Thomas B. Interview with author. 1 August 1984. SNL Archives.

Cotter, Don. Interview with Frederick Alexander. 22 December 1961. SNL Archives.

Crawford, Rosalie Franey. Interview with author. 2 July 1984. SNL Archives.

———. Interview with author. 24 February 1989. SNL Archives.

Critchfield, Charles. Interview with Lillian Hoddeson. 5 August 1980. Los Alamos Archives.

DeMontmollin, Jim. Interview with author. 14 August 1988. SNL Archives.

Duggin, Frank, Jr. Interview with George Kupper. 18 July 1986. SNL Archives.

Ferguson, Waylon. Interview with George Kupper. 18 July 1986. SNL Archives.

Folley, Harold. Interview with author. Albuquerque, New Mexico. 20 November 1986.

Fowler, Glenn A. Interview with author. 17 May 1983. SNL Archives.

———. Interview with Bruce Hawkinson. Sandia *Lab News*. 12 June 1983. SNL Archives.

———. Interview with author. 20 December 1983. SNL Archives.

———. Interview with author. 31 January 1984. SNL Archives.

Fuller, Lloyd. Interview with Ted Alexander. 6 October 1961. SNL Archives.

Greher, Larry, and Kurt Olsen. Interviews with author. Albuquerque, New Mexico. 7 October 1986. SNL Archives.

BIBLIOGRAPHY

Gunn, Hal F. Interview with Sandia *Lab News*. No date. SNL Archives.

Hamilton, Ira. Interview with author. 16 November 1984. SNL Archives.

Harris, Virgil. Interview with Frederick C. Alexander, Jr. 26 September 1960. SNL Archives.

Heilman, Luther J. Interview with author. 22 June 1986. SNL Archives.

Henderson, Robert W. Interview by Hank Willis. No date. SNL Archives.

──────────. Interview with *Lab News*. "Bob Henderson to Take Early Retirement." Sandia *Lab News*. 18 January 1974.

──────────. Interview with author. 23 January 1984. SNL Archives.

──────────. Interview with author. 7 March 1985. SNL Archives.

──────────. Interview with author. 30 October 1985. SNL Archives.

──────────. Interview with author. 26 September 1986. SNL Archives.

──────────. Interview with author. 29 May 1987. SNL Archives.

──────────. Interview with author. 26 April 1988. SNL Archives.

Herrity, Ed. Interview with George Kupper. 28 July 1986. SNL Archives.

Hertford, Kenner F. Interview with author. 9 February 1989. SNL Archives.

Hildebrandt, George. Interview with author. 5 May 1984. SNL Archives.

Hinson, James. Interview with author. 7 March 1988. SNL Archives.

Hirsch, F. G. Interview. January 1962. SNL Archives.

Hollowwa, G. C. Interview with author. 15 December 1983. SNL Archives.

──────────. Interview with author. 23 January 1984. SNL Archives.

──────────. Interview with author. 10 September 1984. SNL Archives.

──────────. Interview with author. 23 June 1986. SNL Archives.

Hopkins, Louis A. Interview with author. 13 April 1987. SNL Archives.

BIBLIOGRAPHY

―――――. Telephone interview with author. 3 February 1989. SNL Archives.

Howard, W. J. Interview with author. 5 September 1984. SNL Archives.

Hudson, Frank. Interview with author. 31 March 1989. SNL Archives.

Hughes, Jay. Interview with George Kupper. 2 December 1986. SNL Archives.

Jacobvitz, Leonard. Interview with author. Albuquerque, New Mexico. 1 October 1985. SNL Archives.

Jacot, Louis. Interview in Sandia *Lab News*. 15 July 1955.

Jercinovic, Leo M. "Jerry." Interview with author. No date. SNL Archives.

―――――. Interview with author. 21 January 1982. SNL Archives.

Johnson, Sam. Interview with author. 2 October 1982. SNL Archives.

Kistiakowsky, G. B. Interview (typescript). No date. L. A. Archives.

Kraft, Bill. Interview with author. 24 August 1984. SNL Archives.

Lenander, Harlan E. Interview with author. 19 August 1986. SNL Archives.

Lieber, Andrew. Interview with author. 8 June 1988. SNL Archives.

Little, W. A. Interview with Beryl Hefley. No date. SNL Archives.

Machen, Arthur B. Interview with author. Rogue River, Oregon. 4 March 1984. SNL Archives.

Management interviews with author. "Ethos" Corporate Culture Study. October 1985. SNL Archives.

Mark, J. Carson. Interview with Arthur Lawrence Norberg. 29 March 1979. Bancroft Library, University of California, Berkeley.

Mathison, Colonel Charles G. and William A. Gardner. Interview with author. 7 December 1984. SNL Archives.

Mauldin, G. Howard. Interview with author. 18 December 1984. SNL Archives.

Maydew, Randy. Interview with author. 22 August 1985. SNL Archives.

McDonald, G. Corry. Interview with author. 11 February 1986. SNL Archives.

BIBLIOGRAPHY

Merritt, Mel. Interview with author. 2 March 1988. SNL Archives.

Miller, Gordon. Interview with author. 11 January 1983. SNL Archives.

Moeding, Henry. Inteview with George Kupper. 18 September 1986. SNL Archives.

Neibel, Larry and William O'Neill. Interview with author. 11 February 1985. SNL Archives.

Olson, Del. Interview with author. 19 August 1985. SNL Archives.

Pace, Tom and Howard Austin. Interview with author. 19 October 1983. SNL Archives.

Paddison, Louis. Interview with author. 16 April 1985. SNL Archives.

Peurifoy, R. L., Jr. Interview with author. 16 October 1985. SNL Archives.

Powell, Ray B. Interview with Beryl Hefley. No date. SNL Archives.

——————. Interview with author. 23 February 1984. SNL Archives.

Rowe, James Les. Interview with author. Livermore, California. 18 September 1984. SNL Archives.

——————. Interview with Sam Johnson. 7 February 1985. SNL Archives.

Schrader, Barry. Interview with author. 15 September 1988. SNL Archives.

Schulz, Lee. Interview with Fred Duimstra. 24 February 1985. SNL Archives.

Scott, Jim. Interview with author. 9 April 1987. SNL Archives.

Shuster, Don. Taped Commentary. "Notes on H–Bomb Development, Pre–Greenhouse and Ranger." December 1988. SNL Archives.

Sherwin, Ted B. Interview conducted by Hank Willis with Bill Carstens and Jim Mitchell. 13 February 1985. SNL Archives.

Sims, Ed. Interview with Sandia *Lab News*. No date. SNL Archives.

Smith, Leon. Interview with author. 19 January 1984. SNL Archives.

―――――. Interview with author. 7 January 1986. SNL Archives.

―――――. Telephone interview with author. 17 March 1989. SNL Archives.

Stewart, Robert L. Interview with Sandia *Lab News*. No date. SNL Archives.

Stromberg, Robert P. Telephone interview with author. 26 January 1988. SNL Archives.

Sundberg, John. Interview with author. 30 April 1985. SNL Archives.

Vortman, Luke. Interview with author. 14 March 1984. SNL Archives.

Wesson, Phillip D. Interview with Ted Alexander. 17 October 1961. SNL Archives.

Whelchel, Lyle. Interview with George Kupper. 20 September 1986. SNL Archives.

Willis, Hank. Interview with author. 13 February 1985. SNL Archives.

―――――. Interview with author. 17 February 1985. SNL Archives.

―――――. Interview with author. 5 March 1985. SNL Archives.

Winter, Charlie. Interview with author. 16 April 1985. SNL Archives.

Correspondence

Alexander, Frederick C., Jr. Letter to Hartley Rowe. Boston, Massachusetts. 21 November 1961. SNL Archives.

Anderson, H. C., Jr. Memo to Walter L. Brown. "Atomic Energy Commission Contract." 7 September 1949. SNL Archives.

Ashworth, Capt. F. L. and R. B. Brode. Letter to William S. Parsons. 6 November 1944. *Daily Log #2*. SNL Archives.

Atomic Energy Commission, Washington, D.C. Memo to Atomic Energy Commission, Sandia Base. 12 August 1951. SNL Archives.

Bainbridge, K. T. Memo to Joseph Hirschfelder. "Fragment Sizes, Velocity, Ranges." 29 August 1944. L. A. Archives.

BIBLIOGRAPHY

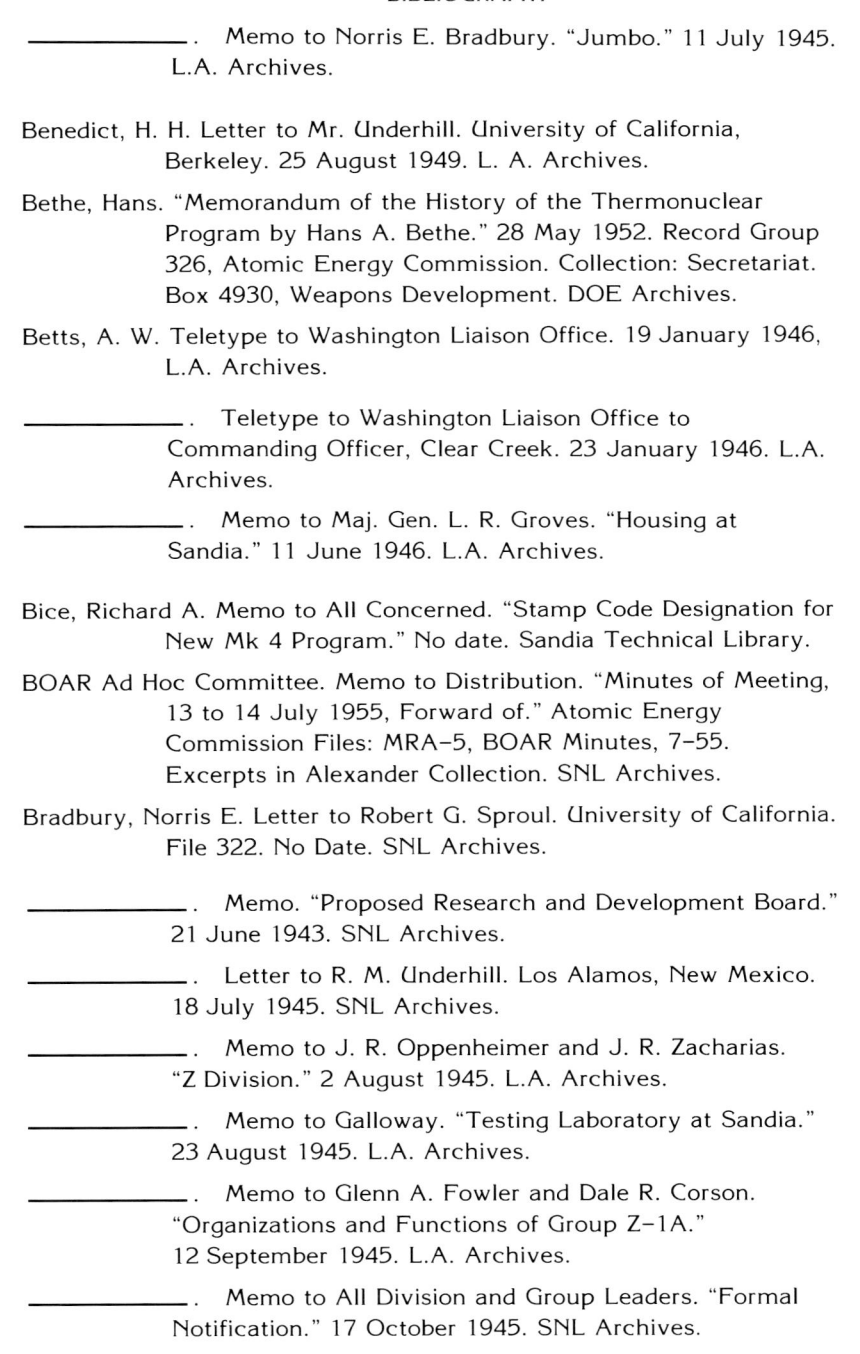

———. Memo to Norris E. Bradbury. "Jumbo." 11 July 1945. L.A. Archives.

Benedict, H. H. Letter to Mr. Underhill. University of California, Berkeley. 25 August 1949. L. A. Archives.

Bethe, Hans. "Memorandum of the History of the Thermonuclear Program by Hans A. Bethe." 28 May 1952. Record Group 326, Atomic Energy Commission. Collection: Secretariat. Box 4930, Weapons Development. DOE Archives.

Betts, A. W. Teletype to Washington Liaison Office. 19 January 1946, L.A. Archives.

———. Teletype to Washington Liaison Office to Commanding Officer, Clear Creek. 23 January 1946. L.A. Archives.

———. Memo to Maj. Gen. L. R. Groves. "Housing at Sandia." 11 June 1946. L.A. Archives.

Bice, Richard A. Memo to All Concerned. "Stamp Code Designation for New Mk 4 Program." No date. Sandia Technical Library.

BOAR Ad Hoc Committee. Memo to Distribution. "Minutes of Meeting, 13 to 14 July 1955, Forward of." Atomic Energy Commission Files: MRA-5, BOAR Minutes, 7-55. Excerpts in Alexander Collection. SNL Archives.

Bradbury, Norris E. Letter to Robert G. Sproul. University of California. File 322. No Date. SNL Archives.

———. Memo. "Proposed Research and Development Board." 21 June 1943. SNL Archives.

———. Letter to R. M. Underhill. Los Alamos, New Mexico. 18 July 1945. SNL Archives.

———. Memo to J. R. Oppenheimer and J. R. Zacharias. "Z Division." 2 August 1945. L.A. Archives.

———. Memo to Galloway. "Testing Laboratory at Sandia." 23 August 1945. L.A. Archives.

———. Memo to Glenn A. Fowler and Dale R. Corson. "Organizations and Functions of Group Z-1A." 12 September 1945. L.A. Archives.

———. Memo to All Division and Group Leaders. "Formal Notification." 17 October 1945. SNL Archives.

BIBLIOGRAPHY

―――――. Memo to Dale R. Corson. "Organization of Z Division While Naval Test is in Progress." 27 March 1946. L.A. Archives.

―――――. Memo to All Division Leaders. "B Division." 18 October 1946. SNL Archives.

―――――. Memo to Paul J. Larsen. 10 November 1947. L.A. Archives.

―――――. Memo to Distribution. "Z Organization." 4 December 1947. SNL Archives.

―――――. Memo to All Group and Division Leaders. 26 July 1948. SNL Archives.

―――――. Memo to P. J. Larsen. "Sandia Branch of Los Alamos Scientific Laboratory." 7 October 1948. L.A. Archives.

―――――. Letter to Robert G. Sproul. 18 November 1948. SNL Archives.

―――――. Letter to R. E. Poole. "A Priori Reliability of Atomic Bomb." 10 October 1950. SNL Archives.

―――――. "TR Hot Run." Enclosure to author from Arthur B. Machen. Rogue River, Oregon. 4 March 1984. SNL Archives.

―――――. Teletype to Gordon Dean, Chairman, Atomic Energy Commission. 26 February 1951. Atomic Energy Commission Record Group 326. Collection: Secretariat. Box 4930. Folder: Weapon Development. DOE Archives.

Brode, R. B. Memo to Group O-3. "Immediate Program." 17 August 1945. SNL Archives.

Brown, Walter L. Letter to H. C. Anderson. 21 July 1949. SNL Archives.

Buckley, Dr. O. E. to Harry Truman. Washington, D.C. 17 May 1949. Atomic Energy Commission Files. Series 199. ALO. Copy in SNL Archives.

Bunker, Howard P. Letter to Chief of Staff, USAF, Washington, D.C. Folder. "Documents on the Air Force Tactical and Liaison Committee, 1946–48." No date. Kirtland Air Force Base, NM.

―――――. Memo to Maj. Gen. Schlatter. "Sandia Lab Proposed Contractor Changes." KAFB Files. General Correspondence. Office of Atomic Energy. 12 August 1948. SNL Archives.

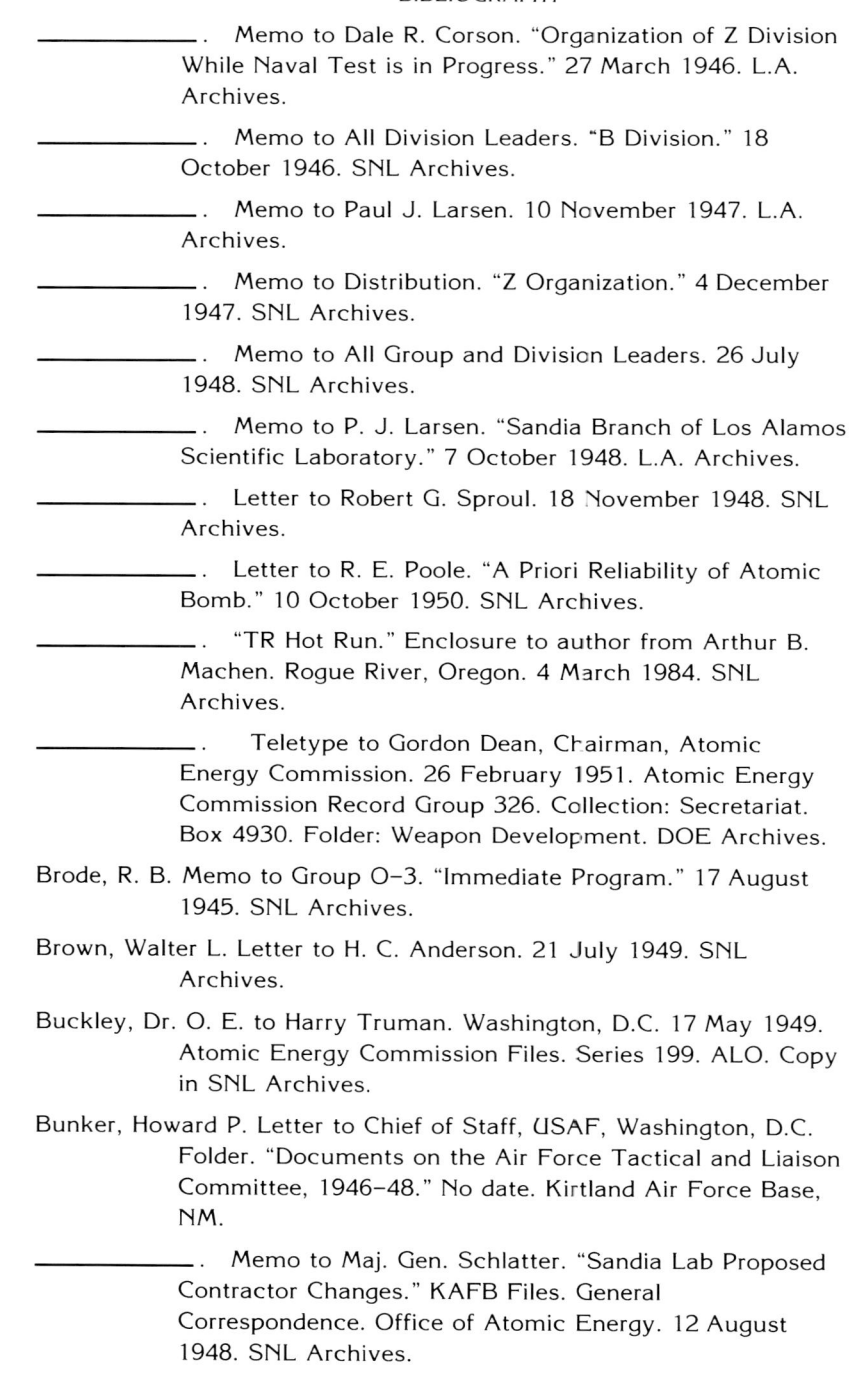

BIBLIOGRAPHY

Burke, William. Memo Enclosure to JCS. "Custody of Atomic Weapons." 13 July 1948. National Archives. Record Group 218. CCS 471.6. 15 August 1945. Section 11. Washington, D.C.

Carlson, Roy W. Letter to author. San Jose, California. 2 July 1984. SNL Archives.

Cheeseman, E. L. Memo to All Groups Concerned. "Nomenclature Conference." No date. SNL Archives.

Chief of Armed Forces Special Weapons Project. Letter to Chairman, Atomic Energy Commission. 4 December 1948. SNL Archives.

Chief of Field Office Atomic Energy Commission. Letter to Col. A. A. Fickel. "Military Characteristics for FM Type Weapons." No date. SNL Archives.

Chief of Naval Operations. Memo to Distribution. "Penetrating Atomic Warhead Program for the Regulus I Guided Missile, Termination of." No date. Atomic Energy Commission Files: MRA-5. Regulus I-55. Excerpt in Alexander Collection. SNL Archives.

——————. Memo to Bureau of Aeronautics. "Regulus XW-5 Warhead Fuzing Systems, Change of." March 1956. Atomic Energy Commission Files: MRA-5, Regulus. Excerpt in Alexander Collection. SNL Archives.

Chief of Staff, USAF. Memo to Joint Chiefs of Staff. "Atomic Bomb Assembly Teams." 27 July 1948. With Enclosure: James Forrestal to Joint Chiefs of Staff. "Custody of Atomic Weapons." 28 July 1948. National Archives. Record Group 218. CCS 471.6. 15 August 1945. Section 11. Washington, D.C.

——————. Memo to Joint Chiefs of Staff. "Review of Military Participation in the Atomic Energy Program." 25 November 1949. File CCS 471.6. National Archives and Records Administration.

Claassen, Dick. Memo to Glenn Fowler. "Meeting with Dr. Kelly on March 16, 1957." 20 March 1957. R. S. Claassen Collection. SNL Archives.

——————. Memo to Frank Hudson. Shirley, New York. 19 September 1987. Frank Hudson Collection. SNL Archives.

BIBLIOGRAPHY

Coiner, R. T., Jr., Colonel, United States Air Force, Acting Atomic Energy Commission/Division of Military Affairs. Memo to C. L. Tyler, Manager, Atomic Energy Commission/Santa Fe Operations. "Guided Missile Fuzing." 14 June 1950. Central Technical File. Sandia Technical Library.

Colby, R. L. Memo to R. S. Warner. "Plans for Z-3 and Z-5 Merger." 2 March 1946. Brode File. SNL Archives.

Commander Joint Task Force One. Memo to Chief of Naval Operations et al. "General Information on Atomic Bomb Tests." 28 January 1946. L.A. Archives.

——————. Mailgram to Public. 19 July 1946. L.A. Archives.

Conant, James B. and Leslie R. Groves. Letter to Dr. J. Robert Oppenheimer. OSRD. Washington, D.C. 25 February 1943. L.A. Archives.

CORPORAL/TX-7 Ad Hoc Working Group to Distribution. Excerpt. "Minutes of Meetings Held June 12 and 13." 13 June 1951. SNL Archives.

Corson, Dale R. Memo to N. E. Bradbury. "Future Z Division Program." 11 April 1946. L.A. Archives.

Dean, Gordon, Chairman, Atomic Energy Commission. Memo to Robert LeBaron, Chairman of the Military Liaison Committee. 13 July 1950. Record Group 326 Atomic Energy Commission. Collection: Secretariat. Box 1228. Folder: Greenhouse. DOE Archives.

——————. Memo to Honorable Rovert A. Lovett, Secretary of Defense, Washington, D.C. 6 May 1952.

——————. Memo to R. L. Gilpatrick, Washington, D.C. 15 May 1952. Atomic Energy Commission. AEC 485/16. Box 1264. Folder MRA 9-1, Design and Development. DOE Archives.

——————. Letter to General Eisenhower. Augusta, Georgia. November 1952. Historical Document Number 356. DOE Archives.

——————. Memo to Robert LeBaron. Washington, D.C. 6 January 1953. Atomic Energy Commission. Circulated as AEC 485/23. DOE Archives.

——————. Memo to Chairman, Military Liaison Committee. "Missile and Rocket Responsibilities." Atomic Energy Commission. AEC 564/1. 22 January 1953. DOE Archives.

BIBLIOGRAPHY

Demson, E. J., R. A. Larkin, and J. A. Muncy. Memo to Roger S. Warner, Jr. Untitled draft response to Memo: "Responsibilities of the Personnel Office to B Division." 30 January 1946. L.A. Archives.

Director, Los Alamos Scientific Laboratory. Memo to Distribution. "Comments and Suggestions on Transfer of the Sandia Laboratory from the University of California to a New Contractor." 8 August 1949. SNL Archives.

District Engineer to Col. L. E. Seeman. "Appointment." No date. L.A. Archives.

Division of Military Affairs. Memo to Circulation. No Date. AEC Files. Series 199. Department of Energy/Albuquerque Operations Office. Copy in SNL Archives.

―――――――. Memo to Sandia Laboratory. 29 June 1949. Atomic Energy Commission Files: MRA-5. Volume V. 1949-1950.

Douglas Aircraft Corporation to Sandia Corporation. "TX-7 Rocket Propelled." Excerpt. 29 June 1951. Microfilm Reel MF-SF-SC-74. Central Technical File. Sandia Technical Library.

Dorland, G. M. Memo to R. W. Henderson. "Road Canning and Nitrogenizing Program." 29 April 1947. SNL Archives.

Draper, E. H. Memo to J. P. Molnar. "History and Status of the Ballistic Missile Program." Excerpt. 7 July 1959. Central Technical File. Sandia Technical Library.

Eisenhower, Dwight D. Letter to Mr. Chairman, Washington, D.C. 8 August 1956. Copy in SNL Archives.

Erickson, K. W. Memo to R. S. Claassen. "Future Research Program." 7 August 1956. Frank Hudson Collection. SNL Archives.

Field Command. Memo to Santa Fe Operations Office. "Early Emergency Capability XW-5/MATADOR." 17 March 1954. Atomic Energy Commission Files: MRA-5, MATADOR. Volume III. Excerpt in Alexander Collection. SNL Archives.

Fields, Kenneth E. Memo to Carroll Tyler. "Development of Honest John and Corporal." Atomic Energy Commission Files: MRA-5, 7/53-8/53. Excerpt in Alexander Collection. SNL Archives.

Fite, A. W. Memo to Distribution. "Nike Meeting Held at Sandia Corporation 22 April 1953." 22 April 1953. SNL Archives.

BIBLIOGRAPHY

Fowler, Glenn A. and Dale Corson. Memo to N. E. Bradbury. "Organization and Functions of Group Z–1A." No date. L.A. Archives.

Fowler, Glenn A. Memo to Roger Warner. 15 November 1945. SNL Archives.

―――――――. Memo to Distribution. 18 February 1957. Including draft of memo to McRae. "The Role of Research at Sandia." 4 February 1957. Frank Hudson Collection. SNL Archives.

Frolich, A. J. Memo to Commanding Officer, Kirtland AFB. "Use of Facilities During Closing of Kirtland Field." 11 December 1945.

―――――――. Memo to All Concerned. "Housing at Kirtland for Sandia Personnel." 21 December 1945. L.A. Archives.

Gilpatrick, R. L., Undersecretary, Department of the Air Force. Memo to Gordon E. Dean, Chairman, Atomic Energy Commission. Washington, D.C. 3 June 1952. Atomic Energy Commission. AEC 485. Box 1264. Folder: MRA–9–1, Design and Development. DOE Archives.

Groves, Leslie R. Memo to Norris Bradbury. "Atomic Warhead in a Guided Missile." 12 December 1946. SNL Archives.

―――――――. Cover Memo to Gen. Omar N. Bradley et al. "Note By the Secretaries to the Joint Chiefs of Staff on the Position of the Military Establishment in the Development of Atomic Weapons: . . . " 28 February 1948. National Archives. Record Group 218. Records of the United States Joint Chiefs of Staff. CCS 471.6. 15 August 1945. Section 9. Washington, D.C.

―――――――. Letter to Commanding Officer. "Construction at Sandia Project." 17 October 1945. L.A. Archives.

―――――――. Letter to Frederick C. Alexander, Jr. Stamford, Connecticut. 6 June 1961. SNL Archives.

Guided Missiles Committee. Memo to Special Weapons Development Board. "Initial Recommendations on REGULUS/TX–5 Arming and Fuzing." Excerpt. Addendum to the 55th Meeting of the Board. 26 August 1951. SNL Archives.

Harris, Sylvan. Memo to L. G. Hawkins. "Sandia Laboratory News." 25 August 1948. 322, Sandia Corporation Correspondence. Bay 9, Drawer 58. L.A. Archives.

BIBLIOGRAPHY

Hawkins, L. G. Letter to A. E. Dyhre. "Sandia Laboratory News." No date. L.A. Archives.

──────. Memo to A. E. Dyhre. "History of Sandia Business Office, July–December 1948." SNL Archives.

Heilman, Luther J. Letter to N. S. Furman. 22 June 1986. SNL Archives.

Henderson, Robert W. (Signing for R. E. Poole) Memo to D. A. Quarles. 17 November 1952. McRae Committee Collection. SNL Archives.

──────. Errata to *Enola Gay* by Gordon Thomas and Max Morgan Witts. New York: Pocket Books. 1978.

Hertford, Kenner F. Memo to Pat, Camp Mercury, Nevada (Las Vegas). 15 May 1952. SNL Archives.

Hollowwa, G. C. Memo to N. S. Furman. 23 June 1986. SNL Archives.

Hudson, Frank. Memo to K. W. Erickson. "A Systematic Program for Component Studies." 26 August 1954. Frank Hudson Collection. SNL Archives.

──────. Memo to Glenn Fowler. "Research Planning, Phase II." 18 February 1957. Frank Hudson Collection. SNL Archives.

Hunter, R. P. Captain, United States Navy, Executive Secretary Department of Energy/Military Liaison Committee. Memo to Director, Atomic Energy Commission/Division of Miitary Affairs. "Responsibility for Development and Procurement of Adaption Kits for Missiles." 7 August 1952. SNL Archives.

Jewett, F. B. Letter to J. B. Conant. 17 July 1941. Folder 5: "Historical File." Record Group No. 222. National Archives.

Johnson, A. C. Memo to Major Wolf. "Assignment of Buildings." 14 July 1945. SNL Archives.

Joint Strategic Plans Group. Memo to Col. C. V. Clifton. 2 May 1952. National Archives. Washington, D.C.

Kelly, M. J. Testimony "In the Matter of J. Robert Oppenheimer." Transcript of Hearing Before Personnel Security Board. Washington, D.C. 12 April 1954 through May 1954.

──────. Letter to D. P. Severance. 25 July 1961. SNL Archives.

Kistiakowsky, George. Memo to Robert Oppenheimer and William Parsons. 13 June 1944. L. A. Archives.

BIBLIOGRAPHY

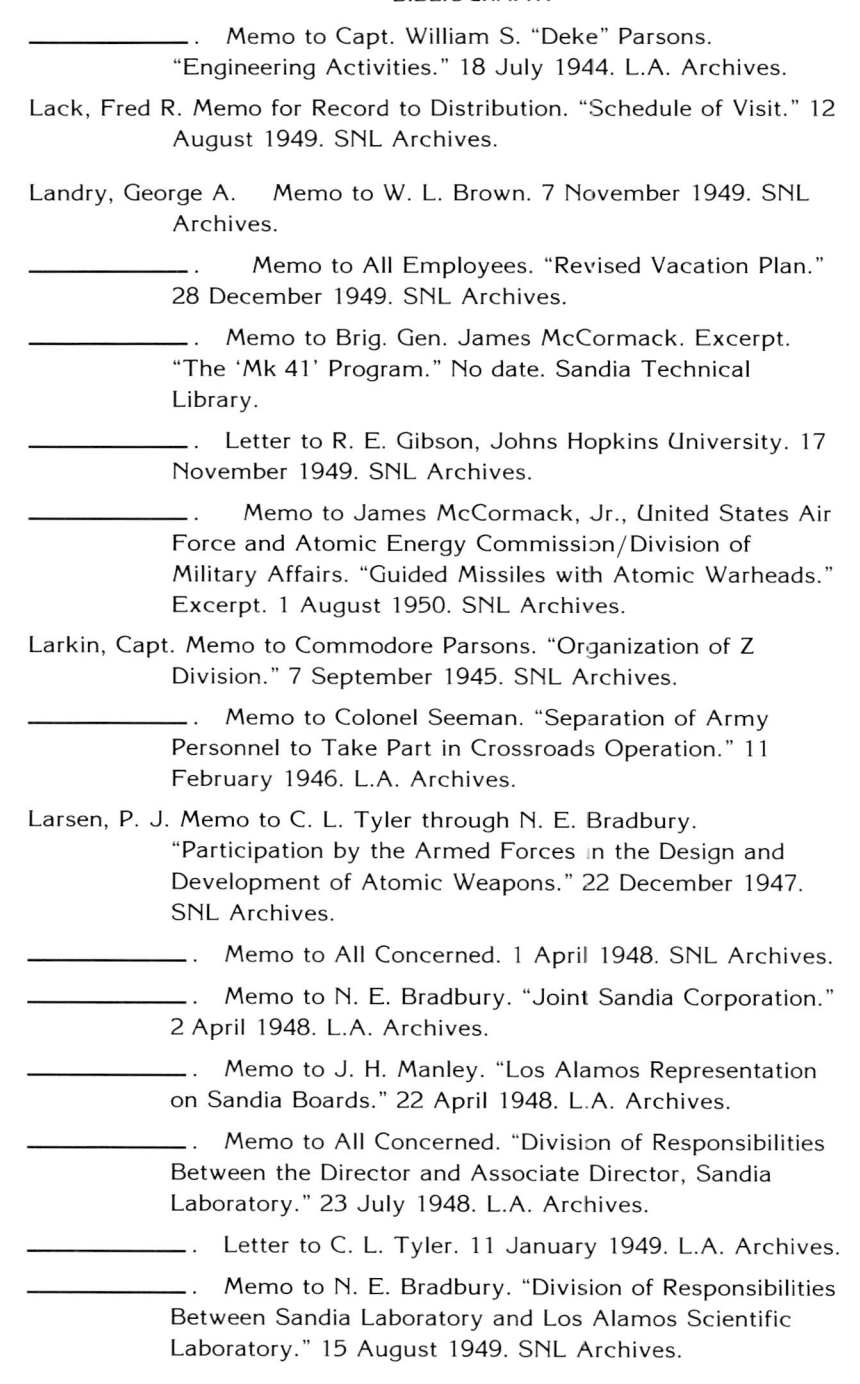

⸺. Memo to Capt. William S. "Deke" Parsons. "Engineering Activities." 18 July 1944. L.A. Archives.

Lack, Fred R. Memo for Record to Distribution. "Schedule of Visit." 12 August 1949. SNL Archives.

Landry, George A. Memo to W. L. Brown. 7 November 1949. SNL Archives.

⸺. Memo to All Employees. "Revised Vacation Plan." 28 December 1949. SNL Archives.

⸺. Memo to Brig. Gen. James McCormack. Excerpt. "The 'Mk 41' Program." No date. Sandia Technical Library.

⸺. Letter to R. E. Gibson, Johns Hopkins University. 17 November 1949. SNL Archives.

⸺. Memo to James McCormack, Jr., United States Air Force and Atomic Energy Commission/Division of Military Affairs. "Guided Missiles with Atomic Warheads." Excerpt. 1 August 1950. SNL Archives.

Larkin, Capt. Memo to Commodore Parsons. "Organization of Z Division." 7 September 1945. SNL Archives.

⸺. Memo to Colonel Seeman. "Separation of Army Personnel to Take Part in Crossroads Operation." 11 February 1946. L.A. Archives.

Larsen, P. J. Memo to C. L. Tyler through N. E. Bradbury. "Participation by the Armed Forces in the Design and Development of Atomic Weapons." 22 December 1947. SNL Archives.

⸺. Memo to All Concerned. 1 April 1948. SNL Archives.

⸺. Memo to N. E. Bradbury. "Joint Sandia Corporation." 2 April 1948. L.A. Archives.

⸺. Memo to J. H. Manley. "Los Alamos Representation on Sandia Boards." 22 April 1948. L.A. Archives.

⸺. Memo to All Concerned. "Division of Responsibilities Between the Director and Associate Director, Sandia Laboratory." 23 July 1948. L.A. Archives.

⸺. Letter to C. L. Tyler. 11 January 1949. L.A. Archives.

⸺. Memo to N. E. Bradbury. "Division of Responsibilities Between Sandia Laboratory and Los Alamos Scientific Laboratory." 15 August 1949. SNL Archives.

BIBLIOGRAPHY

Latta, M. C. To David E. Lilienthal with Enclosure: Harry S. Truman, Executive Order "Providing for the Transfer of Properties and Personnel to the Atomic Energy Commission, The White House, 31 December 1946." National Archives, Record Group 326, Records of the Atomic Energy Commission, Office Files of David E. Lilienthal, Folder "Manhattan District, Transfer of."

LeBaron, Robert. Letter to George A. Landry. 3 May 1951. Washington, D.C. SNL Archives.

Lehr, H. M. Letter to Marshall Holloway. Kwajalein, Marshall Islands. 27 March 1946. L.A. Archives.

Lieber, Andrew. Memo to author. 31 August 1988. SNL Archives.

―――――. "Commentary on Father John." September 1988. SNL Archives.

Lilienthal, David E. Memo to the President. Atomic Energy Commission. Washington, D.C. 13 May 1949. Copy in SNL Archives.

―――――. Letter to Leroy A. Wilson. New York. 6 July 1949. Copy in SNL Archives.

―――――. Letter to Brien McMahon. Washington, D.C. 8 July 1949. Copy in SNL Archives.

Lockridge, Robert W. Memo to All Concerned. "Transfer of Group Z-2A to Sandia." 26 September 1945. L.A. Archives.

Machen, Art. Letter to Manual Control Group Z-3. 14 February 1947. L.A. Archives.

―――――. Letter to author. Rogue River, Oregon. 19 January 1985. SNL Archives.

Manager, Santa Fe Operations and the Director of Military Applications. "Atomic Energy Commission Legislative Jurisdiction Over AEC Properties at Sandia Base." Note by the Secretary Roy B. Snapp. 25 September 1951. AEC 161/1. DOE Archives. Record Group 326. Atomic Energy Commission Collection: Secretariat. Germantown, Maryland.

McCormack, Brig. Gen. James. Memo to Atomic Energy Commission, Santa Fe Office. "Change of Laboratory." No Date. SNL Archives.

―――――. Letter to Carroll L. Wilson. "Sandia Laboratory Forms Attachment to AEC." No Date. SNL Archives.

BIBLIOGRAPHY

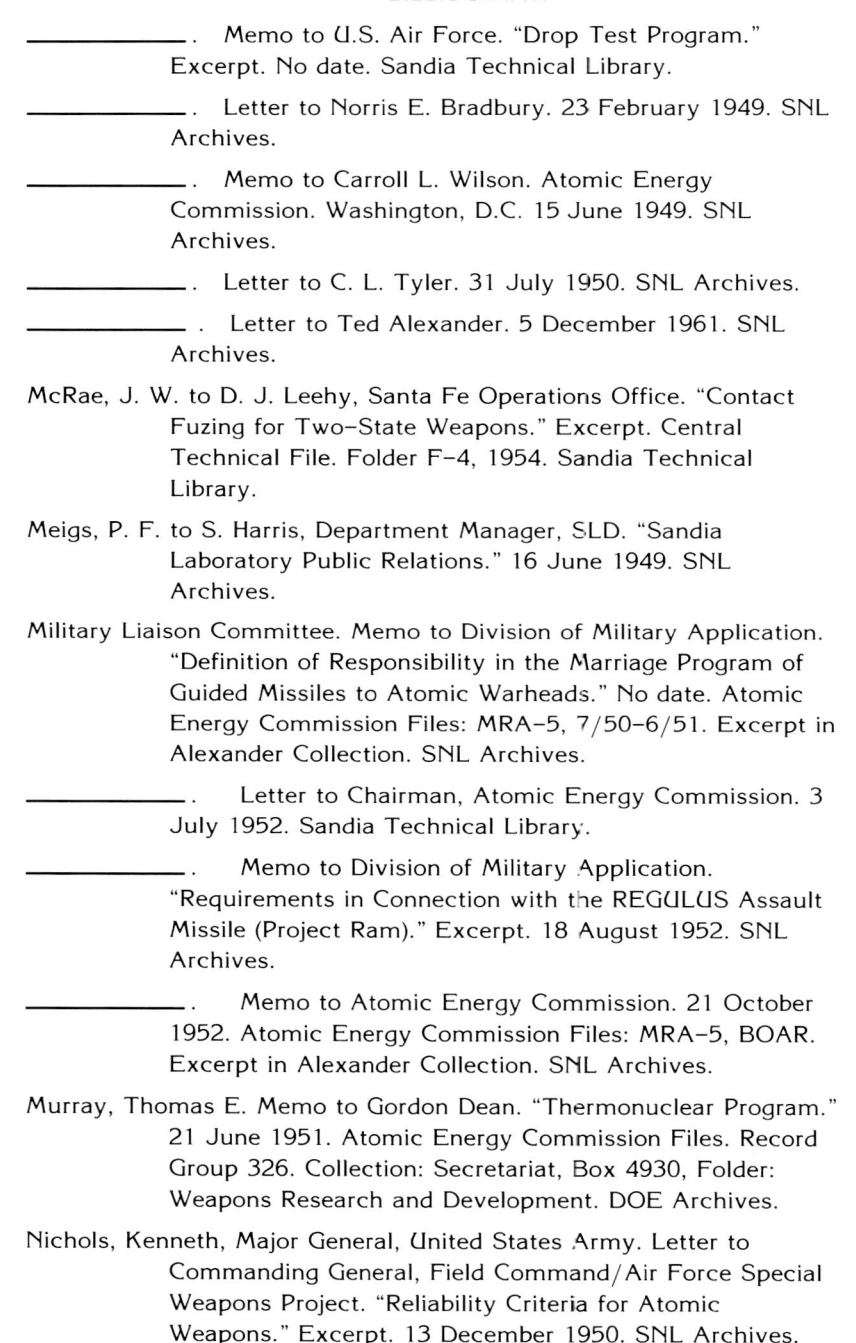

———. Memo to U.S. Air Force. "Drop Test Program." Excerpt. No date. Sandia Technical Library.

———. Letter to Norris E. Bradbury. 23 February 1949. SNL Archives.

———. Memo to Carroll L. Wilson. Atomic Energy Commission. Washington, D.C. 15 June 1949. SNL Archives.

———. Letter to C. L. Tyler. 31 July 1950. SNL Archives.

———. Letter to Ted Alexander. 5 December 1961. SNL Archives.

McRae, J. W. to D. J. Leehy, Santa Fe Operations Office. "Contact Fuzing for Two-State Weapons." Excerpt. Central Technical File. Folder F-4, 1954. Sandia Technical Library.

Meigs, P. F. to S. Harris, Department Manager, SLD. "Sandia Laboratory Public Relations." 16 June 1949. SNL Archives.

Military Liaison Committee. Memo to Division of Military Application. "Definition of Responsibility in the Marriage Program of Guided Missiles to Atomic Warheads." No date. Atomic Energy Commission Files: MRA-5, 7/50-6/51. Excerpt in Alexander Collection. SNL Archives.

———. Letter to Chairman, Atomic Energy Commission. 3 July 1952. Sandia Technical Library.

———. Memo to Division of Military Application. "Requirements in Connection with the REGULUS Assault Missile (Project Ram)." Excerpt. 18 August 1952. SNL Archives.

———. Memo to Atomic Energy Commission. 21 October 1952. Atomic Energy Commission Files: MRA-5, BOAR. Excerpt in Alexander Collection. SNL Archives.

Murray, Thomas E. Memo to Gordon Dean. "Thermonuclear Program." 21 June 1951. Atomic Energy Commission Files. Record Group 326. Collection: Secretariat, Box 4930, Folder: Weapons Research and Development. DOE Archives.

Nichols, Kenneth, Major General, United States Army. Letter to Commanding General, Field Command/Air Force Special Weapons Project. "Reliability Criteria for Atomic Weapons." Excerpt. 13 December 1950. SNL Archives.

BIBLIOGRAPHY

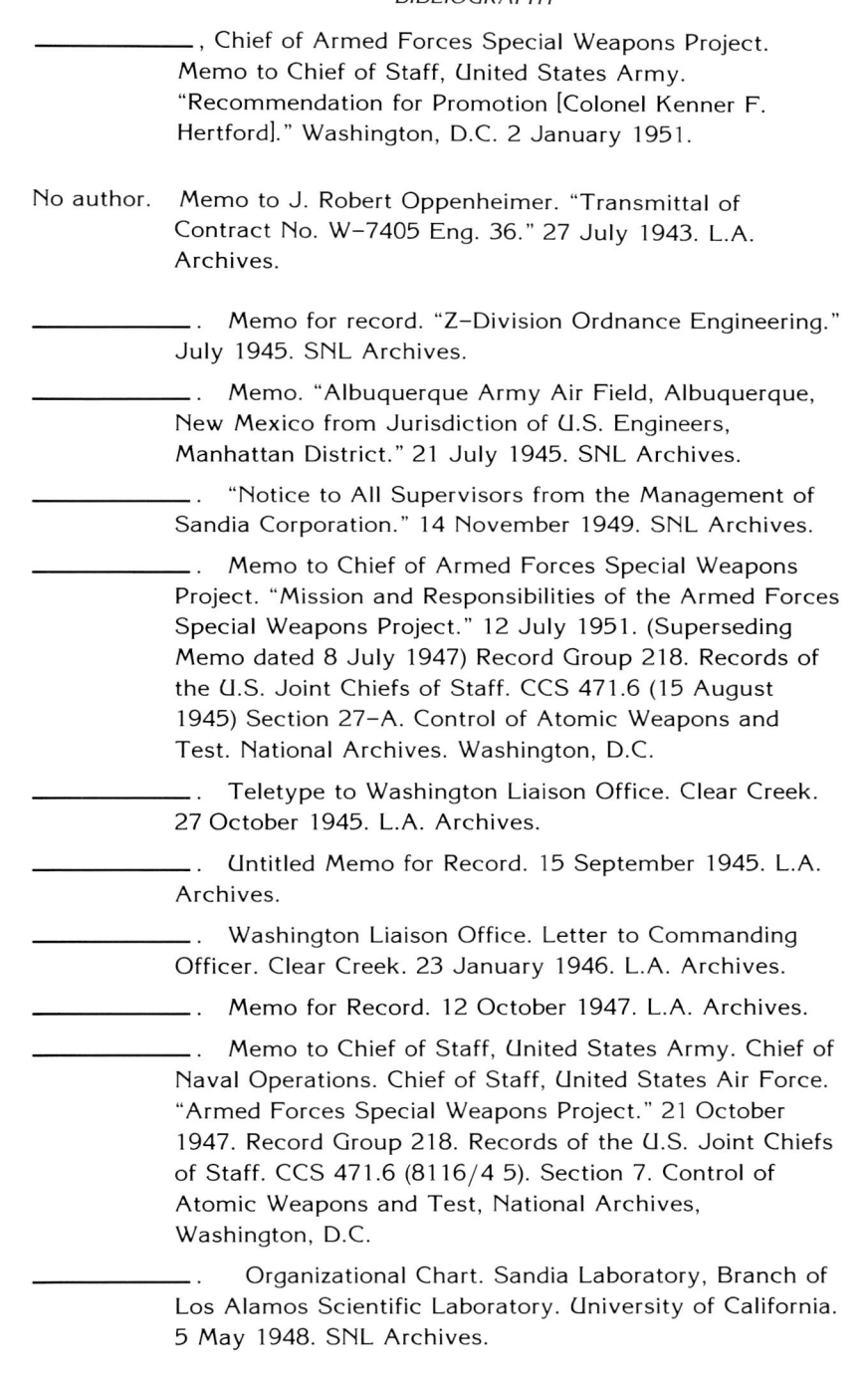

——————, Chief of Armed Forces Special Weapons Project. Memo to Chief of Staff, United States Army. "Recommendation for Promotion [Colonel Kenner F. Hertford]." Washington, D.C. 2 January 1951.

No author. Memo to J. Robert Oppenheimer. "Transmittal of Contract No. W-7405 Eng. 36." 27 July 1943. L.A. Archives.

——————. Memo for record. "Z-Division Ordnance Engineering." July 1945. SNL Archives.

——————. Memo. "Albuquerque Army Air Field, Albuquerque, New Mexico from Jurisdiction of U.S. Engineers, Manhattan District." 21 July 1945. SNL Archives.

——————. "Notice to All Supervisors from the Management of Sandia Corporation." 14 November 1949. SNL Archives.

——————. Memo to Chief of Armed Forces Special Weapons Project. "Mission and Responsibilities of the Armed Forces Special Weapons Project." 12 July 1951. (Superseding Memo dated 8 July 1947) Record Group 218. Records of the U.S. Joint Chiefs of Staff. CCS 471.6 (15 August 1945) Section 27-A. Control of Atomic Weapons and Test. National Archives. Washington, D.C.

——————. Teletype to Washington Liaison Office. Clear Creek. 27 October 1945. L.A. Archives.

——————. Untitled Memo for Record. 15 September 1945. L.A. Archives.

——————. Washington Liaison Office. Letter to Commanding Officer. Clear Creek. 23 January 1946. L.A. Archives.

——————. Memo for Record. 12 October 1947. L.A. Archives.

——————. Memo to Chief of Staff, United States Army. Chief of Naval Operations. Chief of Staff, United States Air Force. "Armed Forces Special Weapons Project." 21 October 1947. Record Group 218. Records of the U.S. Joint Chiefs of Staff. CCS 471.6 (8116/4 5). Section 7. Control of Atomic Weapons and Test, National Archives, Washington, D.C.

——————. Organizational Chart. Sandia Laboratory, Branch of Los Alamos Scientific Laboratory. University of California. 5 May 1948. SNL Archives.

795

BIBLIOGRAPHY

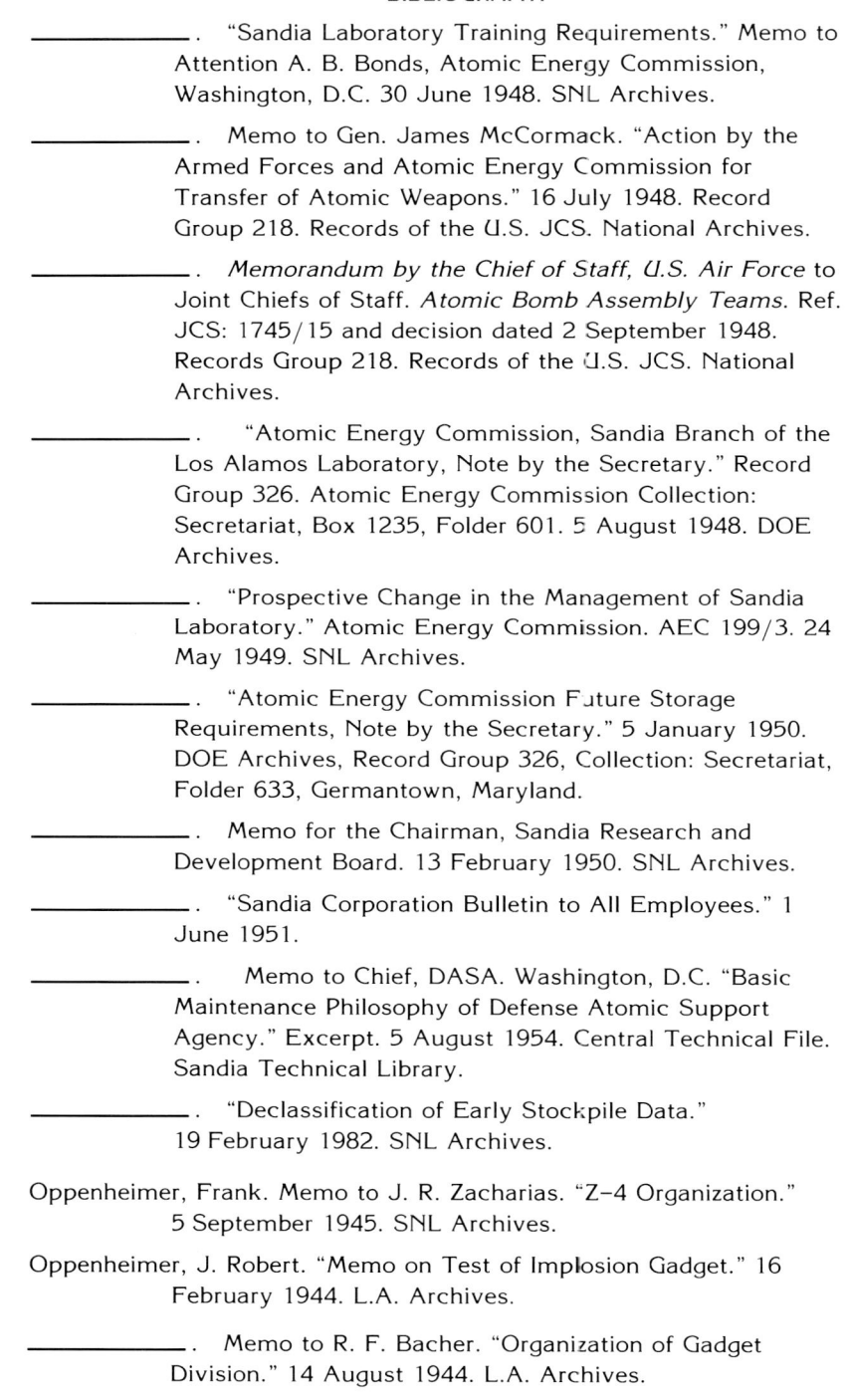

———. "Sandia Laboratory Training Requirements." Memo to Attention A. B. Bonds, Atomic Energy Commission, Washington, D.C. 30 June 1948. SNL Archives.

———. Memo to Gen. James McCormack. "Action by the Armed Forces and Atomic Energy Commission for Transfer of Atomic Weapons." 16 July 1948. Record Group 218. Records of the U.S. JCS. National Archives.

———. *Memorandum by the Chief of Staff, U.S. Air Force* to Joint Chiefs of Staff. *Atomic Bomb Assembly Teams*. Ref. JCS: 1745/15 and decision dated 2 September 1948. Records Group 218. Records of the U.S. JCS. National Archives.

———. "Atomic Energy Commission, Sandia Branch of the Los Alamos Laboratory, Note by the Secretary." Record Group 326. Atomic Energy Commission Collection: Secretariat, Box 1235, Folder 601. 5 August 1948. DOE Archives.

———. "Prospective Change in the Management of Sandia Laboratory." Atomic Energy Commission. AEC 199/3. 24 May 1949. SNL Archives.

———. "Atomic Energy Commission Future Storage Requirements, Note by the Secretary." 5 January 1950. DOE Archives, Record Group 326, Collection: Secretariat, Folder 633, Germantown, Maryland.

———. Memo for the Chairman, Sandia Research and Development Board. 13 February 1950. SNL Archives.

———. "Sandia Corporation Bulletin to All Employees." 1 June 1951.

———. Memo to Chief, DASA. Washington, D.C. "Basic Maintenance Philosophy of Defense Atomic Support Agency." Excerpt. 5 August 1954. Central Technical File. Sandia Technical Library.

———. "Declassification of Early Stockpile Data." 19 February 1982. SNL Archives.

Oppenheimer, Frank. Memo to J. R. Zacharias. "Z-4 Organization." 5 September 1945. SNL Archives.

Oppenheimer, J. Robert. "Memo on Test of Implosion Gadget." 16 February 1944. L.A. Archives.

———. Memo to R. F. Bacher. "Organization of Gadget Division." 14 August 1944. L.A. Archives.

BIBLIOGRAPHY

————. Memo to G. B. Kistiakowsky. "Organization of Gadget Division." 14 August 1944. L.A. Archives.

————. Memo to K. T. Bainbridge. "Gadget Testing Using Water for Recovery and Control." 22 December 1944. L.A. Archives.

————. Memo to J. R. Zacharias. "In Reference to Organization of Z Division. . . . " 6 August 1945. L.A. Archives.

————. Memo to All Concerned. "Additions to Organization of the Lab." 12 September 1945. L.A. Archives.

————. Letter to Frederick C. Alexander, Jr. Princeton, New Jersey. 25 May 1961. SNL Archives.

Parsons, William S. Memo to Maj. Gen. T. F. Farrell. "Bombing Range at Salton Sea." No date. SNL Archives.

————. "Paraphrased Teletype Reference TA-2050 dated 15 July." L.A. Archives.

————. TWX to Leslie R. Groves. 25 October 1944. *Daily Log #2*. SNL Archives.

———— to N. E. Bradbury. "Paraphrased Teletype Reference TA-2461 dated 28 August [1945]." L.A. Archives.

Patterson, Robert T. "Statement of Secretary of War Patterson on Transfer of Manhattan District." War Department. 31 December 1946. The National Archives. Record Group 326. "Records of the Atomic Energy Commission Office Files of David E. Lilienthal." Folder: "Manhattan District, Transfer of."

Pike, Sumner T. Letter to Lindsay C. Warren. Washington, D.C. 30 October 1949. SNL Archives.

Poole, R. E. Memo to Worth. "Project Alias/Betty Complete Design Release." Excerpt. 16 August 1954. SNL Archives.

Powell, Ray B. Memo to Distribution. "Proposed Public Relations Guide." December 1955. SNL Archives.

————. Letter to Necah Furman. 27 February 1984. SNL Archives.

Purnell, Admiral. Letter to William S. Parsons. 9 November 1944. *Daily Log #2*. SNL Archives.

Quarles, D. A. Memo to Colonel K. E. Fields. "Reevaluation of Arming and Fuzing Responsibilities for Atomic Warheads." 4 March 1952. SNL Archives.

BIBLIOGRAPHY

———. Memo to T. E. Shea, R. E. Poole, F. Schmidt, and W. A. MacNair. "Transition—Manufacturing Engineering." 29 November 1952. McRae Committee Collection. SNL Archives.

Ramsey, N. F. Memo to Major Peer de Silva. List of Personnel to Destination. 2 April 1945. SNL Archives.

Research and Development Board. Memo to the Military Liaison Committee. "Guided Missiles With Atomic Warheads. 27 January 1950. Atomic Energy Commission Files: MRA-5, 4/49-6/50. Excerpt in SNL Archives.

———. Memo to Military Liaison Committee. "Subsurface Atomic Weapons of Implosion Type." 6 July 1951. Atomic Energy Commission Files: MRA-5, Alias. Volume I. 6/51-6/53. Excerpts in the Alexander Collection. SNL Archives.

Roper, H. K., Colonel, U.S. Army. Memo to Roy B. Snapp, Secretary. "Military Guidance to the Atomic Energy Commission in the Atomic Weapons Field." 19 July 1951. Contains attachment: "Procedure for Providing Military Guidance to the Atomic Energy Commission in the Atomic Weapons Field." Atomic Energy Commission. 453. Box 1264. Folder: Design and Development. DOE Archives.

Rowe, Hartley. Letter to Frederick C. Alexander, Jr. Newton Centre, Massachusetts. 24 November 1961. SNL Archives.

Roy, M. F. Memo to Technical Personnel Office. 6 December 1945. L.A. Archives.

Russell, J. S. Captain, United States Navy, Acting Director Atomic Energy Commission/Division of Military Affairs. Memo to C. L. Tyler, Manager, Atomic Energy Commission/Santa Fe Operations Office. "Guided Missiles With Atomic Warheads." Excerpt. 13 November 1950. SNL Archives.

Sandia Corporation. File memorandum. "Proposed Nuclear Arming System for Atomic Warheads Used with Guided Missiles." Excerpt. No date. Central Technical File. Sandia Technical Library.

———. Memo to Distribution. "First Meeting of the BOAR Committee on July 22 and 23, 1952." Excerpt. No date. Central Technical File. Sandia Technical Library.

———. Memo to Division of Military Affairs. "Proposed Ordnance Characteristics of the XW-7/HJ-XI Honest

BIBLIOGRAPHY

John Atomic Warhead Installation." No date. Atomic Energy Commission Files: MRA-5, Honest John. Volume I. Excerpt. SNL Archives.

——————. Memo to Distribution. 15 February 1951. Excerpt. Central Technical File. Sandia Technical Library.

——————. Memo to Distribution. "Arming and Fuzing for the CORPORAL/TX-7 Marriage." Excerpt. 11 July 1951. Central Technical File. Sandia Technical Library.

——————. Memo to Division of Military Affairs. "Sandia Corporation Proposal for XW-5/RASCAL Fuzing System." 23 March 1953. Atomic Energy Commission Files: MRA-5, RASCAL. Excerpts in Alexander Collection. SNL Archives.

——————. Memo to Santa Fe Operations Office. "Application of Universal XW-7 Warhead to Various Programs." Excerpt. 13 April 1953. Central Technical File. Sandia Technical Library.

——————. Memo to Sandia Field Office, Atomic Energy Commission. "Project Alias Betty Complete Design Release." 16 August 1954. Atomic Energy Commission Files: MRA-5, LULU, 7-54-12-54. Excerpt in Alexander Collection. SNL Archives.

Sandia Research. Memo to Engineering. "Proposed Conference on Guided Bomb." Excerpt. 26 August 1949. Central Technical File. Microfilm Reel 745. Sandia Technical Library.

Sandia Weapons Development Board. Memo to Distribution. "Minutes of the 94th Meeting, Part 1" Excerpt. No date. SNL Archives.

——————. Memo to Distribution. "Minutes of the 45th Meeting." Excerpt. 18 October 1950. SNL Archives.

Santa Fe Operations Office. Memo to Sandia Corporation. "XW-7/HJ-XI Program." Excerpt. Atomic Energy Commission Files: MRA-5, 5/54-6/54. No date. SNL Archives.

——————. Memo to Sandia Corporation. "Development of an Adaption Kit for *Demi-John*." Atomic Energy Commission Files: MRA-5, Demi-John, 7/54. Excerpts in Alexander Collection. SNL Archives.

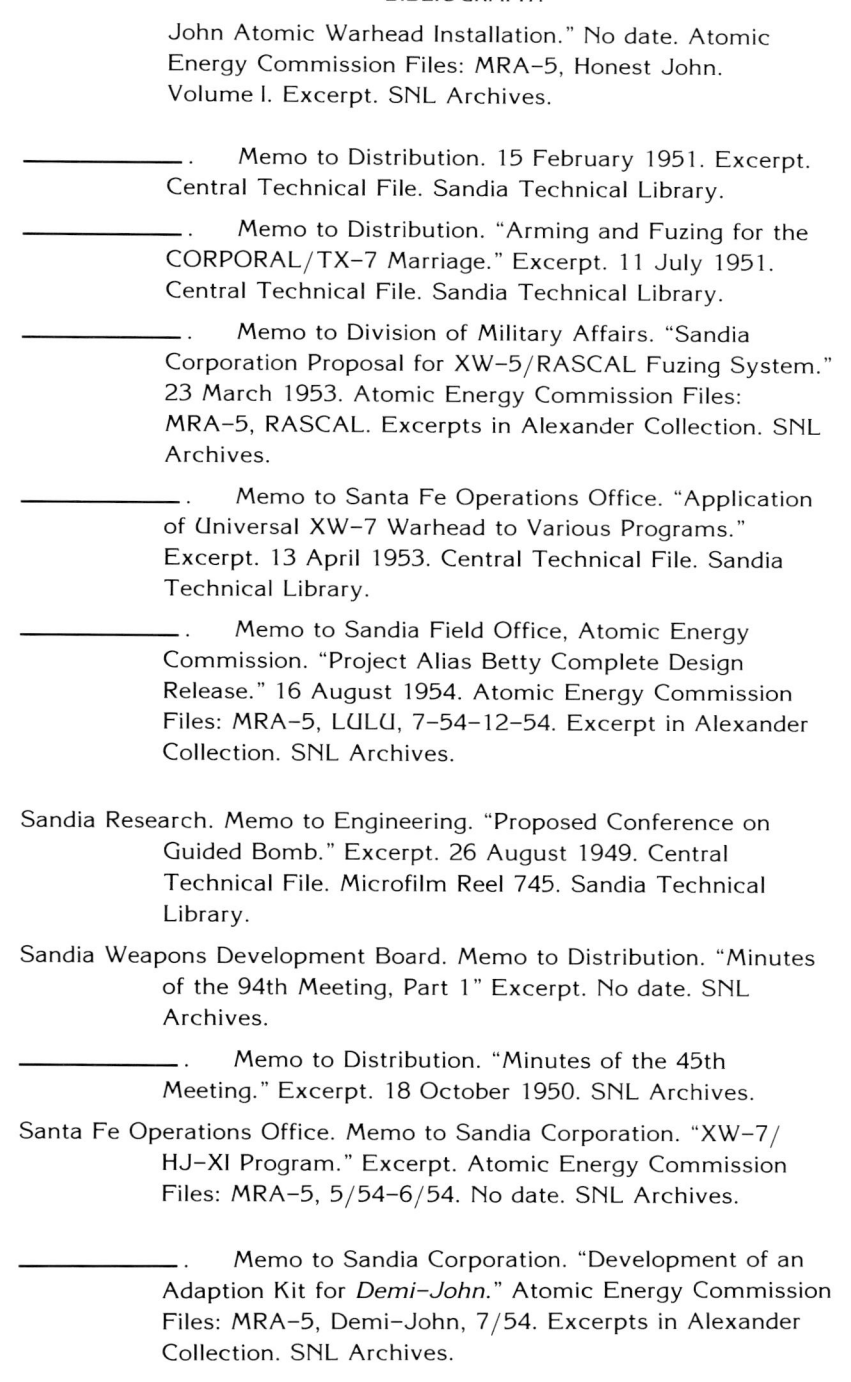

BIBLIOGRAPHY

_____. Memo to Los Alamos Scientific Laboratory and Sandia Corporation. "Implosion Type Weapon Program." 19 January 1951. Excerpt. Central Technical File. Sandia Technical Library.

_____. TWX to Sandia Field Office, Atomic Energy Commission. 9 April 1953. Atomic Energy Commission Files: MRA-5, MATADOR. Volume II. Excerpts in Alexander Collection. SNL Archives.

Santa Fe Operations Office. Memo to Sandia Field Office, Atomic Energy Commission, and Sandia Corporation. "XW-5/RASCAL Fuzing Development." Excerpt. 19 January 1953. Atomic Energy Commission Files: MRA-5, January–February 1953. SNL Archives.

Schaffer, B. H. Memo to Commanding Officer, W. S. Engineer Office. "Sanitation of Los Lunas Range." 20 November 1946. Copy in "Los Lunas Bombing Range, New Mexico." SNL Archives.

Shaffer, W. F., Jr. Letter to Frederick Alexander. Knoxville, Tennessee. 10 May 1962. SNL Archives.

Scott, James H. to author. Albuquerque, New Mexico. 12 April 1989. SNL Archives.

Seeman, Col. L. E. Memo to Col. Johnson. (List of buildings transferred to Albuquerque Air Field, Albuquerque, New Mexico) Santa Fe, New Mexico. 2 July 1945. SNL Archives.

_____. Memo to Col. G. R. Tyler. "Construction at Sandia." 2 September 1945.

_____. Teletype to Col. Nichols and Col. Derry. "Housing at Kirtland for Sandia Personnel." 8 February 1946. L.A. Archives.

Severance, D. P. "Comments on the Sandia History by Frederick C. Alexander, Jr." 25 July 1961. SNL Archives.

Shea, T. E. Memo to D. A. Quarles. 26 November 1952. McRae Committee Collection. SNL Archives.

Sherwin, T. B. to P. F. Meigs, Division Leader. "Public Relations." 13 June 1949. SNL Archives.

_____. "Notes for File Re: Public Relations Program." 15 October 1954. SNL Archives.

BIBLIOGRAPHY

———. "Editorial Policy—Sandia *Lab News*." 2 February 1955. Ted Sherwin Collection. SNL Archives.

Smith, F. B. Memo to All Supervisors. "Union Activities at Sandia Corporation." 21 March 1950. SNL Archives.

Smith, Ralph Carlisle. Memo to Ted Sherwin. Proposed Release. 19 June 1951. L.A. Archives.

Smyth, H. D., Thomas Murray and T. Keith Glennan. "Weapons Research and Development." 15 August 1951. Atomic Energy Commission, Record Group 326. Collection: Secretariat. Box 4930. Folder: Weapons Development. DOE Archives.

Snapp, Roy B. Memo for Information to Brig. Gen. James McCormack. 18 December 1947. Legal Files. SNL Archives.

———. Memo to Brig Gen. James McCormack, Jr. "Commission Action on AEC 199/1–Prospective Change in Management of the Sandia Laboratory." 12 May 1949. SNL Archives.

Stevens, W. L. Commentary. 9 November 1988. SNL Archives.

Strauss, Lewis, Chairman of the Atomic Energy Commission. Memo to Mr. President, the White House. 1 August 1956. Copy in SNL Archives.

Teller, Edward. Memo to Norris E. Bradbury. "Plans for Setting Up a Separate Thermonuclear Division." 24 March 1951. Record Group 326. Atomic Energy Commission. Collection: Secretariat. Box 1235. Folder: Los Alamos. DOE Archives.

Theis, W. T. Memo. "Function and Responsibility of Group Z-6." 1 October 1945. L.A. Archives.

———. Memo to J. R. Zacharias. "Personnel Assigned to Group Z-6. 8 October 1945. L.A. Archives.

Treibel, W. E. to Beryl Hefley. "Sandia Laboratories History." 29 March 1982. SNL Archives.

Truman, Harry. Drafts to Leroy Wilson and Dr. O. E. Buckley. Washington, D.C. 13 May 1949. Copies in SNL Archives.

Tyler, Carroll L. Letter to George Landry. 18 November 1949. SNL Archives.

Underhill, R. M. to N. E. Bradbury. 24 June 1947. SNL Archives.

———. Letter to Carroll L. Tyler. 31 December 1948. SNL Archives.

BIBLIOGRAPHY

Warner, Roger S., Jr. Memo to Commander N. E. Bradbury. "Group II–'Destination'." No date.

——————. Memo to E. J. Demson. "Responsibilities of the Personnel Office to B Division." 30 January 1946. L.A. Archives.

——————. Memo to Capt. R. A. Larkin. "Inspection Party to Crossroads Site." 12 February 1946. L.A. Archives.

——————. Memo to E. B. Doll. "Plans for Survey Trip to Marshall Islands." 8 March 1946. L.A. Archives.

——————. Memo to N. E. Bradbury. 29 November 1946. SNL Archives.

Wesson, P. D. Letter to John E. Hall. Albuquerque, New Mexico. 19 September 1949. SNL Archives.

——————. Memo to W. L. Brown. "Sandia Contract–Unlimited Indemnity Agreement." 26 September 1949. SNL Archives.

——————. Memo to J. A. Dempsey. "Certification of Incorporation." New York. 11 November 1949. SNL Archives.

Wilhoyt, Col. E. E. Memo to W. O. McCord, Jr. "Reclassification of Military Personnel, B Division." 12 March 1946. L.A. Archives.

Willis, Guy. Letter to Hank Willis. 9 May 1985. SNL Archives.

Wilson, Carroll L. Memo to the Commissioners. "Transfer Orders." United States Atomic Energy Commission, Washington, D.C. 17 December 1946. Record Group 326. Records of the Atomic Energy Commission. Office Files of David E. Lilienthal, Folder: "Manhattan District, Transfer of." National Archives.

Wilson, Leroy A. Letter to Harry Truman. Washington, D.C. 17 May 1949. Atomic Energy Commission Files. Series 199. ALO. Copy in SNL Archives.

——————. Letter to David E. Lilienthal. New York. 1 July 1949. Copy in SNL Archives.

——————. Memo for Record AT&T. 15 August 1949. Copy in SNL Archives.

XW-5/MATADOR Ad Hoc Working Group. Memo to Distribution. Minutes of the 6th Meeting. Excerpt. Atomic Energy Commission Files: MRA-5, Matador. Volume I.

BIBLIOGRAPHY

Zacharias, J. R. Memo to J. R. Oppenheimer. "In Reference to Organization of Z Division. . . . " 6 August 1945. L.A. Archives.

Annual Reports and Minutes of Meetings

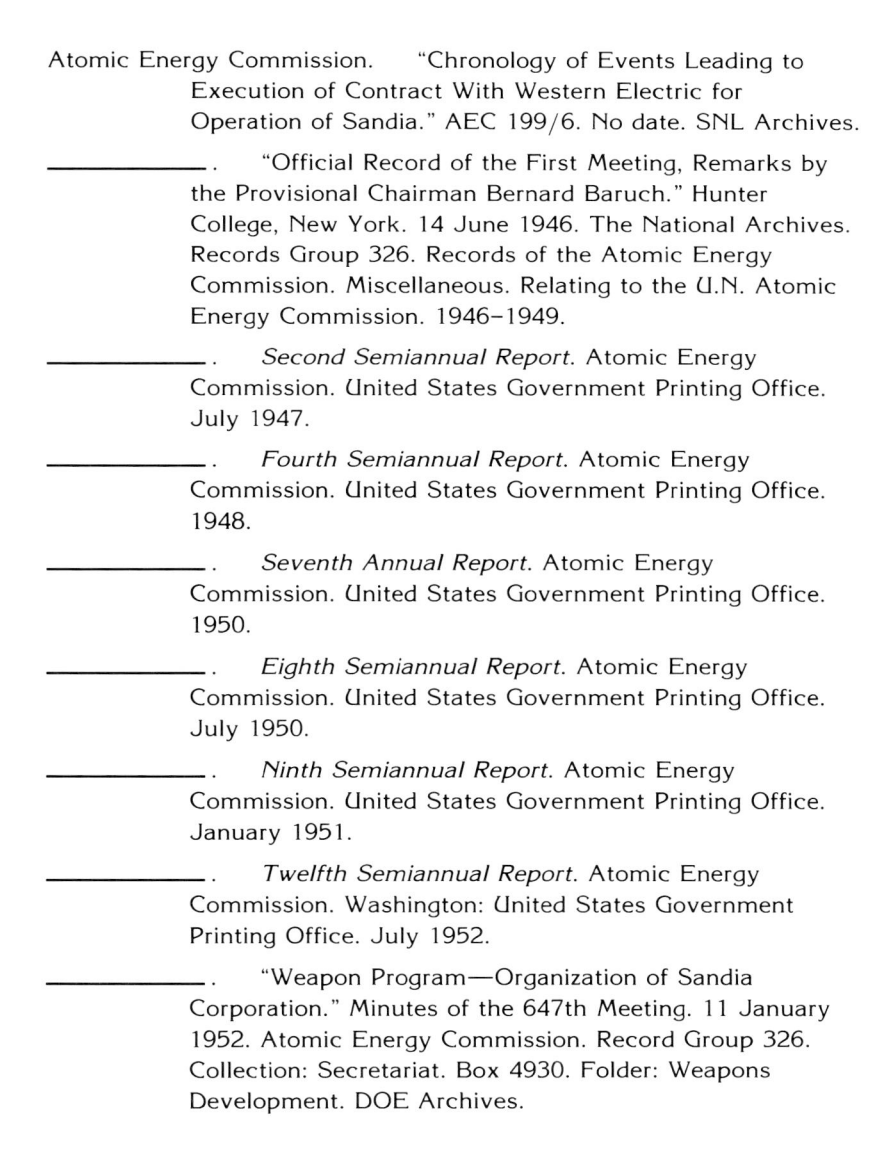

Atomic Energy Commission. "Chronology of Events Leading to Execution of Contract With Western Electric for Operation of Sandia." AEC 199/6. No date. SNL Archives.

———. "Official Record of the First Meeting, Remarks by the Provisional Chairman Bernard Baruch." Hunter College, New York. 14 June 1946. The National Archives. Records Group 326. Records of the Atomic Energy Commission. Miscellaneous. Relating to the U.N. Atomic Energy Commission. 1946–1949.

———. *Second Semiannual Report.* Atomic Energy Commission. United States Government Printing Office. July 1947.

———. *Fourth Semiannual Report.* Atomic Energy Commission. United States Government Printing Office. 1948.

———. *Seventh Annual Report.* Atomic Energy Commission. United States Government Printing Office. 1950.

———. *Eighth Semiannual Report.* Atomic Energy Commission. United States Government Printing Office. July 1950.

———. *Ninth Semiannual Report.* Atomic Energy Commission. United States Government Printing Office. January 1951.

———. *Twelfth Semiannual Report.* Atomic Energy Commission. Washington: United States Government Printing Office. July 1952.

———. "Weapon Program—Organization of Sandia Corporation." Minutes of the 647th Meeting. 11 January 1952. Atomic Energy Commission. Record Group 326. Collection: Secretariat. Box 4930. Folder: Weapons Development. DOE Archives.

BIBLIOGRAPHY

Mc Cormack, Brig. Gen. James. "Notes for Meeting at Los Alamos with Dr. Bradbury." 15 June 1949. SNL Archives.

McRae, James. Minutes of the McRae Committee, 1953–1954. SNL Archives.

——————. "3200 Labor Relations Report." 31 August 1950. SNL Archives.

No author. Minutes of the Governing Board, 17 June 1943 and 23 September 1943. L.A. Archives.

——————. *Los Alamos Daily Log.* 13 October 1943 to 13 September 1945 (Daily Logs #1 and #2). SNL Archives.

——————. Extract from Minutes of the 22nd AEC–MLC Meeting, 17 December 1947. Legal Files. SNL Archives.

——————. "Annual Report of the Military Liaison Committee." Atomic Energy Commission. 11 October 1948. Record Group 218. Records of the United States Joint Chiefs of Staff. National Archives. Washington, D.C.

——————. "Report to the President by the Special Committee of the National Security Council on 'The Proposed Acceleration of the Atomic Energy Program.' " 10 October 1949. Record Group 218. Records of the United States Joint Chiefs of Staff. National Archives. Washington, D.C.

——————. Report to the Joint Chiefs of Staff by the Joint Strategic Survey Committee. No date. Record Group 218. Records of the United States Joint Chiefs of Staff. National Archives. Washington, D.C.

——————. "Minutes of the Sandia Research and Development Board." Sandia Base, Albuquerque. 1948. SNL Archives.

——————. Report by the Manager. Office of Santa Fe Directed Operations and Chief of Engineering. "Sandia Branch of Los Alamos Laboratory." 5 August 1948. Record Group 326. Atomic Energy Commission Collection. Secretariat. Box 1235. Folder 601 LASL. DOE Archives.

——————. "Minutes of the Ad Hoc Committee for Alias Betty." 11 June 1952. AEC Files: MRA-5, Alias. Volume 1. 6–51 to 6–53. Excerpt in the Alexander Collection, SNL Archives.

——————. Minutes of Four O'Clock Group. February 1957. Frank Hudson Collection. SNL Archives.

BIBLIOGRAPHY

Oppenheimer, J. Robert. "Outline of Present Knowledge." No date. L.A. Archives.

Prince, Kimball, with General James McCormack and Carroll Wilson. "Conference Notes Re: Genesis of Sandia Corporation." 29 September 1959. SNL Archives.

RASCAL/XW-5 Ad Hoc Working Group. Memo to Distribution. Minutes of Meeting. Excerpt. 7 November 1951. Central Technical File. Sandia Technical Library.

——————. "Minutes of the 6th Meeting of the Sandia Research and Development Board." Excerpt. 6 May 1948. SNL Archives.

——————. Minutes: "Division Leaders Meeting." 16 November 1948. SNL Archives.

——————. Report of 264th Atomic Energy Commission Meeting. 4 May 1949. SNL Archives.

——————. "Unclassified Minutes of the First Meeting of the Board of Directors of Sandia Corporation." 6 October 1949. SNL Archives.

Sandia Corporation. *Annual Report of the Board of Directors to the Stockholders*. 1950, 1951, 1952, 1953, 1954, 1955. SNL Archives.

Schaffer, B. H. Memo to All Concerned. "Minutes of the Meeting on the Mk 4." 16 April 1947. SNL Archives.

Smith, F. B. Bulletin. "Significant Labor Relations Developments at Sandia Laboratory, American Federation of Labor." 15 February 1950. SNL Archives.

Special Weapons Development Board. Minutes of the 39th Meeting. Excerpt. 21 June 1950. SNL Archives.

——————. Minutes of the 42nd Meeting. Excerpt. 9 August 1950. SNL Archives.

——————. Minutes of Meeting. Excerpt. 22 November 1950. SNL Archives.

——————. "Report of the 2nd Meeting of the Guided Missiles Committee of the Sandia Weapons Development Board." Excerpt. 26 February 1951. SNL Archives.

BIBLIOGRAPHY

——————. Minutes of the 56th Meeting. Excerpt. 12 October 1951. SNL Archives.

——————. Minutes of the 63rd Meeting. Excerpt. 25 June 1952. SNL Archives.

——————. *Report of the Weapons Reliability Committee of the Special Weapons Development Board.* 4 December 1952. SNL Archives.

——————. Minutes of the 89th Meeting. XW-8/REGULUS. Part 1. Excerpt. 1 December 1954. SNL Archives.

——————. Minutes of the 89th Meeting. Excerpt. 12 January 1955. SNL Archives.

——————. "Special Planning Meeting of Small Staff." Excerpt. 24-25 September 1958. SNL Archives.

——————. "AEC-Albuquerque Operations Major Assembly Release and Hold Order Systems." May 1958. SNL Archives.

——————. "Transfer Study." ERDA 97. 1975.

——————. "Report of the President's Blue Ribbon Task Group on Nuclear Weapons Program Management." July 1985. SNL Archives.

TX-N Steering Committee. Minutes of the 40th Meeting of the TX-N Steering Committee. Excerpt. 18 April 1952. Central Technical File. Sandia Technical Library.

Tyler, Caroll L. Memo to George P. Kraker. "Report on Flood at Igloo Area." 10 August 1950. SNL Archives.

Tyler, Carroll L., Manager SFOO, and Daniel F. Worth, Jr., Field Manager, AEC-Sandia. "Three Year Report for Sandia Field Office." 12 October 1953. Folder: Old Sandia Field Office Files—Policy and Historical Documents. DOE AL Archives. Kirtland Air Force Base. Albuquerque, New Mexico.

United Nations Atomic Energy Commission. "Official Record of the First Meeting, Remarks by the Provisional Chairman Bernard Baruch." Hunter College, New York, 14 June 1946. Record Group 326, Records of the Atomic Energy Commission, Miscellaneous, Relating to the United Nations. Atomic Energy Commission, 1946-1949. The National Archives. Washington, D.C.

BIBLIOGRAPHY

Autobiographical Accounts

Anderson, George. Diary, 1956–1957. Harry Saxton Component Collection. SNL Archives.

Ashworth, Capt. F. L. "The Atomic Bombing of Nagasaki." Ft. Belvoir, Virginia. 23 September 1946.

Badash, Lawrence, et al., eds. *Reminiscences of Los Alamos 1943–1945.* Boston: D. Reidel Publishing Company. 1980.

Bainbridge, K. T. "All in Our Time: Prelude to Trinity." *Bulletin of the Atomic Scientists.* Volume 31 (April 1975): 43.

Compton, Arthur Holly. *Atomic Quest: A Personal Narrative.* New York: Oxford University Press, Inc. 1956.

Dean, Gordon. "Dean Diary." Record Group 326. Collection: Dean Diary. DOE Archives.

Fowler, Glenn A. "Logbook." 13 October 1943 to 1951. SNL Archives.

Groves, Leslie R. *Now It Can Be Told: The Story of the Manhattan Project.* New York: Harper and Brothers, 1962.

Hubbard, Dr. Jack. "Los Alamos 1945, Journal of J. Hubbard Meteorologist." L.A. Archives.

Hudson, Frank. "Biographical Material." No date. Frank Hudson Collection. SNL Archives.

Lilienthal, David E. *The Journals of David E. Lilienthal.* Volume II. New York: Harper and Row Publishers. 1964.

McCormack, Brig. Gen. James, and M. J. Kelly. Interview with Kimball Prince. 20 December 1958. "History of Formation." Manuscript in SNL Archives.

Rowe, James Les. *Project W-47.* Livermore, California: Ja A Ro Publishing. 1978.

Stimson, Henry L., and McGeorge Bundy. *On Active Service in Peace and War.* New York: Harper and Brothers, 1947.

_____. "The Decision to Use the Atomic Bomb." *Harper's Magazine* (February 1947).

Upper Management Biographical File. No date. SNL Archives.

Wilson, Jane, ed. *All in Our Time: The Reminiscences of Twelve Nuclear Pioneers.* Chicago: University of Chicago Press, 1975.

BIBLIOGRAPHY

Legal and Government Documents

Eisenhower, Dwight D. *Public Papers of the Presidents of the United States: Dwight D. Eisenhower.* 20 January–31 December 1953.

Kelly, Mervin J. Testimony: "In the Matter of J. Robert Oppenheimer," Transcript of Hearing Before Personnel Security Board, Washington, D.C., April 12, 1954, through May 6, 1954.

No author. *Congressional Quarterly.* Volume I, 1945. Volume II, 1946.

United States Congress. S 1463. 3 October 1945. Copy in "Atomic Energy Bills" File. SNL Archives.

_____. *Atomic Energy Act of 1946.* Public Law 585. In Laws of 79th Congress. 2nd Session. 1 August 1946.

_____. House Committee on Armed Services. *Atomic Energy Legislation Through 95th Congress.* 2nd Session. 1979.

_____. "Report to the President of the United States from the Atomic Energy Commission, January 1–April 1, 1947" and Appendix C, 2 April 1947." Collection: Energy History. DOE Archives. Germantown, Maryland.

_____. License Between the Secretary of War and the Defense Plant Corporation. 18 May 1945. SNL Archives.

_____. *Atomic Energy Act of 1946 and Amendments.* Washington, D.C.: United States Government Printing Office. 1975.

_____. "Takeover Agreement." AT&T-Western Electric. 29 August 1949. SNL Archives.

_____. "Contract AT-(29-1)-789 between United States Atomic Energy Commission and Western Electric Company, Incorporated." 6 October 1949. SNL Archives.

_____. Report Submitted to Senate Government Operations Committee. *Science and Technology Act of 1958, Analysis and Summary.* 85th Congress. Second Session. 1958.

Oppenheimer, J. Robert. *Oppenheimer Papers.* 15 June 1944. Library of Congress. Government File Supplement.

BIBLIOGRAPHY

Roosevelt, Franklin D. *Papers of Franklin D. Roosevelt.* Franklin D. Roosevelt Library. Hyde Park, New York.

Reports

Alexander, Frederick C. "Thermonuclear Weapons Development." Excerpt. May 1969. SNL Archives.

Atomic Energy Commission. "Administration of Sandia Laboratory—Report of Mr. Mervin J. Kelly." Notes by the Secretary, Roy B. Snapp. No date. SNL Archives.

_____. "Executive Secretary on United States Objectives and Programs for National Security." *A Report to the National Security Council.* 14 April 1950. Washington, D.C.

_____. *Agreement Between the AEC and DOD for the Development, Production, and Standardization of Atomic Weapons.* Drafts include Appendix C: "Sandia Corporation Comments on Agreement . . ." 18 November 1952. AEC 485/20. Box 1264. Folder: MRA 9-1, Design and Development. DOE Archives.

_____. *Agreement Between the AEC and DOD . . .* 21 January 1953. AEC 485/24. Box 1264. Folder MRA 9-1, Design and Development. DOE Archives.

_____. "Atomic Energy Commission History of Expansion of AEC Production Facilities." 16 August 1963. AEC 1140. Record Group 326. Collection: Secretariat. DOE Archives.

_____. *Announced United States Nuclear Tests, July 1945 Through December 1982.* NVO-209. January 1983.

Bailey, John Wendell. "History of Sandia Base." Manuscript excerpt. SNL Archives.

Bainbridge, K. T. "Test Site." Typescript in Trinity Site Construction Folder. L.A. Archives.

Bell Telephone Laboratories. Report of the Nike B. 1 February 1954. AEC Files: MRA-5. Excerpt in Alexander Collection. SNL Archives.

Blanchard, T., Jr. "Bimonthly Progress Report of LLD Parachute Research Drop Test Program (September–October 1956)." Excerpt. Central Technical File. Sandia Technical Library.

BIBLIOGRAPHY

Bradbury, Norris E. "Los Alamos Activity." Report Prepared for Technical Director, Joint Task Force 1. 8 July 1946. L.A. Archives.

—————. "Weapon Program of the Los Alamos Laboratory." Excerpt. 27 January 1947. L.A. Archives.

—————. "Scope of the Inter-Related Work of the Sandia Laboratory and the Los Alamos Scientific Laboratory." 29 August 1949. SNL Archives.

Brodie, R. N. "Nuclear Weapon Safety, 1945–1986." Excerpt. SNL Archives.

Bush, Vannevar. "Statement to the National Policy Committee." 28 May 1945. S-1 (Uranium Committee). Folder 36. "Bush, V., 1944." Record Group 227. National Archives.

Calvin, Mark. "On-Roll Employees Who Have an E Number Beginning With a 9." Statistical Data Sheet. SNL Archives.

Carlson, Roy W. "Study of a Modified Jumbo Utilizing Ductility and Inertial Metal." No date. L.A. Archives.

Claassen, R. S. "New Weapon Proposal: Towed Laydown Weapon." Excerpt. 21 June 1957. Central Technical File. SNL Archives.

Clark, A. J., Jr. "History of Field Engineering, Test, and Manuals." No date. Manuscript in SNL Archives.

Cocke, Jim. "Quest for the Wooden Bomb." 29 September 1988. SNL Archives.

Cook, Thomas B., and Carter D. Broyles. "Curves of Atomic Weapons Effects for Various Burst Altitudes (Sea Level to 100,000 feet)." Excerpt. 9 March 1954. SNL Archives.

Dean, Gordon. Memo to Chairman, Military Liaison Committee. *AEC Missile and Rocket Responsibilities*. AEC 564/1. 22 January 1953. DOE Archives.

—————. *Operation Sandstone: 1948*. DNA Technical Report. 1948. SNL Archives.

Defense Nuclear Agency. *The First Five Tests of the Upshot-Knothole Series, 17 March–11 April 1953*. DNA 6017 F. No date. Sandia Corporate File.

—————. *Shots Encore to Climax, The Final Four Tests of the Upshot-Knothole Series, 8 May–4 June 1953*. DNA 6018 F. No date. Sandia Corporate File.

BIBLIOGRAPHY

―――――――. *Shots Able, Baker, Charlie, and Dog, The First Tests of the Tumbler-Snapper Series, 1 April–May 1952*. DNA 6020F. No date. Sandia Corporate File.

―――――――. *Shots Easy, Fox, George, and How, The Final Tests of the Tumbler-Snapper Series, 7 May–5 June 1952*. DNA 6021 F. No date. Sandia Corporate File.

―――――――. *Shots Able, Baker, Easy, Barker, and Fox, 2 January–6 February 1951*. DNA 6022 F. No date. Sandia Corporate File.

―――――――. *Operation Buster-Jangle 1951*. DNA 6023 F. No date. Sandia Corporate File.

―――――――. *Operation Greenhouse*. DNA 6034 F. No date. Sandia Corporate File.

―――――――. *Castle Series, 1954*. DNA 6035 F. No date. Sandia Corporate File.

―――――――. *IVY*. DNA 6036 F. No date. Sandia Corporate File.

Department of Energy. DOE Bulletin WNP-46A. "Declassification of Early Stockpile Data." 19 February 1982. SNL Archives.

Dike, Sheldon. "Atomic Bomb Project Aircraft." *Nuclear Weapons*. Excerpt. SNL Archives.

Director of Military Applications. "Thermonuclear Research at the University of California Radiation Laboratory." 13 June 1952. Record Group 326. Collection: Secretariat. DOE Archives.

Fowler, Glenn. Memo to Roger Warner. "Geological Investigation of the Salton Sea Test Base Area, California." SAND Report No. SC-1408 (TR). SNL Archives.

Hatch, Carl A. Report to President Truman. In Press Release, 11 July 1946. L.A. Archives.

―――――――. "Report of the Second Bikini Atom Test." In Press Release, 2 August 1946.

Henderson, Robert W. Memo to N. E. Bradbury. "Congressional Report." 25 November 1947. Microfilm in SNL Archives.

Holloway, M. G., and R. E. Schreiber. "Design and Assembly of Pit for Model 1561 Fat Man." *Nuclear Weapons*. Excerpt. SNL Archives.

Howard, W. J., and R. D. Jones. *Operation Jangle, Blast and Shock Measurements II*. Excerpt. Central Technical File. Sandia Technical Library.

BIBLIOGRAPHY

Information Research Division. "Mk 4 History." Excerpt. SNL Archives.

—————. "Mk 7 History." Excerpt. SNL Archives.

—————. "TX-9 History." Excerpt. SNL Archives.

—————. "Thermonuclear Weapons." Excerpt. No date. SNL Archives.

Joint Task Force One. "Final Report for Tests Able and Baker." *Operation Crossroads*. 18 October 1946. SNL Archives.

Kelsey, H. P., and Jean La Paz. *Quality Assurance Program for Atomic Weapons*. April 1967. SNL Archives.

Kerr, Donald M. "Forty Years of Service to the Nation." LASL Annual Report, 1982.

Lauritson, James. "Report to Vannevar Bush." 11 July 1941. Draft by MAUD. Technical Committee on the Release of Atomic Energy for Uranium.

Lenander, Harlan E. "Structure Instrumentation, Jangle Project 3.28." Armed Forces Special Weapons Project. WT-406. October 1952. Central Technical File. Sandia Technical Library.

Machen, A. B. "Mechanical Design of Model 1561 Fat Man." *Nuclear Weapons*. Excerpt. SNL Archives.

Manley, John. "Report on Z-Division (Prepared for the General Advisory Committee." 10 November 1947. L.A. Archives.

—————. "Experiment Results and Description of Available Equipment." *Los Alamos Science*. Winter/Spring 1983. L.A. Archives.

Mark, J. Carson. "A Short Account of Los Alamos Theoretical Work on Thermonuclear Weapons, 1946-1950." L.A. Archives.

Maydew, R. C. "Chronology of Parachute R&D at SNL." Draft Report. Aerodynamics Department Files. Sandia National Laboratories.

—————. "Feasibility of Parachutes for Low-Level Delivery." Excerpt. 19 January 1954. SNL Archives.

McDonald, G. C. "Weapon: Birth—Life—Death, Sandia Corporation Controlled Items." 16 November 1959. SC-TM-358-59. SNL Archives.

Merillat, John K. "Geological Investigation of the Salton Sea Test Base Area, California." SAND Report No. SC-1408 (TR).

BIBLIOGRAPHY

Miller, V. A. "Assembly of the 1561 Fat Man." *Nuclear Weapons.* Excerpt. SNL Archives.

Moffat, W. T. to R. W. Henderson. "Report on 'The Case of the Mysterious Niland Bomb.' " 6 June 1949. SNL Archives.

Newhouse Committee. "Technical and Cost Aspects of the Radar Fuze Countermeasures, An Appendix to a Systems Study of Strategic Bombing With Atomic Weapons." Excerpt. 13 July 1950. SNL Archives.

No author. *Sandstone Report 41, Scientific Director's Report of Atomic Weapon Tests.* Annex 17. Parts II and III: Sandia Laboratory Group and Forward Area Administration. Excerpt. No date. SNL Archives.

──────. *Operation Crossroads.* Commemorative album. No date.

──────. "Emergency Capability of Thermonuclear Weapons." Excerpt. Alexander Collection. SNL Archives.

──────. "The Thermonuclear Weapons Program at the Air Force Special Weapons Center." July 1952–July 1954. Excerpt in Alexander Collection. SNL Archives.

──────. "Summary of C. N. Hickman's Participation in Developing Warhead Installations for Guided Missiles." 27 March 1953. SNL Archives.

──────. "Report on Status of Ordnance Work at Y." As of 1 March 1943. L.A. Archives.

──────. "Brief Narrative and Administrative History of the 509th Composite Group." Volume 1. December 1944–April 1946. SNL Archives.

──────. "Report to the President of the United States from the Atomic Energy Commission, January 1–April 1, 1947" and "Appendix C, 2 April 1947." DOE Archives. Collection: Energy History.

──────. "Z Division Progress Reports." 29 May 1947 and 20 October 1947. SNL Archives.

──────. *Atomic Weapons Tests, Operation Sandstone, 1948.* Excerpt. Report of the Commander, Joint Task Force Seven, to the Joint Chiefs of Staff. 16 June 1948. SNL Archives.

BIBLIOGRAPHY

―――――――. AEC "Status of Underground Storage Sites, Report by the Director of Military Applications." 15 January 1949. Record Group 326. Atomic Energy Commission Collection. Secretariat. Folder 633. DOE Archives.

―――――――. "Prospective Change in the Management of Sandia Laboratory." AEC 199/3. 24 May 1949. SNL Archives.

―――――――. "Atomic Energy Commission Future Storage Requirements, Report by the Director of Military Applications." 12 May 1950. Record Group 326. Atomic Energy Commission Collection. Secretariat. Box 4943. Folder 633. DOE Archives.

―――――――. "An Analysis of the Strategic Uses of the Air-Burst Atomic Bomb with Special Reference to the Fuzing Problem." Excerpt. 24 May 1951. SNL Archives.

―――――――. AEC. "Preliminary Examination of the Work Load at Santa Fe Operations." 7 December 1951. Atomic Energy Commission Collection. Secretariat. MR AND A9-1 Weapons Development. DOE Archives.

―――――――. *A Summary of Sandia Corporation Field Testing of Development Atomic Weapons, 1946-1954*. 13 May 1955. SNL Archives.

―――――――. "Personnel Status Report, Departments 5110 and 5130." 16 January 1956. Frank Hudson Collection. SNL Archives.

Owens, Phillip R. "History of Salton Sea Test Base." SAND Report. SNL Archives.

Oppenheimer, J. Robert, John Manley, Edward Teller, et al. "Secret Report on Status of Ordnance Work at Y." No date. L.A. Archives.

Quesada, Lt. Elwood R. *History of Operation Greenhouse 1948-1951*. Joint Task Force Three. No date. Central Technical File. Sandia Technical Library.

Ramsey, Norman F. "History of Project A at Tinian." *Nuclear Weapons*. Excerpt. SNL Archives.

VanDomelen, B. H., and R. D. Wehrle. "A Review of Thermal Battery Technology." No date. Offprint in SNL Archives.

Warren, Leslie G. Statement Issued to "The Chairman, Atomic Energy Commission." Washington, D.C. 6 October 1949.

Wehrle, R. D., and R. K. Quinn. "Long-Life Thermal Battery Development." *Sandia Technical Review.* Vol. 9. January 1979. SNL Archives.

Williams, J. H. "Trip Report to Bikini Area." 29 March 1946. L.A. Archives.

Technical Articles and Unpublished Manuscripts

Bowen, Lee. "Project Silverplate, 1943–1946." In "A History of the Air Force Atomic Energy Program 1943–1953." Volume 1. Unpublished manuscript. USAF History Division, Maxwell Air Force Base Archives, Alabama. Copy in SNL Archives, Atomic Museum, and Sandia Technical Library.

Chambers, Marjorie Bell. "Technically Sweet Los Alamos. The Development of a Federally Sponsored Scientific Community." Ph.D. dissertation. 1974. University of New Mexico.

Deininger, D. Joy. "Unions—Yesterday and Today." 24 April 1986. Manuscript in SNL Archives.

Furman, N. S. "Historical Background for the Evolution of Fuzing and Firing Technology." No date. Manuscript in SNL Archives.

Hurt, D. E. "A History of Quality Assurance at Sandia National Laboratories, Albuquerque." No Date. Manuscript in SNL Archives.

Merritt, Mel. "Weapons Effects Research at Sandia." No date. Manuscript in SNL Archives.

──────────. "Weapons Test Experience." 3 January 1985. Manuscript in SNL Archives.

No author. "History of Design Services." No Date. Manuscript in SNL Archives.

──────────. "Los Lunas Bombing Range, New Mexico." No date. Manuscript in SNL Archives.

──────────. "Working Draft of Production and Manufacturing." No date. SNL Archives.

──────────. "Sandia Corporation's Job in the Nuclear Weapons Program." No date. Manuscript in SNL Archives.

Olson, D. M. "A History of the Coronado Club." No date. Manuscript in SNL Archives.

Owens, Phillip R. "Component Development at Sandia." Excerpt. 7 October 1964. Draft in SNL Archives.

Pope, Alan Y. "Some Experiences With Atomic Weapons." 26 March 1956. Manuscript in SNL Archives.

Shelton, Frank H. "Reflections of a Weaponeer." Manuscript in progress.

Stevens, W. L. "A Brief History of Early Component Development Activities." No date. Draft working paper in SNL Archives.

——————. "Research History." Manuscript in SNL Archives.

——————. "A Brief History of Early Nuclear Weapons Activities at Sandia National Laboratories 1947–1951." Manuscript in SNL Archives.

——————. "On the Characteristics of Nuclear Weapons." No date. Manuscript in SNL Archives.

——————. "On the Division of Responsibilities Between the AEC and DOD for Fuzing of Nuclear Warheads Used on Guided Missiles and Rockets 1950–1953." No date. Manuscript in SNL Archives.

——————. "On the Evolution of the AEC/DOD Agreement on Development, Production, and Standardization of Atomic Weapons of March, 1953." No date. Manuscript in SNL Archives.

Thomas, William P. "Birth and Demise of the Manufacturing Development Engineering Activities at Sandia National Laboratories 1947–1951." Manuscript in SNL Archives.

Wright, Art. "Overview of Component History." Summer 1988. Manuscript in SNL Archives.

Speeches, Films

Claassen, R. S. "Presentation to M. J. Kelly." 16 March 1957. Claassen Collection. SNL Archives.

Durham, Carl T. Speech presented at House of Representatives. 24 June 1952. Photocopy in National Archives. Washington, D.C.

BIBLIOGRAPHY

Eisenhower, Dwight D. "Address Before the General Assembly of the United Nations on Peaceful Uses of Atomic Energy." New York City, 8 December 1953.

Henderson, Robert W. "The Development of the Hydrogen Bomb." Speech presented at the National Atomic Museum. 20 February 1985. SNL Archives.

Landry, George A. "The Responsibilities of Sandia Corporation." Speech presented at Skytop Management Conference. 1950. SNL Archives.

_____. Speech to Sandia Supervisors at the Santa Fe Rotary Club. 10 January 1952. SNL Archives.

Larsen, Paul J. "Information to Sandia Laboratory Department Managers, Division Leaders and Staff Assistants." Talk given at Sandia Base Theatre. 12 May 1949.

Machen, Arthur. Film entitled "Crossroads." No date. SNL Archives.

Quarles, Donald A. "The Public Relations Aspect of Our Job." Speech to Sandia supervisors at Annual Dinner Meeting. 17 February 1953. SNL Archives.

Smith, Fred. "Industrial and Public Relations Program, Sandia Corporation." Speech presented at New Mexico Public Relations Conference. 24 January 1953. SNL Archives.

Newspapers

Bethe, Hans A. "Comments on the History of the H-Bomb." *Los Alamos Science*. Fall 1982. L.A. Archives.

Commander, Joint Task Force One. Mailgram to Public. 19 July 1946. L.A. Archives.

DeSilva, Bruce. "Military Bases: Are They Untouchable?" Albuquerque *Journal*. 28 April 1985.

Graves, Alvin C. Press Release. 13 June 1951. L.A. Archives.

Laurence, William L. "Desert Blast Held 'Baby Bombs' Debut, Atomic Explosion . . . Termed Start of 'New Era' for Tactical Weapons," *New York Times* special. Dateline 23 October 1951, Chicago. L.A. Archives.

Mitchell, Jim. News Release. Public Information Division. 16 January 1974. SNL Archives.

BIBLIOGRAPHY

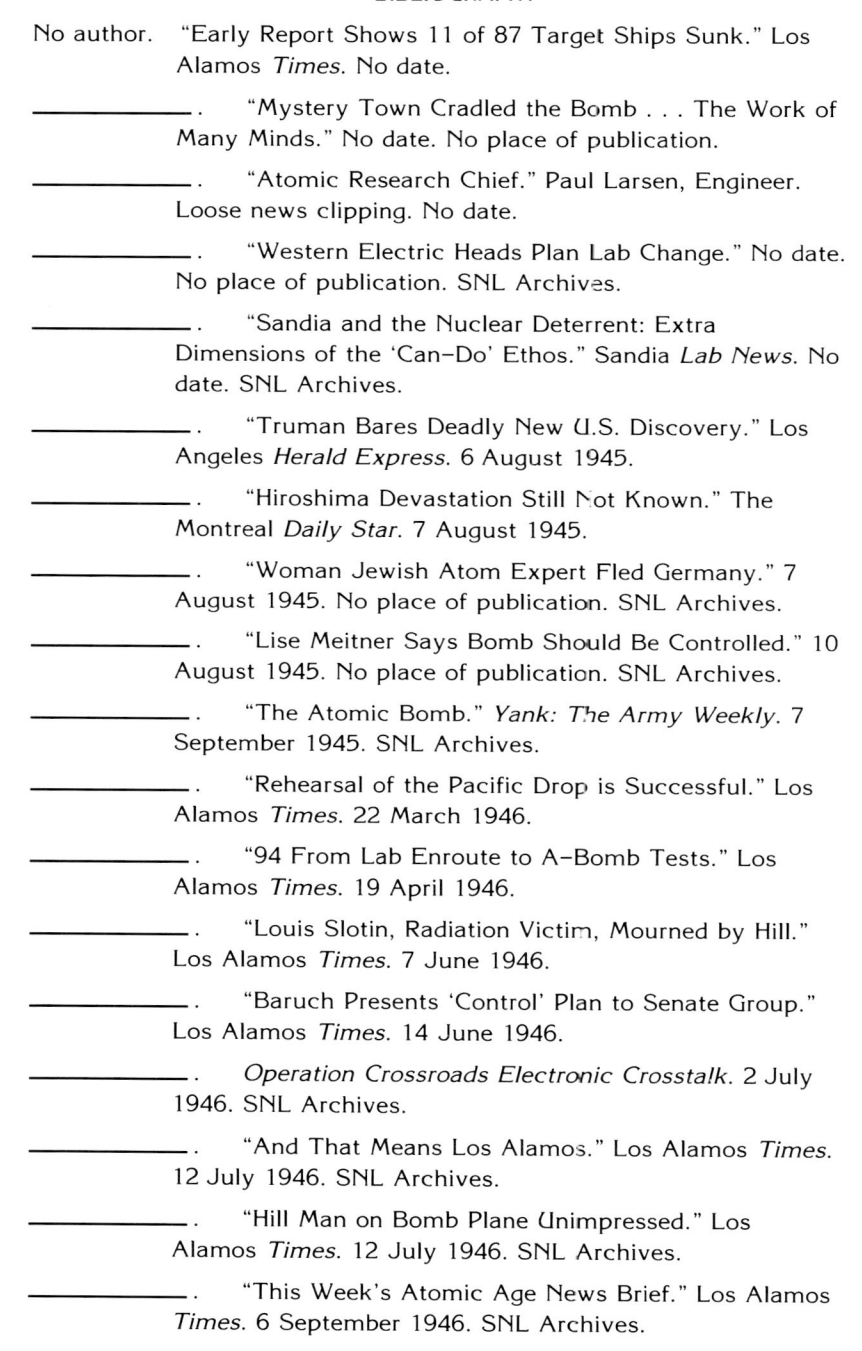

No author. "Early Report Shows 11 of 87 Target Ships Sunk." Los Alamos *Times*. No date.

——————. "Mystery Town Cradled the Bomb . . . The Work of Many Minds." No date. No place of publication.

——————. "Atomic Research Chief." Paul Larsen, Engineer. Loose news clipping. No date.

——————. "Western Electric Heads Plan Lab Change." No date. No place of publication. SNL Archives.

——————. "Sandia and the Nuclear Deterrent: Extra Dimensions of the 'Can-Do' Ethos." Sandia *Lab News*. No date. SNL Archives.

——————. "Truman Bares Deadly New U.S. Discovery." Los Angeles *Herald Express*. 6 August 1945.

——————. "Hiroshima Devastation Still Not Known." The Montreal *Daily Star*. 7 August 1945.

——————. "Woman Jewish Atom Expert Fled Germany." 7 August 1945. No place of publication. SNL Archives.

——————. "Lise Meitner Says Bomb Should Be Controlled." 10 August 1945. No place of publication. SNL Archives.

——————. "The Atomic Bomb." *Yank: The Army Weekly*. 7 September 1945. SNL Archives.

——————. "Rehearsal of the Pacific Drop is Successful." Los Alamos *Times*. 22 March 1946.

——————. "94 From Lab Enroute to A-Bomb Tests." Los Alamos *Times*. 19 April 1946.

——————. "Louis Slotin, Radiation Victim, Mourned by Hill." Los Alamos *Times*. 7 June 1946.

——————. "Baruch Presents 'Control' Plan to Senate Group." Los Alamos *Times*. 14 June 1946.

——————. *Operation Crossroads Electronic Crosstalk*. 2 July 1946. SNL Archives.

——————. "And That Means Los Alamos." Los Alamos *Times*. 12 July 1946. SNL Archives.

——————. "Hill Man on Bomb Plane Unimpressed." Los Alamos *Times*. 12 July 1946. SNL Archives.

——————. "This Week's Atomic Age News Brief." Los Alamos *Times*. 6 September 1946. SNL Archives.

BIBLIOGRAPHY

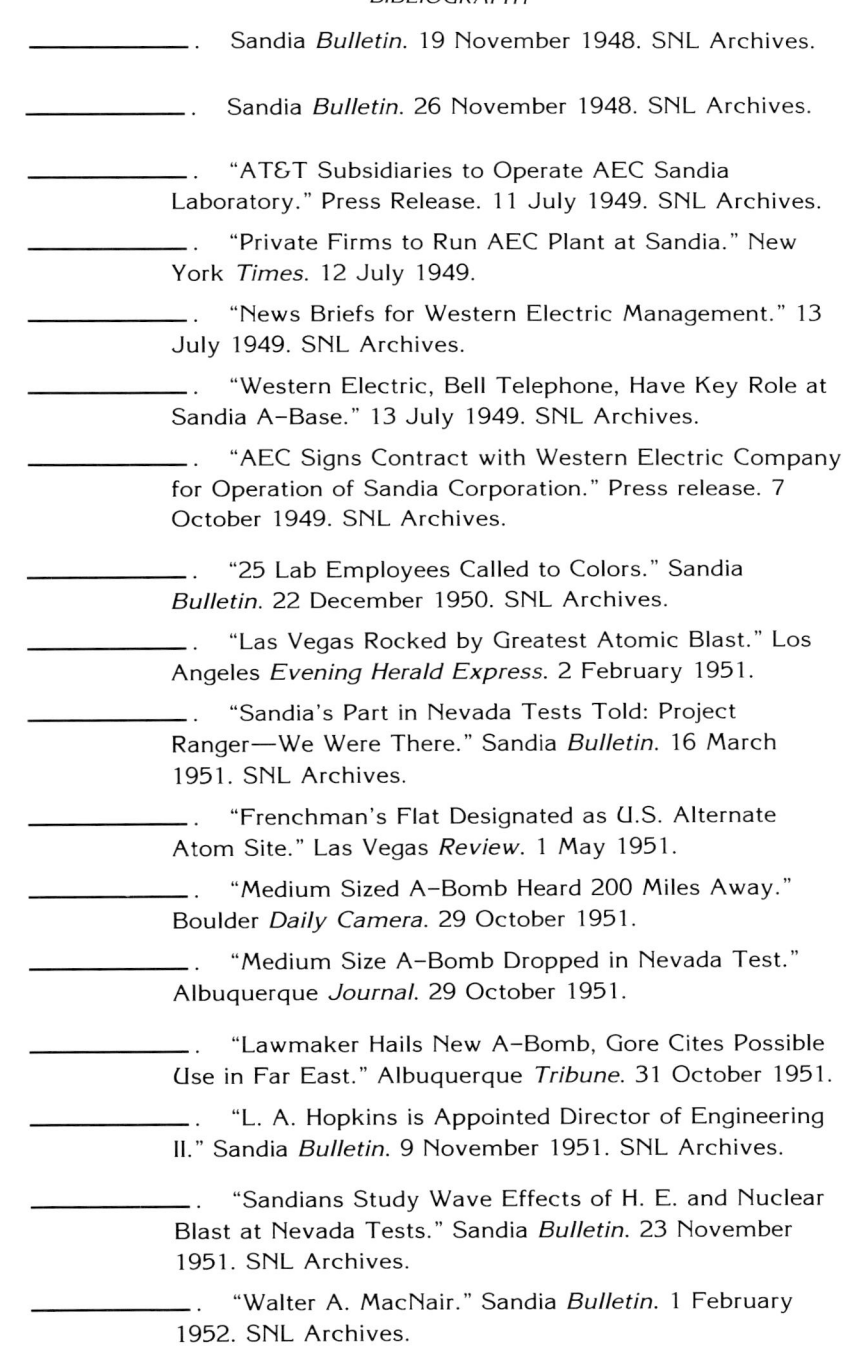

_____. Sandia *Bulletin*. 19 November 1948. SNL Archives.

_____. Sandia *Bulletin*. 26 November 1948. SNL Archives.

_____. "AT&T Subsidiaries to Operate AEC Sandia Laboratory." Press Release. 11 July 1949. SNL Archives.

_____. "Private Firms to Run AEC Plant at Sandia." New York *Times*. 12 July 1949.

_____. "News Briefs for Western Electric Management." 13 July 1949. SNL Archives.

_____. "Western Electric, Bell Telephone, Have Key Role at Sandia A-Base." 13 July 1949. SNL Archives.

_____. "AEC Signs Contract with Western Electric Company for Operation of Sandia Corporation." Press release. 7 October 1949. SNL Archives.

_____. "25 Lab Employees Called to Colors." Sandia *Bulletin*. 22 December 1950. SNL Archives.

_____. "Las Vegas Rocked by Greatest Atomic Blast." Los Angeles *Evening Herald Express*. 2 February 1951.

_____. "Sandia's Part in Nevada Tests Told: Project Ranger—We Were There." Sandia *Bulletin*. 16 March 1951. SNL Archives.

_____. "Frenchman's Flat Designated as U.S. Alternate Atom Site." Las Vegas *Review*. 1 May 1951.

_____. "Medium Sized A-Bomb Heard 200 Miles Away." Boulder *Daily Camera*. 29 October 1951.

_____. "Medium Size A-Bomb Dropped in Nevada Test." Albuquerque *Journal*. 29 October 1951.

_____. "Lawmaker Hails New A-Bomb, Gore Cites Possible Use in Far East." Albuquerque *Tribune*. 31 October 1951.

_____. "L. A. Hopkins is Appointed Director of Engineering II." Sandia *Bulletin*. 9 November 1951. SNL Archives.

_____. "Sandians Study Wave Effects of H. E. and Nuclear Blast at Nevada Tests." Sandia *Bulletin*. 23 November 1951. SNL Archives.

_____. "Walter A. MacNair." Sandia *Bulletin*. 1 February 1952. SNL Archives.

BIBLIOGRAPHY

―――――. "Sandia Corporation Expands." Albuquerque *Tribune*. 1 February 1952.

―――――. "Mr. Quarles Will Speak at Meeting of Engineer Group." Sandia *Bulletin*. 9 May 1952. SNL Archives.

―――――. "D. A. Quarles Succeeds Mr. Landry as President of Sandia Corporation." Sandia *Bulletin*. 29 February 1952. SNL Archives.

―――――. "James W. McRae, Bell Telephone Laboratories, Inc." Biographical data. July 1953. AT&T Archives.

―――――. "Defense Post to D. Quarles; J. McRae New Sandia Head." *Kearnygram*. August 1953. AT&T Archives.

―――――. "Dr. James W. McRae Transferred to Atomic Development Project." Madison *Eagle*. 6 August 1953. AT&T Archives.

―――――. "James W. McRae Heads Corp.; Mr. Quarles Takes Defense Post." Sandia *Bulletin*. 14 August 1953. SNL Archives.

―――――. "James W. McRae Takes Up New Duties as Laboratory President." Sandia *Bulletin*. 25 September 1953. SNL Archives.

―――――. "How Can You Be Taxed if You Can't Vote? And Why Can't You Vote?" Sandia *Lab News*. 12 March 1954. SNL Archives.

―――――. "A Newspaper Goes to Press." Sandia *Lab News*. 18 June 1954. SNL Archives.

―――――. Sandia *Lab News*. 17 December 1954.

―――――. "Sandia Personnel Preparing for AEC Nevada Atomic Tests." Sandia *Lab News*. 28 January 1955. SNL Archives.

―――――. "Baby Atom Blast Set Off in Nevada." Los Angeles *Times*. 20 February 1955.

―――――. "Atomic Blast Jolts Cities for 135 Miles." Norfolk-Virginia *Pilot*. 23 February 1955.

―――――. "Sandians View First Shot of AEC's Operation Teapot." Sandia *Lab News*. 25 February 1955. SNL Archives.

―――――. "Sandia Corporation Role in A-Tests Big One." Albuquerque *Tribune*. 15 March 1955.

BIBLIOGRAPHY

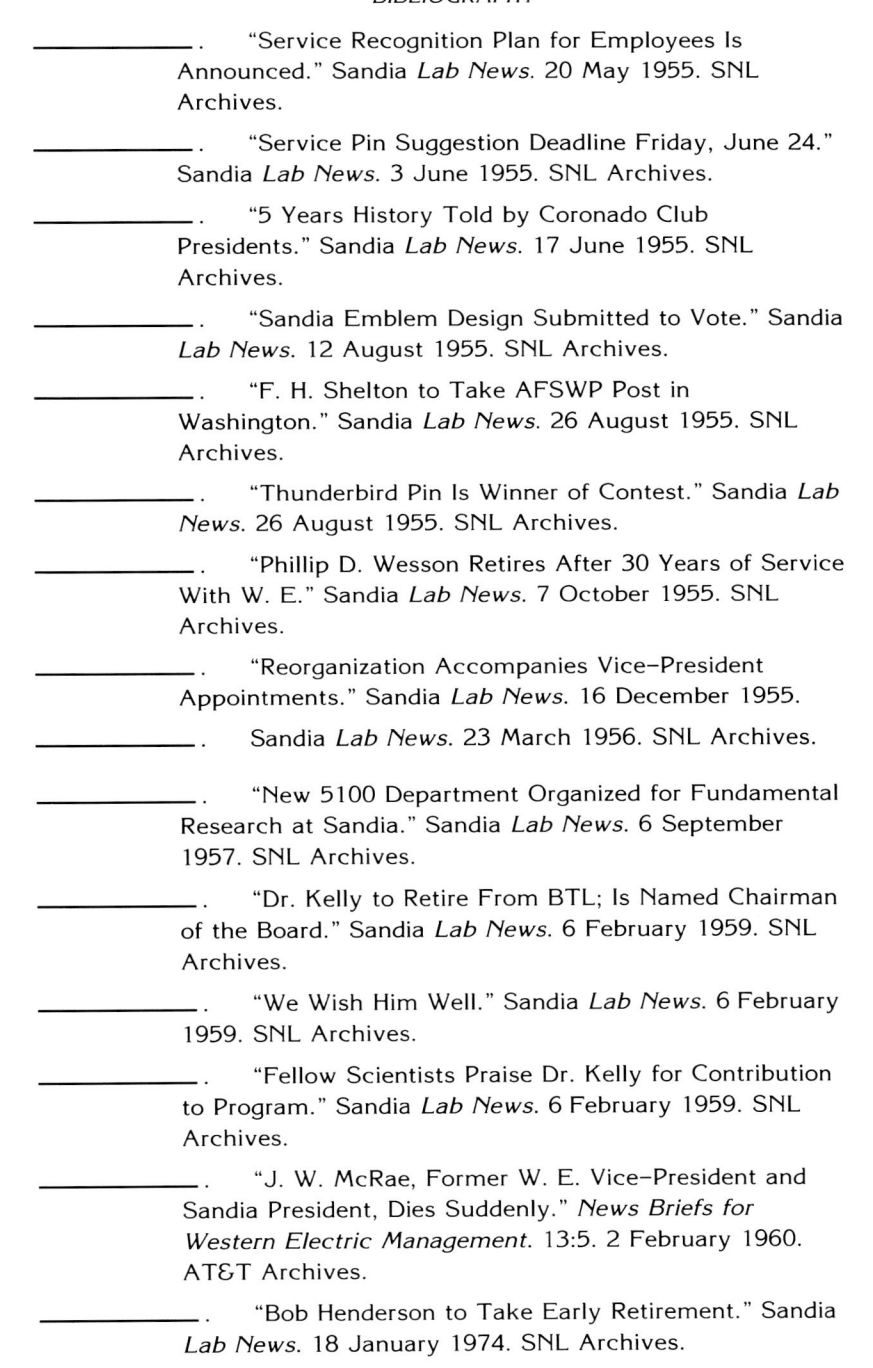

_____. "Service Recognition Plan for Employees Is Announced." Sandia *Lab News*. 20 May 1955. SNL Archives.

_____. "Service Pin Suggestion Deadline Friday, June 24." Sandia *Lab News*. 3 June 1955. SNL Archives.

_____. "5 Years History Told by Coronado Club Presidents." Sandia *Lab News*. 17 June 1955. SNL Archives.

_____. "Sandia Emblem Design Submitted to Vote." Sandia *Lab News*. 12 August 1955. SNL Archives.

_____. "F. H. Shelton to Take AFSWP Post in Washington." Sandia *Lab News*. 26 August 1955. SNL Archives.

_____. "Thunderbird Pin Is Winner of Contest." Sandia *Lab News*. 26 August 1955. SNL Archives.

_____. "Phillip D. Wesson Retires After 30 Years of Service With W. E." Sandia *Lab News*. 7 October 1955. SNL Archives.

_____. "Reorganization Accompanies Vice-President Appointments." Sandia *Lab News*. 16 December 1955.

_____. Sandia *Lab News*. 23 March 1956. SNL Archives.

_____. "New 5100 Department Organized for Fundamental Research at Sandia." Sandia *Lab News*. 6 September 1957. SNL Archives.

_____. "Dr. Kelly to Retire From BTL; Is Named Chairman of the Board." Sandia *Lab News*. 6 February 1959. SNL Archives.

_____. "We Wish Him Well." Sandia *Lab News*. 6 February 1959. SNL Archives.

_____. "Fellow Scientists Praise Dr. Kelly for Contribution to Program." Sandia *Lab News*. 6 February 1959. SNL Archives.

_____. "J. W. McRae, Former W. E. Vice-President and Sandia President, Dies Suddenly." *News Briefs for Western Electric Management*. 13:5. 2 February 1960. AT&T Archives.

_____. "Bob Henderson to Take Early Retirement." Sandia *Lab News*. 18 January 1974. SNL Archives.

———. "Weapon Longevity is Engineered." Sandia *Lab News*. 10 October 1986. SNL Archives.

———. "Working Toward the Maintenance-Free Weapon." Sandia *Lab News*. 10 October 1986. SNL Archives.

Prince, Kimball. "History of Formation." Albuquerque *Journal*. 14 July 1949.

Shalett, Sidney. "The Buck Rogers of the Navy." New York *Times*. 10 January 1946.

Journals and Articles

Campbell, Robert, et al. "Field Testing, the Physical Proof of Design Principles." *Los Alamos Science*. Winter/Spring 1983.

Graybar, Lloyd J. "Bikini Revisited." *Military Affairs 44*. October 1980.

———. "The 1946 Atomic Bomb Tests: Atomic Diplomacy or Bureaucratic Infighting." *Journal of American History* 72. March 1986.

Hahn, Otto. "The Discovery of Fission." *The Scientific American*. February 1958.

Hoddeson, Lillian. "The Emergence of Basic Research in the Bell Telephone System, 1875–1915." *Technology and Culture* 22. 1981.

Lilienthal, David E. "The Kind of Nation We Want." *Colliers*. 14 June 1952.

Miller, Byron S. "A Law is Passed—The Atomic Energy Act of 1946." *The University of Chicago Law Review* 15. Summer 1948.

Napier, Arch. "Sandia Corporation: On the Frontier of Engineering." *Sandia Corporation*. No date. SNL Archives.

Needell, Allan A. "Nuclear Reactors and the Founding of Brookhaven National Laboratory." *Historical Studies in the Physical and Biological Sciences* 14, 1.

No author. *Records of the Manhattan Engineer District, 1942–1948*. World War II Records Division, National Archives and Records Service. No date. Alexandria, Virginia.

BIBLIOGRAPHY

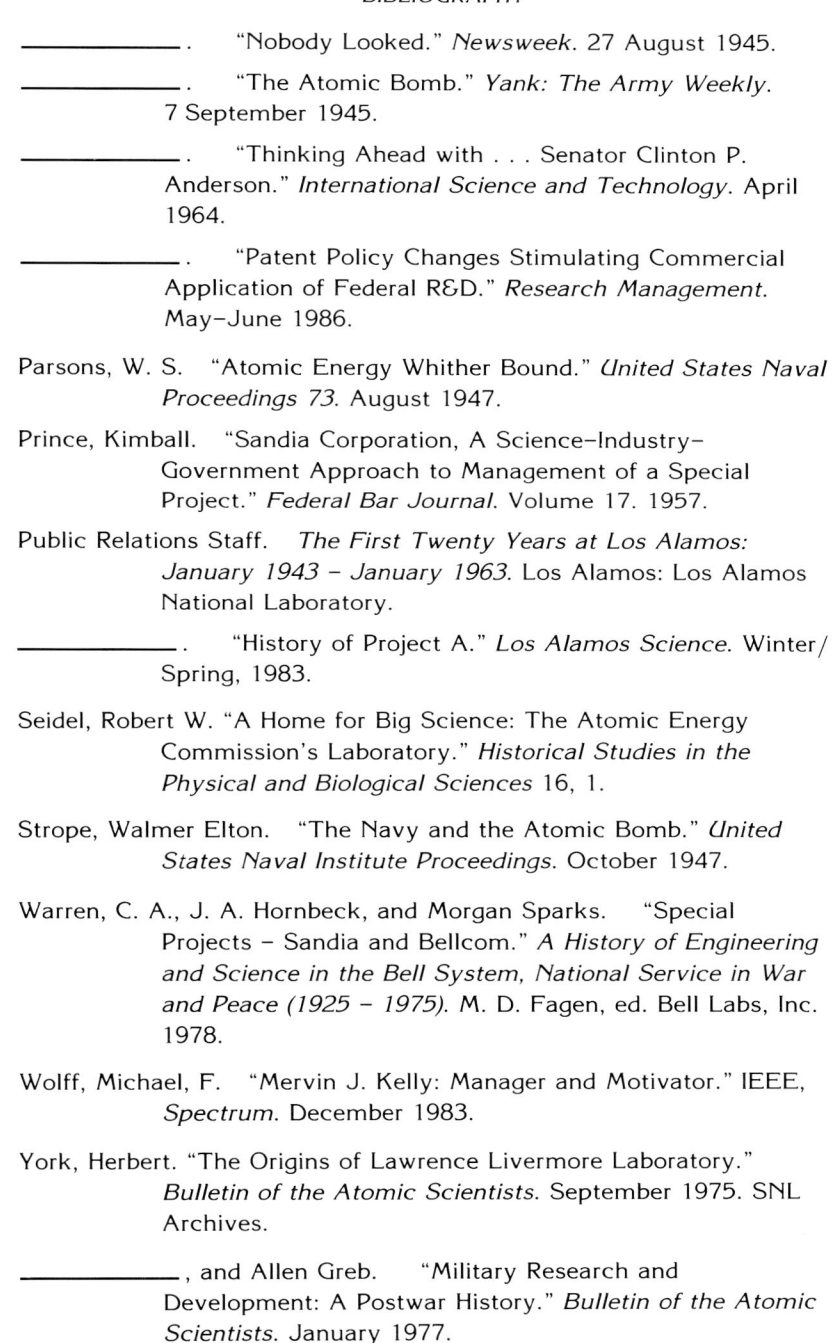

――――――. "Nobody Looked." *Newsweek*. 27 August 1945.

――――――. "The Atomic Bomb." *Yank: The Army Weekly*. 7 September 1945.

――――――. "Thinking Ahead with . . . Senator Clinton P. Anderson." *International Science and Technology*. April 1964.

――――――. "Patent Policy Changes Stimulating Commercial Application of Federal R&D." *Research Management*. May–June 1986.

Parsons, W. S. "Atomic Energy Whither Bound." *United States Naval Proceedings 73*. August 1947.

Prince, Kimball. "Sandia Corporation, A Science–Industry–Government Approach to Management of a Special Project." *Federal Bar Journal*. Volume 17. 1957.

Public Relations Staff. *The First Twenty Years at Los Alamos: January 1943 – January 1963*. Los Alamos: Los Alamos National Laboratory.

――――――. "History of Project A." *Los Alamos Science*. Winter/Spring, 1983.

Seidel, Robert W. "A Home for Big Science: The Atomic Energy Commission's Laboratory." *Historical Studies in the Physical and Biological Sciences* 16, 1.

Strope, Walmer Elton. "The Navy and the Atomic Bomb." *United States Naval Institute Proceedings*. October 1947.

Warren, C. A., J. A. Hornbeck, and Morgan Sparks. "Special Projects – Sandia and Bellcom." *A History of Engineering and Science in the Bell System, National Service in War and Peace (1925 – 1975)*. M. D. Fagen, ed. Bell Labs, Inc. 1978.

Wolff, Michael, F. "Mervin J. Kelly: Manager and Motivator." IEEE, *Spectrum*. December 1983.

York, Herbert. "The Origins of Lawrence Livermore Laboratory." *Bulletin of the Atomic Scientists*. September 1975. SNL Archives.

――――――, and Allen Greb. "Military Research and Development: A Postwar History." *Bulletin of the Atomic Scientists*. January 1977.

BIBLIOGRAPHY

Directorate Histories Relating to Postwar Decade

Banister, John. "History of the Field Engineering Directorate." SNL Archives.

Bickel, Dave. "Rocket Powered Centrifuge." Environmental Test Collection. SNL Archives.

Bohr, Donald, Barbara Kerr, Captain Allen Strouphauer, United States Army, and Donald Gregson. *History 8100 Directorate*. SNL Archives.

Brown, Marvin H. "A History of Local No. 27, International Guards Union of America." Manuscript in SNL Archives.

Clark, A. J. "History of Field Engineering, Test and Manuals." SNL Archives.

Cocke, Jim. "Weapons Directorate History.' SNL Archives.

──────. "5100 Weapons Development." Manuscript in SNL Archives.

Crumley, H. R., and George Kupper. "Purchasing at Sandia." SNL Archives.

Eiffert, Art. "History of Design Engineering." SNL Archives.

Graham, Bob. "Shock Wave Physics History." SNL Archives.

Hogan, Ann. "History of Directorate 3400." Manuscript in SNL Archives.

Hurt, D. E. "History of Quality Assurance at Sandia National Laboratories." SNL Archives.

Johnson, Sam. "Building Construction 1946 through 1956." *History of Plant Engineering*. SNL Archives.

Kupper, G. T. "Purchasing History." SNL Archives.

Marron, J. J. to author. 7 January 1985. Saxton Component History Collection. SNL Archives.

Maydew, Randall C. "History of Nuclear Weapons Parachute Responsibilities." SNL Archives.

No author. "Public Relations Highlights." 3100 Information Services Collection. SNL Archives.

──────. "SNL Budget History." 10 November 1950. SNL Archives.

Peterson, Ernest. "History of Labor Unions at Sandia." 1958. Excerpt in SNL Archives.

BIBLIOGRAPHY

Sherwin, Ted. Information and Communication Services Collection. SNL Archives.

Thomas, Bill. "Birth and Demise of the Manufacturing Development Engineering Technical Organization at Sandia." SNL Archives.

Turney, Scott. "Personnel Directorate History." SNL Archives.

Wehrle, R. D. "History of Power Sources at Sandia, Part 1." Component History Collection. SNL Archives.

Willis, Hank. Information Services Collection. SNL Archives.

Books and Monographs

Alberts, Don E., and Allan E. Putnam. *A History of Kirtland Air Force Base, 1928-1982.* Albuquerque: Kirtland Air Force Base, 1982.

──────. *Balloons to Bombers, Aviation in Albuquerque, 1882-1945.* Albuquerque: Albuquerque Museum, 1987.

Alexander, Frederick C. *History of Sandia Corporation Through Fiscal Year 1963.* Albuquerque: Sandia Corporation, 1963.

Alperovitz, Gar. *Atomic Diplomacy: Hiroshima and Potsdam.* New York: Simon and Schuster, 1965.

Amrine, Michael. *The Great Decision: The Secret History of the Atomic Bomb.* New York: G. P. Putnam's Sons, 1959.

Anders, Roger, ed. *Forging the Atomic Shield, Excerpts from the Office Diary of Gordon E. Dean.* Chapel Hill: University of North Carolina Press, 1987.

Armacost, Michael H. *The Politics of Weapons Innovation, the Thor-Jupiter Controversy.* New York: Columbia University Press. 1969.

Atholl, Justin. *How Stalin Knows.* London: Jarrold and Sons, Ltd. No date.

Arnold, Henry H. *Global Mission.* New York: n. p., 1949.

Ball, Howard. *Justice Downwind, America's Atomic Testing Program in the 1950s.* New York: Oxford University Press. 1986.

Barnett, Richard J. *Who Wants Disarmament?* Boston: Beacon Press, 1960.

BIBLIOGRAPHY

Batchelder, Robert C. *The Irreversive Decision*. New York: MacMillan, 1961.

Bauer, Theodore W., and Eston T. White. *Defense Organization and Management. National Security Management Series*. Washington, D.C.: Industrial College of the Armed Forces, VI Defense Agencies, 1975.

Beard, Charles A. *President Roosevelt and the Coming of the War*. New Haven: Yale University Press, 1948.

Bechhoefer, Bernard G. *Postwar Negotiations for Arms Control*. Washington, D.C.: Brookings Institute, 1961.

Bernstein, Barton J. *The Atomic Bomb: The Critical Issues*. Boston: Little, Brown and Company, 1976.

Blumberg, Stanley A., and Gwinn Owens. *The Life and Times of Edward Teller*. New York: Putnam's Sons. 1976.

Bowen, Lee, and Robert D. Little, et al. *A History of the Air Force Atomic Energy Program, 1942–1953*. 1959.

Bower, Tom. *The Paperclip Conspiracy, The Hunt for the Nazi Scientists*. Boston: Little, Brown and Company. 1987.

Bradley, David. *No Place to Hide*. Boston: Little, Brown and Company, 1984.

Bramlett, William, ed.. *International Guards Union of America*. Volume 1. No. 10. August–September 1958. SNL Archives.

Brooks, John. *Telephone, The First Hundred Years*. New York: Harper and Row, 1975.

Buck, Alice L. *A History of the Atomic Energy Commission*. Washington, D.C.: United States Department of Energy, 1982.

Burns, James McGregor. *Roosevelt: The Soldier of Freedom, 1940–1945*. New York: Harcourt, Brace, Jovanovich, 1970.

Byrnes, James F. *Speaking Frankly*. New York: Harper and Brothers Publishers, 1947.

Carter, Ashton B., John D. Steinbruner, Charles A. Zraket, eds. *Managing Nuclear Operations*. Washington, D.C.: Brookings Institute. 1986.

Cave-Brown, Anthony, and Charles B. MacDonald, eds. *The Secret History of the Atomic Bomb*. New York: Dell Publishing Company, Inc., 1977.

Churchill, Winston. *Triumph and Tragedy*. Boston: Houghton Mifflin Company, 1962.

Craven, Frank, and James Lea Cate, eds. *The Army Air Forces in World War II*. Volume 5. Chicago: University of Chicago Press, 1953.

Davis, Nuel Pharr. *Lawrence and Oppenheimer*. New York: Simon and Schuster, 1968.

Emmé, Eugene M., ed. *The History of Rocket Technology*. Detroit: Wayne State University Press, 1964.

Fagen, M. D. *A History of Engineering and Science in the Bell System–National Service in War and Peace (1925–1975)*. Bell Laboratories, 1978.

Fuller, John G. *The Day We Bombed Utah: America's Most Lethal Secret*. New York: New American Library, 1984.

Gartman, Heintz. *The Man Behind the Space Rockets*. London: Weidenfeld and Nicolson, 1955.

Goddard, Robert H. *Rocket Development, Liquid–Fuel Rocket Research, 1929–1941*. Ed. by Esther C. Goddard and G. Edward Pendray. New York: Prentice–Hall, 1948.

Goldman, Eric F. *The Crucial Decade, America, 1945–1955*. New York: Alfred A. Knopf, 1956.

Goodchild, Peter. *J. Robert Oppenheimer, Shatterer of Worlds*. Boston: Houghton Mifflin Company, 1981.

Green, Harold, and Alan Rosenthal. *Government of the Atom*. New York: Atherton Press, 1963.

Groueff, Stephane. *Manhattan Project: The Untold Story of the Making of the Atomic Bomb*. Boston: Little, Brown and Company, 1976.

Hansen, Chuck. *U. S. Nuclear Weapons, The Secret History*. New York: Orion Books, 1988.

Hawkins, David. *Manhattan District History, Project Y*. Washington, D.C.: Office of Technical Services, 1945–1947.

——————. Part I. "Toward Trinity." In *Project Y: Los Alamos Story*, Volume 2. In *The History of Modern Physics, 1800–1950*, #2 in series. Los Angeles: Tomash Publishers, 1983.

Herken, Gregg. *The Winning Weapon, The Atomic Bomb in the Cold War, 1945–1950*. New York: Vintage Books, 1982.

BIBLIOGRAPHY

Hewlett, Richard G., and Oscar E. Anderson, Jr. *The New World, 1939/1946, A History of the United States Atomic Energy Commission.* Volume I. University Park: Pennsylvania State University Press, 1962.

Hewlett, Richard G., and Francis Duncan. *The Atomic Shield, 1947/1952.* Volume II. University Park: Pennsylvania State University Press, 1969.

Hines, Neal O. *Proving Ground, An Account of the Radiobiological Studies in the Pacific, 1946-1961.* Seattle: Washington Press, 1962.

Hodgson, Godfrey. *America in Our Time.* New York: Vintage Books, 1976.

Holl, Jack M., Roger M. Anders, and Alice L. Buck. *United States Civilian Nuclear Power Policy, 1954-1984: A Summary History.* Washington, D.C.: Department of Energy. February 1986.

Johnson, Paul. *Modern Times: The World From the Twenties to the Eighties.* New York: Harper and Row, 1983.

Kaman, Charles H. *Kaman, Our Early Years.* Indianapolis: Curtis Publishing Company, 1985.

Kevles, Daniel J. *The Physicists, The History of a Scientific Community in Modern America.* New York: Alfred A. Knopf, 1978.

Knebel, Fletcher, and Charles W. Bailey. *No High Ground.* New York: Harper and Brothers Publishers, 1960.

Kunetka, James W. *City of Fire, Los Alamos and the Birth of the Atomic Age, 1943-1945.* Englewood Cliffs, N.J.: Prentice-Hall, Inc., 1978.

Lamont, Lansing. *Day of Trinity.* New York: Atheneum, 1965.

Lauren, William. *The General and the Bomb, A Biography of General Leslie R. Groves, Director of the Manhattan Project.* New York: Dodd, Mead and Company, 1988.

Laurence, William L. *Dawn Over Zero: The Story of the Atomic Bomb.* New York: Alfred A. Knopf, 1947.

Lehman, Milton. *This High Man, The Life of Robert H. Goddard.* New York: Farrar, Strauss and Company, 1963.

Libby, Leona Marshall. *The Uranium People.* New York: Crane Rusak and Company, 1979.

BIBLIOGRAPHY

Manchester, William. *The Glory and the Dream: A Narrative History of America, 1932-1972.* Volume I. Boston: Little, Brown and Company, 1973.

Mojtabai, A. G. *Blessed Assurance: At Home With the Bomb in Amarillo, Texas.* Boston: Houghton Mifflin Company, 1986.

Mazuzan, George T., and J. Samuel Walker. *Controlling the Atom, The Beginnings of Nuclear Regulation, 1946-1962.* Berkeley: University of California Press, 1984.

Miller, Richard L. *Under The Cloud, The Decades of Nuclear Testing.* New York: The Free Press, 1986.

Moss, Norman. *Klaus Fuchs, The Man Who Stole the Atom Bomb.* New York: St. Martin's Press. 1987.

No author. *Weaponization: Sandia Corporation and the Nuclear Weapons Program.* No date. SNL Archives.

——————. *509th Pictorial Album.* U.S. Air Force. No date.

No author. *A Summary of 2100: 1949-1951.* Sandia Corporation. No date. SNL Archives.

——————. *Sandia Corporation Telephone Directory.* 1949-1956. SNL Archives.

——————. *The Government and Atomic Energy.* Albuquerque, New Mexico: Kirtland Air Force Base, 1958.

——————. *The Conference of Berlin* (The Potsdam Conference, 1945). Volumes I & II. Washington, D.C., 1961.

——————. *Who Was Who in America.* Volume 4. 1961-1968.

——————. *A Case Study of the Utilization of Federal Laboratory Resources.* Washington, D.C.: United States Government Printing Office, 1966.

——————. *Information on AEC Interface with Concept Formulation and Contract Definition.* Washington, D.C.: United States Government Printing Office, 1970.

——————. *In the Matter of J. Robert Oppenheimer.* Cambridge, Massachusetts: MIT Press, 1971.

——————. *Story of Albuquerque Operations.* Albuquerque: Department of Energy, 1978.

——————. *Atomic Energy Legislation Through 95th Congress.* Washington, D.C.: United States Government Printing Office, 1979.

BIBLIOGRAPHY

Newman, James R. and Byron S. Miller. *The Control of Atomic Energy*. New York: McGraw-Hill Books Company, 1948.

Nogee, Joseph L. *Soviet Policy Toward International Control of Atomic Energy*. Notre Dame: University of Notre Dame Press, 1961.

O'Keefe, Bernard. *Nuclear Hostages*. Boston: Houghton Mifflin Company, 1983.

Quesada, Lt. Elwood R. *History of Operation Greenhouse 1948-1951*. Joint Task Force Three. No date. Central Technical File. Sandia Technical Library.

Orlans, Howard. *Contracting for Atoms*. Westport, Connecticut: Greenwood Press, Publishers, 1967.

Pringle, Peter, and James Spigelman. *The Nuclear Barons*. New York: Holt, Rinehart, and Winston, 1981.

Public Relations Office. *Los Alamos, The Beginning of an Era, 1943-1945*. Los Alamos: Los Alamos National Laboratory, No date.

Radosh, Ronald, and Joyce Milton. *The Rosenberg File*. New York: Holt, Rinehart, and Winston, 1983.

Rhodes, Richard. *The Making of the Atomic Bomb*. New York: Simon and Schuster, 1988.

Schilling, Warner R., Paul Y. Hammond, and Glenn H. Snyder. *Strategy, Politics, and Defense Budgets*. New York: Columbia University Press, 1962.

Science Policy Research Division. *A Case Study of the Utilization of Federal Laboratory Resources*. Washington, D.C.: United States Government Printing Office. 1966.

Seidel, Roger, and John Hielbron. *LLL, Twenty-Five Years*. University of California, Berkeley: Lawrence Berkeley Laboratory, 1981.

Shurcliff, William A. *Bombs at Bikini: The Official Report of Operation Crossroads*. New York: William H. Wise and Company, 1947.

Smith, Alice Kimball. *A Peril and a Hope, The Scientists' Movement in America: 1945-1947*. Chicago: University of Chicago Press, 1965.

Storms, Barbara. *Reach to the Unknown*. Los Alamos: Los Alamos National Laboratory, 1965.

BIBLIOGRAPHY

Strauss, Lewis L. *Men and Decisions.* New York: Doubleday and Company, 1948.

Szasz, Ferenc. *The Day the Sun Rose Twice: The Story of the Trinity Site Nuclear Explosion, July 16, 1945.* Albuquerque: University of New Mexico Press, 1984.

Teller, Edward with Allen Brown. *Legacy of Hiroshima.* Garden City: Doubleday and Company, 1962.

Titus, A. Costandina. *Bombs in the Backyard, Atomic Testing and American Politics.* Reno and Las Vegas: University of Nevada Press, 1986.

Truslow, Edith C., and Ralph Carlisle Smith. "Beyond Trinity," Part 2. In "Project Y: Los Alamos Story," Volume 2. In *The History of Modern Physics, 1800-1950,* #2 in series. Los Angeles: Tomash Publishers, 1983.

Tunstall, W. Brooke. *Disconnecting Parties, Managing the Bell System Break-Up: An Inside View.* New York: McGraw-Hill Books Company, 1985.

Ulam, Stanley. *Adventures of a Mathematician.* New York: Charles Scribners' Sons, 1976.

von Braun, Wernher, and Frederick I. Ordway III. *History of Rocketry and Space Travel.* New York: Thomas Y. Crowell Company, 1975.

Williams, Robert Chadwell. *Klaus Fuchs, Atom Spy.* Cambridge, Massachusetts: Harvard University Press. 1987.

Wyden, Peter. *Day One, Before Hiroshima and After.* New York: Simon and Schuster, 1984.

York, Herbert. *Race to Oblivion: A Participant's View of the Arms Race.* New York: Simon and Schuster, 1970.

──────. *The Advisors, Oppenheimer, Teller, and the Superbomb.* San Francisco: W. H. Freeman and Company. 1976.

PHOTO CREDITS

Acknowledgment is hereby made to all who have kindly granted permission to reprint copyrighted photographs and exhibits. Recognition is given also to the many individuals, agencies, institutions, and organizations who have generously provided photographs and exhibits. Other visuals appearing in this book may be found in the photographic collection of the SNL Archives.

PROLOGUE: (2) UPI/Bettman Newsphotos; (10) G. W. Szilard/American Institute of Physics; (13) University of California, Lawrence Berkeley Laboratory; (19) Los Alamos National Laboratory (LANL); (21) above: Franklin D. Roosevelt Library, below: Harry S. Truman Library; (4) reprint courtesy Albuquerque *Journal.*

CHAPTER 1: (29) Harry S. Truman Library; (34) LANL; (35) LANL; (36) LANL; (37) LANL; (38) LANL; (39) above: UPI/Bettman Newsphotos, below: LANL; (42) Glenn A. Fowler.

CHAPTER 2: (44) LANL; (45) above: LANL, below: Leon D. Smith; (53, 70, 71) LANL; (87) Ben Benjamin.

CHAPTER 3: (88, 91) Leon D. Smith; (95) above: LANL, below: National Atomic Museum (NAM); (110, 111) Leon D. Smith; (112) NAM; (113) LANL; (114) above: Leon D. Smith, below left: Leon D. Smith, below right: LANL; (115) Leon D. Smith.

CHAPTER 4: (120) LANL; (124) Frank Speakman Collection, Albuquerque Museum; (131) above: The MIT Museum, below: LANL; (135) Art Eiffert; (142) Frank Speakman Collection, Albuquerque Museum; (144, 145) J. J. Miller.

PHOTO CREDITS

CHAPTER 5: (149) Glenn A. Fowler; (162–168) Howard Austin; (169) above: Guy Willis, below: Howard Austin; (171) above: Howard Austin, below: B. G. Edwards; (175) Howard Austin; (179) Ben Benjamin.

CHAPTER 6: (193) above: Leon D. Smith; (200) Art Machen; (201) LANL; (202) Ira "Tiny" Hamilton; (204) below: Robert W. Henderson; (205) Robert W. Henderson

CHAPTER 7: (206) Harris and Ewing photo courtesy Harry S. Truman Library; (211) UPI/Bettman Newsphotos; (212) top left: Library of Congress, all others: Department of Energy, Headquarters (DOE); (229) Robert Hopper.

CHAPTER 8: (230, 237) LANL.

CHAPTER 9: (263, 24) Oswald Tjeltweed.

CHAPTER 11: (316) Louis J. Paddison.

CHAPTER 12: (328, 343) AT&T Archives.

CHAPTER 14: (364, 376) AT&T Archives; (387) Robert Hopper; (388) Sam Johnson.

CHAPTER 16: (445) Art Jimenez; (446) lower right: Jim Hinson; (447) Jim Hinson; (448) William Bramlett.

CHAPTER 17: (479) Ted Sherwin; (481) H. Clyde Walker.

CHAPTER 18: (487) Kenner Hertford; (500, 510–513) Andrew A. Lieber.

CHAPTER 19: (526) AT&T Archives; (532) Rosalie Crawford.

CHAPTER 20: (566, 568) LANL; (573) Don Shuster; (586) Robert W. Henderson; (587) above: LANL, below: Lawrence Livermore National Laboratory (LLNL); (588) above: Glenn Fowler, below: Robert W. Henderson; (589) Robert W. Henderson; (593) LANL.

PHOTO CREDITS

CHAPTER 21: (596, 598, 599) Defense Nuclear Agency; (606) Kenner Hertford; (610) Carter Broyles.

CHAPTER 22: (628) AT&T Archives; (632) DOE; (666) Fred Duimstra; (670, 671) Frank Hudson.

INDEX

Abee radar, 272
Abeyta, Greg, 158
Able Day, Crossroads Test, 191–96,
"Able" storage site, 324, 326
Abraham, L. G., 528
Acheson, Dean: as Secretary of State, 209, 565
Ad Hoc Committee on Guided Missiles, 489, 497, 499
Adams, Paul, 517
Adamson, Colonel Keith F., 12
Adaption Kit: for guided missile warheads, 539–40
Advisory Board to Operation Crossroads, 200
Advisory Committee on Uranium, 12
AEC Missile and Rocket Responsibilities, 540
aerodynamicists, 159, 410, 656
Agnew, Harold M., 276
Ahl, H. V., 469
Air Depot Training Station, at Sandia Base, 123
Aircraft Holdings, Inc., 122
"Alarm clock": thermonuclear concept, 563, 577
"Alberta" Project: to deliver first airborne atomic weapon, 89, 119, 125. See also Tinian.
Albuquerque Municipal Airport, 121, 124
Albuquerque Operations Office, of AEC, 382
Alexander, Frederick C., 123, 461, 485, 611

Allen, Jimmy, 167
Allen, Ted, 302
ALIAS/BETTY depth bomb, 516, 518–21
Allison, Samuel K., 40, 84
Alvarez, Luis, 92, 564, 569
American Car and Foundry (ACF) 379, 380, 406, 410, 636, 640
American Federation of Scientists, 188
American Telephone and Telegraph (AT&T): selected as industrial manager for Sandia, 338–347 passim, 348–362 passim, 683
Amole, Beulah: Secretary to Robert Henderson, 532
analog computer, 620. See also "Ray Pac."
Anchor Ranch Proving Ground: at Los Alamos, 49, 57, 67
Anders, Roger, 592
Anderson, D. T., 456
Anderson, Ensign D. L., 194
Anderson, George: demonstrates exposive-to-electric transducers, 649, 665
Anderson, Herbert, 85
Anderson, Oscar E., Jr., 212
Anderson, Senator Clinton P., 218
Anderson, T. J., 134, 150
Anderson, Ted, 464
Angel, Albert Joe, 438
"Apple": as Colorado production facility, 378, 379, 380, 381
Applied Physics Department: at Sandia Laboratory, 271–73, 351

INDEX

AR-10A radars, 276
"Archies," 104
Arledge vs. Mabry, 465
Armed Forces Special Weapons Project (AFSWP), 158, 217, 231, 274, 287, 338
Army Air Corps Convalescent Center: at Sandia Base, 123
Army Corps of Engineers, 224
Arnold, General Henry H. "Hap," 101, 189
Arthur, Bryan E., 191, 202, 637
Ashley Pond: Los Alamos, New Mexico, 18–19
Ashworth, Commander Frederick L., 90, 108
Askania phototheodolites, 158
Associated Universities, 355
ATLAS intercontinental ballistic missile, 486, 490, 525
atomic bomb: decision to use against Japan, 22, 23; device first detonated at Trinity site, 27–43 passim; development of, 45–86 passim; dropped on Hiroshima, 106, 113, 126; dropped on Nagasaki, 108, 114; impact on Sandia, 126–38 passim. See also "Fat Man" and "Little Boy."
Atomic Energy Act, 207–22 passim, 630, 632
Atomic Energy Commission: establishment of: 206, 308–22 passim; organization of (1946), 215; 172, 214, 223, 324–85; Commissioners, Chairmen, General Managers, 220
Atomic Project and Production Workers, Metal Trades Council, AF of L: formation of union, 432
"Atoms for Peace": Eisenhower policy, 630–31

Austin, Howard, 154, 157, 160, 167, 168, 175, 191, 588
autocatalytic: theory of atomic bomb assembly, 57
Ayers, Alan N., 50, 234, 237

B-43, recognized as first laydown weapon, 659
Babcock and Wilcox: manufacturer of Jumbo containment vessel, 78, 80
Bacher, Robert, 30, 39, 66, 84, 208, 233, 565, 566
Bailey, Milton E., 524
Bain, Col. James, 493
Bainbridge, Kenneth T., 39, 40, 46, 49, 66, 79, 80–81, 84
"Baker" storage site, 323–24
Baker test: of Operation Crossroads, 186, 187, 204–5
Ballard, Douglas A., 422, 552
Banister, John, 608, 619
Barncord, Charles R., 28, 679, 682
Barnes, Philip, 90, 99, 105, 108, 115, 280, 294, 299
barometric pressure sensing switches: for Mk 4, 409
barometric switch: as safety device, 270
Barr, H. C., 150
Baruch, Bernard: plan for control of atomic energy, 188; as U.S. delegate to United Nations, 209–10
battery technology: evolution of, 420, 424, 664
Bauman, Les, 450
Baxter, Hal, 388
Beahan, Kermit K., 106, 108
Beal, H. C., 356, 357, 358, 426
B Division: of Los Alamos, established for Operation Crossroads, 183, 185

838

Beeson, Bernice, 228
Begg, Captain Charles, 100
Bell, Alexander Graham, 339
Bell Telephone Laboratories: concept of auditing quality, 317, 332, 334, 338–59 passim, 428, 527, 530
Bendix Aviation Corporation ("Royal"), 312
Benjamin, Ben, 35, 85, 179, 659
Bergson, Herbert, 344
Bethe, Hans, 32, 561, 577, 585
Betts, A. W. "Cy," 258
Bice, Richard A., 130, 139, 176, 181, 187, 207, 234, 271, 274, 278, 301, 337, 389, 410, 416, 496, 619, 624, 679
Biehl, Arthur T., 592
Biggs, Bernard S., 391
Bikini Atoll, 181, 198
Birch, Alvin F., 84, 126
Biskner, L. J., 314, 383
Blanchard, Colonel W. J. "Butch," 194
Blandy, Adm. William Henry Parnell, 184, 188
Bledsoe, W. W, 674
Bliss, Dr. S. P., 455
Blythe, Bill, 581
BOAR Ad Hoc Committee: for missile development, 521
BOAR missile, 521
Bock's Car: designated atomic bomb aircraft carrier for Nagasaki, 106–8, 114
Bode, Hendrik, W., 400–1
Bode–McNair Report, 397, 399
Bohr, Niels, 4–5, 84
Bollinger, Norm, 613
Bolstad, Milo M., 103
Bonaparte, Napoleon, 7
Bonbrake, L. D. 50
Booster Technique, 563

Boskey, Bennett, 344, 354
Boyes, William E., 401, 498, 667
Bracken, Stanley, 358, 426
Bradbury, Norris, 30, 61–62, 64, 84, 90, 126–140 passim, 185–258 passim, 269, 274, 287, 303, 312, 329–47 passim, 350, 354, 356, 370, 401, 491, 567, 569, 577, 591, 635, 647–49
Bradford, R. J., 385
Bradley, Gen. Omar N., 243, 564, 597
Bramlett, William, 443, 448, 452
Brandenburg, Don, 158, 169
Brass Ring: Air Force Operation, 563
Bravo Event: of Operation Castle, 618
Brawley, Ed, 262, 288
Breit, Gregory, 561
Brereton, Gen. Lewis H., 200, 221, 238, 244
Brett, Roy, 450
Bridges, Al: as President of Kaman Nuclear Division of Kaman Aircraft, 657
Briggs, Lyman, 12
Bright, Ben, 660
Brin, Ray, 67, 103, 108, 491, 679
Brixner, Berlyn, 85, 183, 191
Brode, Robert B., 49, 55, 84, 94, 126–27, 129–30
Brodsky, A. I., 15, 402
Brodsky, Robert F., 656
Brookhaven National Laboratory, 355
Brooks, John, 339, 354
Brown, Bill, 388
Brown, Harold, 525, 592
Brown, Marvin, 441–42, 448
Brown, R., 376
Brown, Walter L., 356–57, 360–61, 426–27, 464

Browne, Dick, 659
Brownell, Atty. Gen. Herbert, 347
Broyles, Carter, 609, 610, 618
Bruda, E. J., 420
Bryson, Jack, 302
Buck Act, 465
Buckley, Oliver E.: as President of Bell Telephone Laboratories, 329–43 passim
Buckner, Bob, 302
Bunker, Bob, 582, 614
Buonagurio, G. C., 388
Burleson: at Operation Crossroads, 187
Burriss, Stanley, 580, 584, 587
Bush, Vannevar, 12, 14, 48, 78, 109, 208
Business Methods Organization, 461
Buster Shots: *Able, Baker, Charlie, Dog,* and *Easy,* 601–3
Buster–Jangle: continental test operation, 597–605
"Button Coordination Committee," 419
Byrd, Nora Bell, 682
Byrd, Robert, 448, 451
Byrnes, James, 11, 28

Cain, S. Gayle, 682
Caldes, William, 32, 134, 318, 480
Caleca, Vincent, 92
Camel Project, 92
Camp Desert Rock: at Nevada Test Site, 601
Camp Mercury: at Nevada Test Site, 599, 620
Campbell, Charles W., 258, 262, 459, 636–38
"Can-Do" ethos: of Sandia Laboratory, 472, 683
Canterbury, Col. William M., 274, 276
Carco Aircraft Company, 158

Carlson, Roy W., 33, 73, 75, 76, 286, 287
Carnahan, C. W., 399
Carpenter, Donald F., 244
Carr, Barney, J., 619
Castle Test Operation, 617, 618
Castle, L. E., 455
"Caucasian": Air Force Program, 563
Cave-Brown, Anthony, 14, 22
"Centerboard": code name for plan to drop atomic bomb, 99
centrifuge: construction of, 516–17
Certificate of Incorporation: of Sandia, 556–57
Chadwick, James, 32, 50
Chambers, Marjorie Bell, 46
Chaney, Mel, 666
"Charlie" storage site, 324, 326
Chenchar, Captain P., Jr. (Army), 194
Chiado, Thomas, 442
"Chicken Pox": flying assembly lab, 252, 253, 254, 255
"Christy Gadget," 58
Christy, Robert F., 58
Church, Allen, 19
Church, Hugh, 19
Church, Pete, 612, 614
Church, Theodore, 19
Churchill, Winston P., 11, 15, 20, 22–23, 27–29, 54, 210
cinetheodolites, 160
Civilian Conservation Corps., 123
Claassen, Richard S., 411, 476, 657–60 passim, 669–79 passim
Clark, Ed, 403
Clark, John, 287, 607
Clark, Tom, 344
"classical" super, ("Runaway Super"), 563
Cline, Grant, 129
Clinton Engineer Works, 14, 16

INDEX

Clyde, Chet, 302
Clyde, Major, 121
Cochran, Jacqueline, 122
Cocke, Jim, 645, 662
Cody, J. P., 498, 499, 647, 658
Coiner, Col. R. T., 354
Colby, R. L., 134, 140, 141
Cole, Ruben, 82
Cole, Sterling, Representative, 632
Collins, R. H., 278
"Committee of Three Principals": for warhead development, 642-43
Component Development, 273, 279-80, 667-68 passim
Composition B, 79
Compton, Arthur H., 14, 46, 561
Compton, K. T., 200
Conant, James, 12, 18, 42, 57, 65, 219, 230, 564
"Cook Book," 618
Cook, Thomas B., 609, 618, 669, 673-74
Cooksey, Donald, 627
Cooney, E. J., 458
Coronado Club, 262-67; Board members, 263
CORPORAL missile, 499-503
Corrigan, Bruce, 105
Corsbie, Robert L., Deputy Director, 611
Corson, Dale, 128, 130, 141, 148, 150-51, 154, 185
Costello, Edward M., 104
Cothran, Tech. Sgt. J. W., 194
Cotter, Donald R., 524, 584, 588
"Cowpuncher" Committee, 84
Cox, Everett, 376, 398-99, 579-80, 600-1, 609, 613, 615, 675
Cox, Otis, 259
Coyote Test Area: of Sandia Laboratory, 515, 581
Crawford, Rosalie Franey, 531

credit rating: of Sandia Laboratory, 464
Critchfield, Charles L., 65, 561
"Crossroads Airline," 186
Crossroads: test operation, 181-99 passim
Crumley, Roy, 469, 471
Csinnjinni, Carl, 612
Cullen, Gen. P. T., 292
Cumberland Sound: at Crossroads, 187, 191
Curry, Warren, 403
Curtis, Roger, 664
Cushman, Capt. J. A., 286

Daghlian, Harry, 69
Dahlgren Naval Proving Grounds, 54
Dahlquist, John A., 620
Dailey, Philip H., 191, 202
Dalbey, Dayton E., 122
Darby, Colonel G. C., 276
Daspit, Admiral L. R., 276
Dave's Dream: aircraft named after pilot Dave Semple, 192; used in Crossroads *Able* test drop, 193, 195
Davies, L. E., 489
Davis, Nuel Pharr, 46
Davis Panel: intercedes in Sandia union negotiations, 432-33, 436
Davis, Thomas V., 183
Dawson, J. J., 491
Dawson, Joe, 271, 299
Dean, Gordon, 212, 220, 534, 555, 567-85 passim, 607, 641
Decker, Leonard, 448
Deeter, Lee, 660
Defense Atomic Support Agency (DASA), 661
Defense Reorganization Act, 556
deHoffman, Frederick, 569, 578
DEMI-JOHN: adaption kit, 519

INDEX

Demler, Colonel M. C., 99
DeMontmollin, Jim, 643
Dempsey, James A., 368, 594
Demson, E. J., 189
Denfield, Admiral Louis E, 243
Denison, Bill, 647
Derry, Lt. Col. J. H., 200
DeSelm, C. H., 150–51, 155, 278–79, 280, 297, 301, 409, 496
Desert Rock IV Exercise: Troop maneuvers at Nevada Test Site, 607
de Silva, Peer, 81
Destination, 90, 110
Detector: test operation planned for Crossroads, 186
Development Engineering, 549
development fabrication, 383
Dewey, Bradley, 200
Dewey, Frank, 462
Dietrich, W. F., 454, 468, 471
Dike, Sheldon, 56, 102, 649
Dillingham, L. A., 660
Dingman, James E., 428
DMA, first Director of, 222
Documents Department (SLD), 473
Dodge, Harold, 390
Doll, E. B., 49–50, 130, 181, 186–87
Domeier, E. J., 434, 458
Donne, John, 46
Doolittle, James, 122
Dorland, Gilbert M., 141, 231, 286
Dornberger, Walter, 486
Douglas, Helen Gahagan, 211
Downing, Jackie, 261
Draper, Eaton H., 409, 412, 420, 495, 643, 661
Dreesen, D. S., 50
Dubose, Major, 187
Dubridge, Lee, 219
Dudley, Major John H., 17
Duff, Robert, 311
Duggin, Frank, 470, 472
Duncan, Francis, 235, 285

Dunn, L. A., 495
DuPont de Nemours, E. I., 342, 355; manufacturer of black powder, 378; as operator of Savannah River Plant, 378–79
Durham, Carl T., 365, 378, 632
Dyhre, A. E., 259, 261

E Division, 49, 51
"Easy" storage site, 320–22
Econnomu, George, 35
Eden, Anthony, 28
Editorial Review Board: of Sandia Laboratory, 475–76
Edwards, George, 390–91
Edgerton, Germeshausen, and Grier (EG&G), 271, 291, 293
Egger, Sam, 322
Eichleay Corporation, 70
Eiffert, Art, 833
Einstein, Albert, 4, 6, 8, 10; letter to President Roosevelt, 12; 564
Eisen, Nathan, 150–51
Eisenhower, General Dwight, 284, 524, 556–57, 592, 629, 632
Ela, Cmdr D. K., 276
Electrical Systems Coordination Group, 549
Electrostatic Generator Group, 83
Ellis, Bo, 451
Elugelab: Pacific island of, 592
"Employee Bulletin," 482
Employee Contribution Plan, 480
Employee Review Committee, 467
Engel, Vic, 652
Engineering Department, 278, 310, 351, 494
Engleman, Cdr. C. L., 286
ENIAC (Electronic Numerical Integrator/Calculator), 563
Enola Gay, 103, 104, 105, 106, 201
Environmental Test Facilities, 273, 514–18 passim

842

ER-12-10 lead acid battery, 269, 424
Erb, Otto, 664
Erickson, Clifford O., 682
Erickson, K. W., 276, 411, 649, 652, 658, 668–75 passim
Esterline, P., 50
Ethos: of Sandia, 147.
Evans, G. Foster, 567
Everett, Cornelius J., 569
Exercises Desert Rock I, II, and III: military troop maneuvers during, 597–601 passim
Explosives Division, 66
Eyster, Gene, 202

fallout: first model for, 86
Farrell, Gen. T. F., 200
Fat Man: code name for atomic bomb, 50, 54–55, 92, 96–7, 101, 104, 106–7, 114, 127, 129, 133, 153–54, 238, 267–71 passim, 414
FATHER JOHN rocket, 508, 513, 514
Favia, M. L., 496, 506
Fay, F. P., 454, 462
Federal Island issue, 465
Federal Mediation and Conciliation Service, 433
Federation of American Scientists, 213
Ferebee, Maj. Thomas, 105
Ferguson, Waylon, 466
Fermi, Enrico, 4–7, 16, 32, 41, 62, 69, 85, 561, 567
Feynman, Richard, 62
Field, Paul, 666
Field Test Group (Z–1A), 151, 176
Field, Vernon M., 682
Fields, Col. Kenneth E., 208, 519, 534

Fifth George Washington Conference on Theoretical Physics, 5
Fireball yield method, 86
Firesets and detonators: for Mk 4 Mod 0, 271; development of, 652
First Engineering Special Batallion, 141
Fisher, Benjamin, F., 682
Fisher, Ralph, 390
Fisk, James B., 330, 334, 376
Fite, A. W., 496, 499
Flugge, Dr. S., 6
Ford, Peyton, 344
Fortine, Frank, 324
Foster, C. E., 383
Foster, Dr. John S., Jr., 592
Four O'Clock Group, 674. See also Research.
Fowler, Glenn A., 41, 43, 84, 126, 128, 130, 141, 147–54 passim, 160–65 passim, 167, 174–77 passim, 186, 196, 202, 237, 294, 463, 482, 493, 529, 572, 575, 588, 607, 623, 625, 638, 669, 672–78 passim
Fowler, Mary Alice, 149, 160, 179
"Fowler's Law," 175
Franklin, William, 121–22
Frantik, Rudy O., 524
Frenchman Flat: of Nevada Test Site, 570–72; 599, 601, 602–21 passim
Frisch, Otto, 5, 69
Froman, Darol K., 285, 287–89
"Fubar," 80, 83. See also Trinity Site.
Fuchs, Klaus, 6, 15, 565, 618
Fulbright, Senator J. William, 188, 577
Full Scale Test Observer Program, 482
Fuller, Lloyd, 462, 464

INDEX

Fulton, Robert, 7
Funk, W. G., 461
Fussell, Louis, 84, 126, 128, 130
fuzing controversy, 535–42

GAFCO (Glenn A. Fowler Company), 174–76
Galbreath, J. W., 461
Gallegos, Tony V., 442
Gamow, George, 561, 569
Garblink, A. M., 496
Gardner, William A., 409, 496, 643–44
Gaskill, W. C., 507
Gee, Col. H. C., 187–88
Geelan, Barney, 302
General Advisory Committee, 215, 218–21 passim; visits Sandia, 231–35; Manley evaluation of Sandia for, 236–38; criticism of, 564–65
George shot: of Operation Greenhouse, 558, 578, 584
Geskell, W. S., 496
Gibson, R. E., 371
Gillespie, Bob, 475
Gillespie, Ken, 666
Given, Frederick J., 528, 636, 639, 664
Glasstone, Samuel, 620
Goddard, Robert H., 486, 492
Goddell, C. A., 259
Gold, Harry, 15, 565
Golden, William T., 208
Gore, Rep. Albert, 603
Gottwald, Klement, 246
Graves, Alvin, 285, 287, 571, 574–75, 578–79, 587, 600, 607, 611
Graves, Charlie, 440
Gray and Barton, 340
Great Artiste, The, 106
Greb, Allen, 556

Green Hornet (C–54), 90, 93, 111
Greenhouse test operation: Sandia's role in, 578–83
Grier, H. E., 286
Grimshaw, Wayne A., 682
Gromyko, Andrei: presents "Gromyko Plan" for control of atomic energy, 210
Groves, Brig. Gen. Leslie, 14, 16–7, 32–49 passim, 56–69 passim, 80, 82–3, 99, 101, 120–21, 125, 132, 140, 187, 188, 206, 216, 221, 242–43, 490, 491
Gruer, Al, 262
Guided Missile Committee, 518; subcommittee of, 536, 537–40 passim
Guided missiles: Sandia's role in warhead development for, 485–525 passim
Gump, Charles, 681, 682
Gun program, 66. See also Little Boy.
Gunn, Harold, 261–62, 442
Gustafson, Gus, 613, 620, 626
Gutierrez, Leo, 302, 411, 495

Hadley, Ross, 122
Hahn, Otto, 4
Hake, Lyle, 176
Hall, Jane, 525
Hall, John H., 357
Hamilton, Ira D. "Tiny", 28, 191, 202, 298, 576
Hampson, Price, 612
Hanford Works of the Manhattan Engineer District, 14, 140
Hansche, George, 272, 376, 649
Hansen, Hans, 612, 622
Hansen, Jack, 262
Harding, John, 613
Harling, Gene, 179
Harris, Lt. Col. John, 493

844

Harris, Sylvan, 260–61
Harris, Virgil, 224, 261, 267–78
Harrison, Captain W. C., Jr., 194
Hart, J. C., 462
Hatch, Senator Carl, 192
Hawk, Robert, 434, 435
Hawkins, David, 56, 58
Hawkins, L. G., 260–63
Hayward, Capt. John T. "Chick," 592
Health and Hazards Division, 466–67
Heaston, Joe, 323
Heckman, G. F., 278, 494, 496, 524
Heilman, Luther J., 223–24, 386, 429, 442, 454
Heinrich, Dr. Helmut, 656
Heins, Omar, 388
Hempleman, Louis H., 39
Henderson, Robert W., 31–45 passim, 54, 67, 72–83 passim, 106, 107, 129, 132, 172, 176, 183, 187–88, 207, 225, 231, 238–39, 240–41, 274, 276, 287–94 passim, 295, 301, 323, 389, 397, 494, 495, 524, 525, 532, 547–53 passim, 579, 582, 584, 588, 639, 641, 643, 658, 675
Hepplewhite, Robert, 162
Hereford, Bill, 302
HERMES missile, 491, 493, 498, 536
Herrity, Ed, 471
Hertford, Kenner F., 276, 487, 595, 605–7, 639, 678
Hewlett, Richard G., 213, 235, 300
Hickman, Clarence N., 485, 492–94, 502, 503, 508
Hight, Stuart C., 276, 528, 529, 637, 657, 677
Hinde, Jack, 179
Hinds, Col. John H., 216, 287
Hinman, Jerry, 660

Hinshaw, Carl, 632
Hinson, James S., 438, 440, 451
Hiroshima, 3, 89, 105–6, 113, 126, 182
Hirsch, Dr. F. G., 466, 468
Hirschfelder, Joseph O., 77, 183
Hitler, Adolf, 11, 486
Hittell, J. L., 50
Hoaglund, Bill, 647, 651
Hoffman, J. A., 278
Hoffman, Joseph G., 39
Hollenbeck, Whitey, 166, 170
Hollingsworth, Lee, 661, 663, 680
Holloway, Marshall, 30, 183, 185, 643
Hollowwa, G. C., 93–4, 96, 133, 317
Holmes and Narver, 571, 580
HONEST JOHN rocket, 493, 497, 509, 510–11, 512–19
Honeyfield, Verne, 437–38, 440, 451
Hooper, J. T., 274
Hoover, Gilbert C., 12
Hopkins, L. A., Jr., 276, 286, 290–91, 492–94, 496, 499, 501, 503, 524, 547, 639
Hopper, Robert, 229, 386–87
"Hotpoint": stockpiled as Mk 34, 659
Hovde, Dr. F. D., 491
Howard, Orville T., 160
Howard, W. J. "Jack," 289, 298, 496, 497, 524, 580, 597, 600, 629, 635, 680
Hubbard, Jack, 40, 82
Hudson, Frank, 667–72, 674
Huffman, Sen. James, 188
Hughes, Grover, 644
Hughes, J. W., 471
Hull, Lt. Gen. John E., 285, 287, 491
Huning, Fred D., 151
Hutson, E. P., 442

INDEX

hydrogen bomb: development of, 559-71, 577-93 passim

Impact fuze, 411
Imperial Valley Irrigation Company, 155
Implosion concept, 57, 66, 79
Indemnity, 360-61
Indianapolis, 103
Indochina: Civil War in, 631
Informer, Fuzing, and Firing Group, 185
Inyokern, 92
Integrated Contractor Complex: origins of, 310-12; expansion of facilities, 378-82
International Chemical Workers, 430
International Guards Union of America (IGUA), 441; officials of, 442
Irons, D. P., 150-51
Item shot: of Operation Greenhouse, 584-85
Ivy test operation, 592-93

J Division: of Los Alamos, 286-87
Jackson, Sen. Henry "Scoop," 586
Jacot, Louis F., 41
Jacquet, Edward, 140
Jameison, W. A., 259
Jeppson, Morris, 90, 99, 105, 115
Jercinovic, Leo M., 28, 30, 32, 278, 287-90, 291-92, 323
Jimenez, Art, 445
"Joe One": detonation of, 366, 559-60
Johnson, Sam, 322
Johnson, Sen. Edwin C., 210, 632
Joint Chiefs of Staff, 182
Joint Committee on Atomic Energy, 217, 221

Joint Operational Planning Committee (JOPC), 643
Joint Research and Development Board, 300
Joint Task Force One (JTF-1), 184
Joint Task Force Seven (JTF-7), 285-87 passim
Joliot-Curie, Frederick and Marie, 4, 7
Jones, J. W., 278, 494, 522, 660
Jones, N. "Pop," 439
Jornada del Muerto, 28, 81
Jumbo, 33, 45, 69-71, 74, 75, 79-80. See also Trinity Site.
JUPITER missile, 484, 525

K-9 patrol, 139
Kai-Shek, Chiang, 23
Kaman, Charles H., 658
Kane, John T., 384
Karo, Jim, 157, 166
Keller, Gladys, 161
Keller, Kaufman, "K. T.," 490, 497
Kelly, Mervin G., 276, 328, 329, 330, 335-38, 350-58, 376, 428, 494, 527-28, 676, 678
Kelly Report, 350-53 passim
Kennedy, President John F., 317
Kent, Noel, 450
Kepner, Maj. Gen. William E., 285
Kidd, Dick, 417
King, Capt. John, 294
"Kingman," 93
Kingsbury Commitment, 340
Kingsley, W. H., 455, 467, 468
Kirkpatrick, Elmer E., 90
Kirtland, Roy S., 125
Kistiakowsky, George, 30, 42, 65-7, 69, 74-5, 81, 84, 525
Knacke, Theodore, 656
Knapp, Bob, 588
Knott, Don, 388

846

Knudsen, R. A., 461, 478
Koebke, Clara, 613
Koester, George A., 183, 191
Korean War, 246, 565
Kraft, William C., 318
Kraker, George P., 261, 300, 336, 353, 354, 370, 377, 521, 600
Krause, Corbett, 441, 442
Krohn, Bob (LASL), 588
Kurchatov, Igor, 15

Labor Management Relations Act (1947), 431
Lack, Frederick R., 354-57, 426, 427
"Laggin Dragon," 104
Lamkin, Lessel E., 397, 409, 495
Lanahan, T. B., 150
Landes, R. H., 461-62
Landing ship mechanized (LSM-60), 187
Landry, George A., 354, 357, 358, 360, 364, 367-75 passim, 382, 395, 426, 427, 429, 443, 464, 466, 475-76, 480, 525, 537
Larsen, Paul J., 250-52, 256, 258, 260, 267-76 passim, 301, 309, 313-15, 333-35, 348, 350-53, 354, 370
Larson, Don, 613
Laurence, William, 100
Lawrence, Ernest O., 14, 16, 32, 74, 564, 569, 587, 627, 679
Laydown bomb, 648-60
LeBaron, Robert: as Chairman of the MLC, 221, 275, 535
Leehy, Gen. D. J., 647
Lehr, Herbert M. 34, 63, 181, 187
LeMay, Gen. Curtis, 182, 635
Lemm, Robert, 594
Lenander, Harlan, 580-84 passim, 600, 607, 613, 623, 624
Lend-lease, 11

Lesicka, Milt, 450
Leupold, A. K., 383
Libby, Commissioner, Willard F., 591
Lieber, Andrew, 501, 514, 660
Liebler, R. A., 402
Lilienthal, David E., 206, 208, 212, 233, 235, 246, 247, 283-84, 318, 335-37, 342, 345-46, 353, 564
Lindbergh, Charles A., 122
Linenburger, Gustave A., 183
Ling, Donald P., 400
Linn, Max K., 276
Linschitz, Harry, 30, 31
List, Dean, 614
Little Boy, 55, 97, 100, 104, 106, 112, 133, 278, 414
Little Red Wagon, 134-35
Little, William A., 515
Lockridge, Lt. Col. Robert W., 128, 130
"Lone Star," 646. See also laydown bomb.
Long, Col. P. J., 276
Long, Earl, 67
Long, Sen. Russell: of Louisiana, 621
Longyear, Frank H., 308, 309, 314, 384, 454
Looney, Trevor C., 612
Loper, Gen. Herbert B. "Doc," 534, 632
Lopez, Fred, 451
Los Alamos Ranch School, 17-19
Los Lunas Test Range, 146, 151-54, 158, 192
Lovelace Foundation, 468
Lower Rio Grande Flood Control Association, 480
Low-Level Delivery (LLD) Flight Test Program, 656
Lucas, Sen. Scott: of Illinois, 188
Lucky Dragon, 614

Luftwaffe, 147
Lyons, Corp. H. B., 194
MacArthur, Douglas A., 20, 247
MacArthur, Robert A., 613
MacDonald, Roderick, 441
Machen, Arthur B., 30–31, 63, 92, 107, 129, 133, 183, 191, 201, 237, 274, 276, 286, 291, 294, 298, 317, 414, 575
Mack, Julian, 85
MacLean, Donald, 15
MacNair, Walter A., 376, 397–401 passim, 480, 494, 527–28, 529, 547, 553, 649, 657
Mahinske, Aggie, 666
Maier, John Wahlen, 442
Major Assembly Release, 552
Malenkov, Georgi, 630
Maloney, Mary Lou, 302
Manchester, William, 6
Mandelkorn, Capt. R. S., 276, 286
Manhattan Engineer District, 14, 16, 18, 141, 155, 184, 310
Manley, John, 33, 62, 86, 230–37, 252, 269
Manual Committee, 278
Manufacturing Engineering, 385
Mark, J. Carson, 560, 568–69, 591
Markos, General, 246
Marks, Herbert S., 208
Marron, Jack, 665
Marrs, Albert T., 613
Marsh, William B., 682
Martin, Frank, 451
Martin, Glenn L. Company, 484, 506
Marshall, George, 56, 209, 284
Marshall Plan, 247
Martin, M. D., 276
Masaryk, Jan, 247
Massachusetts Institute of Technology Radiation Laboratory, 126

"Mastiff," 491
MATADOR guided missile, 410, 484, 490, 493, 505–6
MATADOR Interim Capability Program, 506
Mathison, Col. Charles G. "Moose," 644
Mauldin, George Howard, 105, 396, 406–8, 424, 495, 524
Maxwell, P., 402
May, Alan Nunn, 213
May, Andrew J., 212
May, Mike, 525
Maydew, Randy, 403, 646, 656–60
May-Johnson Bill, 210
MC–193 nickel cadmium battery, 424
McCampbell, Carroll B., 573, 590, 619
McCarthyism, 565
McCartney, Forrest, 659
McCone, John, 490
McCord, W. O. 150, 183, 191, 202, 278, 286, 294, 299, 575
McCormack, Brig. Gen. James M. (U.S.A.F.), 222, 230, 244, 330, 332, 336, 338, 342, 344, 348, 352, 354, 395, 536–37. See also DMA.
McCullough, Guy, 224
McDonald, David, 82
McDonald, G. C. "Corry," 30, 34, 273, 315, 385, 389–90, 415, 495, 551
McGovern, J. E., 469
McKee, Robert E., 224
McMahon Bill (S. 1717), 213, 214
McMahon, Sen. Brien, 182, 198, 211
McMillan, B., 400–1, 649, 657
McMillan, Edwin, 17, 49, 66, 93

McMinn, James, 682
McMurfrey, Mac, 451
McRae Committee, 636–40, 667
McRae, James W., 276, 453, 477, 481, 620, 628, 631–40, 648, 667, 669, 672
Mechanical Test Laboratory, 154
Meitner, Lise, 4–5
Mehlhouse, Harvey G., 636
Merillat, Jack, 167
Merritt, Mel, 601, 604, 674
Metal Trades Council, 432, 435, 441
Metzger, A. "Burt," 468
Michnovicz, Mike, 302
Microbarograph research, 611–14
Miegs, Purdy, 474
Mike Shot: of Operation Ivy, 569, 592–93
Military–Industrial Complex, 347
Military Liaison Committee, 216–21 passim
Military Training Liaison, 314
Miller, Gordon, 158, 178
Miller, J. J., 150
Miller, V. A. 129, 160
Millican, Roland S., 619
Millikan, Clark, 489
Millikan, Robert, 334
Mills, Maj. Gen. John S., 642, 648
MINUTEMAN missile, 490, 525
Mitchum, Billy, 167
Mk 3, 238, 252, 267; x–unit for, 269–71, 275
Mk 4, 129, 132, 160, 258; firing set for, 271, 272; x–unit for, 275, 277, 278, 280–81, 307, 336, 402, 403, 404–5, 406–11 passim; case controversy; 414–15; for use as warhead, 492
Mk 5, 277, 411–15, 419, 492, 497, 499–501; warhead program for RASCAL, 522
Mk 6, 403, 411, 412, 498

Mk 7, 411, 416, 420, 494, 497, 498–501, 518, 653
Mk 8, 411, 417
Mk 11, 662
Mk 12, 418
Mk 15, 650, 662–63
Mk 17/24, 662, 663
Mk 25/Genie, 651
Mk 27, 650
Mk 28: as "Building Block" bomb, 652–53
Mk 34, 659
Mk 37, 505
Mk 39, 651
Mk 43, 652, 659
Mk 91, 662
Modlin, Corp. R. M., 194
Moeding, Henry, 134, 260
Moffat, W. T., 290, 298
Mohler, Fred L., 12
Montague, Commanding General Robert A.: of Sandia Base, 231, 240, 338, 370
Moore, Sam A., 496, 647
Morton, Frank, 434–35
Mound Laboratory: at Miamisburg, Ohio, 311
Mount McKinley, 290
Muench, J. O., 286, 298
Murphey, Byron, 580, 588, 604, 609, 610
Murphree, Eger V., 14
Murray, Commissioner Thomas, 577, 591, 632
Musser, Sam A., 121, 123
Myer, Warren, 403
Myre, William C., 176–77, 179

Naden, Lt. E. T., 276
Nagasaki, 89, 182, 270
Narver, David Jr., 580
National Defense Research Committee (NDRC), 12, 18, 42

National Policy Committee, 109
National Security Act of 1947, 216
National Security Council, 365–66, 565
National Sites, 322–24
National Wage Stabilization Act, 462
Nautilus, 631
NAVAHO cruise missile, 490, 497, 498, 506–7
Neddermeyer, Seth H., 56–58, 65–66
Neilson, Frank, 649, 665
Nellis Bombing and Gunnery Range, 595
Nereson, Norris, 183
Nevada Proving Grounds, 572; location of 598–99. See also Nevada Test Site: Test Observer Program at, 482, 572, 595–97, 600–21 passim
Newberry Frank, 556
Newhouse, R. C., 401
News Nob Observation Point: at Nevada Proving Grounds, 599, 608
Nichols, Gen. K. D., 200, 217, 245, 401, 490
Nichols, Ruth, 122
nickel–cadmium storage battery, 424
Niebel, Lawrence, 272, 323, 325
Nightclub on the Mesa, 122
NIKE–HERCULES, 523
Niland bomb scare, 161
Nooker, Eugene, 92
Nordheim, Lothar W., 563
Norris, George, 659
Norris, M. J., 675
North American Aviation, 506
North, Harvey, 607, 619

North Island (S–2): of Salton Sea, 173
Northrop, J. K., 403, 405
NSC–68, 365–66
NT–6 lead acid battery, 268

O'Beirne, Rear Adm. F., 276
O'Connor, Bob, 299
Office Employees International Union (OEIU), 433, 434–36 passim, 441
Ofstie, Rear Adm. Ralph, 200, 216
Ogden, Gen. David A. "Dad," 288
Ogle, William E., 607
O'Keefe, Bernard J., 107–8, 293, 584
O'Keefe, Charles, 262
Olajos, C., 454
Olson, Del, 504, 651
O'Neill, William, 324
Operation Big Shot: at Yucca Flat, 594
Operation "Black Swan," 506
Operation Centerboard, 104, 106, 108. See also Tinian.
Operation Crossroads, 181–82, 196, 420
Operation Greenhouse, 419, 557–58, 569, 575, 578–85 passim, 587
Operation Ivy, 585, 605
Operation Memo, 320
Operation Milestone, 284
Operation Paperclip, 488
Operation Redwing, 627, 647
Operation Teapot, 618–21
Operation Tumbler–Snapper, 605–10
Operation Wigwam, 621–23
"Operational Requirements for Guided Missiles": restraints on U.S. missile program, 488
operational storage sites, 321–22

INDEX

Oppenheimer, Robert J., 14–18 passim, 33–48 passim, 56–57, 61–69 passim, 74–75, 81, 119–32 passim, 209, 221, 230; hearings mentioned relating to, 337; takes over fission program, 562; position on Super, 563, 564, 578, 592
Ordnance Division: of Los Alamos, 49, 51
Osborne, Carroll: develops SNL rotochute, 658
Overbury, K. G., 497–98
Oxnard Air Field, 118, 121–22, 125

Pace, Tom, 154, 158, 161, 165, 178
Paddison, Louis, 272, 276, 315–16, 318–19, 321, 325, 337, 390–91
Padilla, Felix, 475
Pagenkopf, Walter H., 384, 454
Pantex production facility, 379, 380, 382
parachute technology, 649–60
Parker, Maj. H. S., 130, 493
Parsons, William S., 48–49, 51, 58, 60, 65, 73, 84, 89, 99–100, 120, 187–88, 201, 216, 285, 370
patent provisions: of Sandia contract, 361
Patterson, Hilda, 179
Patterson, Robert T.: as Secretary of War, 209
Pearl Harbor, 12
Peirce, E. W., 461, 477
Pelsor, Gene T., 402
pendulum: construction of, 515
Penny, William, 84
Penzien, Joe, 584
Perlman, Theodore, 92
Perrett, Bill, 609
Personnel organization, 455, 458–61
Petersen, Robert P., 271, 337, 376, 400–1, 575, 579

Peurifoy, Robert L., 541, 642, 662
Phelps, Marcos J., 442
Photooptical group, 85
physical sciences research, 670–71
Pike, Sumner T., 208
Pinkham, Bob, 666
Plagge, H. J., 600
Plant Security Division, 437
Plant Services Department, 386
plutonium: discovery of, 12
Poage, Robert S., 357
POLARIS missile, 525
Pond, Margaret Hallet, 19
Poole, Robert: chairman of SWDB, 275; 276, 374, 376, 390, 459, 492, 494, 527, 529, 553, 600, 627, 680
Pope, Alan Y., 403, 674
Porter, Jim, 325
Potsdam Conference, 20, 23, 27, 29
Potter, Virginia, 532
Powell, Dr. R. C., 466
Powell, Ray, 48, 125, 140, 259, 369, 455, 460, 477
President's Test Evaluation Board, 199
Price, Melvin, 632
Prichett, Robert, 614
Prim, Robert C., 400
Prince, Kimball, 355
Production Engineering, 383–84
Production Planning Division, 384
Project Alberta, 84
Project Elsie, 504
Project Greenfruit, 423–24
Project Mercury, 572
Project Nutmeg, 571
Project Royal, 312, 314, 355, 406. See also Bendix Aviation Corporation.
Project Sugar, 311. See also Burlington, Iowa.
Project Water Supply, 322

INDEX

Project Y, 47, 74, 132, 148
Public Relations Division, 473–83
"Pumpkins," 92, 132
Purchasing, 468–73
Purnell, William R., 99
Putt, Don L., 101
Pyle, Ernie, 27

Quadrapartite Committee: for guided missiles, 643
Quality Assurance, 315, 318, 319, 325, 640
Quarles, Donald A., 354, 376, 426, 428, 453, 526, 529, 530–57 passim, 633
Quesada, Lt. Gen. Elwood, 578, 586
Quinlan, Bob, 666

Rabi, Isidor I., 84, 221, 230
Radar Scanning Link, 522
Radiological Safety Program, 290
Ramey, Gen. R. M., 194
Ramey, James T.: staff director of the Joint Committee, 218
Ramo–Wooldridge Group, 525
Ramsey, Norman, 50, 55, 56, 84, 89, 90, 93, 101, 126, 148
Randle, G. W., 660
Randle, Mavis, 532
Ranger, 571, 572, 575, 576, 577, 603
RASCAL missile, 497, 522
Ratner, S. J., 49
Raytheon Manufacturing Company, 86, 269
Reconstruction Finance Corporation, 123, 139
Redstone Arsenal: at Huntsville, Alabama, 488–89, 524, 645
Reed, J. W., 600
Reese, George, 582
"Refraction–Reflection" phenomenon, 604

REGULUS, 484, 490, 491, 493, 497, 503–5
Research and Development Board, 351, 400
Research Directorate, 398–99, 637, 667–78
Reuben James: destroyed by German U-boat, 11
Rhien, R. E., 259
Richtmyer, Robert Davis, 563
Ridenour, Ralph, 434–44
Riede, John, 129
RIGEL missile, 498.
Road Department, 309–10, 313–14
Roark, Maj. Robert L., 202
Roberts, Richard B, 12
Robertson, Tom, 390
Robinett, Rush, 179
Robinson, C. F., 420, 495
Robinson, Don, 658
Rocky Flats, 378, 379, 380, 381
Roebuck, Kenneth O., 191, 201
Roerkohl, Captain, 138
Roesener, Lt. A. K., 276
Rogers, Col. Benjamin, 323
Roosevelt, Franklin D., 6, 7, 8, 11, 12, 15, 20, 21, 50, 209
Ross Aviation, 158
Ross, Charles: as Presidential Press Secretary, 366
Ross, Hardy, 468
Rotochute, 658. See also laydown bomb.
Rourke, Robert L., 588
Rowe, Hartley, 84, 89, 119, 121, 219
Rowe, James Leslie, 98, 132, 133, 261, 267
Rowe, Paul. 403
Roy, Max, 278
Rubinson, William, 183
Ruiz, George, 583
Runyan, Charlie, 289, 298, 417, 495

INDEX

Rush, Lt. Cmdr. Charles, 493
Russ, Harlow, 92, 129, 134, 181, 183, 187, 278, 294
Russell, Cmdr. James S., 285
Russo, Helen, 532
Rynders, Jerry, 666

Sachs, Alexander, 7, 12
Saidor, 187. See also Crossroads operation.
Salary Committee, 463
Salton Sea Test Site, 154-75
Sampson, D. H., 469
Sandia Apparatus (SA), 412
Sandia Area Federal Credit Union, 122, 265-66
Sandia Base Functions Committee, 278
Sandia Board of Directors (1954), 427, 634
Sandia Branch Laboratory, 251-335 passim
Sandia Corporation: first president of, 367; organization chart (1949), 374-75; moves from production to research and development, 395-425; unionization of, 427-48; working conditions, 446; service pin selected for, 481; loses responsibility for fuzing, 535-42; develops classifications for engineering change orders, 546; 367-683 passim; culture, postwar decade, 683.
Sandia Corporate Instructions (SCIs), 461
Sandia Employees Association, 430, 432
Sandia *Lab News*, 475
Sandia public relations charter, 477
Sandia Purchasing, 468
Sandia Research and Development Planning Board, 274-75

Sandia security guards, 437-51
"Sandia Story," 482
Sandia thunderbird logo, 480, 481.
Sandia West Lab, 272
Sandia's "House of Correction," 582, 583. See also Greenhouse operation.
Sandstone test operation: role of Sandia at, 283-306 passim; assistant technical director for, 243; impact on personnel, 582. See also R. W. Henderson.
Sandy Beach, California, 98
Sawyer, Ralph, 185, 187, 188, 191
Saxton, Curly, 158
Scherr, Ruby (Dr.), 65
Schmidt, Fred, 375, 383, 438, 454, 553
Schreiber, Raemer E., 183, 278, 294
Schufflebarger, F. B., 122
Schulz, Lee, 280
Schultz, Cmdr. P. G., 276
Schultz, Don, 666
Schultz, Raymond H., 134, 496, 508
Schwartz, Siegmund P. "Monk," 360; dissolves Editorial Review Committee, 476
Schriever, Brig. Gen. Bernard A., 525
Scott, Jim, 179, 507, 582, 621, 622, 623, 624
Scoville, M. "Pete," 607
SCP (Skills, Crafts, Production) Plan: for wage and salary, 461
Scroggs, Lt. Col. J. P., 286
Seaborg, Glenn T., 12, 219
sealed pit: leads to development of wooden bomb, 662
Sealey, Clyde, 451
Seeman, Lyle E., 123, 128, 130, 138, 139
Segrè, Emilio, 12, 55
Semple, David, 101, 155, 192

Serber, Robert, 57, 92
Shaffer, Wilbur F., 30, 132, 133, 153, 308, 313–14
Sharp, Harold, 460, 475, 477
Shaw, Brad, 434, 435
Shea, Timothy E., 427, 428, 477, 553–54
Shelton, Frank, 601, 609
Sheppard, Lloyd, 302
Sherman Antitrust Act, 342
Sherwin, Ted, 262, 473–83 passim
Shields, Maj. C. S., 101
Shinn, Kenneth, 434, 435
shock mitigating techniques: for laydown weapons, 657
Shockley, William, 334
Shuster, Don, 573, 575–76, 584, 588, 622
Siglock, Robert L., 682
Silverplate, 54, 93, 101, 102
Simpson, Clarence, 665
Sims, Ed, 437, 440, 451
Site Mercury, 572. See also Nevada Test Site.
Skinner, Col. Leslie, 493
skip zone phenomenon: investigated during Buster-Jangle, 603
Slotin, Louis, 30, 69
Smalley, W. M., 286
Smith, Cyril, 219
Smith, Frederick B., 375, 434, 435, 474–76, 594
Smith, K. A., 455
Smith, Leon, 52, 90, 91, 94, 96, 99, 100, 105, 115, 190, 193, 194, 195, 196, 280, 299, 406, 496, 523–24
Smith, Marie, 52
Smith, Ralph Carlisle, 579
Smith, Wallace T., 176, 177, 179
Smyth, H. J., 454
SNARK missile, 484, 485, 498, 507, 645

Solberg, Rear Adm. Thorvald A., 216
South Island (S–1), 173. See also Salton Sea Test Site.
Soviet A-bomb ("Joe One"), 559.
Spaatz, Gen. Carl, U.S.A.F., 243
Sparks, Bob, 159
Speakman, Frank G., 121, 122
Special Engineering Detachment (SED), 28, 67
Special Weapons Committee, 84
Special Weapons Development Board (SWDB), 275; organized in 1948, 276, 277; effect of, 278, 492, 536, 537, 540
Speer, Charles, 101, 102
Spencer, Robert C., 619
Spoon, Ken, 469, 470
Sprink, T. A., 162
Sproul, R. G., president, University of California, 120, 258
SS *Fowler*, 174
Staff Member Committee, 463. See also Glenn Fowler.
Staley, Capt. C. H., 150
Stalin, Joseph, 22, 27, 29
standardization, 532
Stark, John, 434
Steck, George, 179
Steck, Helen, 179
Stephenson, Bill, 403
Stevens, Lex, 45, 81
Stevenson, Cmdr., 130
Stewart, Robert L., 437, 438
Stewart, William R., 191
Stilwell, Gen. Joseph, 200
Stimson, Henry L., 12, 18, 20, 28, 209
Stockpile Committee, 133, 134, 143
Stockpile Surveillance, 314–18, 337, 351
Stoll, James, 434, 435
Stout, Ed, 165

Stranathan, Gen. Leland S., 642, 648
Strassmann, Fritz, 4
Streiby, Bob, 658
Strauss, Lewis, 212, 220, 245, 564, 577–78, 620, 632
Stravasnik, Luke, 652
Stromberg, Robert P., 409, 666
Sutherland, Col. J. R., 194, 195, 202
Suzuki, Adm. Baron K., 23
Swancutt, Maj. W. P. "Woody," 194
Swartzbaugh, Sid, 614, 615
Sweeney, Charles, 102, 106, 108, 109, 114
Sweeney, Harold G., 292, 299
Sweeney, Henry, 164
"Switchman," 285
Sylvania Company, 407
Systems Analysis: origins of, 401–2
Systems Evaluation Department, 402
Szilard, Leo, 4, 6, 7, 10, 12

Tableleg Committee, 659, 660. See also laydown bomb.
tail warning radar (Archie), 55
Taft–Hartley Act, 431
Takeover Agreement: for Sandia, 356
Tarbox, Dave, 224, 262, 387
Tarnawsky, Nick, 438
Tarrant, Brig. Gen. L. K., 276
Taylor, Tommy, 164
Teapot operation, 618; civil defense requirements for, 616; 661
telemetry, 154
Technical Area III, 515
Technical Facilities Committee, 278
Technical Information Program, 476
Technical Writing organization, 391

Teller, Edward, 5, 6, 7, 12, 32, 566, 560–90 passim, 627, 635, 679
Tennessee Valley Authority, 140
Theis, William T., 129
Theoretical Division: of Los Alamos, 58
thermal battery development: leading to wooden bomb, 662, 664–65
thermonuclear program, 559–93 passim
Thin Man, 50, 54, 55, 101, 102. See also gun program.
Thomas, Earl, 130
Thomas, Frank J., 682
Thompson, Francis E. "Tommy," 582, 612
Thompson, Robert, 614
Thompson, Sanford K. "Tommy," 469, 470
THOR missile, 419, 423
Thornbrough, Dean, 482, 615
Tibbets, Paul W., Jr., 99, 104
Tingley, Maj. Clyde, 122
Tinian, 90, 110. See also "Destination."
TITAN missile, 484
Tolman, Richard, 57, 187, 188
Totem Pole experiment, 619
Townsend, Robert, 390–91, 636, 637, 669
Trades Council of the American Federation of Labor, 434
"Transit Time," 575
Transition Engineering, 390, 553–54, 637
Treibel, Walt, 33, 133, 286, 287, 289, 295, 301, 496, 588, 602, 651
Trepke Construction Company, 173
Trinity Site, 20, 22, 27–43 passim, 45, 79, 84, 89, 119, 125
TRITON missile, 484
Trowbridge, George Fox, 208

Truman, Harry S., 3, 20–29 passim, 214, 216, 219, 235, 244, 246, 247, 329, 341–42, 345, 490, 567
Tuck, James L., 65, 183
Tumbler–Snapper operation, 505–10
TWA Engineering, 388–90
TX–G/Gun Committee, 421
TX–N Steering Committee, 421, 423, 547, 643
TX–Theta Committee, 643
TX–5, 411–19, 492, 501, 504
TX–5 Steering Committee, 275–77, 419
TX–7, 419
TX–8, 419, 420, 421, 492, 504
TX–9, 421, 422
TX–10, 421, 422, 423
TX–11, 421, 423, 505
TX–12, 421, 422, 423
TX–14, 641–45 passim
TX–15, 639, 645, 646
TX–16, 643, 645
TX–17, 643, 645
TX–21, 645
TX–22, 647
TX–27, 647
TX–28, 647, 652, 653, 654–55, 662
TX–39–X1: laydown capability of, 646
Tyler, Carroll, L., 222, 223, 224, 330, 334, 335, 336, 349, 350, 356, 465, 519, 574, 600, 605

Udey, Edwin C., 292, 299, 588
Uehlinger, A. E., 276
Uhl, Edward G., 493
Ulam, Stanislaw M., 567, 569, 578
Underhill, Robert, 331, 356
University of California Radiation Laboratory (UCRL), 679
Upshot–Knothole operation, 609, 611–16, 642

"Urchin," 65
Urey, Harold C., 14, 565
U.S. Army Special Battalion, 186
U.S.S. *Albemarle*, 187. See also Crossroads operation.
U.S.S. *Curtiss*, 291, 296, 303, 305. See also Crossroads operation.
Uszuko, Tony, 451
Valentine, Jim, 612, 614, 621
Van Brocklin, Mary A., 682
Van de Graaff Accelerator, 674
Vandenberg Amendment, 213, 214
Vandenberg, Sen. Arthur H., 213
Van Vessem, Alvin D., 30, 191, 202
Van Zandt, Rep. James E., 632
Varoz, Ted, 437, 450
Verburg, Adrian J., 267
vibration facility, 518
Vincent, Coye, 150
Vlasic, Frank, 658
Volpe, Joseph, Jr., 344
von Braun, Wernher, 486, 487, 524
Von Karman, Theodore, 489. See also RAND Corporation.
Von Neumann Committee, 525
Von Neumann, John, 58, 525, 567, 568, 569, 604
Vortman, Luke, 580, 610, 619
Vytacil, N., 455
V–1 "Buzz Bombs," 486
V–2 "Vengeance Weapons," 486

W–47, 93, 94, 97, 98, 138. See also Wendover, Utah.
Wage and Salary Administration, 461
Waley, Russ, 462
Walker, H. C. "Clyde," 480, 481
Walker, Maj. L. D., 276
Wallen, Orval W., 682
Wallick, Elizabeth, 261
War Powers Act, 342
Ward, Malcolm, 442

Warhead Development Directorate, 491
warheads: for guided missiles, 497–525 passim
Warner, Roger, 30, 89, 92, 97, 126–32 passim, 183–88 passim, 200–2, 225, 230, 267, 294
Warren, Martin, 191
Watson, Gen. Edwin "Pa," 11
Waymack, William W., 208
Weapon Development Phases: Phase I: Concept Formulation, 543, 548; Phase II: Development, 543, 549; Phase III: Development Engineering, 543, 550; Phase IV: Production Engineering, 543, 549–50; Phase V: First Production, 543, 552; Phase VI: Quantity Production, 543, 552–53; Phase VII: Retirement, 553, 543
Weapons Analysis Group, 649. See also Ken Erickson.
Weapons Committee: of Los Alamos, 126, 148
Weapons Component Department, 416, 649
weapons effects experiments: at Operation Greenhouse, 579–84 passim.
Weapons Effect organization: at Greenhouse, 579–82, 584; at Nevada Test Site, 595; at Buster–Jangle, 601–5; at Tumbler–Snapper, 605–10; at Upshot–Knothole, 611–13; at Teapot, 618–23
Weapons Physics Division, 66
Weapons Reliability Committee, 401
Weaver, Charles A. "Buck," 437
Webb, Dave, 411
Webb, James: Budget Director, 247

Webster, Col. G. B., Jr., 276
Wehrle, Bob, 664
Weisskopf, Victor F., 5, 39
Wells, W. M., Jr., 496, 497, 584, 647
Wendover Field, Utah, 56, 84, 93, 94, 99, 134. See also W–47 or Kingman.
Wesson, P. D., 357, 360, 466
West, Richard, 207
Westcott, Hoyt, 276
Western Electric, 334, 340, 346, 356, 359, 360, 458, 464
Whitcomb, D. A., 666
Whitlow, Emory, 614, 615
Whitten, Mildred, 159, 466
Wieboldt, James, 183
Wiesner, Jerome B., 128, 130, 183, 525
Wigner, Eugene, 7, 12
Wigwam test operation, 621
Wilcox, 78, 80
Willard Storage Battery Company, 268
Williams, John, 66, 83, 181–87 passim
Williams Plan, 184
Willis, Guy, 158, 159, 169
Willis, H. M., Jr., 276, 389, 478
Wilson, Carroll L., 208, 212, 244, 342, 344, 345
Wilson, Charles, 347
Wilson, Leroy A., 329, 342, 343, 345, 354
Wilson, Ralph, 418, 495, 679
Wilson, Robert, 33, 66, 99, 101
Windstorm cratering test, 597, 600
Winter, Charlie, 682
Witt, Larry, 582
Wood, Harold, 195
Wood, Maj. H. H., 194
Wood, Walt, 649

wooden bomb concept, 424, 660, 661
Wooldridge, Dean, 525
Worth, Daniel F., Jr., 377
Worthen, D., 462
Worthington, Hood, 219
Wright, O. L., 134, 234
Wriston, Gen. Roscoe, 82

X-unit, 269, 270
XW-5/MATADOR: Sandia completes arming and fuzing subsystems for, 541
XW-5 REGULUS I: Sandia completes arming and fuzing subsystems for, 541
XW-7/ALIAS/BETTY depth bomb, 519, 541
XW-7 CORPORAL: Sandia assigned warhead responsibility for, 538, 541
XW-7/HONEST JOHN: Sandia assigned responsibility for arming and fuzing, 518, 541
XW-15 warhead: for SNARK, REDSTONE, and NAVAHO, 645
XW-25: as emergency capability fission warhead, 647
XW-27: Sandians support UCRL on, 679

Y-12 Plant, 310
Yakovlev, Anatoli, 15
York, Herbert, 489, 556, 591, 627, 679
Yost, Luther F., 385
Young, Capt. D. B., 276

Z Division, 23, 49, 118-45 passim; 153 155, 185, impact of Crossroads on; 199-202 passim; characteristics during postwar decade, 225; evaluation by John Manley and AEC, 231-38; Acting Director of, 238; deficiencies of, 247; becomes Branch Laboratory, 258; technical focus of, 252-53
Zacharias, Jerrold R., 126, 128, 129, 130, 131
Zelenski, Sylvester, 299
Zhukov, Marshall, 22
Zia Corporation, 121

4N program, 408, 409, 421
7N program, 420, 423
38N program, 420, 421
509th Composite Group, 90, 94, 99, 100, 103, 141
1953 AEC-DOD Agreement: implementation of, 542-52 passim